Instructor's Solutions Manual

SEARS AND ZEMANSKY'S

UNIVERSITY PHYSICS

TENTH EDITION

VOLUME 1

YOUNG & FREEDMAN

MARK HOLLABAUGH
NORMANDALE COMMUNITY COLLEGE

NATHAN PALMER
COLORADO SCHOOL OF MINES

Addison-
Wesley

An imprint of Addison Wesley Longman

San Francisco • Boston • New York
Capetown • Hong Kong • London • Madrid • Mexico City
Montreal • Munich • Paris • Singapore • Sydney • Tokyo • Toronto

ISBN: 0-8053-8543-6
1 2 3 4 5 6 7 8 9 10—VG—04 03 02 01 00

Addison Wesley Longman, Inc.
1301 Sansome Street
San Francisco, California 94111

Contents

Preface

This Instructor's Solutions Manual, Volume 1, covers Chapters 1 to 21 of *University Physics,* Tenth Edition, by Hugh Young and Roger Freedman. It contains the solutions to all the problems and exercises in those chapters.

My deepest gratitude goes to Nathan Palmer, Colorado School of Mines, who joined this project as a co-author. Nathan checked all the solutions for accuracy.

In preparing this manual, we assumed that its primary users would be college physics professors; thus the solutions are condensed, and some steps are not shown. Some calculations were carried out to more significant figures than demanded by the input data in order to allow for differences in calculator rounding. In many cases answers were then rounded off. Therefore, you may obtain slightly different results, especially when power or trig functions are involved. All calculations were done with basic scientific calculators.

The new graphs in this edition were drawn with either Microsoft Excel 7.0 (Windows 95) or Mathsoft Mathcad Plus 6.0. Some of the source files for these graphs are included on the Instructor's CD-ROM, enabling you to model the problem situation for different numerical inputs or to include the graphs in examinations. Many of the graphs could also be drawn on graphing calculators. (Note: If you do not have Mathcad and you wish to view these files, Mathcad Explorer can be downloaded from the Mathsoft web site at http://www.mathsoft.com. Mathcad Explorer is a free, read-only version of Mathcad for browsing live mathematical documents on the computer, the network, or the Web. You will find Mathcad Explorer in the "Store" section of Mathsoft's site.)

This manual is based on its predecessor, the Instructor's Solutions Manual, Volume 1, for University Physics, Ninth Edition. I wish to recognize the excellent work of Craig Watkins in preparing that edition. Starting from a superb book made our task much easier. I also wish to thank the helpful staff at Addison Wesley Longman, especially Sami Iwata, Grace Wong, Larry Olsen, Janet Vail, Nancy Gee, and Bridget Biscotti Bradley. Lewis Ford, author of the Student Solutions Manual, offered his opinion on many problems. Hugh Young gave us much encouragement. It is a privilege to work with the textbook that bears his name.

Roger Freedman has prepared some excellent problems for University Physics! My thanks to Roger for asking me to help with this project and for his assistance and encouragement. And yes, I can attest to the existence of at least one airplane problem in each chapter!

I also wish to express my thanks to one of my own students, Christopher Burns. Chris assisted with the recalculation of problems that had different numerical inputs from the Ninth Edition. My wish for all instructors who use University Physics is to teach many capable, diligent, and intelligent students like Chris.

Mark Hollabaugh
Normandale Community College
Bloomington, Minnesota
E-mail: Hollabaugh@aol.com

Chapter 1   Units, Physical Quantities, and Vectors

1-1: 1 mi × (5280 ft/mi) × (12 in/ft) × (2.54 cm/in) × (1 km/10^5 cm) = 1.61 km
Although rounded to three figures, this conversion is exact because the given conversion from inches to centimeters defines the inch.

1-2:
$$0.473 \text{ L} \times \left(\frac{1000 \text{ cm}^3}{1 \text{ L}}\right) \times \left(\frac{1 \text{ in}}{2.54 \text{ cm}}\right)^3 = 28.9 \text{ in}^3.$$

1-3:   The time required for light to travel any distance in a vacuum is the distance divided by the speed of light;
$$\frac{10^3 \text{ m}}{3.00 \times 10^8 \text{ m/s}} = 3.33 \times 10^{-6} \text{ s} = 3.33 \times 10^3 \text{ ns}.$$

1-4:
$$11.3 \frac{\text{g}}{\text{cm}^3} \times \left(\frac{1 \text{ kg}}{1000 \text{ g}}\right) \times \left(\frac{100 \text{ cm}}{1 \text{ m}}\right)^3 = 1.13 \times 10^4 \frac{\text{kg}}{\text{m}^3}.$$

1-5:
$$(327 \text{ in}^3) \times (2.54 \text{ cm/in})^3 \times (1 \text{ L}/1000 \text{ cm}^3) = 5.36 \text{ L}.$$

1-6:
$$1 \text{ m}^3 \times \left(\frac{1000 \text{ L}}{1 \text{ m}^3}\right) \times \left(\frac{1 \text{ gal}}{3.788 \text{ L}}\right) \times \left(\frac{128 \text{ oz.}}{1 \text{ gal}}\right) \times \left(\frac{1 \text{ bottle}}{16 \text{ oz.}}\right)$$
$$= 2111.9 \text{ bottles} \approx 2112 \text{ bottles}.$$

The daily consumption must then be
$$2.11 \times 10^3 \frac{\text{bottles}}{\text{yr}} \times \left(\frac{1 \text{ yr}}{365.24 \text{ da}}\right) = 5.78 \frac{\text{bottles}}{\text{da}}.$$

1-7:   (1450 mi/hr) × (1.61 km/mi) = 2330 km/hr.
   2330 km/hr × (10^3 m/km) × (1 hr/3600 s) = 648 m/s.

1-8: $180{,}000 \dfrac{\text{furlongs}}{\text{fortnight}} \times \left(\dfrac{1 \text{ mile}}{8 \text{ furlongs}}\right) \times \left(\dfrac{1 \text{ fortnight}}{14 \text{ day}}\right) \times \left(\dfrac{1 \text{ day}}{24 \text{ h}}\right) = 67 \dfrac{\text{mi}}{\text{h}}.$

1-9: $15.0 \dfrac{\text{km}}{\text{L}} \times \left(\dfrac{1 \text{ mi}}{1.609 \text{ km}}\right) \times \left(\dfrac{3.788 \text{ L}}{1 \text{ gal}}\right) = 35.3 \dfrac{\text{mi}}{\text{gal}}.$

1-10: $(3.16 \times 10^7 \text{ s} - \pi \times 10^7 \text{ s})/(3.16 \times 10^7 \text{ s}) \times 100 = 0.58\%$

1-11:   a)
$$\frac{10 \text{ m}}{890 \times 10^3 \text{ m}} = 1.1 \times 10^{-3}\%.$$

   b) Since the distance was given as 890 km, the total distance should be 890,000 meters.
   To report the total distance as 890,010 meters, the distance should be given as 890.01 km.

1-12:   a) $(12 \text{ mm}) \times (5.98 \text{ mm}) = 72 \text{ mm}^2$ (two significant figures).

b) $\frac{5.98 \text{ mm}}{12 \text{ mm}} = 0.50$ (also two significant figures).

c) 36 mm (to the nearest millimeter).

d) 6 mm.

e) 2.0.

1-13:   a) If a meter stick can measure to the nearest millimeter, the error will be about 0.13%. b) If the chemical balance can measure to the nearest milligram, the error will be about $8.3 \times 10^{-3}\%$. c) If a handheld stopwatch (as opposed to electric timing devices) can measure to the nearest tenth of a second, the error will be about $2.8 \times 10^{-2}\%$.

1-14:   The area is $10.64 \pm 0.08 \text{ cm}^2$, where the extreme values in the piece's length and width are used to find the uncertainty in the area. The fractional uncertainty in the area is $\frac{0.08 \text{ cm}^2}{10.64 \text{ cm}^2} = 0.75\%$, and the fractional uncertainties in the length and width are $\frac{0.01 \text{ cm}}{5.60 \text{ cm}} = 0.18\%$ and $\frac{0.01 \text{ cm}}{1.9 \text{ cm}} = 0.53\%$.

1-15:   a) The average volume is

$$\pi \frac{(8.50 \text{ cm})^2}{4} (0.050 \text{ cm}) = 2.8 \text{ cm}^3$$

(two significant figures) and the uncertainty in the volume, found from the extreme values of the diameter and thickness, is about 0.3 cm^3, and so the volume of a cookie is $2.8 \pm 0.3 \text{ cm}^3$. (This method does not use the usual form for progation of errors, which is not addressed in the text. The fractional uncertainty in the thickness is so much greater than the fractional uncertainty in the diameter that the fractional uncertainty in the volume is 10%, reflected in the above answer.)

b) $\frac{8.50}{.05} = 170 \pm 20$.

1-16:   (Number of cars $\times$ miles/car·day)/mi/gal = gallons/day

$(2 \times 10^8 \text{ cars} \times 10000 \text{ mi/yr/car} \times 1 \text{ yr/365 days})/(20 \text{ mi/gal}) = 2.75 \times 10^8 \text{ gal/day}$

1-17:   Ten thousand; if it were to contain ten million, each sheet would be on the order of a millionth of an inch thick.

1-18:   If it takes about four kernels to fill 1 cm^3, a 2-L bottle will hold about 8000 kernels.

1-19:   Assuming the two-volume edition, there are approximately a thousand pages, and each page has between 500 and a thousand words (counting captions and the smaller print, such as the end-of-chapter exercise and problems), so an estimate for the number of words is about 10^6.

1-20:   Assuming about 10 breaths per minutes, 24×60 minutes per day, 365 days per year, and a lifespan of fourscore (80) years, the total volume of air breathed in a lifetime is about $2 \times 10^5 \text{ m}^3$. This is the volume of a room $100 \text{ m} \times 100 \text{ m} \times 20 \text{ m}$, which is kind of tight for a major-league baseball game, but it's the same order of magnitude as the volume of the Astrodome.

1-21:   This will vary from person to person, but should be of the order of 1×10^5.

1-22: With a pulse rate of a bit more than one beat per second, a heart will beat 10^5 times per day. With 365 days in a year and the above lifespan of 80 years, the number of beats in a lifetime is about 3×10^9. With $\frac{1}{20}$ L (50 cm^3) per beat, and about $\frac{1}{4}$ gallon per liter, this comes to about 4×10^7 gallons.

1-23: The shape of the pile is not given, but gold coins stacked in a pile might well be in the shape of a pyramid, say with a height of 2 m and a base 3 m $\times$ 3 m. The volume of such a pile is 6 m^3, and the calculations of Example 1-4 indicate that the value of this volume is \$$6 \times 10^8$.

1-24: The surface area of the earth is about $4\pi R^2 = 5 \times 10^{14}$ m^2, where R is the radius of the earth, about 6×10^6 m, so the surface area of all the oceans is about 4×10^{14} m^2. An average depth of about 10 km gives a volume of 4×10^{18} m$^3 = 4 \times 10^{24}$ cm^3. Characterizing the size of a "drop" is a personal matter, but 25 drops/cm^3 is reasonable, giving a total of 10^{26} drops of water in the oceans.

1-25: This will of course depend on the size of the school and who is considered a "student". A school of thousand students, each of whom averages ten pizzas a year (perhaps an underestimate) will total 10^4 pizzas, as will a school of 250 students averaging 40 pizzas a year each.

1-26: The moon is about 4×10^8 m $= 4 \times 10^{11}$ mm away. Depending on age, dollar bills can be stacked with about 2-3 per millimeter, so the number of bills in a stack to the moon would be about 10^{12}. The value of these bills would be \$1 trillion (1 terabuck).

1-27: (Area of USA)/(Area/bill) = number of bills.
(9,372,571 km^2 $\times$ 10^6 m^2/km^2)/(15.6 cm $\times$ 6.7 cm $\times$ 1 m^2/10^4 cm^2) = 9×10^{14} bills
9×10^{14} bills/2.5 10^8 inhabitants = \$3.6 million/inhabitant.

1-28:

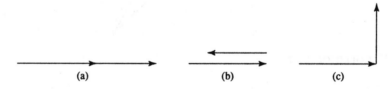

(a)   (b)   (c)

1-29:

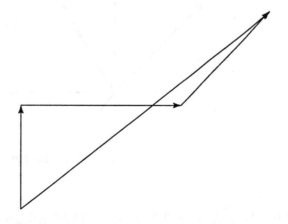

7.8 km, 38° north of east

1-30:

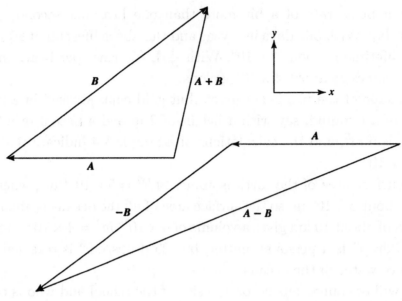

    a) 11.1 m @ 77.6°

    b) 28.5 m @ 202°

    c) 11.1 m @ 258°

    d) 28.5 m @ 22°

1-31:

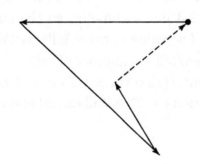

      104 m, 43° north of east.

1-32:

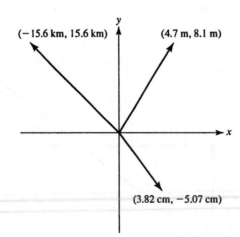

1-33: $\vec{A}$; $A_x = (12.0 \text{ m}) \sin 37.0° = 7.2$ m, $A_y = (12.0 \text{ m}) \cos 37.0° = 9.6$ m.

        $\vec{B}$; $B_x = (15.0 \text{ m}) \cos 40.0° = 11.5$ m, $B_y = -(15.0 \text{ m}) \sin 40.0° = -9.6$ m.

        $\vec{C}$; $C_x = -(6.0 \text{ m}) \cos 60.0° = -3.0$ m, $C_y = -(6.0 \text{ m}) \sin 60.0° = -5.2$ m.

1-34: (The figure is given with the solution to Exercise 1-29).

The net northward displacement is $(2.6 \text{ km}) + (3.1 \text{ km}) \sin 45° = 4.8$ km, and the net eastward displacement is $(4.0 \text{ km}) + (3.1 \text{ km}) \cos 45° = 6.2$ km. The magnitude of the resultant displacement is $\sqrt{(4.8 \text{ km})^2 + (6.2 \text{ km})^2} = 7.8$ km, and the direction is $\arctan\left(\frac{4.8}{6.2}\right) = 38°$ north of east.

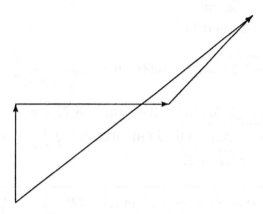

1-35: Using components as a check for any graphical method, the components of $\overrightarrow{B}$ are $B_x = 14.4$ m and $B_y = 10.8$ m, $\overrightarrow{A}$ has one component, $A_x = -12$ m.

a) The x- and y-components of the sum are 2.4 m and 10.8 m, for a magnitude of $\sqrt{(2.4 \text{ m})^2 + (10.8 \text{ m})^2} = 11.1$ m, and an angle of $\arctan\left(\frac{10.8}{2.4}\right) = 77.6°$.

b) The magnitude and direction of $\mathbf{A} + \mathbf{B}$ are the same as $\mathbf{B} + \mathbf{A}$.

c) The x- and y-components of the vector difference are -26.4 m and -10.8 m, for a magnitude of 28.5 m and a direction $\arctan\left(\frac{-10.8}{-26.4}\right) = 202°$. Note that the result of 28.5 m assumes that extra significant figures are kept in the intermediate calculation (*i.e.*, that the imprecise values $\sin 37° = \frac{3}{5}$, $\cos 37° = \frac{4}{5}$ are *not* used), and that 180° must be added to $\arctan\left(\frac{-10.8}{-26.4}\right) = \arctan\left(\frac{10.8}{26.4}\right) = 22°$ in order to give an angle in the third quadrant.

d) $\vec{B} - \vec{A} = 14.4 \text{ m}\hat{i} + 10.8 \text{ m}\hat{j} - 12.0 \text{ m}\hat{i} = 2.4 \text{ m}\hat{i} + 10.8 \text{ m}\hat{j}$.
Magnitude $= \sqrt{(2.4 \text{ m})^2 + (10.8 \text{ m})^2} = 11.1$ m at an angle of $\arctan\left(\frac{10.8}{2.4}\right) = 77.5°$.

1-36: Using Equations (1-8) and (1-9), the magnitude and direction of each of the given vectors is:

a) $\sqrt{(-8.6 \text{ cm})^2 + (5.20 \text{ cm})^2} = 10.05$ cm, $\arctan\left(\frac{5.20}{-8.60}\right) = 328.8°$ (which is $360° - 31.2°$).

b) $\sqrt{(-9.7 \text{ m})^2 + (-2.45 \text{ m})^2} = 10.0$ m, $\arctan\left(\frac{-2.45}{-9.7}\right) = 14° + 180° = 194°$.

c) $\sqrt{(7.75 \text{ km})^2 + (-2.70 \text{ km})^2} = 8.21$ km, $\arctan\left(\frac{-2.7}{7.75}\right) = 340.8°$ (which is $360° - 19.2°$).

1-37:

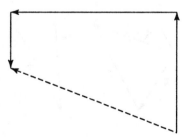

The total northward displacement is 3.25 km − 1.50 km = 1.75 km, and the total westward displacement is 4.75 km. The magnitude of the net displacement is $\sqrt{(1.75 \text{ km})^2 + (4.75 \text{ km})^2} = 5.06$ km. The direction of the net displacement is 69.80° West of North.

1-38: a) The x- and y-components of the sum are 1.30 cm + 4.10 cm = 5.40 cm, 2.25 cm + (−3.75 cm) = −1.50 cm.

b) Using Equations (1-8) and (1-9),

$$\sqrt{(5.40 \text{ cm})^2 + (-1.50 \text{ cm})^2} = 5.60 \text{ cm}, \arctan\left(\frac{-1.50}{+5.40}\right) = 344.5° \text{ ccw}.$$

c) Similarly, 4.10 cm − (1.30 cm) = 2.80 cm, −3.75 cm − (2.25 cm) = −6.00 cm.

d) $\sqrt{(2.80 \text{ cm})^2 + (-6.0 \text{ cm})^2} = 6.62$ cm, $\arctan\left(\frac{-6.00}{2.80}\right) = 295°$ (which is 360° − 65°).

1-39: a) The magnitude of $\vec{A} + \vec{B}$ is

$$\sqrt{\left(\begin{array}{c} ((2.80 \text{ cm}) \cos 60.0° + (1.90 \text{ cm}) \cos 60.0°)^2 \\ + ((2.80 \text{ cm}) \sin 60.0° - (1.90 \text{ cm}) \sin 60.0°)^2 \end{array}\right)} = 2.48 \text{ cm}$$

and the angle is

$$\arctan\left(\frac{(2.80 \text{ cm}) \sin 60.0° - (1.90 \text{ cm}) \sin 60.0°}{(2.80 \text{ cm}) \cos 60.0° + (1.90 \text{ cm}) \cos 60.0°}\right) = 18°.$$

b) The magnitude of $\vec{A} - \vec{B}$ is

$$\sqrt{\left(\begin{array}{c} ((2.80 \text{ cm}) \cos 60.0° - (1.90 \text{ cm}) \cos 60.0°)^2 \\ + ((2.80 \text{ cm}) \sin 60.0° + (1.90 \text{ cm}) \sin 60.0°)^2 \end{array}\right)} = 4.10 \text{ cm}$$

and the angle is

$$\arctan\left(\frac{(2.80 \text{ cm}) \sin 60.0° + (1.90 \text{ cm}) \sin 60.0°}{(2.80 \text{ cm}) \cos 60.0° - (1.90 \text{ cm}) \cos 60.0°}\right) = 84°.$$

c) $\vec{B} - \vec{A} = -\left(\vec{A} - \vec{B}\right)$; the magnitude is 4.10 cm and the angle is 84° + 180° = 264°.

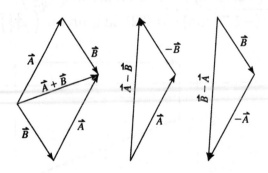

1-40: $\vec{A} = (-12.0 \text{ m})\hat{i}$. More precisely,

$$\vec{A} = (12.0 \text{ m})(\cos 180°)\hat{i} + (12.0 \text{ m})(\sin 180°)\hat{j}.$$
$$\vec{B} = (18.0 \text{ m})(\cos 37°)\hat{i} + (18.0 \text{ m})(\sin 37°)\hat{j} = (14.4 \text{ m})\hat{i} + (10.8 \text{ m})\hat{j}.$$

1-41:
$$\vec{A} = (12.0 \text{ m})\sin 37.0°\hat{i} + (12.0 \text{ m})\cos 37.0°\hat{j} = (7.2 \text{ m})\hat{i} + (9.6 \text{ m})\hat{j}$$
$$\vec{B} = (15.0 \text{ m})\cos 40.0°\hat{i}v - (15.0 \text{ m})\sin 40.0°\hat{j} = (11.5 \text{ m})\hat{i} - (9.6 \text{ m})\hat{j}$$
$$\vec{C} = -(6.0 \text{ m})\cos 60.0°\hat{i} - (6.0 \text{ m})\sin 60.0°\hat{j} = -(3.0 \text{ m})\hat{i} - (5.2 \text{ m})\hat{j}$$

1-42: a)
$$\vec{A} = (3.60 \text{ m})\cos 70.0°\hat{i} + (3.60 \text{ m})\sin 70.0°\hat{j} = (1.23 \text{ m})\hat{i} + (3.38 \text{ m})\hat{j}$$
$$\vec{B} = -(2.40 \text{ m})\cos 30.0°\hat{i} - (2.40 \text{ m})\sin 30.0°\hat{j} = (-2.08 \text{ m})\hat{i} + (-1.20 \text{ m})\hat{j}$$

b) $\vec{C} = (3.00)\vec{A} - (4.00)\vec{B}$

$$= (3.00)(1.23 \text{ m})\hat{i} + (3.00)(3.38 \text{ m})\hat{j} - (4.00)(-2.08 \text{ m})\hat{i} - (4.00)(-1.20 \text{ m})\hat{j}$$
$$= (12.01 \text{ m})\hat{i} + (14.94 \text{ m})\hat{j}$$

(Note that in adding components, the fourth figure becomes significant.)

c) From Equations (1-8) and (1-9),

$$c = \sqrt{(12.01 \text{ m})^2 + (14.94 \text{ m})^2} = 19.17 \text{ m}, \quad \arctan\left(\frac{14.94 \text{ m}}{12.01 \text{ m}}\right) = 51.2°.$$

1-43: a) $A = \sqrt{(4.00)^2 + (3.00)^2} = 5.00, \quad B = \sqrt{(5.00)^2 + (2.00)^2} = 5.39$

b) $\quad \vec{A} - \vec{B} = (4.00 - 3.00)\hat{i} + (5.00 - (-2.00))\hat{j} = (-1.00)\hat{i} + (5.00)\hat{j}$

c) $\quad \sqrt{(1.00)^2 + (5.00)^2} = 5.10, \quad \arctan\left(\frac{5.00}{-1.00}\right) = 101.3°$

d)

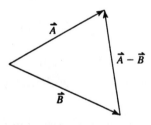

1-44: Method 1: (Product of magnitudes $\times \cos\theta$)

$AB\cos\theta = (12 \text{ m} \times 15 \text{ m})\cos 93° = -9.4 \text{ m}^2$
$BC\cos\theta = (15 \text{ m} \times 6 \text{ m})\cos 80° = 15.6 \text{ m}^2$
$AC\cos\theta = (12 \text{ m} \times 6 \text{ m})\cos 187° = -71.5 \text{ m}^2$

Method 2: (Sum of products of components)

$\mathbf{A} \cdot \mathbf{B} = (7.22)(11.49) + (9.58)(-9.64) = -9.4 \text{ m}^2$
$\mathbf{B} \cdot \mathbf{C} = (11.49)(-3.0) + (-9.64)(-5.20) = 15.6 \text{ m}^2$
$\mathbf{A} \cdot \mathbf{C} = (7.22)(-3.0) + (9.58)(-5.20) = -71.5 \text{ m}^2$

1-45: a) From Eq. (1-21),

$$\vec{A} \cdot \vec{B} = (4.00)(5.00) + (3.00)(-2.00) = 14.00.$$

b) $\mathbf{A} \cdot \mathbf{B} = AB\cos\theta$, so $\theta = \arccos[(14.00)/(5.00 \times 5.39)] = \arccos(.5195) = 58.7°$.

1-46: For all of these pairs of vectors, the angle is found from combining Equations (1-18) and (1-21), to give the angle ϕ as

$$\phi = \arccos\left(\frac{\vec{A} \cdot \vec{B}}{AB}\right) = \arccos\left(\frac{A_x B_x + A_y B_y}{AB}\right).$$

In the intermediate calculations given here, the significant figures in the dot products and in the magnitudes of the vectors are suppressed.

a) $\vec{A} \cdot \vec{B} = -22$, $A = \sqrt{40}$, $B = \sqrt{13}$, and so

$$\phi = \arccos\left(\frac{-22}{\sqrt{40}\sqrt{13}}\right) = 165°.$$

b) $\vec{A} \cdot \vec{B} = 60$, $A = \sqrt{34}$, $B = \sqrt{136}$, $\phi = \arccos\left(\frac{60}{\sqrt{34}\sqrt{136}}\right) = 28°$.

c) $\vec{A} \cdot \vec{B} = 0$, $\phi = 90$.

1-47: Use of the right-hand rule to find cross products gives (a) out of the page and b) into the page.

1-48: a) From Eq. (1-22), the magnitude of the cross product is

$$(12.0 \text{ m})(18.0 \text{ m})\sin(180° - 37°) = 130 \text{ m}^2.$$

The right-hand rule gives the direction as being into the page, or the $-z$-direction. Using Eq. (1-27), the only non-vanishing component of the cross product is

$$C_z = A_x B_y = (-12 \text{ m})((18.0 \text{ m})\sin 37°) = -130 \text{ m}^2.$$

b) The same method used in part (a) can be used, but the relation given in Eq. (1-23) gives the result directly: same magnitude (130 m²), but the opposite direction ($+z$-direction).

1-49: In Eq. (1-27), the only non-vanishing component of the cross product is

$$C_z = A_x B_y - A_y B_x = (4.00)(-2.00) - (3.00)(5.00) = -23.00,$$

so $\vec{A} \times \vec{B} = -(23.00)\hat{k}$, and the magnitude of the vector product is 23.00.

1-50: a) From the right-hand rule, the direction of $\vec{A} \times \vec{B}$ is into the page (the $-z$-direction). The magnitude of the vector product is, from Eq. (1-22),

$$AB\sin\phi = (2.80 \text{ cm})(1.90 \text{ cm})\sin 120° = 4.61 \text{ cm}^2.$$

Or, using Eq. (1-27) and noting that the only non-vanishing component is

$$C_z = A_x B_y - A_y B_x$$
$$= (2.80 \text{ cm}) \cos 60.0°(-1.90 \text{ cm}) \sin 60.0°$$
$$- (2.80 \text{ cm}) \sin 60.0°(1.90 \text{ cm}) \cos 60.0°$$
$$= -4.61 \text{ cm}^2$$

gives the same result.

b) Rather than repeat the calculations, Eq. (1-23) may be used to see that $\vec{B} \times \vec{A}$ has magnitude 4.61 cm² and is in the +z-direction (out of the page).

1-51: a) The area of one acre is $\frac{1}{8}$ mi $\times \frac{1}{80}$ mi $= \frac{1}{640}$ mi², so there are 640 acres to a square mile.

b)
$$(1 \text{ acre}) \times \left(\frac{1 \text{ mi}^2}{640 \text{ acre}} \right) \times \left(\frac{5280 \text{ ft}}{1 \text{ mi}} \right)^2 = 43,560 \text{ ft}^2$$

(all of the above conversions are exact).

c)
$$(1 \text{ acre-foot}) = (43,560 \text{ ft}^3) \times \left(\frac{7.477 \text{ gal}}{1 \text{ ft}^3} \right) = 3.26 \times 10^5 \text{ gal},$$

which is rounded to three significant figures.

1-52: a) ($4,950,000/102 acres) × (1 acre/43560 ft²) × (10.77 ft²/m²) = $12/m².

b) ($12/m²) × (2.54 cm/in)² × (1 m/100 cm)² = $.008/in².
$.008/in² × (1 in × 7/8 in) = $.007 for a postage stamp sized parcel.

1-53: a) To three significant figures, the time for one cycle is

$$\frac{1}{1.420 \times 10^9 \text{ Hz}} = 7.04 \times 10^{-10} \text{ s.}$$

b)
$$\left(1.420 \times 10^9 \frac{\text{cycles}}{\text{s}} \right) \times \left(\frac{3600 \text{ s}}{1 \text{ hr}} \right) = 5.11 \times 10^{12} \frac{\text{cycles}}{\text{hr}}$$

c) Using the conversion from years to seconds given in Appendix F,

$$(1.42 \times 10^9 \text{ Hz}) \times \left(\frac{3.156 \times 10^7 \text{ s}}{1 \text{ yr}} \right) \times (4.600 \times 10^9 \text{ yr}) = 2.06 \times 10^{26}.$$

d) 4.600×10^9 yr $= (4.60 \times 10^4)(1.00 \times 10^5$ yr), so the clock would be off by 4.60×10^4 s.

1-54: Assume a 70-kg person, and the human body is mostly water. Use Appendix D to find the mass of one H_2O molecule: 18.015 u × 1.661 × 10^{-27} kg/u = 2.992 × 10^{-26} kg/molecule. (70 kg/2.992 × 10^{-26} kg/molecule) = 2.34 × 10^{27} molecules. (Assuming carbon to be the most common atom gives 3 × 10^{27} molecules)

1-55: Assume each person sees the dentist twice a year for checkups, for 2 hours. Assume 2 more hours for restorative work. Assuming most dentists work less than 2000 hours per year, this gives 2000 hours/4 hours per patient = 500 patients per dentist. Assuming only half of the people who should go to a dentist do, there should be about 1 dentist per 1000 inhabitants. Note: A dental assistant in an office with more than one treatment room could increase the number off patients seen in a single dental office.

1-56:   a) $(6.0 \times 10^{24} \text{ kg}) \times \left(\dfrac{6.0 \times 10^{23} \frac{\text{atoms}}{\text{mole}}}{14 \times 10^{-3} \frac{\text{kg}}{\text{mole}}} \right) = 2.6 \times 10^{50}$ atoms.

b) The number of neutrons is the mass of the neutron star divided by the mass of a neutron:

$$\frac{(2)(2.0 \times 10^{30} \text{ kg})}{(1.7 \times 10^{-27} \text{ kg/neutron})} = 2.4 \times 10^{57} \text{ neutrons.}$$

c) The average mass of a particle is essentially $\frac{2}{3}$ the mass of either the proton or the neutron, 1.7×10^{-27} kg. The total number of particles is the total mass divided by this average, and the total mass is the volume times the average density. Denoting the density by ρ (the notation introduced in Chapter 14),

$$\frac{M}{m_{\text{ave}}} = \frac{\frac{4}{3}\pi R^3 \rho}{\frac{2}{3}m_{\text{p}}} = \frac{(2\pi)(1.5 \times 10^{11} \text{ m})^3(10^{18} \text{ kg/m}^3)}{(1.7 \times 10^{-27} \text{ kg})} = 1.2 \times 10^{79}.$$

Note the conversion from g/cm^3 to kg/m^3.

1-57:   a)

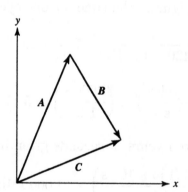

b) Algebraically, $\vec{A} = \vec{C} - \vec{B}$, and so the components of $\vec{A}$ are

$$A_x = C_x - B_x = (6.40 \text{ cm}) \cos 22.0° - (6.40 \text{ cm}) \cos 63.0° = 3.03 \text{ cm}$$
$$A_y = C_y - B_y = (6.40 \text{ cm}) \sin 22.0° + (6.40) \sin 63.0° = 8.10 \text{ cm.}$$

c) $A = \sqrt{(3.03 \text{ cm})^2 + (8.10 \text{ cm})^2} = 8.65 \text{ cm},$ $\arctan \left(\dfrac{8.10 \text{ cm}}{3.03 \text{ cm}} \right) = 69.5°$

1-58:　a) $R_x = A_x + B_x + C_x$

$$= (12.0 \text{ m})\cos(90° - 37°) + (15.00 \text{ m})\cos(-40°) + (6.0 \text{ m})\cos(180° + 60°)$$

$$= 15.7 \text{ m}, \quad \text{and}$$

$$R_y = A_y + B_y + C_y$$

$$= (12.0 \text{ m})\sin(90° - 37°) + (15.0 \text{ m})\sin(-40°) + (6.0 \text{ m})\sin(180° + 60°)$$

$$= -5.3 \text{ m}.$$

The magnitude of the resultant is $R = \sqrt{R_x^2 + R_y^2} = 16.6$ m, and the direction from the positive x-axis is $\arctan\left(\frac{-5.3}{15.7}\right) = -18.6°$. Keeping extra significant figures in the intermediate calculations gives an angle of $-18.49°$, which when considered as a positive counterclockwise angle from the positive x-axis and rounded to the nearest degree is 342°.

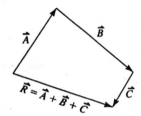

b) $S_x = -3.00 \text{ m} - 7.22 \text{ m} - 11.49 \text{ m} = -21.71 \text{ m};$

$\quad S_y = -5.20 \text{ m} - (-9.64 \text{ m}) - 9.58 \text{ m} = -5.14 \text{ m}$

$$\theta = \arctan\left[\frac{(-5.14)}{(-21.71)}\right] = 13.3°$$

$$S = \sqrt{(-21.71 \text{ m})^2 + (-5.14 \text{ m})^2} = 22.3 \text{ m}$$

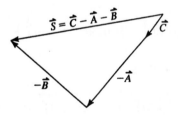

1-59:

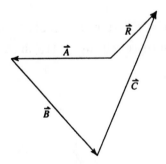

As in Problem 1-64, take the east direction to be the x-direction and the north direction to be the y-direction. The x- and y-components of the resultant displacement of the first three displacements are then

$$(-180 \text{ m}) + (210 \text{ m})\sin 45° + (280 \text{ m})\sin 30° = 108 \text{ m},$$
$$-(210 \text{ m})\cos 45° + (280 \text{ m})\cos 30° = +94.0 \text{ m},$$

keeping an extra significant figure. The magnitude and direction of this net displacement are

$$\sqrt{(108 \text{ m})^2 + (94.0 \text{ m})^2} = 144 \text{ m}, \quad \arctan\left(\frac{94 \text{ m}}{108 \text{ m}}\right) = 40.9°.$$

The fourth displacement must then be 144 m in a direction 40.9° south of west.

1-60:

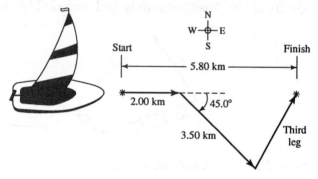

The third leg must have taken the sailor east a distance

$$(5.80 \text{ km}) - (3.50 \text{ km})\cos 45° - (2.00 \text{ km}) = 1.33 \text{ km}$$

and a distance north

$$(3.5 \text{ km})\sin 45° = 2.47 \text{ km}.$$

The magnitude of the displacement is

$$\sqrt{(1.33 \text{ km})^2 + (2.47 \text{ km})^2} = 2.81 \text{ km},$$

and the direction is $\arctan\left(\frac{2.47}{1.33}\right) = 62°$ north of east, which is $90° - 62° = 28°$ east of north. A more precise answer will require retaining extra significant figures in the intermediate calculations.

1-61: a)

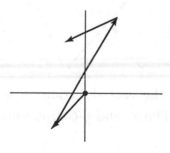

b) The net east displacement is $-(2.80 \text{ km}) \sin 45° + (7.40 \text{ km}) \cos 30° - (3.30 \text{ km}) \cos 22° = 1.37 \text{ km}$, and the net north displacement is

$$-(2.80 \text{ km}) \cos 45.0° + (7.40 \text{ km}) \sin 30.0° - (3.30 \text{ km}) \sin 22.0° = 0.48 \text{ km},$$

and so the distance traveled is

$$\sqrt{(1.37 \text{ km})^2 + (0.48 \text{ km})^2} = 1.45 \text{ km}.$$

1-62: The eastward displacement of Manhattan from Lincoln is

$$(147 \text{ km}) \sin 85° + (106 \text{ km}) \sin 167° + (166 \text{ km}) \sin 235° = 34.3 \text{ km}$$

and the northward displacement is

$$(147 \text{ km}) \cos 85° + (106 \text{ km}) \cos 167° + (166 \text{ km}) \cos 235° = -185.7 \text{ km}.$$

(A negative northward displacement is a southward displacement, as indicated in Fig. (1-30). Extra figures have been kept in the intermediate calculations.)

a) $$\sqrt{(34.3 \text{ km})^2 + (185.7 \text{ km})^2} = 189 \text{ km}$$

b) The direction from Lincoln to Manhattan, relative to the north, is

$$\arctan\left(\frac{34.3 \text{ km}}{-185.7 \text{ km}}\right) = 169.5°$$

and so the direction to fly in order to return to Lincoln is $169.5° + 180° = 349.5°$.

1-63: a) Angle of first line is $\theta = \tan^{-1}\left(\frac{200-20}{210-10}\right) = 42°$. Angle of second line is $42° + 30° = 72°$. Therefore

$$X = 10 + 250 \cos 72° = 87$$
$$Y = 20 + 250 \sin 72° = 258$$

for a final point of (87,258).

b) The computer screen now looks something like this:

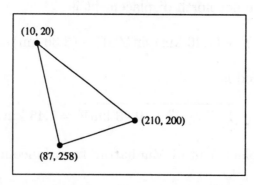

The length of the bottom line is $\sqrt{(210-87)^2 + (200-258)^2} = 136$ and its direction is $\tan^{-1}\left(\frac{258-200}{210-87}\right) = 25°$ below straight left.

1-64: a)

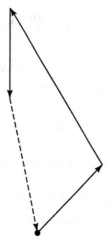

b) To use the method of components, let the east direction be the x-direction and the north direction be the y-direction. Then, the explorer's net x-displacement is, in units of his step size,

$$(40)\cos 45° - (80)\cos 60° = -11.7$$

and the y-displacement is

$$(40)\sin 45° + (80)\sin 60° - 50 = 47.6.$$

The magnitude and direction of the displacement are

$$\sqrt{(-11.7)^2 + (47.6)^2} = 49, \quad \arctan\left(\frac{47.6}{-11.7}\right) = 104°.$$

(More precision in the angle is not warranted, as the given measurements are to the nearest degree.) To return to the hut, the explorer must take 49 steps in a direction $104° - 90° = 14°$ east of south.

1-65: a) With $A_z = B_z = 0$, Eq. (1-22) becomes

$$A_x B_x + A_y B_y = (A \cos \theta_A)(B \cos \theta_B) + (A \sin \theta_A)(B \sin \theta_B)$$
$$= AB \left(\cos \theta_A \cos \theta_B + \sin \theta_A \sin \theta_B \right)$$
$$= AB \cos \left(\theta_A - \theta_B \right)$$
$$= AB \cos \phi,$$

where the expression for the cosine of the difference between two angles has been used (see Appendix B).

b) With $A_z = B_z = 0$, $\vec{C} = C_z \hat{k}$ and $C = |C_z|$. From Eq. (1-27),

$$|C| = |A_x B_y - A_y B_x|$$
$$= |(A \cos \theta_A)(B \sin \theta_B) - (A \sin \theta_A)(B \cos \theta_A)|$$
$$= AB \left| \cos \theta_A \sin \theta_B - \sin \theta_A \cos \theta_B \right|$$
$$= AB \left| \sin(\theta_B - \theta_A) \right|$$
$$= AB \sin \phi.$$

1-66: a) The angle between the vectors is $210° - 70° = 140°$, and so Eq. (1-18) gives $\vec{A} \cdot \vec{B} = (3.60 \text{ m})(2.40 \text{ m}) \cos 140° = -6.62 \text{ m}^2$. Or, Eq. (1-21) gives

$$\vec{A} \cdot \vec{B} = A_x B_x + A_y B_y$$
$$= (3.60 \text{ m}) \cos 70° (2.4 \text{ m}) \cos 210° + (3.6 \text{ m}) \sin 70° (2.4 \text{ m}) \sin 210°$$
$$= -6.62 \text{ m}^2.$$

b) From Eq. (1-22), the magnitude of the cross product is

$$(3.60 \text{ m})(2.40 \text{ m}) \sin 140° = 5.55 \text{ m}^2,$$

and the direction, from the right-hand rule, is out of the page (the $+z$-direction). From Eq. (1-30), with the z-components of $\vec{A}$ and $\vec{B}$ vanishing, the z-component of the cross product is

$$A_x B_y - A_y B_x = (3.60 \text{ m}) \cos 70° (2.40 \text{ m}) \sin 210°$$
$$- (3.60 \text{ m}) \sin 70° (2.40 \text{ m}) \cos 210°$$
$$= 5.55 \text{ m}^2.$$

1-67: a) Parallelogram area $= 2 \times$ area of triangle ABC

Triangle area $= 1/2$ (base)(height) $= 1/2$ (B)(A $\sin \theta$)

Paralellogram area $= BA \sin \theta$

b) 90°

1-68: With the $+x$-axis to the right, $+y$-axis toward the top of the page, and $+z$-axis out of the page, $(\vec{A} \times \vec{B})_x = 87.8 \text{ cm}^2$, $(\vec{A} \times \vec{B})_y = 68.9 \text{ cm}^2$, $(\vec{A} \times \vec{B})_z = 0$.

1-69: a)
$$A = \sqrt{(2.00)^2 + (3.00)^2 + (4.00)^2} = 5.39.$$
$$B = \sqrt{(3.00)^2 + (1.00)^2 + (3.00)^2} = 4.36.$$

b)
$$\vec{A} - \vec{B} = (A_x - B_x)\hat{i} + (A_y - B_y)\hat{j} + (A_z - B_z)\hat{k}$$
$$= (-5.00)\hat{i} + (2.00)\hat{j} + (7.00)\hat{k}$$

c)
$$\sqrt{(5.00)^2 + (2.00)^2 + (7.00)^2} = 8.83,$$

and this will be the magnitude of $\vec{B} - \vec{A}$ as well.

1-70: The direction vectors each have magnitude $\sqrt{3}$, and their dot product is $(1)(1) + (1)(-1) + (1)(-1) = -1$, so from Eq. (1-18) the angle between the bonds is $\arccos\left(\frac{-1}{\sqrt{3}\sqrt{3}}\right) = \arccos\left(-\frac{1}{3}\right) = 109°$.

1-71: The best way to show these results is to use the result of part (a) of Problem 1-65, a restatement of the law of cosines. We know that

$$C^2 = A^2 + B^2 + 2AB\cos\phi,$$

where ϕ is the angle between $\vec{A}$ and $\vec{B}$.

a) If $C^2 = A^2 + B^2$, $\cos\phi = 0$ and the angle between $\vec{A}$ and $\vec{B}$ is 90° (the vectors are perpendicular).

b) If $C^2 < A^2 + B^2$, $\cos\phi < 0$, and the angle between $\vec{A}$ and $\vec{B}$ is greater than 90°.

c) If $C^2 > A^2 + B^2$, $\cos\phi > 0$, and the angle between $\vec{A}$ and $\vec{B}$ is less than 90°.

1-72: a) This is a statement of the law of cosines, and there are many ways to derive it. The most straightforward way, using vector algebra, is to assume the linearity of the dot product (a point used, but not explicitly mentioned in the text) to show that the square of the magnitude of the sum $\vec{A} + \vec{B}$ is

$$(\vec{A} + \vec{B}) \cdot (\vec{A} + \vec{B}) = \vec{A} \cdot \vec{A} + \vec{A} \cdot \vec{B} + \vec{B} \cdot \vec{A} + \vec{B} \cdot \vec{B}$$
$$= \vec{A} \cdot \vec{A} + \vec{B} \cdot \vec{B} + 2\vec{A} \cdot \vec{B}$$
$$= A^2 + B^2 + 2\vec{A} \cdot \vec{B}$$
$$= A^2 + B^2 + 2AB\cos\phi.$$

Using components, if the vectors make angles θ_A and θ_B with the x-axis, the components of the vector sum are $A\cos\theta_A + B\cos\theta_B$ and $A\sin\theta_A + B\sin\theta_B$, and the square of the magnitude is

$$(A\cos\theta_A + B\cos\theta_B)^2 + \left(A\sin\theta_A + B\sin\theta_B\right)^2$$
$$= A^2(\cos^2\theta_A + \sin^2\theta_A) + B^2\left(\cos^2\theta_B + \sin^2\theta_B\right)$$
$$+ 2AB(\cos\theta_A\cos\theta_B + \sin\theta_A\sin\theta_B)$$
$$= A^2 + B^2 + 2AB\cos(\theta_A - \theta_B)$$
$$= A^2 + B^2 + 2AB\cos\phi,$$

where $\phi = \theta_A - \theta_B$ is the angle between the vectors.

b) A geometric consideration shows that the vectors $\vec{A}$, $\vec{B}$ and the sum $\vec{A} + \vec{B}$ must be the sides of an equilateral triangle. The angle *between* $\vec{A}$ and $\vec{B}$ is 120°, since one vector must shifted to add head-to-tail. Using the result of part (a), with $A = B$, the condition is that $A^2 = A^2 + A^2 + 2A^2 \cos\phi$, which solves for $1 = 2 + 2\cos\phi$, $\cos\phi = -\frac{1}{2}$, and $\phi = 120°$.

c) Either method of derivation will have the angle ϕ replaced by $180° - \phi$, so the cosine will change sign, and the result is $\sqrt{A^2 + B^2 - 2AB\cos\phi}$.

d) Similar to what is done in part (b), when the vector *difference* has the same magnitude, the angle between the vectors is 60°. Algebraically, ϕ is obtained from $1 = 2 - 2\cos\phi$, so $\cos\phi = \frac{1}{2}$ and $\phi = 60°$.

1-73: Take the length of a side of the cube to be L, and denote the vectors from a to b, a to c and a to d as $\vec{B}$, $\vec{C}$ and $\vec{D}$. In terms of unit vectors,

$$\vec{B} = L\hat{k}, \quad \vec{C} = L\left(\hat{j} + \hat{k}\right), \quad \vec{D} = L\left(\hat{i} + \hat{j} + \hat{k}\right).$$

Using Eq. (1-18),

$$\arccos\left(\frac{\vec{B}\cdot\vec{D}}{BD}\right) = \arccos\left(\frac{L^2}{(L)(L\sqrt{3})}\right) = 54.7°,$$

$$\arccos\left(\frac{\vec{C}\cdot\vec{D}}{CD}\right) = \arccos\left(\frac{2L^2}{(L\sqrt{2})(L\sqrt{3})}\right) = 35.3°.$$

1-74: From Eq. (1-27), the cross product is

$$(-13.00)\hat{i} + (6.00)\hat{j} + (-11.00)\hat{k} = 13\left[-(1.00)\hat{i} + \left(\frac{6.00}{13.00}\right)\hat{j} - \frac{11.00}{13.00}\hat{k}\right].$$

The magnitude of the vector in square brackets is $\sqrt{1.93}$, and so a unit vector in this direction (which is necessarily perpendicular to both $\vec{A}$ and $\vec{B}$) is

$$\frac{[-(1.00)\hat{i} + (6.00/13.00)\hat{j} - (11/13)\hat{k}]}{\sqrt{1.93}}.$$

The negative of this vector,

$$\frac{[(1.00)\hat{i} - (6.00/13.00)\hat{j} + (11/13)\hat{k}]}{\sqrt{1.93}},$$

is also a unit vector perpendicular to $\vec{A}$ and $\vec{B}$.

1-75: a) Using Equations (1-21) and (1-27), and recognizing that the vectors $\vec{A}$, $\vec{B}$ and $\vec{C}$ do not have the same meanings as they do in those equations,

$$(\vec{A} \times \vec{B}) \cdot \vec{C} = \left((A_y B_z - A_z B_y)\hat{i} + (A_z B_x - A_x B_z)\hat{j} + (A_x B_y - A_y B_x)\hat{k}\right) \cdot \vec{C}$$
$$= A_y B_z C_x - A_z B_y C_x + A_z B_x C_y - A_x B_z C_y + A_x B_y C_z - A_y B_x C_z.$$

A similar calculation shows that

$$\vec{A} \cdot (\vec{B} \times \vec{C}) = A_x B_y C_z - A_x B_z C_y + A_y B_z C_x - A_y B_x C_z + A_z B_x C_y - A_z B_y C_x$$

and a comparison of the expressions shows that they are the same.

b) Although the above expression could be used, the form given allows for ready compuation of $\vec{A} \times \vec{B}$; the magnitude is $AB \sin \phi = (20.00) \sin 37.0°$ and the direction is, from the right-hand rule, in the $+z$-direction, and so

$$(\vec{A} \times \vec{B}) \cdot \vec{C} = +(20.00) \sin 37.0° (6.00) = +72.2.$$

1-76: a) The maximum and minimum areas are

$$(L + l)(W + w) = LW + lW + Lw, \quad (L - l)(W - w) = LW - lW - Lw,$$

where the common terms wl have been omitted. The area and its uncertainty are then $WL \pm (lW + Lw)$, so the uncertainty in the area is $a = lW + Lw$.

b) The fractional uncertainty in the area is

$$\frac{a}{A} = \frac{lW + Wl}{WL} = \frac{l}{L} + \frac{w}{W},$$

the sum of the fractional uncertainties in the length and width.

c) The simlilar calculation to find the uncertainty v in the volume will involve neglecting the terms lwH, lWh and Lwh as well as lwh; the uncertainty in the volume is $v = lWH + LwH + LWh$, and the fractional uncertainty in the volume is

$$\frac{v}{V} = \frac{lWH + LwH + LWh}{LWH} = \frac{l}{L} + \frac{w}{W} + \frac{h}{H},$$

the sum of the fractional uncertainties in the length, width and height.

1-77: The receiver's position is

$$(+1.0 + 9.0 - 6.0 + 12.0)\hat{i} + (-5.0 + 11.0 + 4.0 + 18.0)\hat{j} = (16.0)\hat{i} + (28.0)\hat{j}.$$

The vector from the quarterback to the receiver is the receiver's position minus the quarterback's position, or $(16.0)\hat{i} + (35.0)\hat{j}$, a vector with magnitude $\sqrt{(16.0)^2 + (35.0)^2} = 38.5$, given as being in yards. The angle is $\arctan\left(\frac{16.0}{35.0}\right) = 24.6°$ to the right of downfield.

1-78: a)

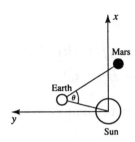

b) i) In AU, $\sqrt{(0.3182)^2 + (0.9329)^2} = 0.9857$.

ii) In AU, $\sqrt{(1.3087)^2 + (-.4423)^2 + (-.0414)^2} = 1.3820$

iii) In AU,

$$\sqrt{(0.3182 - (1.3087))^2 + (0.9329 - (-.4423))^2 + (0.0414)^2} = 1.695.$$

c) The angle between the directions from the Earth to the Sun and to Mars is obtained from the dot product. Combining Equations (1-18) and (1-21),

$$\phi = \arccos\left(\frac{(-0.3182)(1.3087 - 0.3182) + (-0.9329)(-0.4423 - 0.9329) + 0}{(0.9857)(1.695)}\right) = 54.6°.$$

d) Mars could not have been visible at midnight, because the Sun-Mars angle is less then 90°.

1-79: a)

The law of cosines (see Problem 1-72) gives the distance as

$$\sqrt{(138 \text{ ly})^2 + (77 \text{ ly})^2 + 2(138 \text{ ly})(77 \text{ ly}) \cos 154.4°} = 76.2 \text{ ly},$$

where the supplement of 25.6° has been used for the angle between the direction vectors.

b) Although the law of cosines could be used again, it's far more convenient to use the law of sines (Appendix B), and the angle is given by

$$\arcsin\left(\frac{\sin 25.6°}{76.2 \text{ ly}} \, 138 \text{ ly}\right) = 51.5°, 180° - 51.5° = 129°,$$

where the appropriate angle in the second quadrant is used.

1-80: Define $\vec{S} = A\hat{i} + B\hat{j} + C\hat{k}$

$$\vec{r} \cdot \vec{S} = (x\hat{i} + y\hat{j} + z\hat{k}) \cdot (A\hat{i} + B\hat{j} + C\hat{k})$$
$$= Ax + By + Cz$$

If the points satisfy $Ax + By + Cz = 0$, then $\vec{r} \cdot \vec{S} = 0$ and all points $\vec{r}$ are perpendicular to $\vec{S}$.

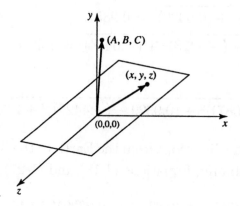

Chapter 2 Motion Along a Straight Line

2-1: a) During the later 4.75-s interval, the rocket moves a distance 1.00×10^3 m-63 m, and so the magnitude of the average velocity is

$$\frac{1.00 \times 10^3 \text{ m} - 63 \text{ m}}{4.75 \text{ s}} = 197 \text{ m/s}.$$

b) $\frac{1.00 \times 10^3 \text{ m}}{5.90 \text{ s}} = 169$ m/s.

2-2: a) The magnitude of the average velocity on the return flight is

$$\frac{(5150 \times 10^3 \text{ m})}{(13.5 \text{ da})(86,400 \text{ s/da})} = 4.42 \text{ m/s}.$$

The direction has been defined to be the $-x$-direction $(-\hat{\imath})$.

b) Because the bird ends up at the starting point, the average velocity for the round trip is **0**.

2-3: Although the distance could be found, the intermediate calculation can be avoided by considering that the time will be inversely proportional to the speed, and the extra time will be

$$(140 \text{ min})\left(\frac{105 \text{ km/hr}}{70 \text{ km/hr}} - 1\right) = 70 \text{ min}.$$

2-4: The eastward run takes $(200 \text{ m}/5.0 \text{ m/s}) = 40.0$ s and the westward run takes $(280 \text{ m}/4.0 \text{ m/s}) = 70.0$ s. a) $(200 \text{ m} + 280 \text{ m})/(40.0 \text{ s} + 70.0 \text{ s}) = 4.4$ m/s to two significant figures. b) The net displacement is 80 m west, so the average velocity is $(80 \text{ m}/110.0 \text{ s}) = 0.73$ m/s in the $-x$-direction $(-\hat{\imath})$.

2-5: a) The van will travel 480 m for the first 60 s and 1200 m for the next 60 s, for a total distance of 1680 m in 120 s and an average speed of 14.0 m/s. b) The first stage of the journey takes $\frac{240 \text{ m}}{8.0 \text{ m/s}} = 30$ s and the second stage of the journey takes $(240 \text{ m}/20 \text{ m/s}) = 12$ s, so the time for the 480-m trip is 42 s, for an average speed of 11.4 m/s. c) The first case (part (a)); the average speed will be the numerical average only if the time intervals are the same.

2-6: From the expression for $x(t)$, $x(0) = 0$, $x(2.00 \text{ s}) = 5.60$ m and $x(4.00 \text{ s}) = 20.8$ m.
a) $\frac{5.60 \text{ m} - 0}{2.00 \text{ s}} = 2.80$ m/s   b) $\frac{20.8 \text{ m} - 0}{4.00 \text{ s}} = 5.2$ m/s   c) $\frac{20.8 \text{ m} - 5.60 \text{ m}}{2.00 \text{ s}} = 7.6$ m/s

2-7: a) At $t_1 = 0$, $x_1 = 0$, so Eq. (2-2) gives

$$v_{\text{av}} = \frac{x_2}{t_2} = \frac{(2.4 \text{ m/s}^2)(10.0 \text{ s})^2 - (0.120 \text{ m/s}^3)(10.0 \text{ s})^3}{(10.0 \text{ s})} = 12.0 \text{ m/s}.$$

b) From Eq. (2-3), the instantaneous velocity as a function of time is

$$v = 2bt - 3ct^2 = (4.80 \text{ m/s}^2)t - (0.360 \text{ m/s}^3)t^2,$$

so i) $v(0) = 0$,

ii) $v(5.0 \text{ s}) = (4.80 \text{ m/s}^2)(5.0 \text{ s}) - (0.360 \text{ m/s}^3)(5.0 \text{ s})^2 = 15.0 \text{ m/s}$,
and iii) $v(10.0 \text{ s}) = (4.80 \text{ m/s}^2)(10.0 \text{ s}) - (0.360 \text{ m/s}^3)(10.0 \text{ s})^2 = 12.0 \text{ m/s}$.

c) The car is at rest when $v = 0$. Therefore $(4.80 \text{ m/s}^2)t - (0.360 \text{ m/s}^3)t^2 = 0$. The only time after $t = 0$ when the car is at rest is $t = \frac{4.80 \text{ m/s}^2}{0.360 \text{ m/s}^3} = 13.3 \text{ s}$

2-8: a) IV: The curve is horizontal; this corresponds to the time when she stops. b) I: This is the time when the curve is most nearly straight and tilted upward (indicating postive velocity). c) V: Here the curve is plainly straight, tilted downward (negative velocity). d) II: The curve has a postive slope that is increasing. e) III: The curve is still tilted upward (positive slope and positive velocity), but becoming less so.

2-9:   Time (s)   0   2   4   6   8   10   12   14   16
      Acceleration (m/s²)   0   1   2   2   3   1.5   1.5   0

a) The acceleration is not constant, but is approximately constant between the times $t = 4$ s and $t = 8$ s.

b)

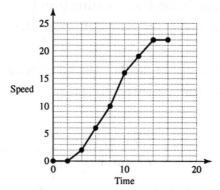

At $t = 9$ s, $a = 3 \text{ m/s}^2$, at $t = 13$ s, $a = 1.5 \text{ m/s}^2$, at $t = 15$ s, $a = 0$.

2-10: The cruising speed of the car is 60 km/hr = 16.7 m/s. a) $\frac{16.7 \text{ m/s}}{10 \text{ s}} = 1.7 \text{ m/s}^2$ (to two significant figures). b) $\frac{0 - 16.7 \text{ m/s}}{10 \text{ s}} = -1.7 \text{ m/s}^2$ c) No change in speed, so the acceleration is zero. d) The final speed is the same as the initial speed, so the average acceleration is zero.

2-11: a) The plot of the velocity seems to be the most curved upward near $t = 5$ s. b) The only negative acceleration (downward-sloping part of the plot) is between $t = 30$ s and $t = 40$ s. c) At $t = 20$ s, the plot is level, and in Exercise 2-10 the car is said to be cruising at constant speed, and so the acceleration is zero. d) The plot is very nearly a straight line, and the acceleration is that found in part (b) of Exercise 2-10, -1.7 m/s^2.

e)

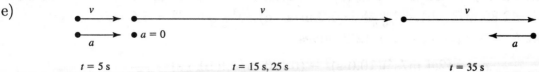

2-12: Use of Eq. (2-5), with $\Delta t = 10$ s in all cases,

a) $((5.0 \text{ m/s}) - (15.0 \text{ m/s}))/(10 \text{ s}) = -1.0 \text{ m/s}^2$,

b) $((-15.0 \text{ m/s}) - (-5.0 \text{ m/s}))/(10 \text{ s}) = -1.0 \text{ m/s}^2$,

c) $((-15.0 \text{ m/s}) - (15.0 \text{ m/s}))/(10 \text{ s}) = -3.0 \text{ m/s}^2$.

In all cases, the negative acceleration indicates an acceleration to the left.

2-13: a) Assuming the car comes to rest from 65 mph (29 m/s) in 4 seconds, a = (29 m/s − 0)/(4 s) = 7.25 m/s^2.

b) Since the car is coming to a stop, the acceleration is in the direction opposite to the velocity. If the velocity is in the positive direction, the acceleration is negative; if the velocity is in the negative direction, the acceleration is positive.

2-14: a) The velocity at $t = 0$ is

$$(3.00 \text{ m/s}) + (0.100 \text{ m/s}^3)(0) = 3.00 \text{ m/s},$$

and the velocity at $t = 5.00$ s is

$$(3.00 \text{ m/s}) + (0.100 \text{ m/s}^3)(5.00 \text{ s})^2 = 5.50 \text{ m/s},$$

so Eq. (2-4) gives the average acceleration as v

$$\frac{(5.50 \text{ m/s}) - (3.00 \text{ m/s})}{(5.00 \text{ s})} = .50 \text{ m/s}^2.$$

b) The instantaneous acceleration is obtained by using Eq. (2-5),

$$a = \frac{dv}{dt} = 2\beta t = (0.2 \text{ m/s}^3)t.$$

Then, i) at $t = 0$, $a = (0.2 \text{ m/s}^3)(0) = 0$, and

ii) at $t = 5.00$ s, $a = (0.2 \text{ m/s}^3)(5.00 \text{ s}) = 1.0 \text{ m/s}^2$.

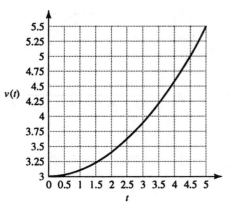
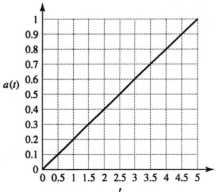

2-15: a)

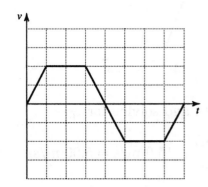
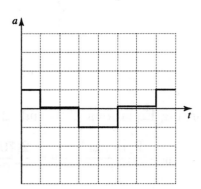

b)

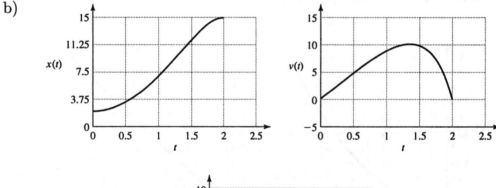

t = 2.5 s

a = 0
t = 10 s

t = 20 s
v = 0

a = 0
t = 30 s

t = 37.5 s

2-16: a) The bumper's velocity and acceleration are given as functions of time by

$$v = \frac{dx}{dt} = (9.60 \text{ m/s}^2)t - (0.600 \text{ m/s}^6)t^5$$

$$a = \frac{dv}{dt} = (9.60 \text{ m/s}^2) - (3.000 \text{ m/s}^6)t^4.$$

There are two times at which $v = 0$ (three if negative times are considered), given by $t = 0$ and $t^4 = 16$ s^4. At $t = 0$, $x = 2.17$ m and $a = 9.60$ m/s^2. When $t^4 = 16$ s^4,

$$x = (2.17 \text{ m}) + (4.80 \text{ m/s}^2)\sqrt{(16 \text{ s}^4)} - (0.100 \text{ m/s}^6)(16 \text{ s}^4)^{3/2} = 14.97 \text{ m},$$

$$a = (9.60 \text{ m/s}^2) - (3.000 \text{ m/s}^6)(16 \text{ s}^4) = -38.4 \text{ m/s}^2.$$

b)

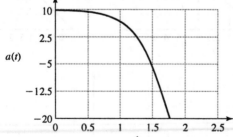

2-17: a) Equating Equations (2-9) and (2-10) and solving for v_0,

$$v_0 = \frac{2(x - x_0)}{t} - v = \frac{2(70 \text{ m})}{7.00 \text{ s}} - 15.0 \text{ m/s} = 5.00 \text{ m/s}.$$

b) The above result for v_0 may be used to find

$$a = \frac{v - v_0}{t} = \frac{15.0 \text{ m/s} - 5.00 \text{ m/s}}{7.00 \text{ s}} = 1.43 \text{ m/s}^2,$$

or the intermediate calculation can be avoided by using the result of Exercise 2-18; solving for a,

$$a = 2\left(\frac{v}{t} - \frac{x - x_0}{t}\right) = 2\left(\frac{15.0 \text{ m/s}}{7.00 \text{ s}} - \frac{70.0 \text{ m}}{(7.00 \text{ s})^2}\right) = 1.43 \text{ m/s}^2.$$

2-18:   a) The acceleration is found from Eq. (2-13), with $v_0 = 0$;

$$a = \frac{v^2}{2(x - x_0)} = \frac{\left((173 \text{ mi/hr})\left(\frac{0.4770 \text{ m/s}}{1 \text{ mi/hr}}\right)\right)^2}{2\left((307 \text{ ft})\left(\frac{1 \text{ m}}{3.281 \text{ ft}}\right)\right)} = 36.4 \text{ m/s}^2,$$

where the conversions are from Appendix E.

b) The time can be found from the above acceleration,

$$t = \frac{v}{a} = \frac{(173 \text{ mi/hr})\left(\frac{0.4770 \text{ m/s}}{1 \text{ mi/hr}}\right)}{36.4 \text{ m/s}^2} = 2.27 \text{ s}.$$

The intermediate calculation may be avoided by using Eq. (2-14), again with $v_0 = 0$,

$$t = \frac{2(x - x_0)}{v} = \frac{2\left((307 \text{ ft})\left(\frac{1 \text{ m}}{3.281 \text{ ft}}\right)\right)}{(173 \text{ mi/hr})\left(\frac{0.4770 \text{ m/s}}{1 \text{ mi/hr}}\right)} = 2.27 \text{ s}.$$

2-19:   From Eq. (2-13), with $v = 0$, $a = \frac{v_0^2}{2(x - x_0)} < a_{max}$. Taking $x_0 = 0$,

$$x > \frac{v_0^2}{2a_{max}} = \frac{((105 \text{ km/hr})(1 \text{ m/s})/(3.6 \text{ km/hr}))^2}{2(250 \text{ m/s}^2)} = 1.70 \text{ m}.$$

2-20:   In Eq. (2-14), with $x - x_0$ being the length of the runway, and $v_0 = 0$ (the plane starts from rest), $v = 2\frac{x - x_0}{t} = 2\frac{280 \text{ m}}{8 \text{ s}} = 70.0 \text{ m/s}$.

2-21:   a) From Eq. (2-13), with $v_0 = 0$,

$$a = \frac{v_0^2}{2(x - x_0)} = \frac{(20 \text{ m/s})^2}{2(120 \text{ m})} = 1.67 \text{ m/s}^2.$$

b) Using Eq. (2-14), $t = 2(x - x_0)/v = 2(120 \text{ m})/(20 \text{ m/s}) = 12 \text{ s}$.
c) $(12 \text{ s})(20 \text{ m/s}) = 240 \text{ m}$.

2-22: a) $x_0 < 0, v_0 < 0, a < 0$

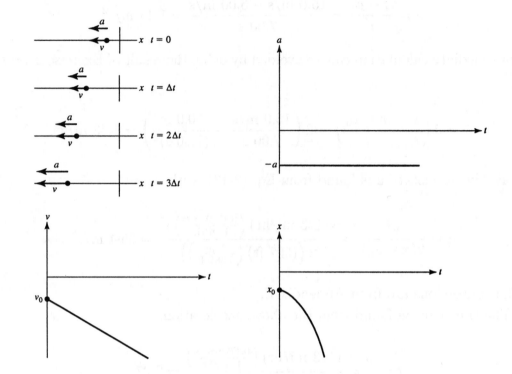

b) $x_0 > 0, v_0 < 0, a > 0$

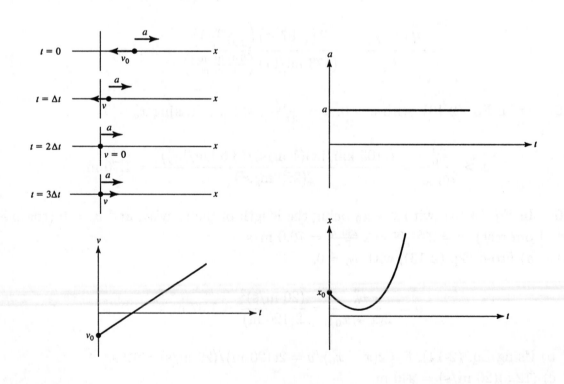

c) $x_0 > 0, v_0 > 0, a < 0$

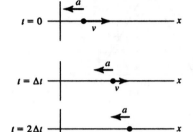

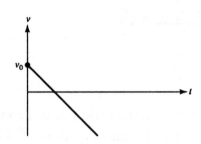

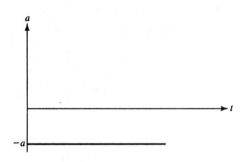
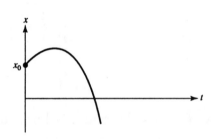

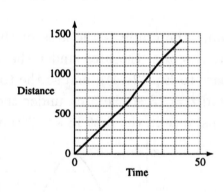
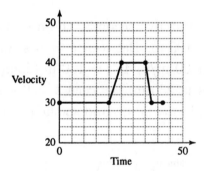

2-23:   a)

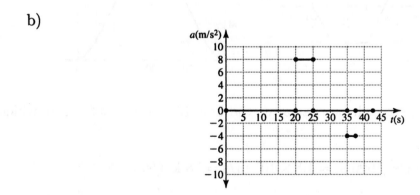

b)

2-24:   a)

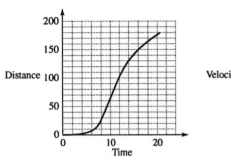
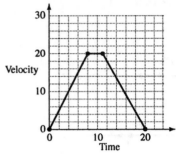

b)

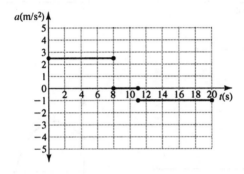

2-25:   a) At $t = 3$ s the graph is horizontal and the acceleration is 0. From $t = 5$ s to $t = 9$ s, the acceleration is constant (from the graph) and equal to $\frac{45 \text{ m/s} - 20 \text{ m/s}}{4 \text{ s}} = 6.3 \text{ m/s}^2$. From $t = 9$ s to $t = 13$ s the acceleration is constant and equal to $\frac{0 - 45 \text{ m/s}}{4 \text{ s}} = -11.2 \text{ m/s}^2$.

b) In the first five seconds, the area under the graph is the area of the rectangle, $(20 \text{ m})(5 \text{ s}) = 100$ m. Between $t = 5$ s and $t = 9$ s, the area under the trapezoid is $(1/2)(45 \text{ m/s} + 20 \text{ m/s})(4 \text{ s}) = 130$ m (compare to Eq. (2-14)), and so the total distance in the first 9 s is 230 m. Between $t = 9$ s and $t = 13$ s, the area under the triangle is $(1/2)(45 \text{ m/s})(4 \text{ s}) = 90$ m, and so the total distance in the first 13 s is 320 m.

2-26:

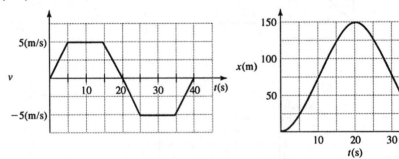

2-27:   a) The maximum speed will be that after the first 10 min (or 600 s), at which time the speed will be

$$(20.0 \text{ m/s}^2)(900 \text{ s}) = 1.8 \times 10^4 \text{ m/s} = 18 \text{ km/s}.$$

b) During the first 15 minutes (and also during the last 15 minutes), the ship will travel $(1/2)(18 \text{ km/s})(900 \text{ s}) = 8100$ km, so the distance traveled at non-constant speed is 16,200 km and the fraction of the distance traveled at constant speed is

$$1 - \frac{16{,}200 \text{ km}}{384{,}000 \text{ km}} = 0.958,$$

keeping an extra significant figure.

c) The time spent at constant speed is $\frac{384{,}000 \text{ km} - 16{,}200 \text{ km}}{18 \text{ km/s}} = 2.04 \times 10^4$ s and the time spent during both the period of acceleration and deceleration is 900 s, so the total time required for the trip is 2.22×10^4 s, about 6.2 hr.

2-28: After the initial acceleration, the train has traveled

$$\frac{1}{2}(1.60 \text{ m/s}^2)(14.0 \text{ s})^2 = 156.8 \text{ m}$$

(from Eq. (2-12), with $x_0 = 0$, $v_0 = 0$), and has attained a speed of

$$(1.60 \text{ m/s}^2)(14.0 \text{ s}) = 22.4 \text{ m/s}.$$

During the 70-second period when the train moves with constant speed, the train travels (22.4 m/s)(70 s) = 1568 m. The distance traveled during deceleration is given by Eq. (2-13), with $v = 0$, $v_0 = 22.4$ m/s and $a = -3.50$ m/s^2, so the train moves a distance $x - x_0 = \frac{-(22.4 \text{ m/s})^2}{2(-3.50 \text{ m/s}^2)} = 71.68$ m. The total distance covered is then 156.8 m + 1568 m + 71.7 m = 1.8 km.

In terms of the initial acceleration a_1, the initial acceleration time t_1, the time t_2 during which the train moves at constant speed and the magnitude a_2 of the final acceleration, the total distance x_T is given by

$$x_T = \frac{1}{2}a_1 t_1^2 + (a_1 t_1)t_2 + \frac{1}{2}\frac{(a_1 t_1)^2}{|a_2|} = \left(\frac{a_1 t_1}{2}\right)\left(t_1 + 2t_2 + \frac{a_1 t_1}{|a_2|}\right),$$

which yields the same result.

2-29: a)

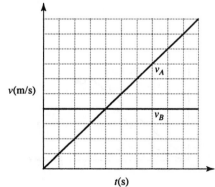

b) From the graph (Fig. (2-30)), the curves for A and B intersect at $t = 1$ s and $t = 3$ s.

c)

   d) From Fig. (2-30), the graphs have the same slope at $t = 2$ s. e) Car A passes car B when they have the same position and the slope of curve A is greater than that of curve B in Fig. (2-30); this is at $t = 3$ s. f) Car B passes car A when they have the same position and the slope of curve B is greater than that of curve A; this is at $t = 1$ s.

2-30:   a) The truck's position as a function of time is given by $x_T = v_T t$, with v_T being the truck's constant speed, and the car's position is given by $x_C = (1/2)a_C t^2$. Equating the two expressions and dividing by a factor of t (this reflects the fact that the car and the truck are at the same place at $t = 0$) and solving for t yields

$$t = \frac{2v_T}{a_C} = \frac{2(20.0 \text{ m/s})}{3.20 \text{ m/s}^2} = 12.5 \text{ s}$$

and at this time

$$x_T = x_C = 250 \text{ m}.$$

   b) $a_C t = (3.20 \text{ m/s}^2)(12.5 \text{ s}) = 40.0 \text{ m/s}$ (See Exercise 2-31 for a discussion of why the car's speed at this time is twice the truck's speed.)

   c)

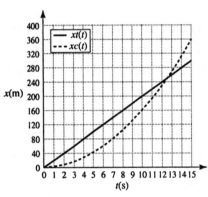

   d)

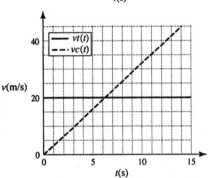

2-31:   a)

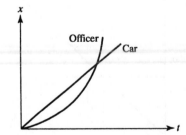

The car and the motorcycle have gone the same distance during the same time, so their average speeds are the same. The car's average speed is its constant speed v_C, and for constant acceleration from rest, the motorcycle's speed is always twice its average, or $2v_C$. b) From the above, the motorcyle's speed will be v_C after half the time needed to catch the car. For motion from rest with constant acceleration, the distance traveled is proportional to the square of the time, so for half the time one-fourth of the total distance has been covered, or $d/4$.

2-32: a) An initial height of 200 m gives a speed of 60 m/s when rounded to one significant figure. This is approximately 200 km/hr or approximately 150 mi/hr. (Different values of the approximate height will give different answers; the above may be interpreted as slightly better than order of magnitude answers.) b) Personal experience will vary, but speeds on the order of one or two meters per second are reasonable. c) Air resistance may certainly not be neglected.

2-33: a) From Eq. (2-13), with $v = 0$ and $a = -g$,

$$v_0 = \sqrt{2g(y - y_0)} = \sqrt{2(9.80 \text{ m/s}^2)(0.440 \text{ m})} = 2.94 \text{ m/s},$$

which is probably too precise for the speed of a flea; rounding down, the speed is about 2.9 m/s.

b) The time the flea is rising is the above speed divided by g, and the total time is twice this; symbolically,

$$t = 2\frac{\sqrt{2g(y - y_0)}}{g} = 2\sqrt{\frac{2(y - y_0)}{g}} = 2\sqrt{\frac{2(0.440 \text{ m})}{(9.80 \text{ m/s}^2)}} = 0.599 \text{ s},$$

or about 0.60 s.

2-34: Using Eq. (2-13), with downward velocities and accelerations being positive, $v^2 = (0.8 \text{ m/s})^2 + 2(1.6 \text{ m/s}^2)(5.0 \text{ m}) = 16.64 \text{ m}^2/\text{s}^2$ (keeping extra significant figures), so $v = 4.1 \text{ m/s}$.

2-35: a) If the meter stick is in free fall, the distance d is related to the reaction time t by $d = (1/2)gt^2$, so $t = \sqrt{2d/g}$. If d is measured in centimeters, the reaction time is

$$t = \sqrt{\frac{2}{g}}\sqrt{d} = \sqrt{\frac{2}{980 \text{ cm/s}^2}}\sqrt{d} = (4.52 \times 10^{-2} \text{ s})\sqrt{d/(1 \text{ cm})}.$$

b) Using the above result, $(4.52 \times 10^{-2} \text{ s})\sqrt{17.6} = 0.190 \text{ s}$.

2-36: a) $(1/2)gt^2 = (1/2)(9.80 \text{ m/s}^2)(2.5 \text{ s})^2 = 30.6 \text{ m}$.

b) $gt = (9.80 \text{ m/s}^2)(2.5 \text{ s}) = 24.5 \text{ m/s}$.

c)

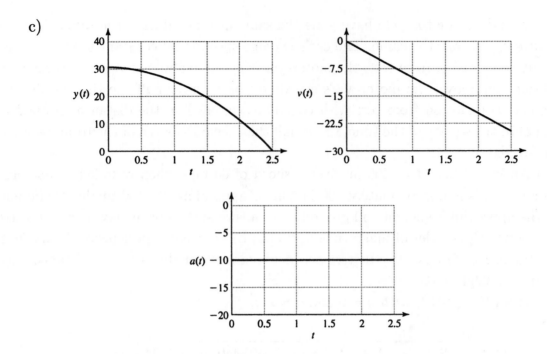

2-37: a) Using the method of Example 2-8, the time the ring is in the air is

$$t = \frac{v_0 + \sqrt{v_0^2 - 2g(y - y_0)}}{g}$$

$$= \frac{(5.00 \text{ m/s}) + \sqrt{(5.00 \text{ m/s})^2 - 2(9.80 \text{ m/s}^2)(-12.0 \text{ m})}}{(9.80 \text{ m/s}^2)}$$

$$= 2.156 \text{ s},$$

keeping an extra significant figure. The average velocity is then $\frac{12.0 \text{ m}}{2.156 \text{ s}} = 5.57 \text{ m/s}$, down.

As an alternative to using the quadratic formula, the speed of the ring when it hits the ground may be obtained from $v^2 = v_0^2 - 2g(y - y_0)$, and the average velocity found from $\frac{v + v_0}{2}$; this is algebraically identical to the result obtained by the quadratic formula.

b) While the ring is in free fall, the average acceleration is the constant acceleration due to gravity, 9.80 m/s^2 down.

c)
$$y = y_0 + v_0 t - \frac{1}{2}gt^2$$

$$0 = 12.0 \text{ m} + (5.00 \text{ m/s})t - \frac{1}{2}(9.8 \text{ m/s}^2)t^2$$

Solve this quadratic as in part a) to obtain $t = 2.156$ s.

d)
$$v^2 = v_0^2 - 2g(y - y_0) = (5.00 \text{ m/s})^2 - 2(9.8 \text{ m/s}^2)(-12.0 \text{ m})$$

$$|v| = 16.1 \text{ m/s}$$

e)

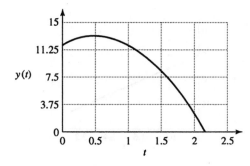
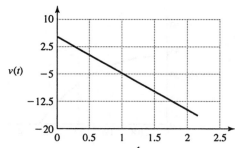
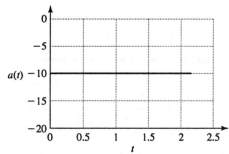

2-38: a) Using $a = -g$, $v_0 = 5.00$ m/s and $y_0 = 40.0$ m in Eqs. (2-8) and (2-12) gives i) at $t = 0.250$ s,

$$y = (40.0 \text{ m}) + (5.00 \text{ m/s})(0.250 \text{ s}) - (1/2)(9.80 \text{ m/s}^2)(0.250 \text{ s})^2 = 40.9 \text{ m},$$
$$v = (5.00 \text{ m/s}) - (9.80 \text{ m/s}^2)(0.250 \text{ s}) = 2.55 \text{ m/s}$$

and ii) at $t = 1.00$ s,

$$y = (40.0 \text{ m}) + (5.00 \text{ m/s})(1.00 \text{ s}) - (1/2)(9.80 \text{ m/s}^2)(1.00 \text{ s})^2 = 40.1 \text{ m},$$
$$v = (5.00 \text{ m/s}) - (9.80 \text{ m/s}^2)(1.00 \text{ s}) = -4.80 \text{ m/s}.$$

b) Using the result derived in Example 2-8, the time is

$$t = \frac{(5.00 \text{ m/s}) + \sqrt{(5.00 \text{ m/s})^2 - 2(9.80 \text{ m/s}^2)(0 - 40.0 \text{ m})}}{(9.80 \text{ m/s}^2)} = 3.41 \text{ s}.$$

c) Either using the above time in Eq. (2-8) or avoiding the intermediate calculation by using Eq. (2-13),

$$v^2 = v_0^2 - 2g(y - y_0) = (5.00 \text{ m/s})^2 - 2(9.80 \text{ m/s}^2)(-40.0 \text{ m}) = 809 \text{ m}^2/\text{s}^2,$$
$$v = 28.4 \text{ m/s}.$$

d) Using $v = 0$ in Eq. (2-13) gives

$$y = \frac{v_0^2}{2g} + y_0 = \frac{(5.00 \text{ m/s})^2}{2(9.80 \text{ m/s}^2)} + 40.0 \text{ m} = 41.2 \text{ m}.$$

e)

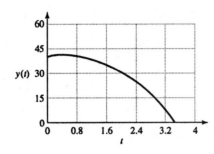

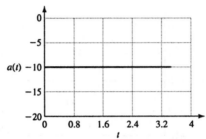

2-39: a) $v = v_0 - gt = (-6.00 \text{ m/s}) - (9.80 \text{ m/s}^2)(2.00 \text{ s}) = -25.6 \text{ m/s}$, so the speed is 25.6 m/s.

b) $y = v_0 t - \frac{1}{2}gt^2 = (-6.00 \text{ m/s})(2.00 \text{ s}) - \frac{1}{2}(9.80 \text{ m/s}^2)(2.00 \text{ s})^2 = -31.6 \text{ m}$,

with the minus sign indicating that the balloon has indeed fallen.

c) $v^2 = v_0^2 - 2g(y_0 - y) = (6.00 \text{ m/s})^2 - 2(9.80 \text{ m/s}^2)(-10.0 \text{ m}) = 232 \text{ m}^2/\text{s}^2$,

so $v = 15.2 \text{ m/s}$.

d)

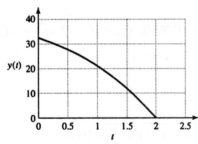

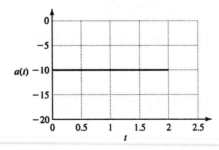

2-40: a) The vertical distance from the initial position is given by

$$y = v_0 t - \frac{1}{2}gt^2;$$

solving for v_0,

$$v_0 = \frac{y}{t} + \frac{1}{2}gt = \frac{(-50.0 \text{ m})}{(5.00 \text{ s})} + \frac{1}{2}(9.80 \text{ m/s}^2)(5.00 \text{ s}) = 14.5 \text{ m/s}.$$

b) The above result could be used in $v^2 = v_0^2 - 2g(y - y_0)$, with $v = 0$, to solve for $y - y_0 = 10.7$ m (this requires retention of two extra significant figures in the calculation for v_0). c) 0 d) 9.8 m/s^2, down.

e) Assume the top of the building is 50 m above the ground for purposes of graphing:

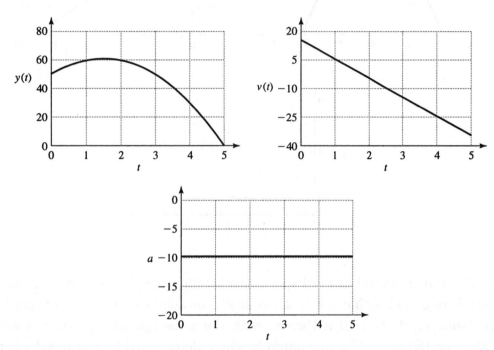

2-41: a) $(224 \text{ m/s})/(0.9 \text{ s}) = 249 \text{ m/s}^2$. b) $\frac{249 \text{ m/s}^2}{9.80 \text{ m/s}^2} = 25.4$. c) The most direct way to find the distance is $v_{\text{ave}}t = ((224 \text{ m/s})/2)(0.9 \text{ s}) = 101$ m.

d) $(283 \text{ m/s})/(1.40 \text{ s}) = 202 \text{ m/s}^2$ but $40g = 392 \text{ m/s}^2$, so the figures are not consistent.

2-42: a) From Eq. (2-8), solving for t gives $(40.0 \text{ m/s} - 20.0 \text{ m/s})/9.80 \text{ m/s}^2 = 2.04$ s.

b) Again from Eq. (2-8),

$$\frac{40.0 \text{ m/s} - (-20.0 \text{ m/s})}{9.80 \text{ m/s}^2} = 6.12 \text{ s}.$$

c) The displacement will be zero when the ball has returned to its original vertical position, with velocity opposite to the original velocity. From Eq. (2-8),

$$\frac{40 \text{ m/s} - (-40 \text{ m/s})}{9.80 \text{ m/s}^2} = 8.16 \text{ s}.$$

(This ignores the $t = 0$ solution.)

d) Again from Eq. (2-8), $(40 \text{ m/s})/(9.80 \text{ m/s}^2) = 4.08$ s. This is, of course, half the time found in part (c).

e) 9.80 m/s^2, down, in all cases.

f)

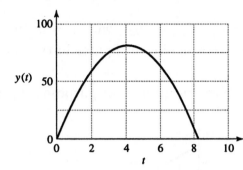
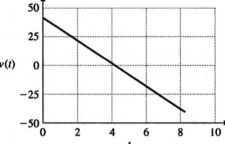

2-43: a) For a given initial upward speed, the height would be inversely proportional to the magnitude of g, and with g one-tenth as large, the height would be ten times higher, or 7.5 m. b) Similarly, if the ball is thrown with the same upward speed, it would go ten times as high, or 180 m. c) The maximum height is determined by the speed when hitting the ground; if this speed is to be the same, the maximum height would be ten times as large, or 20 m.

2-44: a) From Eq. (2-15), the velocity v_2 at time t

$$v_2 = v_1 + \int_{t_1}^{t} \alpha t \, dt$$
$$= v_1 + \frac{\alpha}{2}\left(t^2 - t_1^2\right)$$
$$= v_1 - \frac{\alpha}{2}t_1^2 + \frac{\alpha}{2}t^2$$
$$= (5.0 \text{ m/s}) - (0.6 \text{ m/s}^3)(1.0 \text{ s})^2 + (0.6 \text{ m/s}^3)t^2$$
$$= (4.40 \text{ m/s}) + (0.6 \text{ m/s}^3)t^2.$$

At $t_2 = 2.0$ s, the velocity is $v_2 = (4.40 \text{ m/s}) + (0.6 \text{ m/s}^3)(2.0 \text{ s})^2 = 6.80$ m/s, or 6.8 m/s to two significant figures.

b) From Eq. (2-16), the position x_2 as a function of time is

$$x_2 = x_1 + \int_{t_1}^{t} v \, dt$$

$$= (6.0 \text{ m}) + \int_{t_1}^{t} ((4.40 \text{ m/s}) + (0.6 \text{ m/s}^3)t^2) \, dt$$

$$= (6.0 \text{ m}) + (4.40 \text{ m/s})(t - t_1) + \frac{(0.6 \text{ m/s}^3)}{3} \left(t^3 - t_1^3 \right).$$

At $t = 2.0$ s, and with $t_1 = 1.0$ s,

$$x = (6.0 \text{ m}) + (4.40 \text{ m/s}) \left((2.0 \text{ s}) - (1.0 \text{ s}) \right) + (0.20 \text{ m/s}^3)((2.0 \text{ s})^3 - (1.0 \text{ s})^3)$$

$$= 11.8 \text{ m}.$$

c)

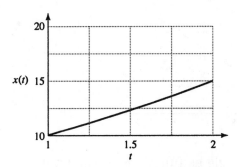
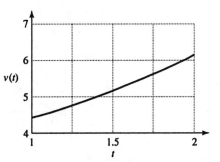
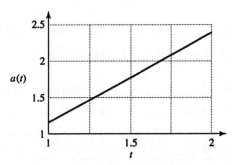

2-45: a) From Eqs. (2-17) and (2-18), with $v_0 = 0$ and $x_0 = 0$,

$$v = \int_{0}^{t} \left(At - Bt^2 \right) dt = \frac{A}{2}t^2 - \frac{B}{3}t^3 = (0.75 \text{ m/s}^3)t^2 - (0.040 \text{ m/s}^4)t^3$$

$$x = \int_{0}^{t} \left(\frac{A}{2}t^2 - \frac{B}{3}t^3 \right) dt = \frac{A}{6}t^3 - \frac{B}{12}t^4 = (0.25 \text{ m/s}^3)t^3 - (0.010 \text{ m/s}^4)t^4.$$

b) For the velocity to be a maximum, the acceleration must be zero; this occurs at $t = 0$ and $t = \frac{A}{B} = 12.5$ s. At $t = 0$ the velocity is a minimum, and at $t = 12.5$ s the velocity is

$$v = (0.75 \text{ m/s}^3)(12.5 \text{ s})^2 - (0.040 \text{ m/s}^4)(12.5 \text{ s})^3 = 39.1 \text{ m/s}.$$

2-46: a) To average 4 mi/hr, the total time for the twenty-mile ride must be five hours, so the second ten miles must be covered in 3.75 hours, for an average of 2.7 mi/hr. b) To average 12 mi/hr, the second ten miles must be covered in 25 minutes and the average speed must be 24 mi/hr. c) After the first hour, only ten of the twenty miles have been covered, and 16 mi/hr is not possible as the average speed.

2-47: a)

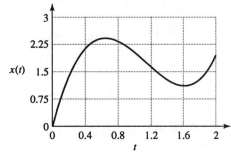 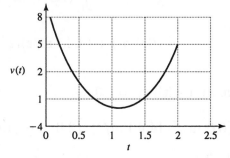

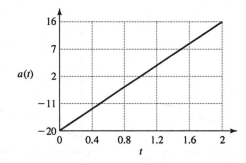

The velocity and acceleration of the particle as functions of time are

$$v(t) = (9.00 \text{ m/s}^3)t^2 - (20.00 \text{ m/s}^2)t + (9.00 \text{ m/s})$$
$$a(t) = (18.0 \text{ m/s}^3)t - (20.00 \text{ m/s}^2).$$

b) The particle is at rest when the velocity is zero; setting $v = 0$ in the above expression and using the quadratic formula to solve for the time t,

$$t = \frac{(20.0 \text{ m/s}^3) \pm \sqrt{(20.0 \text{ m/s}^3)^2 - 4(9.0 \text{ m/s}^3)(9.0 \text{ m/s})}}{2(9.0 \text{ m/s}^3)}$$

and the times are 0.63 s and 1.60 s. c) The acceleration is negative at the earlier time and positive at the later time. d) The velocity is instantaneously not changing when the acceleration is zero; solving the above expression for $a(t) = 0$ gives

$$\frac{20.00 \text{ m/s}^2}{18.00 \text{ m/s}^3} = 1.11 \text{ s}.$$

Note that this time is the numerical average of the times found in part (c). e) The greatest distance is the position of the particle when the velocity is zero and the acceleration is negative; this occurs at 0.63 s, and at that time the particle is at

$$(3.00 \text{ m/s}^3)(0.63 \text{ s})^3 - (10.0 \text{ m/s}^2)(0.63 \text{ s})^2 + (9.00 \text{ m/s})(0.63 \text{ s}) = 2.45 \text{ m}.$$

(In this case, retaining extra significant figures in evaluating the roots of the quadratic equation does not change the answer in the third place.) f) The acceleration is negative at $t = 0$ and is increasing, so the particle is speeding up at the greatest rate at $t = 2.00$ s and slowing down at the greatest rate at $t = 0$. This is a situation where the extreme values of a function (in the case the acceleration) occur not at times when $\frac{da}{dt} = 0$ but at the endpoints of the given range.

2-48: a) $\frac{25.0 \text{ m}}{20.0 \text{ s}} = 1.25$ m/s.

b) $\frac{25 \text{ m}}{15 \text{ s}} = 1.67$ m/s.

c) Her net displacement is zero, so the average velocity has zero magnitude.

d) $\frac{50.0 \text{ m}}{35.0 \text{ s}} = 1.43$ m/s. Note that the answer to part (d) is the *harmonic* mean, not the arithmetic mean, of the answers to parts (a) and (b). (See Exercise 2-5.)

2-49: Denote the times, speeds and lengths of the two parts of the trip as t_1 and t_2, v_1 and v_2, and l_1 and l_2.

a) The average speed for the whole trip is

$$\frac{l_1 + l_2}{t_1 + t_2} = \frac{l_1 + l_2}{(l_1/v_1) + (l_2/v_2)} = \frac{(76 \text{ km}) + (34 \text{ km})}{\left(\frac{76 \text{ km}}{88 \text{ km/hr}}\right) + \left(\frac{34 \text{ km}}{72 \text{ km/hr}}\right)} = 82 \text{ km/hr},$$

or 82.3 km/hr, keeping an extra significant figure.

b) Assuming nearly straight-line motion (a common feature of Nebraska highways), the total distance traveled is $l_1 - l_2$ and

$$|v_{\text{ave}}| = \frac{l_1 - l_2}{t_1 + t_2} = \frac{(76 \text{ km}) - (34 \text{ km})}{\left(\frac{76 \text{ km}}{88 \text{ km/hr}}\right) + \left(\frac{34 \text{ km}}{72 \text{ km/hr}}\right)} = 31 \text{ km/hr}.$$

(31.4 km/hr to three significant figures.)

2-50: a) The space per vehicle is the speed divided by the frequency with which the cars pass a given point;

$$\frac{96 \text{ km/h}}{2400 \text{ vehicles/h}} = 40 \text{ m/vehicle}.$$

An average vehicle is given to be 4.6 m long, so the average spacing is 40.0 m − 4.6 m = 35.4 m.

b) An average spacing of 9.2 m gives a space per vehicle of 13.8 m, and the traffic flow rate is

$$\frac{96000 \text{ m/h}}{13.8 \text{ m/vehicle}} = 6960 \text{ vehicle/h}.$$

2-51: (a) Denote the time for the acceleration (4.0 s) as t_1 and the time spent running at constant speed (5.1 s) as t_2. The constant speed is then at_1, where a is the unknown acceleration. The total l is then given in terms of a, t_1 and t_2 by

$$l = \frac{1}{2}at_1^2 + at_1 t_2,$$

and solving for a gives

$$a = \frac{l}{(1/2)t_1^2 + t_1 t_2} = \frac{(100 \text{ m})}{(1/2)(4.0 \text{ s})^2 + (4.0 \text{ s})(5.1 \text{ s})} = 3.5 \text{ m/s}^2.$$

(b) During the 5.1 s interval, the runner is not accelerating, so a = 0.

(c) $\Delta v / \Delta t = [(3.5 \text{ m/s}^2)(4 \text{ s})]/(9.1 \text{ s}) = 1.54 \text{ m/s}$.

(d) The runner was moving at constant velocity for the last 5.1 s.

2-52: a) Simple subtraction and division gives average speeds during the 2-second intervals as 5.6, 7.2 and 8.8 m/s.

b) The average speed increases by 1.6 m/s during each 2-second interval, so the acceleration is 0.8 m/s^2.

c) From Eq. (2-13), with $v_0 = 0$, $v = \sqrt{2(0.8 \text{ m/s}^2)(14.4 \text{ m})} = 4.8$ m/s. Or, recognizing that for constant acceleration the average speed of 5.6 m/s is the speed one second after passing the 14.4-m mark, 5.6 m/s $- (0.8 \text{ m/s}^2)(1.0 \text{ s}) = 4.8$ m/s.

d) With both the acceleration and the speed at the 14.4-m known, either Eq. (2-8) or Eq. (2-12) gives the time as 6.0 s.

e) From Eq. (2-12), $x - x_0 = (4.8 \text{ m/s})(1.0 \text{ s}) + \frac{1}{2}(0.8 \text{ m/s}^2)(1.0 \text{ s})^2 = 5.2$ m. This is also the average velocity $(1/2)(5.6 \text{ m/s} + 4.8 \text{ m/s})$ times the time interval of 1.0 s.

2-53: If the driver steps on the gas, the car will travel

$$(20 \text{ m/s})(3.0 \text{ s}) + (1/2)(2.3 \text{ m/s}^2)(3.0 \text{ s})^2 = 70.4 \text{ m}.$$

If the brake is applied, the car will travel

$$(20 \text{ m/s})(3.0 \text{ s}) + (1/2)(-3.8 \text{ m/s}^2)(3.0 \text{ s})^2 = 42.9 \text{ m},$$

so the driver should apply the brake.

2-54: a) The simplest way to do this is to go to a frame in which the freight train (which moves with constant velocity) is stationary. Then, the passenger train has an initial relative velocity of $v_{\text{rel},0} = 10$ m/s. This relative speed would be decreased to zero after the relative separation had decreased to $\frac{v_{\text{rel},0}^2}{2a_{\text{rel}}} = +500$ m. Since this is larger in magnitude than the original relative separation of 200 m, there will be a collision.   b) The time at which the relative separation goes to zero (*i.e.*, the collision time) is found by solving a quadratic (see Problems 2-29 & 2-30 or Example 2-8). The time is given by

$$t = \frac{1}{a}\left(v_{\text{rel},0} - \sqrt{v_{\text{rel},0}^2 + 2ax_{\text{rel},0}}\right)$$
$$= (10 \text{ s}^2/\text{m})(10 \text{ m/s} - \sqrt{100 \text{ m}^2/\text{s}^2 - 40 \text{ m}^2/\text{s}^2})$$
$$= (100 \text{ s})(1 - \sqrt{0.6}).$$

Substitution of this time into Eq. (2-12), with $x_0 = 0$, yields 538 m as the distance the passenger train moves before the collision.

2-55: The total distance you cover is $1.20 \text{ m} + 0.90 \text{ m} = 2.10 \text{ m}$ and the time available is $\frac{1.20 \text{ m}}{1.50 \text{ m/s}} = 0.80 \text{ s}$. Solving Eq. (2-12) for a,

$$a = 2\frac{(x - x_0) - v_0 t}{t^2} = 2\frac{(2.10 \text{ m}) - (0.80 \text{ m/s})(0.80 \text{ s})}{(0.80 \text{ s})^2} = 4.56 \text{ m/s}^2.$$

2-56: One convenient way to do the problem is to do part (b) first; the time spent accelerating from rest to the maximum speed is $\frac{20 \text{ m/s}}{2.5 \text{ m/s}^2} = 8.0 \text{ s}$. At this time, the officer is

$$x_1 = \frac{v_1^2}{2a} = \frac{(20 \text{ m/s})^2}{2(2.5 \text{ m/s}^2)} = 80.0 \text{ m}.$$

This could also be found from $(1/2)a_1 t_1^2$, where t_1 is the time found for the acceleration. At this time the car has moved $(15 \text{ m/s})(8.0 \text{ s}) = 120 \text{ m}$, so the officer is 40 m behind the car.

a) The remaining distance to be covered is $300 \text{ m} - x_1$ and the average speed is $(1/2)(v_1 + v_2) = 17.5 \text{ m/s}$, so the time needed to slow down is

$$\frac{360 \text{ m} - 80 \text{ m}}{17.5 \text{ m/s}} = 16.0 \text{ s},$$

and the total time is 24.0 s.

c) The officer slows from 20 m/s to 15 m/s in 16.0 s (the time found in part (a)), so the acceleration is -0.31 m/s^2.

d), e)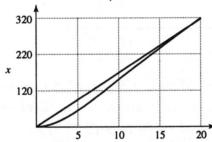

2-57: a) $x_T = (1/2)a_T t^2$, and with $x_T = 40.0 \text{ m}$, solving for the time gives $t = \sqrt{\frac{2(40.0 \text{ m})}{(2.10 \text{ m/s}^2)}} = 6.17 \text{ s}$.

b) The car has moved a distance

$$\frac{1}{2}a_C t^2 = \frac{a_C}{a_T}x_1 = \frac{3.40 \text{ m/s}^2}{2.10 \text{ m/s}^2} 40.0 \text{ m} = 64.8 \text{ m},$$

and so the truck was initially 24.8 m in front of the car.

c) The speeds are $a_T t = 13 \text{ m/s}$ and $a_C t = 21 \text{ m/s}$.

d)

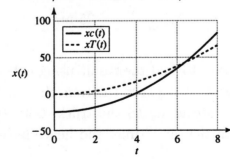

2-58: The position of the cars as functions of time (taking $x_1 = 0$ at $t = 0$) are

$$x_1 = \frac{1}{2}at^2, \qquad x_2 = D - v_0 t.$$

The cars collide when $x_1 = x_2$; setting the expressions equal yields a quadratic in t,

$$\frac{1}{2}at^2 + v_0 t - D = 0,$$

the solutions to which are

$$t = \frac{1}{a}\left(\sqrt{v_0^2 + 2aD} - v_0\right), \qquad t = \frac{1}{a}\left(-\sqrt{v_0^2 + 2aD} - v_0\right).$$

The second of these times is negative and does not represent the physical situation.

     b) $$v_1 = at = \left(\sqrt{v_0^2 + 2aD} - v_0\right).$$

     c)

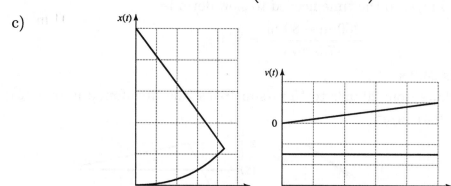

2-59: a) Travelling at 20 m/s, Juan is $x_1 = 37\text{ m} - (20\text{ m/s})(0.80\text{ s}) = 21\text{ m}$ from the spreader when the brakes are applied, and the magnitude of the acceleration will be $a = \frac{v_1^2}{2x_1}$. Travelling at 25 m/s, Juan is $x_2 = 37\text{ m} - (25\text{ m/s})(0.80\text{ s}) = 17\text{ m}$ from the spreader, and the speed of the car (and Juan) at the collision is obtained from

$$v^2 = v_0^2 - 2ax_2 = v_0^2 - 2\left(\frac{v_1^2}{2x_1}\right)x_2 = v_0^2 - v_1^2\left(\frac{x_2}{x_1}\right) = (25\text{ m/s})^2 - (20\text{ m/s})^2\left(\frac{17\text{ m}}{21\text{ m}}\right)$$

$$= 301\text{ m}^2/\text{s}^2$$

and so $v = 17.4$ m/s.

     b) The time is the reaction time plus the magnitude of the change in speed $(v_0 - v)$ divided by the magnitude of the acceleration, or

$$t_{\text{flash}} = t_{\text{reaction}} + 2\frac{v_0 - v}{v_0^2}x_1 = (0.80\text{ s}) + 2\frac{25\text{ m/s} - 17.4\text{ m/s}}{(20\text{ m/s})^2}(21\text{ m}) = 1.60\text{ s}.$$

2-60: a) There are many ways to find the result using extensive algebra, but the most straightforward way is to note that between the time the truck first passes the police car and the time the police car catches up to the truck, both the truck and the car have travelled the same distance in the same time, and hence have the same average velocity over that time. Since the truck had initial speed $\frac{3}{2}v_p$ and the average speed is v_p, the truck's final speed must be $\frac{1}{2}v_p$.

b)

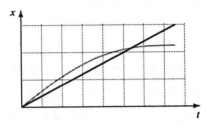

2-61: a) The most direct way to find the time is to consider that the truck and the car are initially moving at the same speed, and the time of the acceleration must be that which gives a difference between the truck's position and the car's position as 24 m + 21 m + 26 m + 4.5 m = 75.5 m, or $t = \sqrt{2(75.5 \text{ m})/(0.600 \text{ m/s}^2)} = 15.9$ s.

b) $v_0 t + (1/2)at^2 = (20.0 \text{ m/s})(15.9 \text{ s}) + (1/2)(0.600 \text{ m/s}^2)(15.9 \text{ s})^2 = 394$ m.

c) $v_0 + at = (20.0 \text{ m/s}) + (0.600 \text{ m/s}^2)(15.9 \text{ s}) = 29.5$ m/s.

2-62: a) From Eq. (2-17), $x(t) = \alpha t - \frac{\beta}{3}t^3 = (4.00 \text{ m/s})t - (0.667 \text{ m/s}^3)t^3$. From Eq. (2-5), the acceleration is $a(t) = -2\beta t = (-4.00 \text{ m/s}^3)t$.

b) The velocity is zero at $t = \pm\sqrt{\frac{\alpha}{\beta}}$ ($a = 0$ at $t = 0$, but this is an inflection point, not an extreme). The extreme values of x are then

$$x = \pm\left(\alpha\sqrt{\frac{\alpha}{\beta}} - \frac{\beta}{3}\sqrt{\frac{\alpha^3}{\beta^3}}\right) = \pm\frac{2}{3}\sqrt{\frac{\alpha^3}{\beta}}.$$

The positive value is then

$$x = \frac{2}{3}\left(\frac{(4.00 \text{ m/s})^3}{2.00 \text{ m/s}^3}\right)^{\frac{1}{2}} = \frac{2}{3}\sqrt{32 \text{ m}^2} = 3.77 \text{ m}.$$

2-63: a) The particle's velocity and position as functions of time are

$$v(t) = v_0 + \int_0^t \left((-2.00 \text{ m/s}^2) + (3.00 \text{ m/s}^3)t\right) dt$$

$$= v_0 - (2.00 \text{ m/s}^2)t + \left(\frac{3.00 \text{ m/s}^3}{2}\right)t^2,$$

$$x(t) = \int_0^t v(t)\, dt = v_0 t - (1.00 \text{ m/s}^2)t^2 + (0.50 \text{ m/s}^3)t^3$$

$$= t(v_0 - (1.00 \text{ m/s}^2)t + (0.50 \text{ m/s}^3)t^2),$$

where x_0 has been set to 0. Then, $x(0) = 0$, and to have $x(4 \text{ s}) = 0$,

$$v_0 - (1.00 \text{ m/s}^2)(4.00 \text{ s}) + (0.50 \text{ m/s}^3)(4.00 \text{ s})^2 = 0,$$

which is solved for $v_0 = -4.0$ m/s. b) $v(4 \text{ s}) = 12.0$ m/s.

2-64: The time needed for the egg to fall is

$$t = \sqrt{\frac{2\Delta h}{g}} = \sqrt{\frac{2(46.0 \text{ m} - 1.80 \text{ m})}{(9.80 \text{ m/s}^2)}} = 3.00 \text{ s},$$

and so the professor should be a distance $vt = (1.20 \text{ m/s})(3.00 \text{ s}) = 3.60 \text{ m}$.

2-65: Let t_1 be the fall for the watermelon, and t_2 be the travel time for the sound to return. The total time is $T = t_1 + t_2 = 2.5 \text{ s}$. Let y be the height of the building, then, $y = \frac{1}{2}gt_1^2$ and $y = v_s t_2$. There are three equations and three unknowns. Eliminate t_2, solve for t_1, and use the result to find y. A quadratic results: $\frac{1}{2}gt_1^2 + v_s t_1 - v_s T = 0$.

If $at^2 + bt + c = 0$, then $t = \frac{-b \pm \sqrt{b^2 - 4ac}}{2a}$. Here, $t = t_1$, $a = 1/2 \ g = 4.9 \text{ m/s}^2$, $b = v_s = 340 \text{ m/s}$, and $c = -v_s T = -(340 \text{ m/s})(2.5 \text{ s}) = -850 \text{ m}$. Then upon substituting these values into the quadratic formula,

$$t_1 = \frac{-(340 \text{ m/s}) \pm \sqrt{(340 \text{ m/s})^2 - 4(4.9 \text{ m/s}^2)(-850 \text{ m})}}{2(4.9 \text{ m/s}^2)}$$

$t_1 = \frac{-(340 \text{ m/s}) \pm (363.7 \text{ m/s})}{2(4.9 \text{ m/s}^2)} = 2.42 \text{ s}$. The other solution, -71.8 s has no real physical meaning. Then, $y = \frac{1}{2} \ gt_1^2 = \frac{1}{2}(9.8 \text{ m/s}^2)(2.42 \text{ s})^2 = 28.6 \text{ m}$. Check: $(28.6 \text{ m})/(340 \text{ m/s}) = .08 \text{ s}$, the time for the sound to return.

2-66: The elevators to the observation deck of the Sears Tower in Chicago move from the ground floor to the 103rd floor observation deck in about 70 s. Estimating a single floor to be about 3.5 m (11.5 ft), the average speed of the elevator is $\frac{(103)(3.5 \text{ m})}{70 \text{ s}} = 5.15 \text{ m/s}$. Estimating that the elevator must come to rest in the space of one floor, the acceleration is about $\frac{0^2 - (5.15 \text{ m/s})^2}{2(3.5 \text{ m})} = -3.80 \text{ m/s}^2$.

2-67: a) $v = \sqrt{2gh} = \sqrt{2(9.80 \text{ m/s}^2)(21.3 \text{ m})} = 20.4 \text{ m/s}$; the announcer is mistaken.

b) The required speed would be

$$v_0 = \sqrt{v^2 + 2g(y - y_0)} = \sqrt{(25 \text{ m/s})^2 + 2(9.80 \text{ m/s}^2)(-21.3 \text{ m})} = 14.4 \text{ m/s},$$

which is not possible for a leaping diver.

2-68: If the speed of the flowerpot at the top of the window is v_0, height h of the window is

$$h = v_{\text{ave}}t = v_0 t + (1/2)gt^2, \quad \text{or} \quad v_0 = \frac{h}{t} - (1/2)gt.$$

The distance l from the roof to the top of the window is then

$$l = \frac{v_0^2}{2g} = \frac{((1.90 \text{ m})/(0.420 \text{ s}) - (1/2)(9.80 \text{ m/s}^2)(0.420 \text{ s}))^2}{2(9.80 \text{ m/s}^2)} = 0.310 \text{ m}.$$

An alternative but more complicated algebraic method is to note that t is the difference between the times taken to fall the heights $l + h$ and h, so that

$$t = \sqrt{\frac{2(l+h)}{g}} - \sqrt{\frac{2l}{g}}, \quad \sqrt{gt^2/2} + \sqrt{l} = \sqrt{l+h}.$$

Squaring the second expression allows cancelation of the l terms,

$$(1/2)gt^2 + 2\sqrt{gt^2l/2} = h,$$

which is solved for

$$l = \frac{1}{2g}\left(\frac{h}{t} - (1/2)gt\right)^2,$$

which is the same as the previous expression.

2-69: a) The football will go an additional $\frac{v^2}{2g} = \frac{(5.00 \text{ m/s})^2}{2(9.80 \text{ m/s}^2)} = 1.27$ m above the window, so the greatest height is 13.27 m or 13.3 m to the given precision.

b) The time needed to reach this height is $\sqrt{2(13.3 \text{ m})/(9.80 \text{ m/s}^2)} = 1.65$ s.

2-70: a)

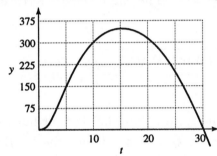
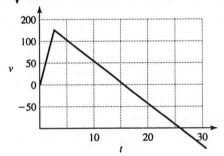

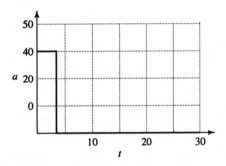

b) From the speed found in part (c), the maximum height is

$$\frac{(111.6 \text{ m/s})^2}{2(9.80 \text{ m/s}^2)} = 635 \text{ m}.$$

c) After the fuel is burned, the rocket has speed $(40.0 \text{ m/s}^2)(2.50 \text{ s}) = 100$ m/s and has reached a height $(1/2)(40.0 \text{ m/s}^2)(2.50 \text{ s})^2 = 125$ m. The speed of the rocket just before it hits the ground is then

$$v = \sqrt{v_0^2 - 2g(y - y_0)} = \sqrt{(100 \text{ m/s})^2 - 2(9.80 \text{ m/s}^2)(-125 \text{ m})} = 111.6 \text{ m/s},$$

or 112 m/s to three significant figures.

d) The time from launch to the highest point is not the same as the time from the highest point back to the ground due to the upward acceleration of the engine in the first 2.5 s.

2-71: a) From Eq. (2-14), with $v_0 = 0$,

$$v = \sqrt{2a(y - y_0)} = \sqrt{2(45.0 \text{ m/s}^2)(0.640 \text{ m})} = 7.59 \text{ m/s}.$$

b) The height above the release point is also found from Eq. (2-14), with $v_0 = 7.59$ m/s, $v = 0$ and $a = -g$,

$$h = \frac{v_0^2}{2g} = \frac{(7.59 \text{ m/s})^2}{2(9.80 \text{ m/s}^2)} = 2.94 \text{ m}$$

(Note that this is also $(64.0 \text{ cm})\frac{45 \text{ m/s}^2}{g}$). The height above the ground is then 5.14 m.

c) See Problems 2-40 & 2-42 or Example 2-8: The shot moves a total distance 2.20 m − 1.83 m = 0.37 m, and the time is

$$\frac{(7.59 \text{ m/s}) + \sqrt{(7.59 \text{ m/s})^2 + 2(9.80 \text{ m/s}^2)(0.37 \text{ m})}}{(9.80 \text{ m/s}^2)} = 1.60 \text{ s}.$$

2-72: a) Suppose that Superman falls for a time t, and that the student has been falling for a time t_0 before Superman's leap (in this case, $t_0 = 5$ s). Then, the height h of the building is related to t and t_0 in two different ways:

$$-h = v_0 t - \frac{1}{2}gt^2$$

$$= -\frac{1}{2}g(t + t_0)^2,$$

where v_0 is Superman's initial velocity. Solving the second for t gives $t = \sqrt{\frac{2h}{g}} - t_0$. Solving the first for v_0 gives $v_0 = -\frac{h}{t} + \frac{g}{2}t$, and substitution of numerical values gives $t = 1.06$ s and $v_0 = -165$ m/s, with the minus sign indicating a downward initial velocity.

b)

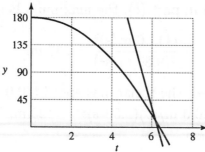

c) If the skyscraper is so short that the student is already on the ground, then $h = \frac{1}{2}gt_0^2 = 123$ m.

2-73: a) The final speed of the first part of the fall (free fall) is the same as the initial speed of the second part of the fall (with the Rocketeer supplying the upward acceleration), and assuming the student is a rest both at the top of the tower and at the ground, the distances fallen during the first and second parts of the fall are $\frac{v_1^2}{2g}$ and $\frac{v_1^2}{10g}$, where v_1 is the student's speed when the Rocketeer catches him. The distance fallen in free fall is then five times the distance from the ground when caught, and so the distance above the ground when caught is one-sixth of the height of the tower, or 92.2 m. b) The student falls a distance $5H/6$ in time $t = \sqrt{5H/3g}$, and the Rocketeer falls the same distance in time $t - t_0$, where $t_0 = 5.00$ s (assigning three significant figures to t_0 is more or less arbitrary). Then,

$$\frac{5H}{6} = v_0(t - t_0) + \frac{1}{2}g(t - t_0)^2, \quad \text{or}$$

$$v_0 = \frac{5H/6}{(t - t_0)} - \frac{1}{2}g(t - t_0).$$

At this point, there is no great advantage in expressing t in terms of H and g algebraically; $t - t_0 = \sqrt{5(553 \text{ m})/29.40 \text{ m/s}^2} - 5.00$ s $= 4.698$ s, from which $v_0 = 75.1$ m/s.

c)

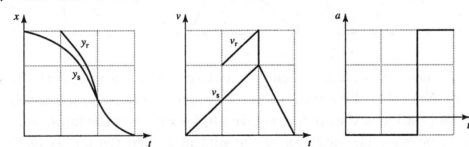

2-74: a) The time is the initial separation divided by the initial relative speed, H/v_0. More precisely, if the positions of the balls are described by

$$y_1 = v_0 t - (1/2)gt^2, \qquad y_2 = H - (1/2)gt^2,$$

setting $y_1 = y_2$ gives $H = v_0 t$. b) The first ball will be at the highest point of its motion if at the collision time t found in part (a) its velocity has been reduced from v_0 to 0, or $gt = gH/v_0 = v_0$, or $H = v_0^2/g$.

2-75: The velocities are $v_A = \alpha + 2\beta t$ and $v_B = 2\gamma t - 3\delta t^2$. a) Since v_B is zero at $t = 0$, car A takes the early lead. b) The cars are both at the origin at $t = 0$. The non-trivial solution is found by setting $x_A = x_B$, cancelling the common factor of t, and solving the quadratic for

$$t = -\frac{1}{2\delta}\left[(\beta - \gamma) \pm \sqrt{(\beta - \gamma)^2 - 4\alpha\delta}\right].$$

Substitution of numerical values gives 2.27 s, 5.73 s. The use of the term "starting point" can be taken to mean that negative times are to be neglected. c) Setting $v_A = v_B$ leads to a different quadratic, the positive solution to which is

$$t = -\frac{1}{6\delta}\left[(2\beta - 2\gamma) - \sqrt{(2\beta - 2\gamma)^2 - 12\alpha\delta}\right].$$

Substitution of numerical results gives 1.00 s and 4.33 s.

d) Taking the second derivative of x_A and x_B and setting them equal, yields, $2\beta = 2\gamma - 6\delta t$. Solving, $t = 2.67$ s.

2-76: a) The speed of any object falling a distance $H - h$ in free fall is $\sqrt{2g(H-h)}$.
b) The acceleration needed to bring an object from speed v to rest over a distance h is $\frac{v^2}{2h} = \frac{2g(H-h)}{2h} = g\left(\frac{H}{h} - 1\right)$.

c)

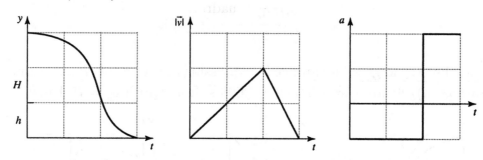

2-77: For convenience, let the student's (constant) speed be v_0 and the bus's initial position be x_0. Note that these quantities are for separate objects, the student and the bus. The initial position of the student is taken to be zero, and the initial velocity of the bus is taken to be zero. The positions of the student x_1 and the bus x_2 as functions of time are then

$$x_1 = v_0 t, \qquad x_2 = x_0 + (1/2)at^2.$$

a) Setting $x_1 = x_2$ and solving for the times t gives

$$t = \frac{1}{a}\left(v_0 \pm \sqrt{v_0^2 - 2ax_0}\right)$$

$$= \frac{1}{(0.170 \text{ m/s}^2)}\left((5.0 \text{ m/s}) \pm \sqrt{(5.0 \text{ m/s})^2 - 2(0.170 \text{ m/s}^2)(40.0 \text{ m})}\right)$$

$$= 9.55 \text{ s}, \quad 49.3 \text{ s}.$$

The student will be likely to hop on the bus the first time she passes it (see part (d) for a discussion of the later time). During this time, the student has run a distance $v_0 t = (5 \text{ m/s})(9.55 \text{ s}) = 47.8 \text{ m}$.

b) The speed of the bus is $(0.170 \text{ m/s}^2)(9.55 \text{ s}) = 1.62 \text{ m/s}$.

c) The results can be verified by noting that the x lines for the student and the bus intersect at two points:

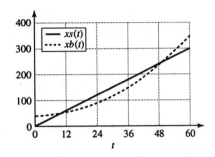

d) At the later time, the student has passed the bus, maintaining her constant speed, but the accelerating bus then catches up to her. At this later time the bus's velocity is $(0.170 \text{ m/s}^2)(49.3 \text{ s}) = 8.38 \text{ m/s}$.

e) No; $v_0^2 < 2ax_0$, and the roots of the quadratic are imaginary. When the student runs at 3.5 m/s, the two lines do *not* intersect:

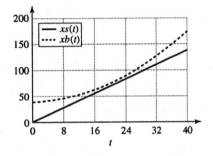

f) For the student to catch the bus, $v_0^2 > 2ax_0$, and so the minimum speed is $\sqrt{2(0.170 \text{ m/s}^2)(40 \text{ m/s})} = 3.688 \text{ m/s}$. She would be running for a time $\frac{3.69 \text{ m/s}}{0.170 \text{ m/s}^2} = 21.7 \text{ s}$, and cover a distance $(3.688 \text{ m/s})(21.7 \text{ s}) = 80.0 \text{ m}$.

However, when the student runs at 3.688 m/s, the lines intersect at *one* point $(x = 80 \text{ m})$:

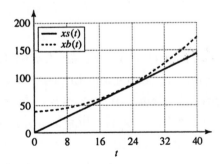

2-78: The time spent above $y_{\max}/2$ is $\frac{1}{\sqrt{2}}$ the total time spent in the air, as the time is proportional to the square root of the change in height. Therefore the ratio is

$$\frac{1/\sqrt{2}}{1 - 1/\sqrt{2}} = \frac{1}{\sqrt{2} - 1} \approx 2.4.$$

2-79: For the purpose of doing all four parts with the least repetition of algebra, quantities will be denoted symbolically. That is, let $y_1 = h + v_0 t - \frac{1}{2}gt^2$, $y_2 = h - \frac{1}{2}g(t - t_0)^2$. In this case, $t_0 = 1.00$ s. Setting $y_1 = y_2 = 0$, expanding the binomial $(t - t_0)^2$ and eliminating the common term $\frac{1}{2}gt^2$ yields $v_0 t = g t_0 t - \frac{1}{2}g t_0^2$, which can be solved for t;

$$t = \frac{\frac{1}{2}g t_0^2}{g t_0 - v_0} = \frac{t_0}{2}\frac{1}{1 - \frac{v_0}{g t_0}}.$$

Substitution of this into the expression for y_1 and setting $y_1 = 0$ and solving for h as a function of v_0 yields, after some algebra,

$$h = \frac{1}{2}g t_0^2 \frac{(\frac{1}{2}g t_0 - v_0)^2}{(g t_0 - v_0)^2}.$$

a) Using the given value $t_0 = 1.00$ s and $g = 9.80$ m/s^2,

$$h = 20.0 \text{ m} = (4.9 \text{ m})\left(\frac{4.9 \text{ m/s} - v_0}{9.8 \text{ m/s} - v_0}\right)^2.$$

This has two solutions, one of which is unphysical (the first ball is still going up when the second is released; see part (c)). The physical solution involves taking the negative square root before solving for v_0, and yields 8.2 m/s.

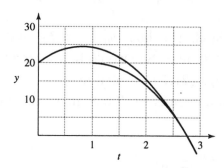

b) The above expression gives for i), 0.411 m and for ii) 1.15 km. c) As v_0 approaches 9.8 m/s, the height h becomes infinite, corresponding to a relative velocity at the time the second ball is thrown that approaches zero. If $v_0 > 9.8$ m/s, the first ball can never catch the second ball. d) As v_0 approaches 4.9 m/s, the height approaches zero. This corresponds to the first ball being closer and closer (on its way down) to the top of the roof when the second ball is released. If $v_0 < 4.9$ m/s, the first ball will already have passed the roof on the way down before the second ball is released, and the second ball can never catch up.

2-80: Let the height be h and denote the 1.30-s interval as Δt; the simultaneous equations $h = \frac{1}{2}gt^2$, $\frac{2}{3}h = \frac{1}{2}g(t - \Delta t)^2$ can be solved for t. Eliminating h and taking the square root, $\frac{t}{t - \Delta t} = \sqrt{\frac{3}{2}}$, and $t = \frac{\Delta t}{1 - \sqrt{2/3}}$, and substitution into $h = \frac{1}{2}gt^2$ gives $h = 246$ m.

This method avoids use of the quadratic formula; the quadratic formula is a generliza-tion of the method of "completing the square", and in the above form, $\frac{2}{3}h = \frac{1}{2}g(t - \Delta t)^2$, the square is already completed.

b) The above method assumed that $t > 0$ when the square root was taken. The neg-ative root (with $\Delta t = 0$) gives an answer of 2.51 m, clearly not a "cliff". This would correspond to an object that was initially near the bottom of this "cliff" being thrown upward and taking 1.30 s to rise to the top and fall to the bottom. Although physically possible, the conditions of the problem preclude this answer.

Chapter 3 Motion in Two or Three Dimensions

3-1: a)
$$v_{x,\,\text{ave}} = \frac{(5.3 \text{ m}) - (1.1 \text{ m})}{(3.0 \text{ s})} = 1.4 \text{ m/s},$$

$$v_{y,\,\text{ave}} = \frac{(-0.5 \text{ m}) - (3.4 \text{ m})}{(3.0 \text{ s})} = -1.3 \text{ m/s}.$$

b) $v_{\text{ave}} = \sqrt{(1.4 \text{ m/s})^2 + (-1.3 \text{ m/s})^2} = 1.91 \text{ m/s}$, or 1.9 m/s to two significant figures, $\theta = \arctan\left(\frac{-1.3}{1.4}\right) = -43°$.

3-2: a)
$$x = (v_{x,\,\text{ave}})\Delta t = (-3.8 \text{ m/s})(12.0 \text{ s}) = -45.6 \text{ m} \quad \text{and}$$

$$y = (v_{y,\,\text{ave}})\Delta t = (4.9 \text{ m/s})(12.0 \text{ s}) = 58.8 \text{ m}.$$

b) $r = \sqrt{x^2 + y^2} = \sqrt{(-45.6 \text{ m})^2 + (58.8 \text{ m})^2} = 74.4 \text{ m}.$

3-3: The position is given by $\vec{r} = [4.0 \text{ cm} + (2.5 \text{ cm/s}^2)t^2]\hat{i} + (5.0 \text{ cm/s})t\hat{j}$.
(a) $\mathbf{r}(0) = [4.0 \text{ cm}]\hat{i}$, and $\mathbf{r}(2 \text{ s}) = [4.0 \text{ cm} + (2.5 \text{ cm/s}^2)(2 \text{ s})^2]\hat{i} + (5.0 \text{ cm/s})(2 \text{ s})\hat{j} = (14.0 \text{ cm})\hat{i} + (10.0 \text{ cm})\hat{j}$. Then using the definition of average velocity, $\vec{v}_{\text{ave}} = \frac{(14 \text{ cm} - 4 \text{ cm})\hat{i} + (10 \text{ cm} - 0)\hat{j}}{2 \text{ s}} = (5 \text{ cm/s})\hat{i} + (5 \text{ cm/s})\hat{j}$.

b) $\vec{v} = \frac{d\vec{r}}{dt} = (2)(2.5 \text{ cm/s})t\hat{i} + (5 \text{ cm/s})\hat{j} = (5 \text{ cm/s})t\hat{i} + (5 \text{ cm/s})\hat{j}$. Substituting for $t = 0$, 1 s, and 2 s, gives:
$\vec{v}(0) = (5 \text{ cm/s})\hat{j}$, $\vec{v}(1 \text{ s}) = (5 \text{ cm/s})\hat{i} + (5 \text{ cm/s})\hat{j}$, and $\vec{v}(2 \text{ s}) = (10 \text{ cm/s})\hat{i} + (5 \text{ cm/s})\hat{j}$.

c)

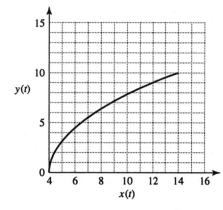

3-4: $\vec{v} = 2bt\hat{i} + 3ct^2\hat{j}$. This vector will make a 45°-angle with both axes when the x- and y-components are equal; in terms of the parameters, this time is $2b/3c$.

3-5: a)

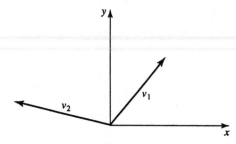

b)

$$a_{x,\,ave} = \frac{(-170 \text{ m/s}) - (90 \text{ m/s})}{(30.0 \text{ s})} = -8.7 \text{ m/s}^2,$$

$$a_{y,\,ave} = \frac{(40 \text{ m/s}) - (110 \text{ m/s})}{(30.0 \text{ s})} = -2.3 \text{ m/s}^2.$$

c) $\sqrt{(-8.7 \text{ m/s}^2)^2 + (-2.3 \text{ m/s}^2)^2} = 9.0 \text{ m/s}^2$, $\arctan\left(\frac{-2.3}{-8.7}\right) = 14.8° + 180° = 195°$.

3-6: a) $a_x = (0.45 \text{ m/s}^2)\cos 31.0° = 0.39 \text{ m/s}^2$, $a_y = (0.45 \text{ m/s}^2)\sin 31.0° = 0.23 \text{ m/s}^2$, so $v_x = 2.6 \text{ m/s} + (0.39 \text{ m/s}^2)(10.0 \text{ s}) = 6.5 \text{ m/s}$ and $v_y = -1.8 \text{ m/s} + (0.23 \text{ m/s}^2)(10.0 \text{ s}) = 0.52 \text{ m/s}$.

b) $v = \sqrt{(0.52 \text{ m/s})^2 + (6.5 \text{ m/s})^2} = 6.48 \text{ m/s}$, at an angle of $\arctan\left(\frac{6.5}{0.52}\right) = 85°$.

c)

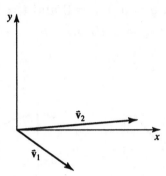

3-7: a)

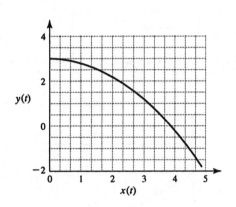

b)

$$\vec{v} = \alpha\hat{i} - 2\beta t\hat{j} = (2.4 \text{ m/s})\hat{i} - [(2.4 \text{ m/s}^2)t]\hat{j}$$

$$\vec{a} = -2\beta\hat{j} = (-2.4 \text{ m/s}^2)\hat{j}.$$

c) At $t = 2.0$ s, the velocity is $\vec{v} = (2.4 \text{ m/s})\hat{i} - (4.8 \text{ m/s})\hat{j}$; the magnitude is $\sqrt{(2.4 \text{ m/s})^2 + (-4.8 \text{ m/s})^2} = 5.4 \text{ m/s}$, and the direction is $\arctan\left(\frac{-4.8}{2.4}\right) = -63°$. The acceleration is constant, with magnitude 2.4 m/s² in the $-y$-direction. d) The velocity vector has a component parallel to the acceleration, so the bird is speeding up. The bird is turning toward the $-x$-direction, which would be to the bird's right (taking the $+z$-direction to be vertical).

3-8:

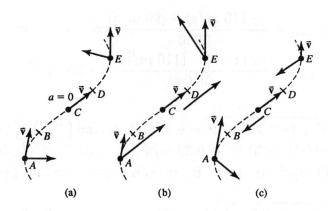

(a)   (b)   (c)

3-9: a) Solving Eq. (3-18) with $y = 0$, $v_{0y} = 0$ and $t = 0.350$ s gives $y_0 = 0.600$ m.

b) $v_x t = 0.385$ m c) $v_x = v_{0x} = 1.10$ m/s, $v_y = -gt = -3.43$ m/s, $v = 3.60$ m/s, $72.2°$ below the horizontal.

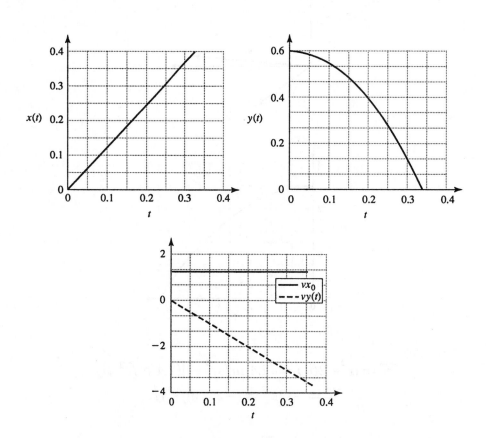

3-10: a) The time t is given by $t = \sqrt{\frac{2h}{g}} = 7.82$ s.

b) The bomb's constant horizontal velocity will be that of the plane, so the bomb travels a horizontal distance $x = v_x t = (60 \text{ m/s})(7.82 \text{ s}) = 470$ m.

c) The bomb's horizontal component of velocity is 60 m/s, and its vertical component is $-gt = -76.7$ m/s.

d)

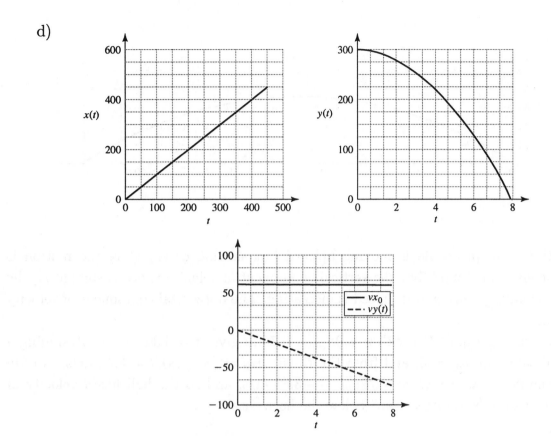

e) Because the airplane and the bomb always have the same x-component of velocity *and* position, the plane will be 300 m above the bomb at impact.

3-11: a) Solving Eq. (3-17) for $v_y = 0$, with $v_{0y} = (15.0 \text{ m/s}) \sin 45.0°$,

$$T = \frac{(15.0 \text{ m/s}) \sin 45°}{9.80 \text{ m/s}^2} = 1.08 \text{ s.}$$

b) Using Equations (3-20) and (3-21) gives at t_1, $(x, y) = (6.18 \text{ m}, 4.52 \text{ m})$: t_2, (11.5 m, 5.74 m): t_3, (16.8 m, 4.52 m). c) Using Equations (3-22) and (3-23) gives at t_1, $(v_x, v_y) = (10.6 \text{ m/s}, 4.9 \text{ m/s})$: t_2, (10.6 m/s, 0) t_3: (10.6 m/s, -4.9 m/s), for velocities, respectively, of 11.7 m/s @ 24.8°, 10.6 m/s @ 0°, and 11.7 m/s @ $-24.8°$. Note that v_x is the same for all times, and that the y-component of velocity at t_3 is negative that at t_1. d) The parallel and perpendicular components of the acceleration are obtained from

$$\vec{a}_\parallel = \frac{(\vec{a} \cdot \vec{v})\vec{v}}{v^2}, \quad \left|\vec{a}_\parallel\right| = \frac{\left|\vec{a} \cdot \vec{v}\right|}{v}, \quad \left|\vec{a}_\perp\right| = \sqrt{\left|\vec{a}\right| - \left|\vec{a}_\parallel\right|}.$$

For projectile motion, $\vec{a} = -g\hat{j}$, so $\vec{a} \cdot \vec{v} = -gv_y$, and the components of acceleration parallel and perpendicular to the velocity are t_1: -4.1 m/s^2, 9.8 m/s^2. t_2: 0, 9.8 m/s^2. t_3: 4.1 m/s^2, 9.8 m/s^2.

e)

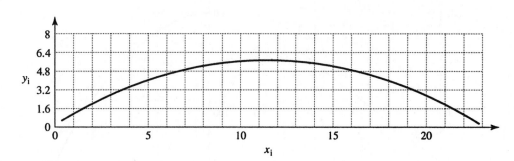

f) At t_1, the projectile is moving upward but slowing down; at t_2 the motion is instantaneously horizontal, but the vertical component of velocity is decreasing; at t_3, the projectile is falling down and its speed is increasing. The horizontal component of velocity is constant.

3-12: a) Solving Eq. (3-18) with $y = 0$, $y_0 = 0.75$ m gives $t = 0.391$ s. b) Assuming a horizontal tabletop, $v_{0y} = 0$, and from Eq. (3-16), $v_{0x} = (x - x_0)/t = 3.58$ m/s. c) On striking the floor, $v_y = -gt = -\sqrt{2gy_0} = -3.83$ m/s, and so the ball has a velocity of magnitude 5.24 m/s, directed 46.9° below the horizontal.

d)

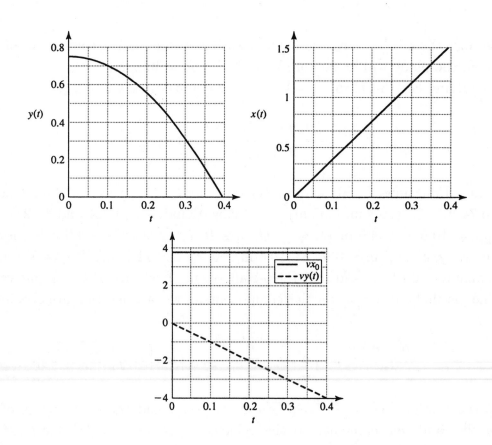

Although not asked for in the problem, this y vs. x graph shows the trajectory of the tennis ball as viewed from the side.

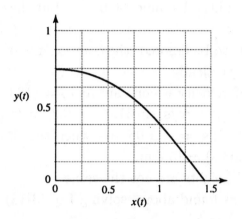

3-13: The range of a projectile is given in Example 3-10, $R = v_0^2 \sin 2\alpha_0 / g$.

a) $(120 \text{ m/s})^2 \sin 110°/(9.80 \text{ m/s}^2) = 1.38$ km.   b) $(120 \text{ m/s})^2 \sin 110°/(1.6 \text{ m/s}^2) = 8.4$ km.

3-14: a) The time t is $\frac{v_{y0}}{g} = \frac{16.0 \text{ m/s}}{9.80 \text{ m/s}^2} = 1.63$ s.   b) $\frac{1}{2}gt^2 = \frac{1}{2}v_{y0}t = \frac{v_{y0}^2}{2g} = 13.1$ m.

c) Regardless of how the algebra is done, the time will be twice that found in part (a), or 3.27 s   d) v_x is constant at 20.0 m/s, so $(20.0 \text{ m/s})(3.27 \text{ s}) = 65.3$ m.

e)

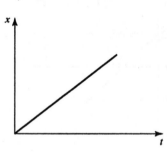
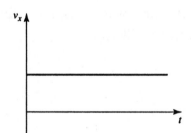
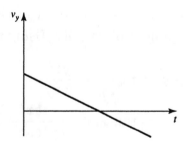

3-15: a) $v_{0y} = (30.0 \text{ m/s}) \sin 36.9° = 18.0 \text{ m/s}$; solving Eq. (3-18) for t with $y_0 = 0$ and $y = 10.0$ m gives

$$t = \frac{(18.0 \text{ m/s}) \pm \sqrt{(18.0 \text{ m/s})^2 - 2(9.80 \text{ m/s}^2)(10.0 \text{ m})}}{9.80 \text{ m/s}^2} = 0.68 \text{ s}, 2.99 \text{ s}.$$

b) The x-component of velocity will be $(30.0 \text{ m/s})\cos 36.9° = 24.0 \text{ m/s}$ at all times. The y-component, obtained from Eq. (3-17), is 11.3 m/s at the earlier time and -11.3 m/s at the later.   c) The magnitude is the same, 30.0 m/s, but the direction is now 36.9° below the horizontal.

3-16:   a) If air resistance is to be ignored, the components of acceleration are 0 horizontally and $-g = -9.80 \text{ m/s}^2$ vertically.

b) The x-component of velocity is constant at $v_x = (12.0 \text{ m/s})\cos 51.0° = 7.55 \text{ m/s}$. The y-component is $v_{0y} = (12.0 \text{ m/s})\sin 51.0° = 9.32 \text{ m/s}$ at release and $v_{0y} - gt = (10.57 \text{ m/s}) - (9.80 \text{ m/s}^2)(2.08 \text{ s}) = -11.06 \text{ m/s}$ when the shot hits.

c) $v_{0x}t = (7.55 \text{ m/s})(2.08 \text{ s}) = 15.7 \text{ m}$.

d) The initial and final heights are not the same.

e) With $y = 0$ and v_{0y} as found above, solving Eq. (3-18) for $y_0 = 1.81$ m.

f)

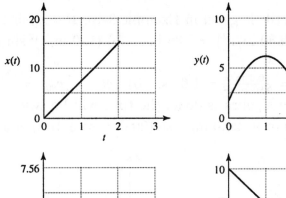

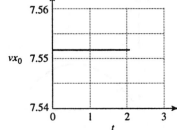

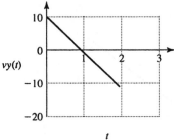

3-17:   a) The time the quarter is in the air is the horizontal distance divided by the horizontal component of velocity. Using this time in Eq. (3-18),

$$y - y_0 = v_{0y}\frac{x}{v_{0x}} - \frac{gx^2}{2v_{0x}^2}$$

$$= \tan \alpha_0 x - \frac{gx^2}{v_0^2 2\cos^2 \alpha_0}$$

$$= \tan 60°(2.1 \text{ m}) - \frac{(9.80 \text{ m/s}^2)(2.1 \text{ m})^2}{2(6.4 \text{ m/s})^2 \cos^2 60°} = 1.53 \text{ m rounded}.$$

b) Using the same expression for the time in terms of the horizontal distance in Eq. (3-17),

$$v_y = v_0 \sin \alpha_0 - \frac{gx}{v_0 \cos \alpha_0} = (6.4 \text{ m/s})\sin 60° - \frac{(9.80 \text{ m/s}^2)(2.1 \text{ m})}{(6.4 \text{ m/s})\cos 60°} = -0.89 \text{ m/s}.$$

3-18: Substituting for t in terms of d in the expression for y_{dart} gives

$$y_{\text{dart}} = d\left(\tan\alpha_0 - \frac{gd}{2v_0^2\cos^2\alpha_0}\right).$$

Using the given values for d and α_0 to express this as a function of v_0,

$$y = (3.00\text{ m})\left(0.90 - \frac{26.62\text{ m}^2/\text{s}^2}{v_0^2}\right).$$

Then, a) $y = 2.14$ m, b) $y = 1.45$ m, c) $y = -2.29$ m. In the last case, the dart was fired with so slow a speed that it hit the ground before traveling the 3-meter horizontal distance.

d)

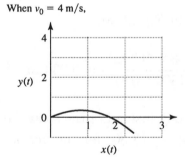

3-19: a) With $v_y = 0$ in Eq. (3-17), solving for t and substituting into Eq. (3-18) gives

$$(y - y_0) = \frac{v_{0y}^2}{2g} = \frac{v_0^2\sin^2\alpha_0}{2g} = \frac{(30.0\text{ m/s})^2\sin^2 33.0^\circ}{2(9.80\text{ m/s}^2)} = 13.6\text{ m}.$$

b) Rather than solving a quadratic, the above height may be used to find the time the rock takes to fall from its greatest height to the ground, and hence the vertical component of velocity, $v_y = \sqrt{2yg} = \sqrt{2(28.6\text{ m})(9.80\text{ m/s}^2)} = 23.7$ m/s, and so the speed of the rock is $\sqrt{(23.7\text{ m/s})^2 + ((30.0\text{ m/s})(\cos 33.0^\circ))^2} = 34.6$ m/s.

c) The time the rock is in the air is given by the change in the vertical component of velocity divided by the acceleration $-g$; the distance is the constant horizontal component of velocity multiplied by this time, or

$$x = (30.0\text{ m/s})\cos 33.0^\circ \frac{(-23.7\text{ m/s} - ((30.0\text{ m/s})\sin 33.0^\circ))}{(-9.80\text{ m/s}^2)} = 103\text{ m}.$$

d)

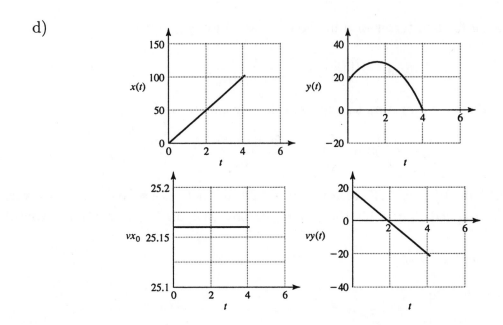

3-20: For any item in the washer, the centripetal acceleration will be proportional to the square of the frequency, and hence inversely proportional to the square of the rotational period; tripling the centripetal acceleration involves decreasing the period by a factor of $\sqrt{3}$, so that the new period T' is given in terms of the previous period T by $T' = T/\sqrt{3}$.

3-21: Using the given values in Eq. (3-30),

$$a_{rad} = \frac{4\pi^2(6.38 \times 10^6 \text{ m})}{((24 \text{ h})(3600 \text{ s/h}))^2} = 0.034 \text{ m/s}^2 = 3.4 \times 10^{-3}g.$$

(Using the time for the siderial day instead of the solar day will give an answer that differs in the third place.)   b) Solving Eq. (3-30) for the period T with $a_{rad} = g$,

$$T = \sqrt{\frac{4\pi^2(6.38 \times 10^6 \text{ m})}{9.80 \text{ m/s}^2}} = 5070 \text{ s} \sim 1.4 \text{ h}.$$

3-22: $550 \text{ rev/min} = 9.17 \text{ rev/s}$, corresponding to a period of 0.109 s. a) From Eq. (3-29), $v = \frac{2\pi R}{T} = 196$ m/s. b) From either Eq. (3-30) or Eq. (3-31), $a_{rad} = 1.13 \times 10^4$ m/s^2 = $1.15 \times 10^3 g$.

3-23: Solving Eq. (3-30) for T in terms of R and a_{rad},
    a) $\sqrt{4\pi^2(7.0 \text{ m})/(3.0)(9.80 \text{ m/s}^2)} = 3.07$ s.   b) 1.68 s.

3-24: a) Using Eq. (3-31), $\frac{2\pi R}{T} = 2.97 \times 10^4$ m/s.   b) Either Eq. (3-30) or Eq. (3-31) gives $a_{rad} = 5.91 \times 10^{-3}$ m/s^2.   c) $v = 4.78 \times 10^4$ m/s, and $a = 3.97 \times 10^{-2}$ m/s^2.

3-25: a) From Eq. (3-31), $a = (7.00 \text{ m/s})^2/(15.0 \text{ m}) = 3.50$ m/s^2. The acceleration at the bottom of the circle is toward the center, up.   b) $a = 3.50$ m/s^2, the same as part (a), but is directed *down*, and still towards the center.   c) From Eq. (3-29), $T = 2\pi R/v = 2\pi(15.0 \text{ m})/(7.00 \text{ m/s}) = 12.6$ s.

3-26:   a) $a_{\text{rad}}=(3 \text{ m/s})^2/(14 \text{ m})=0.643 \text{ m/s}^2$, and $a_{\text{tan}} = 0.5 \text{ m/s}^2$. So, $a=((0.643 \text{ m/s}^2)^2+ (0.5 \text{ m/s}^2)^2)^{1/2} = 0.814 \text{ m/s}^2, 37.9°$ to the right of vertical.

   b)

3-27:   b) No. Only in a circle would a_{rad} point to the center (See planetary motion in Chapter 12).

   c) Where the car is farthest from the center of the ellipse.

3-28:   Repeated use of Eq. (3-33) gives a) $5.0 =$ m/s to the right, b) 16.0 m/s to the left, and c) $13.0 =$ m/s to the left.

3-29:   a) The speed relative to the ground is $1.5 \text{ m/s} + 1.0 \text{ m/s} = 2.5$ m/s, and the time is $35.0 \text{ m}/2.5 \text{ m/s} = 14.0$ s   b) The speed relative to the ground is 0.5 m/s, and the time is 70 s.

3-30:   The walker moves a total distance of 3.0 km at a speed of 4.0 km/h, and takes a time of three fourths of an hour (45.0 min). The boat's speed relative to the shore is 6.8 km/h downstream and 1.2 km/h upstream, so the total time the rower takes is

$$\frac{1.5 \text{ km}}{6.8 \text{ km/h}} + \frac{1.5 \text{ km}}{1.2 \text{ km/h}} = 1.47 \text{ hr} = 88 \text{ min}.$$

3-31:   The velocity components are

$$-0.50 \text{ m/s} + (0.40 \text{ m/s})/\sqrt{2} \text{ east} \quad \text{and} \quad (0.40 \text{ m/s})/\sqrt{2} \text{ south},$$

for a velocity relative to the earth of 0.36 m/s, 52.5° south of east.

3-32:   a) The plane's northward component of velocity relative to the air must be 80.0 km/h, so the heading must be $\arcsin \frac{80.0}{320} = 14°$ north of west.   b) Using the angle found in part (a), $(320 \text{ km/h}) \cos 14° = 310$ km/h.   Equivalently, $\sqrt{(320 \text{ km/h})^2 - (80.0 \text{ km/h})^2} = 310$ km/h.

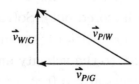

3-33: a) $\sqrt{(2.0 \text{ m/s})^2 + (4.2 \text{ m/s})^2} = 4.7$ m/s, $\arctan \frac{2.0}{4.2} = 25.5°$, south of east.

b) 800 m/4.2 m/s = 190 s.   c) 2.0 m/s × 190 s = 381 m.

3-34: a) The speed relative to the water is still 4.2 m/s; the necessary heading of the boat is $\arcsin \frac{2.0}{4.2} = 28°$ north of east.   b) $\sqrt{(4.2 \text{ m/s})^2 - (2.0 \text{ m/s})^2} = 3.7$ m/s, east.

d) 800 m/3.7 m/s = 217 s, rounded to three significant figures.

3-35: a)

b) x: $-(10 \text{ m/s}) \cos 45° = -7.1$ m/s. y: $-(35 \text{ m/s}) - (10 \text{ m/s}) \sin 45° = -42.1$ m/s.

c) $\sqrt{(-7.1 \text{ m/s})^2 + (-42.1 \text{ m/s})^2} = 42.7$ m/s, $\arctan \frac{-42.1}{-7.1} = 80°$, south of west.

3-36: a) Using generalizations of Equations 2-17 and 2-18, $v_x = v_{0x} + \frac{\alpha}{3}t^3$, $v_y = v_{0y} + \beta t - \frac{\gamma}{2}t^2$, and $x = v_{0x}t + \frac{\alpha}{12}t^4$, $y = v_{0y}t + \frac{\beta}{2}t^2 - \frac{\gamma}{6}t^3$.   b) Setting $v_y = 0$ yields a quadratic in t, $0 = v_{0y} + \beta t - \frac{\gamma}{2}t^2$, which has as the positive solution

$$t = \frac{1}{\gamma}\left[\beta + \sqrt{\beta^2 + 2v_0\gamma}\right] = 13.59 \text{ s},$$

keeping an extra place in the intermediate calculation. Using this time in the expression for $y(t)$ gives a maximum height of 341 m.

c)

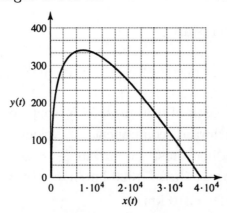

d) The time at which $y = 0$ involves solving another quadratic, $0 = v_{0y} + \frac{\beta}{2}t - \frac{\gamma}{6}t^2$ (note that the root $t = 0$ has been factored out). Solving for t gives $t = 20.73$ s (keeping the extra figure in the intermediate calculation), at which time $x = 38.5$ km.

3-37: a) The $a_x = 0$ and $a_y = -2\beta$, so the velocity and the acceleration will be perpendicular only when $v_y = 0$, which occurs at $t = 0$.

b) The speed is $v = (\alpha^2 + 4\beta^2 t^2)^{1/2}$, $dv/dt = 0$ at $t = 0$. (See part d below.)

c) r and v are perpendicular when their dot product is 0: $(\alpha t)(\alpha) + (15.0 \text{ m} - \beta t^2) \times$
$(-2\beta t) = \alpha^2 t - (30.0 \text{ m})\beta t + 2\beta^2 t^3 = 0$. Solve this for t: $t = \pm\sqrt{\frac{(30.0 \text{ m})(0.500 \text{ m/s}^2) - (1.2 \text{ m/s})^2}{2(0.500 \text{ m/s}^2)^2}} =$
$+5.208$ s, and 0 s, at which times the student is at $(6.25 \text{ m}, 1.44 \text{ m})$ and $(0 \text{ m}, 15.0 \text{ m})$,
respectively.

d) At $t = 5.208$ s, the student is 6.41 m from the origin, at an angle of 13° from
the x-axis. A plot of $d(t) = (x(t)^2 + y(t)^2)^{1/2}$ shows the minimum distance of 6.41 m at
5.208 s:

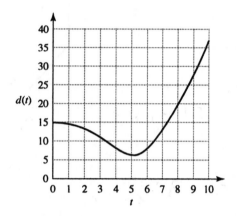

e) In the x-y plane the student's path is:

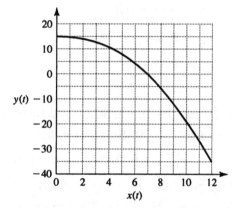

3-38: a) Integrating, $\vec{r} = (\alpha t - \frac{\beta}{3}t^3)\hat{i} + (\frac{\gamma}{2}t^2)\hat{j}$. Differentiating, $\vec{a} = (-2\beta)\hat{i} + \gamma\hat{j}$.
b) The positive time at which $x = 0$ is given by $t^2 = 3\alpha/\beta$. At this time, the y-coordinate is

$$y = \frac{\gamma}{2}t^2 = \frac{3\alpha\gamma}{2\beta} = \frac{3(2.4 \text{ m/s})(4.0 \text{ m/s}^2)}{2(1.6 \text{ m/s}^3)} = 9.0 \text{ m}.$$

3-39: a) The acceleration is

$$a = \frac{v^2}{2x} = \frac{((88 \text{ km/h})(1 \text{ m/s})/(3.6 \text{ km/h}))^2}{2(300 \text{ m})} = 0.996 \text{ m/s}^2 \approx 1 \text{ m/s}^2.$$

b) $\arctan\left(\frac{15 \text{ m}}{460 \text{ m} - 300 \text{ m}}\right) = 5.4°$. c) The vertical component of the velocity is
$(88 \text{ km/h})\left(\frac{1 \text{ m/s}}{3.6 \text{ km/h}}\right)\frac{15 \text{ m}}{160 \text{ m}} = 2.3 \text{ m/s}$. d) The average speed for the first 300 m is 44 km/h,

so the elapsed time is

$$\frac{300 \text{ m}}{(44 \text{ km/h})(1 \text{ m/s})/(3.6 \text{ km/h})} + \frac{160 \text{ m}}{(88 \text{ km/h})(1 \text{ m/s})\cos 5.4°/(3.6 \text{ km/h})} = 31.1 \text{ s},$$

or 31 s to two places.

3-40: a)

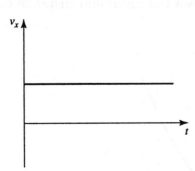

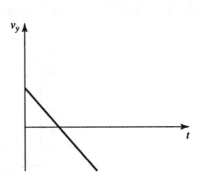

The equations of motions are:

$$y = h + (v_0 \sin \alpha)t - \frac{1}{2}gt^2$$

$$x = (v_0 \cos \alpha)t$$

$$v_y = v_0 \sin \alpha - gt$$

$$v_x = v_0 \cos \alpha$$

Note that the angle of 36.9° results in $\sin 36.9° = 3/5$ and $\cos 36.9° = 4/5$.

b) At the top of the trajectory, $v_y = 0$. Solve this for t and use in the equation for y to find the maximum height: $t = \frac{v_0 \sin \alpha}{g}$. Then, $y = h + (v_0 \sin \alpha)\left(\frac{v_0 \sin \alpha}{g}\right) - \frac{1}{2}g\left(\frac{v_0 \sin \alpha}{g}\right)^2$, which reduces to $y = h + \frac{v_0^2 \sin^2 \alpha}{2g}$. Using $v_0 = \sqrt{25gh/8}$, and $\sin \alpha = 3/5$, this becomes $y = h + \frac{(25gh/8)(3/5)^2}{2g} = h + \frac{9}{16}h$, or $y = \frac{25}{16}h$. Note: This answer assumes that $y_0 = h$. Taking $y_0 = 0$ will give a result of $y = \frac{9}{16}h$.

c) The total time of flight can be found from the y equation by setting $y = 0$, assuming $y_0 = h$, solving the quadratic for t and inserting the total flight time in the x equation to find the range. The quadratic is $\frac{1}{2}gt^2 - \frac{3}{5}v_0 - h = 0$. Using the quadratic formula gives $t = \frac{(3/5)v_0 \pm \sqrt{(-(3/5)v_0)^2 - 4(\frac{1}{2}g)(-h)}}{2(\frac{1}{2}g)}$. Substituting $v_0 = \sqrt{25gh/8}$ gives $t = \frac{(3/5)\sqrt{25gh/8} \pm \sqrt{\frac{9}{25}\cdot\frac{25gh}{8} + \frac{16gh}{8}}}{g}$. Collecting terms gives t: $t = \frac{1}{2}\left(\sqrt{\frac{9h}{2g}} \pm \sqrt{\frac{25h}{2g}}\right) = \frac{1}{2}\left(3\sqrt{\frac{h}{2g}} \pm 5\sqrt{\frac{h}{2g}}\right)$. Only the positive root is meaningful and so $t = 4\sqrt{\frac{h}{2g}}$. Then, using $x = (v_0 \cos \alpha)t$, $x = \sqrt{\frac{25gh}{8}}\left(\frac{4}{5}\right)\left(4\sqrt{\frac{h}{2g}}\right) = 4h$.

3-41: The range for a projectile that lands at the same height from which it was launched is $R = \frac{v_0^2 \sin 2\alpha}{g}$. Assuming $\alpha = 45°$, and $R = 50$ m, $v_0 = \sqrt{gR} = 22$ m/s.

3-42: a) Setting $y = -h$ in Eq. (3-27) (h being the stuntwoman's initial height above the ground) and rearranging gives

$$x^2 - \frac{2v_0^2 \sin \alpha_0 \cos \alpha_0}{g}x - \frac{2v_{0x}^2}{g}h = 0.$$

The easier thing to do here is to recognize that this can be put in the form

$$x^2 - \frac{2v_{0x}v_{0y}}{g}x - \frac{2v_{0x}^2}{g}h = 0,$$

the solution to which is

$$x = \frac{v_{0x}}{g}\left[v_{0y} + \sqrt{v_{0y}^2 + 2gh}\right] = 55.5 \text{ m}.$$

b) The graph of $v_x(t)$ is a horizontal line.

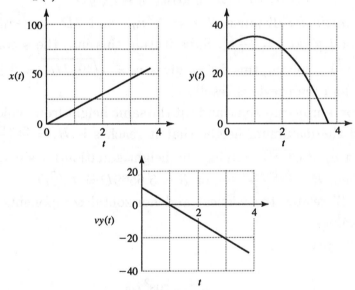

3-43: The distance is the horizontal speed times the time of free fall,

$$v_x\sqrt{\frac{2y}{g}} = (64.0 \text{ m/s})\sqrt{\frac{2(90 \text{ m})}{(9.80 \text{ m/s}^2)}} = 274 \text{ m}.$$

3-44: In terms of the range R and the time t that the balloon is in the air, the car's original distance is $d = R + v_{\text{car}}t$. The time t can be expressed in terms of the range and the horizontal component of velocity, $t = \frac{R}{v_0 \cos \alpha_0}$, so $d = R\left(1 + \frac{v_{\text{car}}}{v_0 \cos \alpha_0}\right)$. Using $R = v_0^2 \sin 2\alpha_0/g$ and the given values yields $d = 29.5$ m.

3-45: a) With $\alpha_0 = 45°$, Eq. (3-27) is solved for $v_0^2 = \frac{gx^2}{x-y}$. In this case, $y = -0.9$ m is the change in height. Substitution of numerical values gives $v_0 = 42.8$ m/s. b) Using the above algebraic expression for v_0 in Eq. (3-27) gives

$$y = x - \left(\frac{x}{188 \text{ m}}\right)^2 (188.9 \text{ m}).$$

Using $x = 116$ m gives $y = 44.1$ m above the initial height, or 45.0 m, which is 42.0 m above the fence.

3-46: The equations of motions are $y = (v_0 \sin \alpha)t - 1/2gt^2$ and $x = (v_0 \cos \alpha)t$, assuming the match starts out at $x = 0$ and $y = 0$. When the match goes in the wastebasket for the *minimum* velocity, $y = 2D$ and $x = 6D$. When the match goes in the wastebasket for the *maximum* velocity, $y = 2D$ and $x = 7D$. In both cases, $\sin \alpha = \cos \alpha = \sqrt{2}/2$.

To reach the *minimum* distance: $6D = \frac{\sqrt{2}}{2}v_0 t$, and $2D = \frac{\sqrt{2}}{2}v_0 t - \frac{1}{2}gt^2$. Solving the first equation for t gives $t = \frac{6D\sqrt{2}}{v_0}$. Substituting this into the second equation gives $2D = 6D - \frac{1}{2}g\left(\frac{6D\sqrt{2}}{v_0}\right)^2$. Solving this for v_0 gives $v_0 = 3\sqrt{gD}$.

To reach the *maximum* distance: $7D = \frac{\sqrt{2}}{2}v_0 t$, and $2D = \frac{\sqrt{2}}{2}v_0 t - \frac{1}{2}gt^2$. Solving the first equation for t gives $t = \frac{7D\sqrt{2}}{v_0}$. Substituting this into the second equation gives $2D = 7D - \frac{1}{2}g\left(\frac{7D\sqrt{2}}{v_0}\right)^2$. Solving this for v_0 gives $v_0 = \sqrt{49gD/5} = 3.13\sqrt{gD}$, which, as expected, is larger than the previous result.

3-47: The range for a projectile that lands at the same height from which it was launched is $R = \frac{v_0^2 \sin 2\alpha}{g}$, and the maximum height that it reaches is $H = \frac{v_0^2 \sin^2 \alpha}{2g}$. We must find R when $H = D$ and $v_0 = \sqrt{6gD}$. Solving the height equation for $\sin \alpha$, $D = \frac{6gD \sin^2 \alpha}{2g}$, or $\sin \alpha = (1/3)^{1/2}$. Then, $R = \frac{6gD \sin(70.72)}{g}$, or $R = 5.6569D = 4\sqrt{2}D$.

3-48: Equation 3-27 relates the vertical and horizontal components of position for a given set of initial values.

a) Solving for v_0 gives

$$v_0^2 = \frac{gx^2/2 \cos^2 \alpha_0}{x \tan \alpha_0 - y}.$$

Insertion of numerical values gives $v_0 = 16.6$ m/s.

b) Eliminating t between Equations 3-20 and 3-23 gives v_y as a function of x,

$$v_y = v_0 \sin \alpha_0 - \frac{gx}{v_0 \cos \alpha_0}.$$

Using the given values yields $v_x = v_0 \cos \alpha_0 = 8.28$ m/s, $v_y = -6.98$ m/s, so $v = \sqrt{(8.28 \text{ m/s})^2 + (-6.98 \text{ m/s})^2} = 10.8$ m/s, at an angle of $\arctan\left(\frac{-6.98}{8.24}\right) = -40.1°$, with the negative sign indicating a direction *below* the horizontal.

c) The graph of $v_x(t)$ is a horizontal line.

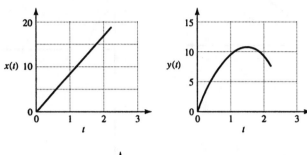

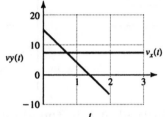

3-49: a) In Eq. (3-27), the change in height is $y = -h$. This gives a quadratic equation in x, the solution to which is

$$x = \frac{v_0^2 \cos^2 \alpha_0}{g} \left[\tan \alpha_0 + \sqrt{\tan^2 \alpha_0 + \frac{2gh}{v_0^2 \cos^2 \alpha_0}} \right]$$

$$= \frac{v_0 \cos \alpha_0}{g} \left[v_0 \sin \alpha_0 + \sqrt{v_0^2 \sin^2 \alpha_0 + 2gh} \right].$$

If $h = 0$, the square root reduces to $v_0 \sin \alpha_0$, and $x = R$.   b) The expression for x becomes $x = (10.2 \text{ m}) \cos \alpha_0 \left[\sin \alpha_0 + \sqrt{\sin^2 \alpha_0 + 0.98} \right]$.

The angle $\alpha_0 = 90°$ corresponds to the projectile being launched straight up, and there is no horizontal motion. If $\alpha_0 = 0$, the projectile moves horizontally until it has fallen the distance h.

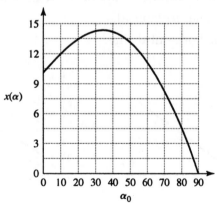

c) The maximum range occurs for an angle less than 45°, and in this case the angle is about 36°.

3-50: a) This may be done by a direct application of the result of Problem 3-49; with $\alpha_0 = -40°$, substitution into the expression for x gives 6.93 m.

b)

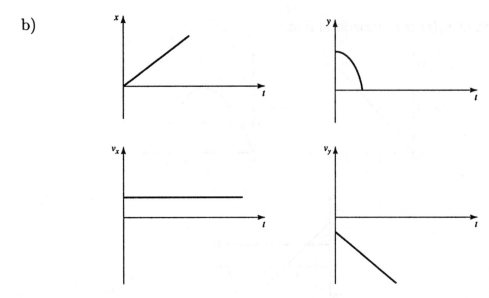

c) Using (14.0 m − 1.9 m) instead of h in the above calculation gives $x = 6.3$ m, so the man will not be hit.

3-51:   a) The expression for the range, as derived in Example 3-9, involves the sine of twice the launch angle, and

$$\sin(2(90° - \alpha_0)) = \sin(180° - 2\alpha_0) = \sin 180° \cos 2\alpha_0 - \cos 180° \sin 2\alpha_0 = \sin 2\alpha_0,$$

and so the range is the same. As an alternative, using $\sin(90° - \alpha_0) = \cos\alpha$ and $\cos(90° - \alpha_0) = \sin\alpha_0$ in the expression for the range that involves the product of the sine and cosine of α_0 gives the same result.

b) The range equation is $R = \frac{v_0^2 \sin 2\alpha}{g}$. In this case, $v_0 = 2.2$ m/s and $R = 0.25$ m. Hence, $\sin 2\alpha = (9.8 \text{ m/s}^2)(0.25 \text{ m})/(2.2 \text{ m/s})^2$, or $\sin 2\alpha = 0.5062$; and $\alpha = 15.2°$ or $74.8°$.

3-52:   a) Using the same algebra as in Problem 3-48(a), $v_0 = 13.8$ m/s.

b) Again, the algebra is the same as that used in Problem 3-48; $v = 8.4$ m/s, at an angle of 9.1°, this time above the horizontal.

c) The graph of $v_x(t)$ is a horizontal line.

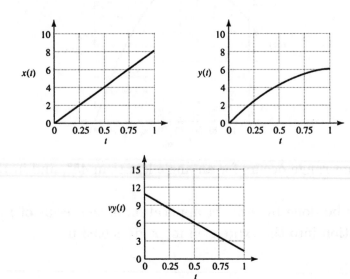

A graph of $y(t)$ vs. $x(t)$ shows the trajectory of Mary Belle as viewed from the side:

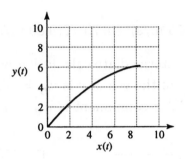

d) In this situation it's convenient to use Eq. (3-27), which becomes $y = (1.327)x - (0.071115\text{ m}^{-1})x^2$. Use of the quadratic formula gives $x = 23.8$ m.

3-53: a) The algebra is the same as that for Problem 3-48,

$$v_0^2 = \frac{gx^2}{2\cos^2\alpha_0(x\tan\alpha_0 - y)}.$$

In this case, the value for y is -15.0 m, the change in height. Substitution of numerical values gives 17.8 m/s. b) 28.4 m from the near bank (i.e., in the water!).

3-54: Combining equations 3-25, 3-22 and 3-23 gives

$$\begin{aligned}
v^2 &= v_0^2\cos^2\alpha_0 + (v_0\sin\alpha_0 - gt)^2 \\
&= v_0^2(\sin^2\alpha_0 + \cos^2\alpha_0) - 2v_0\sin\alpha_0 gt + (gt)^2 \\
&= v_0^2 - 2g\left(v_0\sin\alpha_0 t - \frac{1}{2}gt^2\right) \\
&= v_0^2 - 2gy,
\end{aligned}$$

where Eq. (3-21) has been used to eliminate t in favor of y. This result, which will be seen in the chapter dealing with conservation of energy (Chapter 7), is valid for any y, positive, negative or zero, as long as $v^2 > 0$. For the case of a rock thrown from the roof of a building of height h, the speed at the ground is found by substituting $y = -h$ into the above expression, yielding $v = \sqrt{v_0^2 + 2gh}$, which is independent of α_0.

3-55: a) The height above the player's hand will be $\frac{v_{0y}^2}{2g} = \frac{v_0^2\sin^2\alpha_0}{2g} = 0.40$ m, so the maximum height above the floor is 2.23 m.   b) Use of the result of Problem 3-49 gives 3.84 m.   c) The algebra is the same as that for Problems 3-48 and 3-52. The distance y is 3.05 m $- 1.83$ m $= 1.22$ m, and

$$v_0 = \sqrt{\frac{(9.80\text{ m/s}^2)(4.21\text{ m})^2}{2\cos^2 35°((4.21\text{ m})\tan 35° - 1.22\text{ m})}} = 8.65\text{ m/s}.$$

d) As in part (a), but with the larger speed,

$$1.83\text{ m} + (8.65\text{ m/s})^2\sin^2 35°/2(9.80\text{ m/s}^2) = 3.09\text{ m}.$$

The distance from the basket is the distance from the foul line to the basket, minus half the range, or

$$4.21 \text{ m} - (8.655 \text{ m/s})^2 \sin 70°/2(9.80 \text{ m/s}^2) = 0.62 \text{ m}.$$

Note that an extra figure in the intermediate calculation was kept to avoid roundoff error.

3-56: The initial y-component of the velocity is $v_{0y} = \sqrt{2gy}$, and the time the pebble is in flight is $t = \sqrt{2y/g}$. The initial x-component is $v_{0x} = x/t = \sqrt{x^2 g/2y}$. The magnitude of the initial velocity is then

$$v_0 = \sqrt{2gy + \frac{x^2 g}{2y}} = \sqrt{2gy}\sqrt{1 + \left(\frac{x}{2y}\right)^2},$$

and the angle is $\arctan\left(\frac{v_{0y}}{v_{0x}}\right) = \arctan(2y/x)$.

3-57: a) The acceleration is given as g at an angle of 53.1° to the horizontal. This is a 3-4-5 triangle, and thus, $a_x = (3/5)g$ and $a_y = (4/5)g$ during the "boost" phase of the flight. Hence this portion of the flight is a straight line at an angle of 53.1° to the horizontal. After time T, the rocket is in free flight, the acceleration is $a_x = 0$ and $a_y = g$, and the familiar equations of projectile motion apply. During this coasting phase of the flight, the trajectory is the familiar parabola.

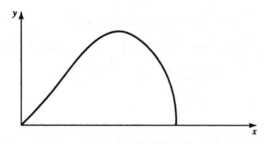

b) During the boost phase, the velocities are: $v_x = (3/5)gt$ and $v_y = (4/5)gt$, both straight lines. After $t = T$, the velocities are $v_x = (3/5)gT$, a horizontal line, and $v_y = (4/5)gT - gt$, a negatively sloping line which crosses the axis at the time of the maximum height.

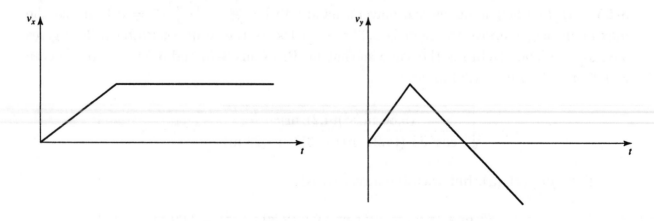

c) To find the maximum height of the rocket, set $v_y = 0$, and solve for t, use this time in the familiar equation for y. Thus, using $t = (4/5)T$ and $y_{max} = y_0 + v_{0y}t - \frac{1}{2}gt^2$, $y_{max} = \frac{2}{5}gT^2 + \frac{4}{5}gT\left(\frac{4}{5}T\right) - \frac{1}{2}g\left(\frac{4}{5}T\right)^2$, $y_{max} = \frac{2}{5}gT^2 + \frac{16}{25}gT^2 - \frac{8}{25}gT^2$. Combining terms, $y_{max} = \frac{18}{25}gT^2$.

d) To find the total horizontal distance, break the problem into three parts: The boost phase, the rise to maximum, and the fall back to earth. The fall time back to earth can be found from the answer to part (c), $(18/25)gT^2 = (1/2)gt^2$, or $t = (6/5)T$. Then, multiplying these times and the velocity, $x = \frac{3}{10}gT^2 + \left(\frac{3}{5}gT\right)\left(\frac{4}{5}T\right) + \left(\frac{3}{5}gT\right)\left(\frac{6}{5}T\right)$, or $x = \frac{3}{10}gT^2 + \frac{12}{25}gT^2 + \frac{18}{25}gT^2$. Combining terms gives $x = \frac{3}{2}gT^2$.

3-58: In the frame of the hero, the range of the object must be the initial separation plus the amount the enemy has pulled away in that time. Symbolically, $R = x_0 + v_{E/H}t = x_0 + v_{E/H}\frac{R}{v_{0x}}$, where $v_{E/H}$ is the velocity of the enemy relative to the hero, t is the time of flight, v_{0x} is the (constant) x-component of the grenade's velocity, as measured by the hero, and R is the range of the grenade, also as measured by the hero. Using Eq. (3-29) for R, with $\sin 2\alpha_0 = 1$ and $v_{0x} = v_0/\sqrt{2}$,

$$\frac{v_0^2}{g} = x_0 + v_{E/H}\frac{v_0}{g}\sqrt{2}, \quad \text{or} \quad v_0^2 - \left(\sqrt{2}v_{E/H}\right)v_0 - gx_0 = 0.$$

This quadratic is solved for

$$v_0 = \frac{1}{2}\left(\sqrt{2}v_{E/H} + \sqrt{2v_{E/H}^2 + 4gx_0}\right) = 61.1 \text{ km/h},$$

where the units for g and x_0 have been properly converted. Relative to the earth, the x-component of velocity is $90.0 \text{ km/h} + (61.1 \text{ km/h})\cos 45° = 133.2 \text{ km/h}$, the y-component, the same in both frames, is $(61.1 \text{ km/h})\sin 45° = 43.2 \text{ km/h}$, and the magnitude of the velocity is then 140 km/h.

3-59: a) $\quad x^2 + y^2 = (R\cos\omega t)^2 + (R\sin\omega t)^2 = R^2(\cos^2\omega t + \sin^2\omega t) = R^2,$

so the radius is R.

b) $$v_x = -\omega R\sin\omega t, \quad v_y = \omega R\cos\omega t,$$

and so the dot product

$$\begin{aligned}\vec{r} \cdot \vec{v} &= xv_x + yv_y \\ &= (R\cos\omega t)(-\omega R\sin\omega t) + (R\sin\omega t)(\omega R\cos\omega t) \\ &= \omega R(-\cos\omega t\sin\omega t + \sin\omega t\cos\omega t) \\ &= 0.\end{aligned}$$

c) $$a_x = -\omega^2 R\cos\omega t = -\omega^2 x, \quad a_y = \omega^2 R\sin\omega t = -\omega^2 y,$$

and so $\vec{a} = -\omega^2\vec{r}$ and $a = \omega^2 R$.

d) $v^2 = v_x^2 + v_y^2 = (-\omega R \sin \omega t)^2 + (\omega R \cos \omega t)^2 = \omega^2 R^2 (\sin^2 \omega t + \cos^2 \omega t) = \omega^2 R^2$,

and so $v = \omega R$.

e)
$$a = \omega^2 R = \frac{(\omega R)^2}{R} = \frac{v^2}{R}.$$

3-60: a)
$$\frac{dv}{dt} = \frac{d}{dt}\sqrt{v_x^2 + v_y^2}$$
$$= \frac{(1/2)\frac{d}{dt}(v_x^2 + v_y^2)}{\sqrt{v_x^2 + v_y^2}}$$
$$= \frac{v_x a_x + v_y a_y}{\sqrt{v_x^2 + v_y^2}}.$$

b) Using the numbers from Example 3-1 and 3-2,

$$\frac{dv}{dt} = \frac{(-1.0 \text{ m/s})(-0.50 \text{ m/s}^2) + (1.3 \text{ m/s})(0.30 \text{ m/s}^2)}{\sqrt{(-1.0 \text{ m/s})^2 + (1.3 \text{ m/s})^2}} = 0.54 \text{ m/s}.$$

The acceleration is due to changing both the magnitude and direction of the velocity. If the direction of the velocity is changing, the magnitude of the acceleration is larger than the rate of change of speed. c) $\vec{v} \cdot \vec{a} = v_x a_x + v_y a_y$, $v = \sqrt{v_x^2 + v_y^2}$, and so the above form for $\frac{dv}{dt}$ is seen to be $\vec{v} \cdot \vec{a}/v$.

3-61: a) The path is a cycloid.

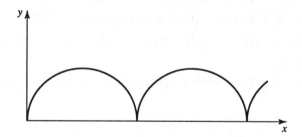

b) To find the velocity components, take the derivative of x and y with respect to time: $v_x = R\omega(1 - \cos \omega t)$, and $v_y = R\omega \sin \omega t$. To find the acceleration components, take the derivative of v_x and v_y with respect to time: $a_x = R\omega^2 \sin \omega t$, and $a_y = R\omega^2 \cos \omega t$.

c) The particle is at rest ($v_x = v_y = 0$) every period, namely at $t = 0, 2\pi/\omega, 4\pi/\omega, \dots$. At that time, $x = 0, 2\pi R, 4\pi R, \dots$; and $y = 0$. The acceleration is $a = R\omega^2$ in the $+y$-direction.

d) No, since $a = [(R\omega^2 \sin \omega t)^2 + (R\omega^2 \cos \omega t)^2]^{1/2} = R\omega^2$.

3-62: A direct way to find the angle is to consider the velocity relative to the air and the velocity relative to the ground as forming two sides of an isosceles triangle. The wind direction relative to north is half of the included angle, or $\arcsin(10/50) = 11.53°$, east of north.

3-63: Finding the infinite series consisting of the times between meeting with the brothers is possible, and even entertaining, but hardly necessary. The relative speed of the brothers is 70 km/h, and as they are initially 42 km apart, they will reach each other in six-tenths of an hour, during which time the pigeon flies 30 km.

3-64: a) The drops are given as falling vertically, so their horizontal component of velocity with respect to the earth is zero. With respect to the train, their horizontal component of velocity is 12.0 m/s, west (as the train is moving eastward). b) The vertical component, in either frame, is $(12.0 \text{ m/s})/(\tan 30°) = 20.8$ m/s, and this is the magnitude of the velocity in the frame of the earth. The magnitude of the velocity in the frame of the train is $\sqrt{(12.0 \text{ m/s})^2 + (20.8 \text{ m/s})^2} = 24$ m/s. This is, of course, the same as $(12.0 \text{ m/s})/\sin 30°$.

3-65: With no wind, the plane would be 110 km west of the starting point; the wind has blown the plane 10 km west and 20 km south in half an hour, so the wind velocity is $\sqrt{(20 \text{ km/h})^2 + (40 \text{ km/h})^2} = 44.7$ km/h at a direction of $\arctan(40/20) = 63°$ south of west. b) $\arcsin(40/220) = 10.5°$ north of west.

3-66: a) $2D/v$ b) $2Dv/(v^2 - w^2)$ c) $2D/\sqrt{v^2 - w^2}$ d) 1.50 h, 1.60 h, 1.55 h.

3-67: a) The position of the bolt is $3.00 \text{ m} + (2.50 \text{ m/s})t - 1/2(9.80 \text{ m/s}^2)t^2$, and the position of the floor is $(2.50 \text{ m/s})t$. Equating the two, $3.00 \text{ m} = (4.90 \text{ m/s}^2)t^2$. Therefore $t = 0.782$ s. b) The velocity of the bolt is $2.50 \text{ m/s} - (9.80 \text{ m/s}^2)(0.782 \text{ s}) = -5.17$ m/s relative to Earth, therefore, relative to an observer in the elevator $v = -5.17 \text{ m/s} - 2.50 \text{ m/s} = -7.67$ m/s. c) As calculated in part (b), the speed relative to Earth is 5.17 m/s. d) Relative to Earth, the distance the bolt travelled is $(2.50 \text{ m/s})t - 1/2(9.80 \text{ m/s}^2)t^2 = (2.50 \text{ m/s})(0.782 \text{ s}) - (4.90 \text{ m/s}^2)(0.782 \text{ s})^2 = -1.04$ m

3-68: a) $v_{0y} = \sqrt{2gh} = \sqrt{2(9.80 \text{ m/s}^2)(4.90 \text{ m})} = 9.80$ m/s. b) $v_{0y}/g = 1.00$ s. c) The speed relative to the man is $\sqrt{(10.8 \text{ m/s})^2 - (9.80 \text{ m/s})^2} = 4.54$ m/s, and the speed relative to the hoop is 13.6 m/s (rounding to three figures), and so the man must be 13.6 m in front of the hoop at release. d) Relative to the flat car, the ball is projected at an angle $\theta = \tan^{-1}\left(\frac{9.80 \text{ m/s}}{4.54 \text{ m/s}}\right) = 65°$. Relative to the ground the angle is $\theta = \tan^{-1}\left(\frac{9.80 \text{ m/s}}{4.54 \text{ m/s} + 9.10 \text{ m/s}}\right) = 35.7°$

3-69: a) $(150 \text{ m/s})^2 \sin 2°/9.80 \text{ m/s}^2 = 80$ m.
b) $1000 \times \frac{\pi(10 \times 10^{-2} \text{ m})^2}{\pi(80 \text{ m})^2} = 1.6 \times 10^{-3}$.

c) The slower rise will tend to reduce the time in the air and hence reduce the radius. The slower horizontal velocity will also reduce the radius. The lower speed would tend to increase the time of descent, hence increasing the radius. As the bullets fall, the friction effect is smaller than when they were rising, and the overall effect is to decrease the radius.

3-70: Write an expression for the square of the distance (D^2) from the origin to the particle, expressed as a function of time. Then take the derivative of D^2 with respect to t, and solve for the value of t when this derivative is zero. If the discriminant is zero or negative, the distance D will never decrease. Following this process, $\sin^{-1}(8/9) = 62.7°$.

3-71: a) The trajectory of the projectile is given by Eq. (3-27), with $\alpha_0 = \theta + \phi$, and the equation describing the incline is $y = x \tan \theta$. Setting these equal and factoring out the

$x = 0$ root (where the projectile is on the incline) gives a value for x_0; the range measured along the incline is

$$x/\cos\theta = \left[\frac{2v_0^2}{g}\right][\tan(\theta + \phi) - \tan\theta]\left[\frac{\cos^2(\theta + \phi)}{\cos\theta}\right].$$

b) Of the many ways to approach this problem, a convenient way is to use the same sort of "trick", involving double angles, as was used to derive the expression for the range along a horizontal incline. Specifically, write the above in terms of $\alpha = \theta + \phi$, as

$$R = \left[\frac{2v_0^2}{g\cos^2\theta}\right][\sin\alpha\cos\alpha\cos\theta - \cos^2\alpha\sin\theta].$$

The dependence on α and hence ϕ is in the second term. Using the identities $\sin\alpha\cos\alpha = (1/2)\sin 2\alpha$ and $\cos^2\alpha = (1/2)(1 + \cos 2\alpha)$, this term becomes

$$(1/2)\left[\cos\theta\sin 2\alpha - \sin\theta\cos 2\alpha - \sin\theta\right] = (1/2)\left[\sin(2\alpha - \theta) - \sin\theta\right].$$

This will be a maximum when $\sin(2\alpha - \theta)$ is a maximum, at $2\alpha - \theta = 2\phi + \theta = 90°$, or $\phi = 45° - \theta/2$. Note that this reduces to the expected forms when $\theta = 0$ (a flat incline, $\phi = 45°$ and when $\theta = -90°$ (a vertical cliff), when a horizontal launch gives the greatest distance).

3-72: As in the previous problem, the horizontal distance x in terms of the angles is

$$\tan\theta = \tan(\theta + \phi) - \left(\frac{gx}{2v_0^2}\right)\frac{1}{\cos^2(\theta + \phi)}.$$

Denote the dimensionless quantity $gx/2v_0^2$ by β; in this case

$$\beta = \frac{(9.80 \text{ m/s}^2)(60.0 \text{ m})\cos 30.0°}{2(32.0 \text{ m/s})^2} = 0.2486.$$

The above relation can then be written, on multiplying both sides by the product $\cos\theta\cos(\theta + \phi)$,

$$\sin\theta\cos(\theta + \phi) = \sin(\theta + \phi)\cos\theta - \frac{\beta\cos\theta}{\cos(\theta + \phi)},$$

and so

$$\sin(\theta + \phi)\cos\theta - \cos(\theta + \phi)\sin\theta = \frac{\beta\cos\theta}{\cos(\theta + \phi)}.$$

The term on the left is $\sin((\theta + \phi) - \theta) = \sin\phi$, so the result of this combination is

$$\sin\phi\cos(\theta + \phi) = \beta\cos\theta.$$

Although this can be done numerically (by iteration, trial-and-error, or other methods), the expansion $\sin a\cos b = \frac{1}{2}(\sin(a + b) + \sin(a - b))$ allows the angle ϕ to be isolated;

specifically, then

$$\frac{1}{2}(\sin(2\phi + \theta) + \sin(-\theta)) = \beta \cos\theta,$$

with the net result that

$$\sin(2\phi + \theta) = 2\beta\cos\theta + \sin\theta.$$

a) For $\theta = 30°$, and β as found above, $\phi = 19.3°$ and the angle above the horizontal is $\theta + \phi = 49.3°$. For level ground, using $\beta = 0.2871$, gives $\phi = 17.5°$. b) For $\theta = -30°$, the same β as with $\theta = 30°$ may be used ($\cos 30° = \cos(-30°)$), giving $\phi = 13.0°$ and $\phi + \theta = -17.0°$.

3-73: In a time Δt, the velocity vector has moved through an angle (in radians) $\Delta\phi = \frac{v\Delta t}{R}$ (see Figure 3-23). By considering the isosceles triangle formed by the two velocity vectors, the magnitude $\left|\Delta\vec{v}\right|$ is seen to be $2v\sin(\phi/2)$, so that

$$\left|\vec{a}_{ave}\right| = 2\frac{v}{\Delta t}\sin\left(\frac{v\Delta t}{2R}\right) = \frac{10 \text{ m/s}}{\Delta t}\sin(1.0/\text{s}\cdot\Delta t)$$

Using the given values gives magnitudes of 9.59 m/s², 9.98 m/s² and 10.0 m/s². The instantaneous acceleration magnitude, $v^2/R = (5.00 \text{ m/s})^2/(2.50 \text{ m}) = 10.0 \text{ m/s}^2$ is indeed approached in the limit at $\Delta t \to 0$. The changes in direction of the velocity vectors are given by $\Delta\theta = \frac{v\Delta t}{R}$ and are, respectively, 1.0 rad, 0.2 rad, and 0.1 rad. Therefore, the angle of the average acceleration vector with the original velocity vector is $\frac{\pi+\Delta\theta}{2} = \pi/2 + 1/2$ rad $(118.6°)$, $\pi/2 + 0.1$ rad $(95.7°)$, and $\pi/2 + 0.05$ rad $(92.9°)$.

3-74:

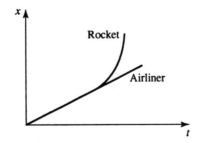

The x-position of the plane is $(236 \text{ m/s})t$ and the x-position of the rocket is $(236 \text{ m/s})t + 1/2(3.00)(9.80 \text{ m/s}^2)\cos 30°(t - T)^2$. The graphs of these two have the form,

If we take $y = 0$ to be the altitude of the airliner, then $y(t) = -1/2gT^2 - gT(t - T) + 1/2(3.00)(9.80 \text{ m/s}^2)(\sin 30°)(t - T)^2$ for the rocket. This graph looks like

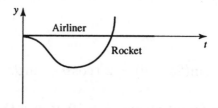

By setting $y = 0$ for the rocket, we can solve for t in terms of T, $0 = -(4.90 \text{ m/s}^2)T^2 - (9.80 \text{ m/s}^2)T(t - T) + (7.35 \text{ m/s}^2)(t - T)^2$. Using the quadratic formula for the variable $x = t - T$, we find $x = t - T = \frac{(9.80 \text{ m/s}^2)T + \sqrt{(9.80 \text{ m/s}^2 T)^2 + (4)(7.35 \text{ m/s}^2)(4.9)T^2}}{2(7.35 \text{ m/s}^2)}$ or $t = 2.72\, T$.
Now, using the condition that $x_{\text{rocket}} - x_{\text{plane}} = 1000$ m, we find $(236 \text{ m/s})t + (12.7 \text{ m/s}^2) \times (t - T)^2 - (236 \text{ m/s})t = 1000$ m, or $(1.72T)^2 = 78.6 \text{ s}^2$. Therefore $T = 5.15$ s.

3-75: a) Taking all units to be in km and hr, we have three equations. We know that heading upstream $v_{c/w} - v_{w/G} = 2$ where $v_{c/w}$ is the speed of the curve relative to water and $v_{w/G}$ is the speed of the water relative to the ground. We know that heading downstream for a time t, $(v_{c/w} + v_{w/G})t = 5$. We also know that for the bottle $v_{w/G}(t + 1) = 3$. Solving these three equations for $v_{w/G} = x$, $v_{c/w} = 2 + x$, therefore $(2 + x + x)t = 5$ or $(2 + 2x)t = 5$. Also $t = 3/x - 1$, so $(2 + 2x)(\frac{3}{x} - 1) = 5$ or $2x^2 + x - 6 = 0$. The positive solution is $x = v_{w/G} = 1.5$ km/hr.

  b) $v_{c/w} = 2$ km/hr $+ v_{w/G} = 3.5$ km/hr.

Chapter 4 Newton's Laws of Motion

4-1: a) For the magnitude of the sum to be the sum of the magnitudes, the forces must be parallel, and the angle between them is zero. b) The forces form the sides of a right isosceles triangle, and the angle between them is 90°. Alternatively, the law of cosines may be used as

$$F^2 + F^2 = \left(\sqrt{2}F\right)^2 - 2F^2 \cos\theta,$$

from which $\cos\theta = 0$, and the forces are perpendicular. c) For the sum to have 0 magnitude, the forces must be antiparallel, and the angle between them is 180°.

4-2: In the new coordinates, the 300-N force acts at an angle of 75° from the $-x$-axis, or 105° from the $+x$-axis, and the 155-N force acts at an angle of 23° from the $-x$-axis, or 203° from the $+x$-axis. a) The components of the net force are

$$R_x = (200 \text{ N}) + (300 \text{ N})\cos 105° + (155 \text{ N})\cos 203° = -20 \text{ N}$$
$$R_y = (200 \text{ N})\sin 0 + (300 \text{ N})\sin 105° + (155 \text{ N})\sin 203° = 229 \text{ N}.$$

b) $R = \sqrt{R_x^2 + R_y^2} = 230$ N, $\arctan\left(\frac{229}{-20}\right) = 95°$. The results have the same magnitude, and the angle has been changed by the amount (30°) that the coordinates have been rotated.

4-3: The horizontal component of the force is $(10 \text{ N})\cos 45° = 7.1$ N to the right and the vertical component is $(10 \text{ N})\sin 45° = 7.1$ N down.

4-4: a) $F_x = F\cos\theta$, where θ is the angle that the rope makes with the ramp ($\theta = 30°$ in this problem), so $F = |\overrightarrow{F}| = \frac{F_x}{\cos\theta} = \frac{60.0 \text{ N}}{\cos 30°} = 69.3$ N.
b) $F_y = F\sin\theta = F_x\tan\theta = 34.6$ N.

4-5: Of the many ways to do this problem, two are presented here.

Geometric: From the law of cosines, the magnitude of the resultant is

$$R = \sqrt{(270 \text{ N})^2 + (300 \text{ N})^2 + 2(270 \text{ N})(300 \text{ N})\cos 60°} = 494 \text{ N}.$$

The angle between the resultant and dog A's rope (the angle opposite the side corresponding to the 250-N force in a vector diagram) is then

$$\arcsin\left(\frac{\sin 120°(300 \text{ N})}{(494 \text{ N})}\right) = 31.7°.$$

Components: Taking the $+x$-direction to be along dog A's rope, the components of the resultant are

$$R_x = (270 \text{ N}) + (300 \text{ N})\cos 60° = 420 \text{ N}$$
$$R_y = (300 \text{ N})\sin 60° = 259.8 \text{ N},$$

so $R = \sqrt{(420 \text{ N})^2 + (259.8 \text{ N})^2} = 494$ N, $\theta = \arctan\left(\frac{259.8}{420}\right) = 31.7°$.

4-6: a) $F_{1x} + F_{2x} = (9.00\text{ N})\cos 120° + (6.00\text{ N})\cos(-126.9°) = -8.10\text{ N}$

$F_{1y} + F_{2y} = (9.00\text{ N})\sin 120° + (6.00\text{ N})\sin(-126.9°) = +3.00\text{ N}.$

 b) $R = \sqrt{R_x^2 + R_y^2} = \sqrt{(8.10\text{ N})^2 + (3.00\text{ N})^2} = 8.64\text{ N}.$

4-7: $a = F/m = (132\text{ N})/(60\text{ kg}) = 2.2\text{ m/s}^2$ (to two places).

4-8: $F = ma = (135\text{ kg})(1.40\text{ m/s}^2) = 189\text{ N}.$

4-9: $m = F/a = (48.0\text{ N})/(3.00\text{ m/s}^2) = 16.00\text{ kg}.$

4-10: a) The acceleration is $a = \frac{2x}{t^2} = \frac{2(11.0\text{ m})}{(5.00\text{ s})^2} = 0.88\text{ m/s}^2$. The mass is then $m = \frac{F}{a} = \frac{80.0\text{ N}}{0.88\text{ m/s}^2} = 90.9\text{ kg}.$

 b) The speed at the end of the first 5.00 seconds is $at = 4.4\text{ m/s}$, and the block on the frictionless surface will continue to move at this speed, so it will move another $vt = 22.0\text{ m}$ in the next 5.00 s.

4-11: a) During the first 2.00 s, the acceleration of the puck is $F/m = 1.563\text{ m/s}^2$ (keeping an extra figure). At $t = 2.00$ s, the speed is $at = 3.13\text{ m/s}$ and the position is $at^2/2 = vt/2 = 3.13\text{ m}$. b) The acceleration during this period is also 1.563 m/s^2, and the speed at 7.00 s is $3.13\text{ m/s} + (1.563\text{ m/s}^2)(2.00\text{ s}) = 6.26\text{ m/s}$. The position at $t = 5.00$ s is $x = 3.13\text{ m} + (3.13\text{ m/s})(5.00\text{ s} - 2.00\text{ s}) = 12.5\text{ m}$, and at $t = 7.00$ s is

$$12.5\text{ m} + (3.13\text{ m/s})(2.00\text{ s}) + (1/2)(1.563\text{ m/s}^2)(2.00\text{ s})^2 = 21.89\text{ m},$$

or 21.9 m to three places.

4-12: a) $a = F/m = 140\text{ N}/32.5\text{ kg} = 4.31\text{ m/s}^2.$

 b) With $v_0 = 0$, $x = \frac{1}{2}at^2 = 215\text{ m}.$

 c) With $v_0 = 0$, $v = at = 2x/t = 43.0\text{ m/s}.$

4-13: a) $\sum \vec{F} = 0$

 b), c), d)

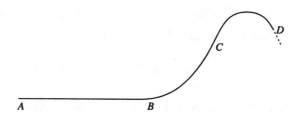

4-14: a) With $v_0 = 0$,

$$a = \frac{v^2}{2x} = \frac{(3.00 \times 10^6\text{ m/s})^2}{2(1.80 \times 10^{-2}\text{ m})} = 2.50 \times 10^{14}\text{ m/s}^2.$$

 b) $t = \frac{v}{a} = \frac{3.00 \times 10^6\text{ m/s}}{2.50 \times 10^{14}\text{ m/s}^2} = 1.20 \times 10^{-8}$ s. Note that this time is also the distance divided by the *average* speed.

 c) $F = ma = (9.11 \times 10^{-31}\text{ kg})(2.50 \times 10^{14}\text{ m/s}^2) = 2.28 \times 10^{-16}\text{ N}.$

4-15: $F = ma = w(a/g) = (2400\text{ N})(12\text{ m/s}^2)/(9.80\text{ m/s}^2) = 2.94 \times 10^3\text{ N}.$

4-16:
$$a = \frac{F}{m} = \frac{F}{w/g} = \frac{F}{w}g = \left(\frac{160}{71.2}\right)(9.80 \text{ m/s}^2) = 22.0 \text{ m/s}^2.$$

4-17: a) $m = w/g = (44.0 \text{ N})/(9.80 \text{ m/s}^2) = 4.49 \text{ kg}$ b) The mass is the same, 4.49 kg, and the weight is $(4.49 \text{ kg})(1.81 \text{ m/s}^2) = 8.13 \text{ N}$.

4-18: a) From Eq. (4-9), $m = w/g = (3.20 \text{ N})/(9.80 \text{ m/s}^2) = 0.327 \text{ kg}$.

   b) $w = mg = (14.0 \text{ kg})(9.80 \text{ m/s}^2) = 137 \text{ N}$.

4-19: $F = ma = (55 \text{ kg})(15 \text{ m/s}^2) = 825 \text{ N}$. The net forward force on the sprinter is exerted by the blocks. (The sprinter exerts a backward force on the blocks.)

4-20: a) the earth (gravity)   b) 4 N, the book   c) no   d) 4 N, the earth, the book, up   e) 4 N, the hand, the book, down   f) second   g) third   h) no   i) no   j) yes   k) yes   l) one (gravity)   m) no

4-21: a) When air resistance is not neglected, the net force on the bottle is the weight of the bottle plus the force of air resistance.   b) The bottle exerts an upward force on the earth, and a downward force on the air.

4-22: The reaction to the upward normal force on the passenger is the downward normal force, also of magnitude 620 N, that the passenger exerts on the floor. The reaction to the passenger's weight is the gravitational force that the passenger exerts on the earth, upward and also of magnitude 650 N. $\frac{\Sigma F}{m} = \frac{620 \text{ N} - 650 \text{ N}}{650 \text{ N}/9.80 \text{ m/s}^2} = -0.452 \text{ m/s}^2$. The passenger's acceleration is 0.452 m/s², downward.

4-23:
$$a_E = \frac{F}{m_E} = \frac{mg}{m_E} = \frac{(45 \text{ kg})(9.80 \text{ m/s}^2)}{(6.0 \times 10^{24} \text{ kg})} = 7.4 \times 10^{-23} \text{ m/s}^2.$$

4-24: a) The force the astronaut exerts on the rope and the force that the rope exerts on the astronaut are an action-reaction pair, so the rope exerts a force of 80.0 N on the astronaut.   b) The cable is under tension.   c) $a = \frac{F}{m} = \frac{80.0 \text{ N}}{105.0 \text{ kg}} = 0.762 \text{ m/s}^2$.   d) There is no net force on the massless rope, so the force that the shuttle exerts on the rope must be 80.0 N (this is *not* an action-reaction pair). Thus, the force that the rope exerts on the shuttle must be 80.0 N.   e) $a = \frac{F}{m} = \frac{80.0 \text{ N}}{9.05 \times 10^4 \text{ kg}} = 8.84 \times 10^{-4} \text{ m/s}^2$.

4-25: When the upward force has its maximum magnitude F_{max} (the breaking strength), the net upward force will be $F_{max} - mg$ and the upward acceleration will be

$$a = \frac{F_{max} - mg}{m} = \frac{F_{max}}{m} - g = \frac{75.0 \text{ N}}{4.80 \text{ kg}} - 9.80 \text{ m/s}^2 = 5.83 \text{ m/s}^2.$$

4-26: a) The net force is upward, so $T - mg = m|\vec{a}|$, and $T = m(g + |\vec{a}|)$.

   b) The net force is downward, so $mg - T = m|\vec{a}|$ and $T = m(g - |\vec{a}|)$.

4-27: (Either part (a) or part (b) may be done first with equal ease.) b) The smaller crate will also accelerate at $a = 2.50 \text{ m/s}^2$, so the tension in the rope is $T = (4.00 \text{ kg})(2.50 \text{ m/s}^2) = 10.0 \text{ N}$.   a) The net force on the crate is $F - T = (6.00 \text{ kg})(2.5 \text{ m/s}^2) = 15.0 \text{ N}$, and so the applied force is $F = 15.0 \text{ N} + 10.0 \text{ N} = 25.0 \text{ N}$.

4-28: a)

b) The crates, with a total mass of 10.00 kg, accelerate together with an acceleration $a = \frac{F}{m} = \frac{50.0 \text{ N}}{10.00 \text{ kg}} = 5.00 \text{ m/s}^2$. c) The tension is the only horizontal force acting on the smaller crate, so $T = ma = (4.00 \text{ kg})(5.00 \text{ m/s}^2) = 20.0 \text{ N}$. As a check, the net force on the larger crate is $F - T$, so $T = 50.0 \text{ N} - (6.00 \text{ kg})(5.00 \text{ m/s}^2) = 20 \text{ N}$.

4-29: Taking the upward direction as positive, the acceleration is

$$a = \frac{F}{m} = \frac{F_{air} - mg}{m} = \frac{F_{air}}{m} - 9.80 \text{ m/s}^2 = \frac{620 \text{ N}}{55.0 \text{ kg}} - 9.80 \text{ m/s}^2 = 1.47 \text{ m/s}^2.$$

4-30: Differentiating twice, the acceleration of the helicopter as a function of time is

$$\vec{a} = (0.120 \text{ m/s}^3)t\hat{i} - (0.12 \text{ m/s}^2)\hat{k},$$

and at $t = 5.0$ s, the acceleration is

$$\vec{a} = (0.60 \text{ m/s}^2)\hat{i} - (0.12 \text{ m/s}^2)\hat{k}.$$

The force is then

$$\vec{F} = m\vec{a} = \frac{w}{g}\vec{a} = \frac{(2.75 \times 10^5 \text{ N})}{(9.80 \text{ m/s}^2)}\left[(0.60 \text{ m/s}^2)\hat{i} - (0.12 \text{ m/s}^2)\hat{k}\right]$$

$$= (1.7 \times 10^4 \text{ N})\hat{i} - (3.4 \times 10^3 \text{ N})\hat{k}.$$

4-31: The velocity as a function of time is $v(t) = A - 3Bt^2$ and the acceleration as a function of time is $a(t) = -6Bt$, and so the Force as a function of time is $F(t) = ma(t) = -6mBt$.

4-32: a) The stopping time is $\frac{x}{v_{avc}} = \frac{x}{(v_0/2)} = \frac{2(0.130 \text{ m})}{350 \text{ m/s}} = 7.43 \times 10^{-4}$ s.

b) $F = ma = (1.80 \times 10^{-3} \text{ kg})\frac{(350 \text{ m/s})}{(7.43 \times 10^{-4} \text{ s})} = 848$ N. (Using $a = v_0^2/2x$ gives the same result.)

4-33: Take the $+x$-direction to be along $\vec{F}_1$ and the $+y$-direction to be along $\vec{R}$. Then $F_{2x} = -1300$ N and $F_{2y} = 1300$ N, so $F_2 = 1838$ N, at an angle of $135°$ from $\vec{F}_1$.

4-34: a) $F - w = F - mg = ma$, so $m = \frac{F}{a+g}$ and

$$w = mg = F\frac{g}{a+g} = (50.0 \text{ N})\frac{(9.80 \text{ m/s}^2)}{(2.45 \text{ m/s}^2 + 9.80 \text{ m/s}^2)} = 40.0 \text{ N}.$$

b) Solving the previous relation for a in terms of F,

$$a = \frac{F}{m} - g = \frac{F}{w/g} - g = g\left(\frac{F}{w} - 1\right) = (9.80 \text{ m/s}^2)\left(\frac{30.0 \text{ N}}{40.0 \text{ N}} - 1\right) = -2.45 \text{ m/s}^2,$$

with the negative sign indicating a downward acceleration.

c) If the cable breaks, $a = -g$ and the force F is zero, so the scale reads zero.

4-35: a) The resultant must have no y-component, and so the child must push with a force with y-component $(140 \text{ N})\sin 30° - (100 \text{ N})\sin 60° = -16.6 \text{ N}$. For the child to exert the smallest possible force, that force will have no x-component, so the smallest possible force has magnitude 16.6 N and is at an angle of 270°, or 90° clockwise from the $+x$-direction. b) $m = \frac{\sum F}{a} = \frac{100 \text{ N}\cos 60° + 140 \text{ N}\cos 30°}{2.0 \text{ m/s}^2} = 85.6 \text{ kg}$. $w = mg = (85.6 \text{ kg})(9.80 \text{ m/s}^2) = 840 \text{ N}$.

4-36: The ship would go a distance

$$\frac{v_0^2}{2a} = \frac{v_0^2}{2(F/m)} = \frac{mv_0^2}{2F} = \frac{(3.6 \times 10^7 \text{ kg})(1.5 \text{ m/s})^2}{2(8.0 \times 10^4 \text{ N})} = 506.25 \text{ m},$$

so the ship would hit the reef. The speed when the tanker hits the reef is also found from

$$v = \sqrt{v_0^2 - (2Fx/m)} = \sqrt{(1.5 \text{ m/s})^2 - \frac{2(8.0 \times 10^4 \text{ N})(500 \text{ m})}{(3.6 \times 10^7 \text{ kg})}} = 0.17 \text{ m/s},$$

so the oil should be safe.

4-37: Let v_0 be the speed with which Griffith leaves the floor. The height h is then $h = v_0^2/2g$, and the acceleration while he is still on the floor is v_0/t. The average force is the weight plus the mass times this acceleration,

$$F_{\text{ave}} = w + ma = w\left(1 + \frac{v_0}{gt}\right) = w\left(1 + \sqrt{\frac{2h}{gt^2}}\right)$$

$$= (890 \text{ N})\left(1 + \sqrt{\frac{2(1.2 \text{ m})}{(9.80 \text{ m/s}^2)(0.300 \text{ s})^2}}\right) = 2.4 \times 10^3 \text{ N}.$$

4-38: $$F = ma = m\frac{v_0^2}{2x} = (850 \text{ kg})\frac{(12.5 \text{ m/s})^2}{2(1.8 \times 10^{-2} \text{ m})} = 3.7 \times 10^6 \text{ N}.$$

4-39: a) $x(0.025 \text{ s}) = (9.0 \times 10^3 \text{ m/s}^2)(0.025 \text{ s})^2 - (8.0 \times 10^4 \text{ m/s}^3)(0.025 \text{ s})^3 = 4.4 \text{ m}$.

b) Differentiating, the velocity as a function of time is

$$v(t) = (1.80 \times 10^4 \text{ m/s}^2)t - (2.40 \times 10^5 \text{ m/s}^3)t^2, \quad \text{so}$$
$$v(0.025 \text{ s}) = (1.80 \times 10^4 \text{ m/s}^2)(0.025 \text{ s}) - (2.40 \times 10^5 \text{ m/s}^3)(0.025 \text{ s})^2$$
$$= 3.0 \times 10^2 \text{ m/s}.$$

c) The acceleration as a function of time is

$$a(t) = 1.80 \times 10^4 \text{ m/s}^2 - (4.80 \times 10^5 \text{ m/s}^3)t,$$

so (i) at $t = 0$, $a = 1.8 \times 10^4 \text{ m/s}^2$, and (ii) $a(0.025 \text{ s}) = 6.0 \times 10^3 \text{ m/s}^2$, and the forces are (i) $ma = 2.7 \times 10^4 \text{ N}$ and (ii) $ma = 9.0 \times 10^3 \text{ N}$.

4-40: Denote the acceleration when the thrust is F_1 by a_1 and the acceleration when the thrust is F_2 by a_2. The forces and accelerations are then related by

$$F_1 - w = ma_1, \quad F_2 - w = ma_2.$$

Dividing the first of these by the second to eliminate the mass gives

$$\frac{F_1 - w}{F_2 - w} = \frac{a_1}{a_2},$$

and solving for the weight w gives

$$w = \frac{a_1 F_2 - a_2 F_1}{a_1 - a_2}.$$

In this form, it does not matter which thrust and acceleration are denoted by 1 and which by 2, and the acceleration due to gravity at the surface of Mercury need not be found. Substituting the given numbers,

$$w = \frac{(1.20 \text{ m/s}^2)(10.0 \times 10^3 \text{ N}) - (-0.80 \text{ m/s}^2)(25.0 \times 10^3 \text{ N})}{1.20 \text{ m/s}^2 - (-0.80 \text{ m/s}^2)} = 16.0 \times 10^3 \text{ N}.$$

In the above, note that the upward direction is taken to be positive, so that a_2 is negative. Also note that although a_2 is known to two places, the sums in both numerator and denominator are known to three places.

4-41:

a) The engine is pulling four cars, and so the force that the engine exerts on the first car is $4m|\vec{a}|$. b), c), d): Similarly, the forces the cars exert on the car behind are $3m|\vec{a}|, 2m|\vec{a}|$ and $-m|\vec{a}|$. e) The direction of the acceleration, and hence the direction of the forces, would change but the magnitudes would not; the answers are the same.

4-42: a) If the gymnast climbs at a constant rate, there is no net force on the gymnast, so the tension must equal the weight; $T = mg$.

b) No motion is no acceleration, so the tension is again the gymnast's weight.

c) $T - w = T - mg = ma = m|\vec{a}|$ (the acceleration is upward, the same direction as the tension), so $T = m(g + |\vec{a}|)$.

d) $T - w = T - mg = ma = -m|\vec{a}|$ (the acceleration is downward, the same opposite as the tension), so $T = m(g - |\vec{a}|)$.

4-43: a) The maximum acceleration would occur when the tension in the cables is a maximum,

$$a = \frac{F}{m} = \frac{T - mg}{m} = \frac{T}{m} - g = \frac{28,000 \text{ N}}{2200 \text{ kg}} - 9.80 \text{ m/s}^2 = 2.93 \text{ m/s}^2.$$

b)
$$\frac{28,000 \text{ N}}{2200 \text{ kg}} - 1.62 \text{ m/s}^2 = 11.1 \text{ m/s}^2.$$

4-44: a) His speed as he touches the ground is

$$v = \sqrt{2gh} = \sqrt{2(9.80 \text{ m/s}^2)(3.10 \text{ m})} = 7.80 \text{ m/s}.$$

b) The acceleration while the knees are bending is

$$a = \frac{v^2}{2y} = \frac{(7.80 \text{ m/s})^2}{2(0.60 \text{ m})} = 50.6 \text{ m/s}^2.$$

c) The net force that the feet exert on the ground is the force that the ground exerts on the feet (an action-reaction pair). This force is related to the weight and acceleration by $F - w = F - mg = ma$, so $F = m(a + g) = (75.0 \text{ kg})(50.6 \text{ m/s}^2 + 9.80 \text{ m/s}^2) = 4532$ N. As a fraction of his weight, this force is $\frac{F}{mg} = \left(\frac{a}{g} + 1\right) = 6.16$ (keeping an extra figure in the intermediate calculation of a). Note that this result is the same algebraically as $\left(\frac{3.10 \text{ m}}{0.60 \text{ m}} + 1\right)$.

4-45: a)

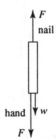

b) The acceleration of the hammer head will be the same as the nail, $a = v_0^2/2x = (3.2 \text{ m/s})^2/2(0.45 \text{ cm}) = 1.138 \times 10^3 \text{ m/s}^2$. The mass of the hammer head is its weight divided by g, $4.9 \text{ N}/9.80 \text{ m/s}^2 = 0.50$ kg, and so the net force on the hammer head is $(0.50 \text{ kg})(1.138 \times 10^3 \text{ m/s}^2) = 570$ N. This is the sum of the forces on the hammer head; the upward force that the nail exerts, the downward weight and the downward 15-N force. The force that the nail exerts is then 590 N, and this must be the magnitude of the force that the hammer head exerts on the nail. c) The distance the nail moves is .12 m, so the acceleration will be 4267 m/s^2, and the net force on the hammer head will be 2133 N. The magnitude of the force that the nail exerts on the hammer head, and hence the magnitude of the force that the hammer head exerts on the nail, is 2153 N, or about 2200 N.

4-46:

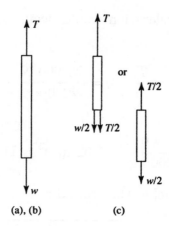

(a), (b)   (c)

a) The net force on a point of the cable at the top is zero; the tension in the cable must be equal to the weight w.

b) The net force on the cable must be zero; the difference between the tensions at the top and bottom must be equal to the weight w, and with the result of part (a), there is no tension at the bottom.

c) The net force on the bottom half of the cable must be zero, and so the tension in the cable at the middle must be half the weight, $w/2$. Equivalently, the net force on the upper half of the cable must be zero. From part (a) the tension at the top is w, the weight of the top half is $w/2$ and so the tension in the cable at the middle must be $w - w/2 = w/2$.

d) A graph of T vs. distance will be a negatively sloped line.

4-47:   a)

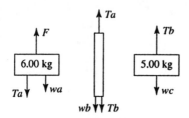

b) The net force on the system is $200 \text{ N} - (15.00 \text{ kg})(9.80 \text{ m/s}^2) = 53.0 \text{ N}$ (keeping three figures), and so the acceleration is $(53.0 \text{ N})/(15.0 \text{ kg}) = 3.53 \text{ m/s}^2$, up.   c) The net force on the 6-kg block is $(6.00 \text{ kg})(3.53 \text{ m/s}^2) = 21.2 \text{ N}$, so the tension is found from $F - T - mg = 21.2 \text{ N}$, or $T = (200 \text{ N}) - (6.00 \text{ kg})(9.80 \text{ m/s}^2) - 21.2 \text{ N} = 120 \text{ N}$. Equivalently, the tension at the top of the rope causes the upward acceleration of the rope and the bottom block, so $T - (9.00 \text{ kg})g = (9.00 \text{ kg})a$, which also gives $T = 120 \text{ N}$. d) The same analysis of part (c) is applicable, but using $6.00 \text{ kg} + 2.00 \text{ kg}$ instead of the mass of the top block, or 7.00 kg instead of the mass of the bottom block. Either way gives $T = 93.3 \text{ N}$.

4-48:   For a given initial velocity, the height that the ball will reach is inversely proportional to its downward acceleration. That is, the acceleration in the presence of the drag force is $a = g\left(\frac{5.0}{3.8}\right) = 1.32g$. Since $mg + F_{\text{drag}} = ma = 1.32 \, mg$, $F_{\text{drag}} = 0.32 \, mg =$

$(0.32)(0.0900 \text{ kg})(9.80 \text{ m/s}^2) = 0.32 \text{ N}$. Note that in this situation (where the upward motion of the ball was considered), the drag force and gravity act in the *same* direction.

4-49:
$$\vec{v}(t) = \frac{1}{m}\int_0^t \vec{a}\, dt = \frac{1}{m}\left(k_1 t\hat{i} + \frac{k_2}{4}t^4\hat{j}\right).$$

4-50: a) The equation of motion, $-Cv^2 = m\frac{dv}{dt}$ cannot be integrated with respect to time, as the unknown function $v(t)$ is part of the integrand. The equation must be *separated* before integration; that is,

$$-\frac{C}{m}dt = \frac{dv}{v^2}$$
$$-\frac{Ct}{m} = -\frac{1}{v} + \frac{1}{v_0},$$

where v_0 is the constant of integration that gives $v = v_0$ at $t = 0$. Note that this form shows that if $v_0 = 0$, there is no motion. This expression may be rewritten as

$$v = \frac{dx}{dt} = \left(\frac{1}{v_0} + \frac{Ct}{m}\right)^{-1},$$

which may be integrated to obtain

$$x - x_0 = \frac{m}{C}\ln\left[1 + \frac{Ctv_0}{m}\right].$$

To obtain x as a function of v, the time t must be eliminated in favor of v; from the expression obtained after the first integration, $\frac{Ctv_0}{m} = \frac{v_0}{v} - 1$, so

$$x - x_0 = \frac{m}{C}\ln\left(\frac{v_0}{v}\right).$$

b) By the chain rule,

$$\frac{dv}{dt} = \frac{dv}{dx}\frac{dv}{dt} = \frac{dv}{dx}v,$$

and using the given expression for the net force,

$$-Cv^2 = \left(v\frac{dv}{dx}\right)m$$
$$-\frac{C}{m}dx = \frac{dv}{v}$$
$$-\frac{C}{m}(x - x_0) = \ln\left(\frac{v}{v_0}\right)$$
$$x - x_0 = \frac{m}{C}\ln\left(\frac{v_0}{v}\right).$$

4-51: In this situation, the x-component of force depends explicitly on the y-component of postion. As the y-component of force is given as an explicit function of time, v_y and

y can be found as functions of time. Specifically, $a_y = (k_3/m)t$, so $v_y = (k_3/2m)t^2$ and $y = (k_3/6m)t^3$, where the initial conditions $v_{0y} = 0$, $y_0 = 0$ have been used. Then, the expressions for a_x, v_x and x are obtained as functions of time:

$$a_x = \frac{k_1}{m} + \frac{k_2 k_3}{6m^2} t^3$$

$$v_x = \frac{k_1}{m} t + \frac{k_2 k_3}{24m^2} t^4$$

$$x = \frac{k_1}{2m} t^2 + \frac{k_2 k_3}{120m^2} t^5.$$

In vector form,

$$\vec{r} = \left(\frac{k_1}{2m} t^2 + \frac{k_2 k_3}{120m^2} t^5 \right) \hat{i} + \left(\frac{k_3}{6m} t^3 \right) \hat{j}$$

$$\vec{v} + \left(\frac{k_1}{m} t + \frac{k_2 k_3}{24m^2} t^4 \right) \hat{i} + \left(\frac{k_3}{2m} t^2 \right) \hat{j}.$$

Chapter 5   Applications of Newton's Laws

5-1:   a) The tension in the rope must be equal to each suspended weight, 25.0 N.   b) If the mass of the light pulley may be neglected, the net force on the pulley is the tension in the chain and the tensions in the two parts of the rope; for the pulley to be in equilibrium, the tension in the chain is twice the tension in the rope, or 50.0 N.

5-2:   In all cases, each string is supporting a weight w against gravity, and the tension in each string is w. Two forces act on each mass: w down and $T(=w)$ up.

5-3:   a) The two sides of the rope each exert a force with vertical component $T\sin\theta$, and the sum of these components is the hero's weight. Solving for the tension T,

$$T = \frac{w}{2\sin\theta} = \frac{(90.0\text{ kg})(9.80\text{ m/s}^2)}{2\sin 10.0°} = 2.54 \times 10^3 \text{ N}.$$

b) When the tension is at its maximum value, solving the above equation for the angle θ gives

$$\theta = \arcsin\left(\frac{w}{2T}\right) = \arcsin\left(\frac{(90.0\text{ kg})(9.80\text{ m/s}^2)}{2(2.50 \times 10^4 \text{ N})}\right) = 1.01°.$$

5-4:   The vertical component of the force due to the tension in each wire must be half of the weight, and this in turn is the tension multiplied by the cosine of the angle each wire makes with the vertical, so if the weight is w, $\frac{w}{2} = \frac{3w}{4}\cos\theta$ and $\theta = \arccos\frac{2}{3} = 48°$.

5-5:   With the positive y-direction up and the positive x-direction to the right, the free-body diagram of Fig. 5-3(b) will have the forces labeled $\mathcal{N}$ and T resolved into x- and y-components, and setting the net force equal to zero,

$$F_x = T\cos\alpha - \mathcal{N}\sin\alpha = 0$$
$$F_y = \mathcal{N}\cos\alpha + T\sin\alpha - w = 0.$$

Solving the first for $\mathcal{N} = T\cot\alpha$ and substituting into the second gives

$$T\frac{\cos^2\alpha}{\sin\alpha} + T\sin\alpha = T\left(\frac{\cos^2\alpha}{\sin\alpha} + \frac{\sin^2\alpha}{\sin\alpha}\right) = \frac{T}{\sin\alpha} = w,$$

and so $\mathcal{N} = T\cot\alpha = w\sin\alpha\cot\alpha = w\cos\alpha$, as in Example 5-3.

5-6: $w\sin\alpha = mg\sin\alpha = (1390\text{ kg})(9.80\text{ m/s}^2)\sin 17.5° = 4.10 \times 10^3 \text{ N}.$

5-7:   a) $T_B\cos\theta = W$, or $T_B = W/\cos\theta = \frac{(4090\text{ kg})(9.8\text{ m/s}^2)}{\cos 40°} = 5.23 \times 10^4 \text{ N}.$

b) $T_A = T_B\sin\theta = (5.23 \times 10^4 \text{ N})\sin 40° = 3.36 \times 10^4 \text{ N}.$

5-8:   a) $T_C = w$, $T_A\sin 30° + T_B\sin 45° = T_C = w$, and $T_A\cos 30° - T_B\cos 45° = 0$. Since $\sin 45° = \cos 45°$, adding the last two equations gives $T_A(\cos 30° + \sin 30°) = w$, and so $T_A = \frac{w}{1.366} = 0.732w$. Then, $T_B = T_A\frac{\cos 30°}{\cos 45°} = 0.897w$.

88

Chapter 5

b) Similar to part (a), $T_C = w$, $-T_A \cos 60° + T_B \sin 45° = w$, and $T_A \sin 60° - T_B \cos 45° = 0$. Again adding the last two, $T_A = \frac{w}{(\sin 60° - \cos 60°)} = 2.73w$, and $T_B = T_A \frac{\sin 60°}{\cos 45°} = 3.35w$.

5-9: The resistive force is $w \sin \alpha = (1600 \text{ kg})(9.80 \text{ m/s}^2)(200 \text{ m}/6000 \text{ m}) = 523$ N.

5-10: The magnitude of the force must be equal to the component of the weight along the incline, or $W \sin \theta = (180 \text{ kg})(9.80 \text{ m/s}^2) \sin 11.0° = 337$ N.

5-11: a) $W = 60 \text{ N}, T \sin \theta = W$, so $T = (60 \text{ N})/\sin 45°$, or $T = 85$ N.

b) $F_1 = F_2 = T \cos \theta$, $F_1 = F_2 = 85 \text{ N} \cos 45° = 60$ N.

5-12: If the rope makes an angle θ with the vertical, then $\sin \theta = \frac{0.110}{1.51} = 0.073$ (the denominator is the sum of the length of the rope and the radius of the ball). The weight is then the tension times the cosine of this angle, or

$$T = \frac{w}{\cos \theta} = \frac{mg}{\cos(\arcsin(.073))} = \frac{(0.270 \text{ kg})(9.80 \text{ m/s}^2)}{0.998} = 2.65 \text{ N.}$$

The force of the pole on the ball is the tension times $\sin \theta$, or $(.055)T = 0.146$ N.

5-13: a)In the absence of friction, the force that the rope between the blocks exerts on block B will be the component of the weight along the direction of the incline, $T = w \sin \alpha$.
b) The tension in the upper rope will be the sum of the tension in the lower rope and the component of block A's weight along the incline, $w \sin \alpha + w \sin \alpha = 2w \sin \alpha$. c) In each case, the normal force is $w \cos \alpha$. d) When $\alpha = 0, N = w$, when $\alpha = 90°$, $N = 0$.

5-14: a) In level flight, the thrust and drag are horizontal, and the lift and weight are vertical. At constant speed, the net force is zero, and so $F = f$ and $w = L$. b) When the plane attains the new constant speed, it is again in equilibrium and so the new values of the thrust and drag, F' and f', are related by $F' = f'$; if $F' = 2F$, $f' = 2f$. c) In order to increase the magnitude of the drag force by a factor of 2, the speed must increase by a factor of $\sqrt{2}$.

5-15: a)

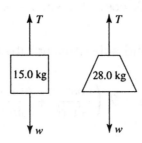

The tension is related to the masses and accelerations by

$$T - m_1 g = m_1 a_1$$
$$T - m_2 g = m_2 a_2.$$

b) For the bricks accelerating upward, let $a_1 = -a_2 = a$ (the counterweight will accelerate down). Then, subtracting the two equations to eliminate the tension gives

$$(m_2 - m_1)g = (m_1 + m_2)a, \quad \text{or}$$

$$a = g\frac{m_2 - m_1}{m_2 + m_1} = 9.80 \text{ m/s}^2 \left(\frac{28.0 \text{ kg} - 15.0 \text{ kg}}{28.0 \text{ kg} + 15.0 \text{ kg}}\right) = 2.96 \text{ m/s}^2.$$

c) The result of part (b) may be substituted into either of the above expressions to find the tension $T = 191$ N. As an alternative, the expressions may be manipulated to eliminate a algebraically by multiplying the first by m_2 and the second by m_1 and adding (with $a_2 = -a_1$) to give

$$T(m_1 + m_2) - 2m_1 m_2 g = 0, \quad \text{or}$$

$$T = \frac{2m_1 m_2 g}{m_1 + m_2} = \frac{2(15.0 \text{ kg})(28.0 \text{ kg})(9.80 \text{ m/s}^2)}{(15.0 \text{ kg} + 28.0 \text{ kg})} = 191 \text{ N}.$$

In terms of the weights, the tension is

$$T = w_1 \frac{2m_2}{m_1 + m_2} = w_2 \frac{2m_1}{m_1 + m_2}.$$

If, as in this case, $m_2 > m_1$, $2m_2 > m_1 + m_2$ and $2m_1 < m_1 + m_2$, so the tension is greater than w_1 and less than w_2; this must be the case, since the load of bricks rises and the counterweight drops.

5-16: Use Second Law and kinematics: $a = g\sin\theta$, $2ax = v^2$, solve for θ. $g\sin\theta = v^2/2x$, or $\theta = \arcsin(v^2/2gx) = \arcsin[(2.5 \text{ m/s})^2/[(2)(9.8 \text{ m/s}^2)(1.5 \text{ m})]]$, $\theta = 12.3°$.

5-17: a)

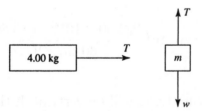

b) In the absence of friction, the net force on the 4.00-kg block is the tension, and so the acceleration will be $(10.0 \text{ N})/(4.00 \text{ kg}) = 2.50 \text{ m/s}^2$. c) The net upward force on the suspended block is $T - mg = ma$, or $m = T/(g + a)$. The block is accelerating downward, so $a = -2.50 \text{ m/s}^2$, and so $m = (10.0 \text{ N})/(9.80 \text{ m/s}^2 - 2.50 \text{ m/s}^2) = 1.37$ kg. d) $T = ma + mg$, so $T < mg$, because $a < 0$.

5-18: The maximum net force on the glider combination is

$$12,000 \text{ N} - 2 \times 2500 \text{ N} = 7000 \text{ N},$$

so the maximum acceleration is $a_{\text{max}} = \frac{7000 \text{ N}}{1400 \text{ kg}} = 5.0 \text{ m/s}^2$.

a) In terms of the runway length L and takeoff speed v, $a = \frac{v^2}{2L} < a_{max}$, so

$$L > \frac{v^2}{2a_{max}} = \frac{(40 \text{ m/s})^2}{2(5.0 \text{ m/s}^2)} = 160 \text{ m}.$$

b) If the gliders are accelerating at a_{max}, from $T - F_{drag} = ma$, $T = ma + F_{drag} = (700 \text{ kg})(5.0 \text{ m/s}^2) + 2500 \text{ N} = 6000 \text{ N}$. Note that this is exactly half of the maximum tension in the towrope between the plane and the first glider.

5-19: Denote the scale reading as F, and take positive directions to be upward. Then,

$$F - w = ma = \frac{w}{g}a, \quad \text{or} \quad a = g\left(\frac{F}{w} - 1\right).$$

a) $a = (9.80 \text{ m/s}^2)((450 \text{ N})/(550 \text{ N}) - 1) = -1.78 \text{ m/s}^2$, down.

b) $a = (9.80 \text{ m/s}^2)((670 \text{ N})/(550 \text{ N}) - 1) = 2.14 \text{ m/s}^2$, up.   c) If $F = 0$, $a = -g$ and the student, scale, and elevator are in free fall. The student should worry.

5-20: Similar to Exercise 5-16, the angle is $\arcsin(\frac{2L}{gt^2})$, but here the time is found in terms of velocity along the table, $t = \frac{x}{v_0}$, x being the length of the table and v_0 the velocity component along the table. Then,

$$\arcsin\left(\frac{2L}{g(x/v_0)^2}\right) = \arcsin\left(\frac{2Lv_0^2}{gx^2}\right)$$

$$= \arcsin\left(\frac{2(2.50 \times 10^{-2} \text{ m})(3.80 \text{ m/s})^2}{(9.80 \text{ m/s}^2)(1.75 \text{ m})^2}\right) = 1.38°.$$

5-21: a) For the car at rest, the string will be vertical. If the car is facing up the incline, the accelerometer will deflect to the left (backwards), and if the car is facing down the incline the accelerometer will deflect to the right (forward).   b) If the sinker is at rest with respect to the car, then both the sinker and car will have an acceleration down the incline, with magnitude $a = g \sin \alpha$, α being the angle of the incline. This is the component of $\overrightarrow{g}$ along the incline, so the string must be perpendicular to the direction of the acceleration; the string is not deflected with respect to the car.

5-22: In all cases, the string will deflect in such a way that the tension provides a component of force in the direction of acceleration.   a) The acceleration is to the left, so the string must pull to the left, and so the accelerometer deflects to the *right*.   b) The acceleration is to the right, and so the accelerometer deflects to the left.   c) The acceleration is to the right, and so the accelerometer deflects to the left.

5-23:

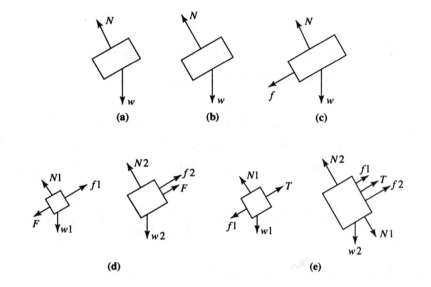

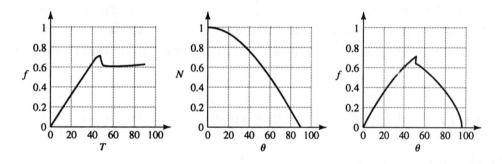

5-24:

5-25: a) For the net force to be zero, the applied force is

$$F = f_k = \mu_k \mathcal{N} = \mu_k mg = (0.20)(11.2 \text{ kg})(9.80 \text{ m/s}^2) = 22.0 \text{ N}.$$

b) The acceleration is $\mu_k g$, and $2ax = v^2$, so $x = v^2/2\mu_k g$, or $x = 3.13$ m.

5-26: a) If there is no applied horizontal force, no friction force is needed to keep the box in equilibrium. b) The maximum static friction force is, from Eq. (5-6), $\mu_s \mathcal{N} = \mu_s w = (0.40)(40.0 \text{ N}) = 16.0 \text{ N}$, so the box will not move and the friction force balances the applied force of 6.0 N. c) The maximum friction force found in part (b), 16.0 N. d) From Eq. (5-5), $\mu_k \mathcal{N} = (0.20)(40.0 \text{ N}) = 8.0 \text{ N}$. e) The applied force is enough to either start the box moving or to keep it moving. The answer to part (d), from Eq. (5-5), is independent of speed (as long as the box is moving), so the friction force is 8.0 N.

5-27: a) At constant speed, the net force is zero, and the magnitude of the applied force must equal the magnitude of the kinetic friction force,

$$|\overrightarrow{F}| = f_k = \mu_k \mathcal{N} = \mu_k mg = (0.12)(6.00 \text{ kg})(9.80 \text{ m/s}^2) = 7 \text{ N}.$$

b) $|\vec{F}| - f_k = ma$, so

$$|\vec{F}| = ma + f_k = ma + \mu_k mg = m(a + \mu_k g)$$
$$= (6.00 \text{ kg})(0.180 \text{ m/s}^2 + (0.12)9.80 \text{ m/s}^2) = 8 \text{ N}.$$

c) Replacing $g = 9.80 \text{ m/s}^2$ with 1.62 m/s^2 gives 1.2 N and 2.2 N.

5-28: The coefficient of kinetic friction is the ratio $\frac{f_k}{N}$, and the normal force has magnitude $85 \text{ N} + 25 \text{ N} = 110 \text{ N}$. The friction force, from $F_H - f_k = ma = w\frac{a}{g}$ is

$$f_k = F_H - w\frac{a}{g} = 20 \text{ N} - 85 \text{ N} \left(\frac{-0.9 \text{ m/s}^2}{9.80 \text{ m/s}^2}\right) = 28 \text{ N}$$

(note that the acceleration is negative), and so $\mu_k = \frac{28 \text{ N}}{110 \text{ N}} = 0.25$.

5-29: As in Example 5-16, the friction force is $\mu_k N = \mu_k w \cos\alpha$ and the component of the weight down the skids is $w \sin\alpha$. In this case, the angle α is $\arcsin(2.00/20.0) = 5.7°$. The ratio of the forces is $\frac{\mu_k \cos\alpha}{\sin\alpha} = \frac{\mu_k}{\tan\alpha} = \frac{0.25}{0.10} > 1$, so the friction force holds the safe back, and another force is needed to move the safe down the skids.

b) The difference between the downward component of gravity and the kinetic friction force is

$$w(\sin\alpha - \mu_k \cos\alpha) = (260 \text{ kg})(9.80 \text{ m/s}^2)(\sin 5.7° - (0.25)\cos 5.7°) = -381 \text{ N}.$$

5-30: a) The stopping distance is

$$\frac{v^2}{2a} = \frac{v^2}{2\mu_k g} = \frac{(28.7 \text{ m/s})^2}{2(0.80)(9.80 \text{ m/s}^2)} = 53 \text{ m}.$$

b) The stopping distance is inversely proportional to the coefficient of friction and proportional to the square of the speed, so to stop in the same distance the initial speed should not exceed

$$v\sqrt{\frac{\mu_{k,wet}}{\mu_{k,dry}}} = (28.7 \text{ m/s})\sqrt{\frac{0.25}{0.80}} = 16 \text{ m/s}.$$

5-31: For a given initial speed, the distance traveled is inversely proportional to the coefficient of kinetic friction. From Table 5-1, the ratio of the distances is then $\frac{0.44}{0.04} = 11$.

5-32: (a) If the block descends at constant speed, the tension in the connecting string must be equal to the hanging block's weight, w_B. Therefore, the friction force $\mu_k w_A$ on block A must be equal to w_B, and $w_B = \mu_k w_A$.

(b) With the cat on board, $a = g(W_B - \mu_k 2W_A)/(W_B + 2W_A)$.

5-33:

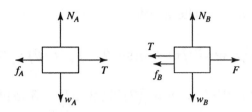

a) For the blocks to have no acceleration, each is subject to zero net force. Considering the horiztonal components,

$$T = f_A, \quad \left|\vec{F}\right| = T + f_B, \quad \text{or}$$

$$\left|\vec{F}\right| = f_A + f_B.$$

Using $f_A = \mu_k g m_A$ and $f_B = \mu_k g m_B$ gives $\left|\vec{F}\right| = \mu_k g \left(m_A + m_B\right)$.

   b) $T = f_A = \mu_k g m_A$.

5-34:

$$\mu_r = \frac{a}{g} = \frac{v_0^2 - v^2}{2Lg} = \frac{v_0^2 - \frac{1}{4}v_0^2}{2Lg} = \frac{3}{8}\frac{v_0^2}{Lg},$$

where L is the distance covered before the wheel's speed is reduced to half its original speed. Low pressure, $L = 18.1$ m; $\frac{3}{8}\frac{(3.50 \text{ m/s})^2}{(18.1 \text{ m})(9.80 \text{ m/s}^2)} = 0.0259$. High pressure, $L = 92.9$ m; $\frac{3}{8}\frac{(3.50 \text{ m/s})^2}{(92.9 \text{ m})(9.80 \text{ m/s}^2)} = 0.00505$.

5-35:   First, determine the acceleration from the freebody diagrams.

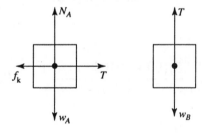

There are two equations and two unknowns, a and T:

$$-\mu_k m_A g + T = m_A a$$

$$m_B g - T = m_B a$$

Add and solve for a: $a = g(m_B - \mu_k m_A)/(m_B + m_A)$, $a = 0.79$ m/s^2.

   (a) $v = (2ax)^{1/2} = 0.22$ m/s.

   (b) Solving either equation for the tension gives $T = 11.7$ N.

5-36:   a) The normal force will be $w \cos\theta$ and the component of the gravitational force along the ramp is $w \sin\theta$. The box begins to slip when $w \sin\theta > \mu_s w \cos\theta$, or $\tan\theta > \mu_s = 0.35$, so slipping occurs at $\theta = \arctan(0.35) = 19.3°$, or $19°$ to two figures.   b) When moving, the friction force along the ramp is $\mu_k w \cos\theta$, the component of the gravitational force along the ramp is $w \sin \theta$, so the acceleration is

$$(w \sin \theta - w\mu_k \cos \theta)/m = g(\sin \theta - \mu_k \cos \theta) = 0.92 \text{ m/s}^2.$$

(c) $2ax = v^2$, so $v = (2ax)^{1/2}$, or $v = [(2)(0.92 \text{ m/s}^2)(5 \text{ m})]^{1/2} = 3 \text{ m}$.

5-37: a) The magnitude of the normal force is $mg + |\overrightarrow{F}| \sin \theta$. The horizontal component of $\overrightarrow{F}$, $|\overrightarrow{F}| \cos \theta$ must balance the frictional force, so

$$|\overrightarrow{F}| \cos \theta = \mu_k \left(mg + |\overrightarrow{F}| \sin \theta \right);$$

solving for $|\overrightarrow{F}|$ gives

$$|\overrightarrow{F}| = \frac{\mu_k mg}{\cos \theta - \mu_k \sin \theta}.$$

b) If the crate remains at rest, the above expression, with μ_s instead of μ_k, gives the force that must be applied in order to start the crate moving. If $\cot \theta < \mu_s$, the needed force is infinite, and so the critical value is $\mu_s = \cot \theta$.

5-38: a) There is no net force in the vertical direction, so $\mathcal{N} + F \sin \theta - w = 0$, or $\mathcal{N} = w - F \sin \theta = mg - F \sin \theta$. The friction force is $f_k = \mu_k \mathcal{N} = \mu_k (mg - F \sin \theta)$. The net horizontal force is $F \cos \theta - f_k = F \cos \theta - \mu_k (mg - F \sin \theta)$, and so at constant speed,

$$F = \frac{\mu_k mg}{\cos \theta + \mu_k \sin \theta}.$$

b) Using the given values,

$$F = \frac{(0.35)(90 \text{ kg})(9.80 \text{ m/s}^2)}{(\cos 25° + (0.35) \sin 25°)} = 293 \text{ N},$$

or 290 N to two figures.

5-39: a)

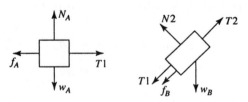

b) The blocks move with constant speed, so there is no net force on block A; the tension in the rope connecting A and B must be equal to the frictional force on block A, $\mu_k = (0.35)(25.0 \text{ N}) = 9 \text{ N}$. c) The weight of block C will be the tension in the rope connecting B and C; this is found by considering the forces on block B. The components of force along the ramp are the tension in the first rope (9 N, from part (a)), the component of the weight along the ramp, the friction on block B and the tension in the second rope.

Thus, the weight of block C is

$$w_C = 9 \text{ N} + w_B(\sin 36.9° + \mu_k \cos 36.9°)$$
$$= 9 \text{ N} + (25.0 \text{ N})(\sin 36.9° + (0.35) \cos 36.9°) = 31.0 \text{ N},$$

or 31 N to two figures. The intermediate calculation of the first tension may be avoided to obtain the answer in terms of the common weight w of blocks A and B,

$$w_C = w(\mu_k + (\sin \theta + \mu_k \cos \theta)),$$

giving the same result.

(d) Applying Newton's Second Law to the remaining masses gives:

$$a = g(W_c - \mu_k W_B \cos \theta - W_B \sin \theta)/(W_B + W_c) = 1.54 \text{ m/s}^2.$$

5-40:   Differentiating Eq. (5-10) with respect to time gives the acceleration

$$a = v_t \left(\frac{k}{m}\right) e^{-(k/m)t} = g\, e^{-(k/m)t},$$

where Eq. (5-9), $v_t = mg/k$ has been used.

Integrating Eq. (5-10) with respect to time with $y_0 = 0$ gives

$$y = \int_0^t v_t \left[1 - e^{-(k/m)t}\right] dt$$
$$= v_t \left[t + \left(\frac{m}{k}\right) e^{-(k/m)t}\right] - v_t \left(\frac{m}{k}\right)$$
$$= v_t \left[t - \frac{m}{k} \left(1 - e^{-(k/m)t}\right)\right].$$

5-41:   a) Solving for D in terms of v_t,

$$D = \frac{mg}{v_t^2} = \frac{(80 \text{ kg})(9.80 \text{ m/s}^2)}{(42 \text{ m/s})^2} = 0.44 \text{ kg/m}.$$

b) $$v_t = \sqrt{\frac{mg}{D}} = \sqrt{\frac{(45 \text{ kg})(9.80 \text{ m/s}^2)}{(0.25 \text{ kg/m})}} = 42 \text{ m/s}.$$

5-42:   At half the terminal speed, the magnitude of the frictional force is one-fourth the weight.   a) If the ball is moving up, the frictional force is down, so the magnitude of the net force is $(5/4)w$ and the acceleration is $(5/4)g$, down.   b) While moving down, the frictional force is up, and the magnitude of the net force is $(3/4)w$ and the acceleration is $(3/4)g$, down.

5-43: Setting F_{net} equal to the maximum tension in Eq. (5-17) and solving for the speed v gives

$$v = \sqrt{\frac{F_{\text{net}}R}{m}} = \sqrt{\frac{(600\ \text{N})(0.90\ \text{m})}{(0.80\ \text{kg})}} = 26.0\ \text{m/s},$$

or 26 m/s to two figures.

5-44: This is the same situation as Example 5-22. Solving for μ_{s} yields

$$\mu_{\text{s}} = \frac{v^2}{Rg} = \frac{(25.0\ \text{m/s})^2}{(220\ \text{m})(9.80\ \text{m/s}^2)} = 0.290.$$

5-45: (a)The magnitude of the force F is given to be equal to $3.8w$. "Level flight" means that the net vertical force is zero, so $F\cos\beta = (3.8)w\cos\beta = w$, and $\beta = \arccos(1/3.8) = 75°$.

(b) The angle does not depend on speed.

5-46: a) The analysis of Example 5-21 may be used to obtain $\tan\beta = (v^2/gR)$, but the subsequent algebra expressing R in terms of L is not valid. Denoting the length of the horizontal arm as r and the length of the cable as l, $R = r + l\sin\beta$. The relation $v = \frac{2\pi R}{T}$ is still valid, so $\tan\beta = \frac{4\pi^2 R}{gT^2} = \frac{4\pi^2(r + l\sin\beta)}{gT^2}$. Solving for the period T,

$$T = \sqrt{\frac{4\pi^2(r + l\sin\beta)}{g\tan\beta}} = \sqrt{\frac{4\pi^2(3.00\ \text{m} + (5.00\ \text{m})\sin 30°)}{(9.80\ \text{m/s}^2)\tan 30°}} = 6.19\ \text{s}.$$

Note that in the analysis of Example 5-22, β is the angle that the support (string or cable) makes, *not* the angle from the mass to the top of the axis (see Figure 5-24(b)). b) To the extent that the cable can be considered massless, the angle will be independent of the rider's weight. The tension in the cable will depend on the rider's mass.

5-47: This is the same situation as Example 5-21, with the lift force replacing the tension in the string. As in that example, the angle β is related to the speed and the turning radius by $\tan\beta = \frac{v^2}{gR}$. Solving for β,

$$\beta = \arctan\left(\frac{v^2}{gR}\right) = \arctan\left(\frac{(240\ \text{km/h} \times ((1\ \text{m/s})/(3.6\ \text{km/h})))^2}{(9.80\ \text{m/s}^2)(1200\ \text{m})}\right) = 20.7°.$$

5-48: a) This situation is equivalent to that of Example 5-22 and Problem 5-44, so $\mu_{\text{s}} = \frac{v^2}{Rg}$. Expressing v in terms of the period T, $v = \frac{2\pi R}{T}$, so $\mu_{\text{s}} = \frac{4\pi^2 R}{T^2 g}$. A platform speed of 40.0 rev/min corresponds to a period of 1.50 s, so

$$\mu_{\text{s}} = \frac{4\pi^2(0.150\ \text{m})}{(1.50\ \text{s})^2(9.80\ \text{m/s}^2)} = 0.269.$$

b) For the same coefficient of static friction, the maximum radius is proportional to the square of the period (longer periods mean slower speeds, so the button may be moved further out) and so is inversely proportional to the square of the speed. Thus, at the higher speed, the maximum radius is $(0.150 \text{ m})\left(\frac{40.0}{60.0}\right)^2 = 0.067$ m.

5-49: a) Setting $a_{\text{rad}} = g$ in Eq. (5-16) and solving for the period T gives

$$T = 2\pi\sqrt{\frac{R}{g}} = 2\pi\sqrt{\frac{400 \text{ m}}{9.80 \text{ m/s}^2}} = 40.1 \text{ s},$$

so the number of revolutions per minute is $(60 \text{ s/min})/(40.1 \text{ s}) = 1.5$ rev/min.

b) The lower acceleration corresponds to a longer period, and hence a lower rotation rate, by a factor of the square root of the ratio of the accelerations, $T' = (1.5 \text{ rev/min}) \times \sqrt{3.70/9.8} = 0.92$ rev/min.

5-50: a) $2\pi R/T = 2\pi(50.0 \text{ m})/(60.0 \text{ s}) = 5.24$ m/s. b) The magnitude of the radial force is $m4\pi^2 R/T^2 = w(4\pi^2 R/gT^2) = 49$ N (to the nearest Newton), so the apparent weight at the top is $882 \text{ N} - 49 \text{ N} = 833$ N, and at the bottom is $882 \text{ N} + 49 \text{ N} = 931$ N. c) For apparent weightlessness, the radial acceleration at the top is equal to g in magnitude. Using this in Eq. (5-16) and solving for T gives

$$T = 2\pi\sqrt{\frac{R}{g}} = 2\pi\sqrt{\frac{50.0 \text{ m}}{9.80 \text{ m/s}^2}} = 14 \text{ s}.$$

d) At the bottom, the apparent weight is twice the weight, or 1760 N.

5-51: a) If the pilot feels weightless, he is in free fall, and $a = g = v^2/R$, so $v = \sqrt{Rg} = \sqrt{(150 \text{ m})(9.80 \text{ m/s}^2)} = 38.3$ m/s, or 138 km/h. b) The apparent weight is the sum of the net inward (upward) force and the pilot's weight, or

$$w + ma = w\left(1 + \frac{a}{g}\right)$$

$$= (700 \text{ N})\left(1 + \frac{(280 \text{ km/h})^2}{(3.6 \text{ (km/h)}/\text{(m/s)})^2(9.80 \text{ m/s}^2)(150 \text{ m})}\right)$$

$$= 3581 \text{ N},$$

or 3580 N to three places.

5-52: a) Solving Eq. (5-14) for R,

$$R = v^2/a = v^2/4g = (95.0 \text{ m/s})^2/(4 \times 9.80 \text{ m/s}^2) = 230 \text{ m}.$$

b) The apparent weight will be five times the actual weight,

$$5mg = 5(50.0 \text{ kg})(9.80 \text{ m/s}^2) = 2450 \text{ N}$$

to three figures.

5-53: For no water to spill, the magnitude of the downward (radial) acceleration must be at least that of gravity; from Eq. (5-14), $v > \sqrt{gR} = \sqrt{(9.80 \text{ m/s}^2)(0.600 \text{ m})} = 2.42$ m/s.

5-54: a) The inward (upward, radial) acceleration will be $\frac{v^2}{R} = \frac{(4.2 \text{ m/s})^2}{(3.80 \text{ m})} = 4.64$ m/s^2. At the bottom of the circle, the inward direction is upward.

b) The forces on the ball are tension and gravity, so $T - mg = ma$,

$$T = m(a + g) = w\left(\frac{a}{g} + 1\right) = (71.2 \text{ N})\left(\frac{4.64 \text{ m/s}^2}{9.80 \text{ m/s}^2} + 1\right) = 105 \text{ N}.$$

The Mathcad simulation, **AirResist.mcd** can be found on the instructor's CD-ROM. Exercises 5-55 through 5-59 can be solved using this simulation of projectile motion *with* air resistance.

5-55: b) The ball travels about 108 m at sea level and about 118 m in Denver, for an increase of 10 m.

5-56: Angles in the range 35° to 40° give the greatest range. Using 37° in the simulation shows the maximum range. The simulation doesn't have enough accuracy to do better than that. Furthermore, the value of the range is quite insensitive to the launch angle, provided it is in this range.

5-57: When the initial velocity is 43 m/s (82 mi/hr) at 40°, the range is about 91 m (300 ft).

5-58: The ball is moving at about 25 m/s = 55 mi/h. (Note: This number was calculated with a different algorithm than in **AirResist.mcd**.)

5-59: For a launch angle of about 30 degrees, the ball goes 14 m. The ping-pong ball travels a shorter distance than a baseball since the force of air resistance is relatively more important for a low-mass ping-pong ball than it is for a more massive baseball. (Note to instructors: If **AirResist.mcd** is used as a laboratory or homework activity, students can work this problem and then see how close they come to their estimated range when they throw a ping-pong ball. They probably will be able to throw a ping-pong ball 12 to 15 m.)

5-60:

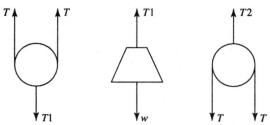

The tension in the lower chain balances the weight and so is equal to w. The lower pulley must have no net force on it, so twice the tension in the rope must be equal to w, and so the tension in the rope is $w/2$. Then, the downward force on the upper pulley due to the rope is also w, and so the upper chain exerts a force w on the upper pulley, and the tension in the upper chain is also w.

5-61: In the absence of friction, the only forces along the ramp are the component of the weight along the ramp, $w \sin \alpha$, and the component of $\vec{F}$ along the ramp, $|\vec{F}| \cos \alpha = F \cos \alpha$. These forces must sum to zero, so $F = w \tan \alpha$.

Considering horizontal and vertical components, the normal force must have horizontal component equal to $\mathcal{N} \sin \alpha$, which must be equal to F; the vertical component must balance the weight, $\mathcal{N} \cos \alpha = w$. Eliminating $\mathcal{N}$ gives the same result.

5-62: The hooks exert forces on the ends of the rope. At each hook, the force that the hook exerts and the force due to the tension in the rope are an action-reaction pair. a) The vertical forces that the hooks exert must balance the weight of the rope, so each hook exerts an upward vertical force of $w/2$ on the rope. Therefore, the downward force that the rope exerts at each end is $T_{end} \sin\theta = w/2$, so $T_{end} = w/(2\sin\theta) = Mg/(2\sin\theta)$.   b) Each half of the rope is itself in equilibrium, so the tension in the middle must balance the horizontal force that each hook exerts, which is the same as the horizontal component of the force due to the tension at the end; $T_{end}\cos\theta = T_{middle}$, so $T_{middle} = Mg\cos\theta/(2\sin\theta) = Mg/(2\tan\theta)$.   (c) Mathematically speaking, $\theta \neq 0$ because this would cause a division by zero in the equation for T_{end} or T_{middle}. Physically speaking, we would need an infinite tension to keep a non-massless rope perfectly straight.

5-63:   Consider a point a distance x from the top of the rope. The forces acting in this point are T up and $\left(M + \frac{m(L-x)}{L}\right)g$ downwards. Newton's Second Law becomes $T - \left(M + \frac{m(L-x)}{L}\right)g = \left(M + \frac{m(L-x)}{L}\right)a$. Since $a = \frac{F-(M+m)g}{M+m}$, $T = \left(M + \frac{m(L-x)}{L}\right)\left(\frac{F}{M+m}\right)$. At $x = 0$, $T = F$, and at $x = L$, $T = \frac{MF}{M+m} = M(a+g)$ as expected.

5-64:   a) The tension in the cord must be m_2g in order that the hanging block move at constant speed. This tension must overcome friction and the component of the gravitational force along the incline, so $m_2g = (m_1g\sin\alpha + \mu_k m_1g\cos\alpha)$ and $m_2 = m_1(\sin\alpha + \mu_k\cos\alpha)$.

b) In this case, the friction force acts in the same direction as the tension on the block of mass m_1, so $m_2g = (m_1g\sin\alpha - \mu_k m_1g\cos\alpha)$, or $m_2 = m_1(\sin\alpha - \mu_k\cos\alpha)$.

c) Similar to the analysis of parts (a) and (b), the largest m_2 could be is $m_1(\sin\alpha + \mu_s\cos\alpha)$ and the smallest m_2 could be is $m_1(\sin\alpha - \mu_s\cos\alpha)$.

5-65:   For an angle of 45.0°, the tensions in the horizontal and vertical wires will be the same.   a) The tension in the vertical wire will be equal to the weight $w = 12.0$ N; this must be the tension in the horizontal wire, and hence the friction force on block A is also 12.0 N.   b) The maximum frictional force is $\mu_s w_A = (0.25)(60.0 \text{ N}) = 15$ N; this will be the tension in both the horizontal and vertical parts of the wire, so the maximum weight is 15 N.

5-66:   a) The most direct way to do part (a) is to consider the blocks as a unit, with total weight 4.80 N. Then the normal force between block B and the lower surface is 4.80 N, and the friction force that must be overcome by the force F is $\mu_k \mathcal{N} = (0.30)(4.80 \text{ N}) = 1.440$ N, or 1.44 N to three figures.   b) The normal force between block B and the lower surface is still 4.80 N, but since block A is moving *relative to block B*, there is a friction force between the blocks, of magnitude $(0.30)(1.20 \text{ N}) = 0.360$ N, so the total friction force that the force F must overcome is 1.440 N + 0.360 N = 1.80 N. (An extra figure was kept in these calculations for clarity.)

5-67:   (Denote $|\overrightarrow{F}|$ by F.)   a) The force normal to the surface is $\mathcal{N} = F\cos\theta$; the vertical component of the applied force must be equal to the weight of the brush plus the friction force, so that $F \sin \theta = w + \mu_k F \cos \theta$, and

$$F = \frac{w}{\sin \theta - \mu_k \cos \theta} = \frac{12.00 \text{ N}}{\sin 53.1° - (0.15) \cos 53.1°} = 16.9 \text{ N},$$

keeping an extra figure.   b) $F \cos \theta = (16.91 \text{ N}) \cos 53.1° = 10.2 \text{ N}.$

5-68:   The key idea in solving this problem is to recognize that if the system is accelerating, the tension that block A exerts on the rope is different from the tension that block B exerts on the rope. (Otherwise the net force on the rope would be zero, and the rope couldn't accelerate.) Also, treat the rope as if it is just another object. Taking the "clockwise" direction to be positive, the Second Law equations for the three different parts of the system are:

Block A (The only horizontal forces on A are tension to the right, and friction to the left):
$-\mu_k m_A g + T_A = m_A a.$

Block B (The only vertical forces on B are gravity down, and tension up): $m_B g - T_B = m_B a.$

Rope (The forces on the rope along the direction of its motion are the tensions at either end and the weight of the portion of the rope that hangs vertically): $m_R(\frac{d}{L})g + T_B - T_A = m_R a.$

To solve for a and eliminate the tensions, add the left hand sides and right hand sides of the three equations: $-\mu_k m_A g + m_B g + m_R(\frac{d}{L})g = (m_A + m_B + m_R)a$, or $a = g\frac{m_B + m_R\frac{d}{L} - \mu_k m_A}{(m_A + m_B + m_R)}.$

   (a) When $\mu_k = 0, a = g\frac{m_B + m_R(\frac{d}{L})}{(m_A + m_B + m_R)}.$ As the system moves, d will increase, approaching L as a limit, and thus the acceleration will approach a maximum value of $a = g\frac{m_B + m_R}{(m_A + m_B + m_R)}.$

   (b) For the blocks to just begin moving, $a > 0$, so solve $0 = [m_B + m_R(\frac{d}{L}) - \mu_s m_A]$ for d. Note that we must use static friction to find d for when the block will *begin* to move. Solving for $d, d = \frac{L}{m_R}(\mu_s m_A - m_B)$, or $d = \frac{1.0 \text{ m}}{.160 \text{ kg}}(.25(2 \text{ kg}) - .4 \text{ kg}) = .63 \text{ m}.$

   (c) When $m_R = .04 \text{ kg}, d = \frac{1.0 \text{ m}}{.04 \text{ kg}}(.25(2 \text{ kg}) - .4 \text{ kg}) = 2.50 \text{ m}.$ This is not a physically possible situation since $d > L$.

5-69:   For a rope of length L, and weight w, assume that a length rL is on the table, so that a length $(1 - r)L$ is hanging. The tension in the rope at the edge of the table is then $(1 - r)w$, and the friction force on the part of the rope on the table is $f_s = \mu_s r w$. This must be the same as the tension in the rope at the edge of the table, so $\mu_s r w = (1 - r)w$ and $r = 1/(1 + \mu_s)$. Note that this result is independent of L and w for a uniform rope. The fraction that hangs over the edge is $1 - r = \mu_s/(1 + \mu_s)$; note that if $\mu_s = 0, r = 1$ and $1 - r = 0$.

5-70:   a) The normal force will be $mg \cos \alpha + F \sin \alpha$, and the net force along (up) the ramp is

$$F \cos \alpha - mg \sin \alpha - \mu_s(mg \cos \alpha + F \sin \alpha) = F(\cos \alpha - \mu_s \sin \alpha) - mg(\sin \alpha + \mu_s \cos \alpha).$$

In order to move the box, this net force must be greater than zero. Solving for F,

$$F > mg \, \frac{\sin \alpha + \mu_s \cos \alpha}{\cos \alpha - \mu_s \sin \alpha}.$$

Since F is the magnitude of a force, F must be positive, and so the denominator of this expression must be positive, or $\cos \alpha > \mu_s \sin \alpha$, and $\mu_s < \cot \alpha$. b) Replacing μ_s with μ_k in the above expression, and making the inequality an equality,

$$F = mg \, \frac{\sin \alpha + \mu_k \cos \alpha}{\cos \alpha - \mu_k \sin \alpha}.$$

5-71: a) The product $\mu_s g = 2.94$ m/s^2 is greater than the magnitude of the acceleration of the truck, so static friction can supply sufficient force to keep the case stationary relative to the truck; the crate accelerates north at 2.20 m/s^2, due to the friction force of $ma = 66.0$ N. b) In this situation, the static friction force is insufficient to maintain the case at rest relative to the truck, and so the friction force is the kinetic friction force, $\mu_k \mathcal{N} = \mu_k mg = 59$ N.

5-72: To answer the question, v_o must be found and compared with 20 m/s (72 km/hr). The kinematics relationship $2ax = -v_o^2$ is useful, but we also need a. The acceleration must be large enough to cause the box to begin sliding, and so we must use the force of static friction in Newton's Second Law: $-\mu_s mg \le ma$, or $a = -\mu_s g$. Then, $2(-\mu_s g)x = -v_o^2$, or $v_o = \sqrt{2\mu_s gx} = \sqrt{2(.30)(9.8 \, \text{m/s}^2)(47 \, \text{m})}$. Hence, $v_o = 16.6$ m/s, which is less than 20 m/s, so do you not go to jail.

5-73: See Exercise 5-40. a) The maximum tension and the weight are related by

$$T_{max} \cos \beta = \mu_k \left(w - T_{max} \sin \beta \right),$$

and solving for the weight w gives

$$w = T_{max} \left(\frac{\cos \beta}{\mu_k} + \sin \beta \right).$$

This will be a maximum when the quantity in parentheses is a maximum. Differentiating with respect to β,

$$\frac{d}{d\beta} \left(\frac{\cos \beta}{\mu_k} + \sin \beta \right) = -\frac{\sin \beta}{\mu_k} + \cos \beta = 0,$$

or $\tan \theta = \mu_k$, where θ is the value of β that maximizes the weight. Substituting for μ_k in terms of θ,

$$w = T_{max} \left(\frac{\cos \theta}{\sin \theta / \cos \theta} + \sin \theta \right)$$

$$= T_{max} \left(\frac{\cos^2 \theta + \sin^2 \theta}{\sin \theta} \right)$$

$$= \frac{T_{max}}{\sin \theta}.$$

b) In the absence of friction, any non-zero horizontal component of force will be enough to accelerate the crate, but slowly.

5-74: a) Taking components along the direction of the plane's descent, $f = w \sin \alpha$ and $L = w \cos \alpha$. b) Dividing one of these relations by the other cancels the weight, so $\tan \alpha = f/L$. c) The distance will be the initial altitude divided by the tangent of α. $f = L \tan \alpha$ and $L = w \cos \alpha$, therefore $\sin \alpha = f/w = \frac{1300 \text{ N}}{12,900 \text{ N}} g$ and so $\alpha = 5.78°$. This makes the horizontal distance $(2500 \text{ m})/\tan(5.78°) = 24.7$ km. d) If the drag is reduced, the angle α is reduced, and the plane goes further.

5-75: If the plane is flying at a constant speed of 36.1 m/s, then $\Sigma F = 0$, or $T - w \sin \alpha - f = 0$. The rate of climb and the speed give the angle α, $\alpha = \arcsin(5/36.1) = 7.96°$. Then, $T = w \sin \alpha + f$. $T = (12,900 \text{ N}) \sin 7.96° + 1300 \text{ N} = 3087$ N. Note that in level flight $(\alpha = 0)$, the thrust only needs to overcome the drag force to maintain the constant speed of 36.1 m/s.

5-76: If the block were to remain at rest relative to the truck, the friction force would need to cause an acceleration of 2.20 m/s^2; however, the maximum acceleration possible due to static friction is $(0.19)(9.80 \text{ m/s}^2) = 1.86$ m/s^2, and so the block will move relative to the truck; the acceleration of the box would be $\mu_k g = (0.15)(9.80 \text{ m/s}^2) = 1.47$ m/s^2. The difference between the distance the truck moves and the distance the box moves (*i.e.*, the distance the box moves relative to the truck) will be 1.80 m after a time

$$t = \sqrt{\frac{2\Delta x}{a_{\text{truck}} - a_{\text{box}}}} = \sqrt{\frac{2(1.80 \text{ m})}{(2.20 \text{ m/s}^2 - 1.47 \text{ m/s}^2)}} = 2.22 \text{ s}.$$

In this time, the truck moves $\frac{1}{2} a_{\text{truck}} t^2 = \frac{1}{2}(2.20 \text{ m/s}^2)(2.221 \text{ s})^2 = 5.43$ m. Note that an extra figure was kept in the intermediate calculation to avoid roundoff error.

5-77: The friction force *on* block A is $\mu_k w_A = (0.30)(1.40 \text{ N}) = 0.420$ N, as in Problem 5-68. This is the magnitude of the friction force that block A exerts on block B, as well as the tension in the string. The force F must then have magnitude

$$F = \mu_k(w_B + w_A) + \mu_k w_A + T = \mu_k(w_B + 3w_A)$$
$$= (0.30)(4.20 \text{ N} + 3(1.40 \text{ N})) = 2.52 \text{ N}.$$

Note that the normal force exerted on block B by the table is the sum of the weights of the blocks.

5-78: The crate will hit the ground after a time

$$t = \sqrt{\frac{2h}{g}} = \sqrt{\frac{2(1200 \text{ m})}{(9.80 \text{ m/s}^2)}} = 15.6 \text{ s}.$$

During this time, the crate will have travelled horizontally a distance

$$x = v_{x0}t - \frac{1}{2}at^2 = v_{x0}t - \frac{1}{2}\frac{F}{m}t^2$$

$$= (70 \text{ m/s})(15.6 \text{ s}) - \frac{1}{2}\frac{(180 \text{ N})}{(30 \text{ kg})}(15.6 \text{ s})^2 = 362 \text{ m}.$$

5-79: Let the tension in the cord attached to block A be T_A and the tension in the cord attached to block C be T_C. The equations of motion are then

$$T_A - m_A g = m_A a$$
$$T_C - \mu_k m_B g - T_A = m_B a$$
$$m_C g - T_C = m_C a.$$

a) Adding these three equations to eliminate the tensions gives

$$a(m_A + m_B + m_C) = g(m_C - m_A - \mu_k m_B),$$

solving for m_C gives

$$m_C = \frac{m_A(a+g) + m_B(a + \mu_k g)}{g - a},$$

and substitution of numerical values gives $m_C = 12.9$ kg.

b) $T_A = m_A(g+a) = 47.2$ N, $T_C = m_C(g-a) = 101$ N.

5-80: Considering positive accelerations to be to the right (up and to the right for the left-hand block, down and to the right for the right-hand block), the forces along the inclines and the accelerations are related by $T - (100 \text{ kg})g \sin 30° = (100 \text{ kg})a$, $(50 \text{ kg})g \sin 53° - T = (50 \text{ kg})a$, where T is the tension in the cord and a the mutual magnitude of acceleration. Adding these relations, $(50 \text{ kg} \sin 53° - 100 \text{ kg} \sin 30°)g = (50 \text{ kg} + 100 \text{ kg})a$, or $a = -0.067g$. a) Since a comes out negative, the blocks will slide to the left; the 100-kg block will slide down. Of course, if coordinates had been chosen so that positive accelerations were to the left, a would be $+0.067g$. b) $0.067(9.80 \text{ m/s}^2) = 0.658 \text{ m/s}^2$.

c) Substituting the value of a (including the proper sign, depending on choice of coordinates) into either of the above relations involving T yields 424 N.

5-81: Denote the magnitude of the acceleration of the block with mass m_1 as a; the block of mass m_2 will descend with acceleration $a/2$. If the tension in the rope is T, the equations of motion are then

$$T = m_1 a$$
$$m_2 g - 2T = m_2 a/2.$$

Multiplying the first of these by 2 and adding to eliminate T, and then solving for a gives

$$a = \frac{m_2 g}{2m_1 + m_2/2} = g \frac{2m_2}{4m_1 + m_2}.$$

The acceleration of the block of mass m_2 is half of this, or $g\, m_2/(4m_1 + m_2)$.

5-82: Denote the common magnitude of the maximum acceleration as a. For block A to remain at rest with respect to block B, $a < \mu_s g$. The tension in the cord is then $T = (m_A + m_B)a + \mu_k g(m_A + m_B) = (m_A + m_B)(a + \mu_k g)$. This tension is related to the mass m_C by $T = m_C(g - a)$. Solving for a yields

$$a = g \frac{m_C - \mu_k(m_A + m_B)}{m_A + m_B + m_C} < \mu_s g.$$

Solving the inequality for m_C yields

$$m_C < \frac{(m_A + m_B)(\mu_s + \mu_k)}{1 - \mu_s}.$$

5-83: See Exercise 5-17 (Atwood's machine). The 2.00-kg block will accelerate upward at $g\,\frac{5.00 \text{ kg} - 2.00 \text{ kg}}{5.00 \text{ kg} + 2.00 \text{ kg}} = 3g/7$, and the 5.00-kg block will accelerate downward at $3g/7$. Let the initial height above the ground be h_0; when the large block hits the ground, the small block will be at a height $2h_0$, and moving upward with a speed given by $v_0^2 = 2ah_0 = 6gh_0/7$. The small block will continue to rise a distance $v_0^2/2g = 3h_0/7$, and so the maximum height reached will be $2h_0 + 3h_0/7 = 17h_0/7 = 1.46$ m, which is 0.860 m above its initial height.

5-84: The floor exerts an upward force $\mathcal{N}$ on the box, obtained from $\mathcal{N} - mg = ma$, or $\mathcal{N} = m(a + g)$. The friction force that needs to be balanced is

$$\mu_k \mathcal{N} = \mu_k m(a + g) = (0.32)(28.0 \text{ kg})(1.90 \text{ m/s}^2 + 9.80 \text{ m/s}^2) = 105 \text{ N}.$$

5-85: The upward friction force must be $f_s = \mu_s \mathcal{N} = m_A g$, and the normal force, which is the only horizontal force on block A, must be $\mathcal{N} = m_A a$, and so $a = g/\mu_s$. An observer on the cart would "feel" a backwards force, and would say that a similar force acts on the block, thereby creating the need for a normal force.

5-86: Since the larger block (the trailing block) has the larger coefficient of friction, it will need to be pulled down the plane; *i.e.*, the larger block will not move faster than the smaller block, and the blocks will have the same acceleration. For the smaller block, $(4.00 \text{ kg})g(\sin 30° - (0.25)\cos 30°) - T = (4.00 \text{ kg})a$, or $11.11 \text{ N} - T = (4.00 \text{ kg})a$, and similarly for the larger, $15.44 \text{ N} + T = (8.00 \text{ kg})a$. a) Adding these two relations, $26.55 \text{ N} = (12.00 \text{ kg})a$, $a = 2.21 \text{ m/s}^2$ (note that an extra figure was kept in the intermediate calculation to avoid roundoff error). b) Substitution into either of the above relations gives $T = 2.27 \text{ N}$. Equivalently, dividing the second relation by 2 and subtracting from the first gives $\frac{3}{2}T = 11.11 \text{ N} - \frac{15.44 \text{ N}}{2}$, giving the same result. c) The upper block (with $\mu_k = 0.25$) will slide down while the lower block will remain at rest until the blocks collide (a subject for a later chapter).

5-87:

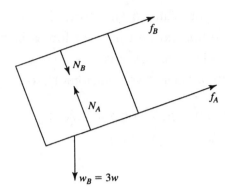

$w_B = 3w$

a)Let $\mathcal{N}_B$ be the normal force between the plank and the block and $\mathcal{N}_A$ be the normal force between the block and the incline. Then, $\mathcal{N}_B = w\cos\theta$ and $\mathcal{N}_A = \mathcal{N}_B + 3w\cos\theta = 4w\cos\theta$. The net frictional force on the block is $\mu_k(\mathcal{N}_A + \mathcal{N}_B) = \mu_k 5w\cos\theta$. To move at constant speed, this must balance the component of the block's weight along the incline, so $3w\sin\theta = \mu_k 5w\cos\theta$, and $\mu_k = \frac{3}{5}\tan\theta = \frac{3}{5}\tan 37° = 0.452$.

5-88: (a) There is a contact force $\mathcal{N}$ between the man (mass M) and the platform (mass m). The equation of motion for the man is $T + \mathcal{N} - Mg = Ma$, where T is the tension in the rope, and for the platform, $T - \mathcal{N} - mg = ma$. Adding to eliminate $\mathcal{N}$, and rearranging, $T = \frac{1}{2}(M+m)(a+g)$. This result could be found directly by considering the man-platform combination as a unit, with mass $m + M$, being pulled upward with a force $2T$ due to the *two* ropes on the combination. The tension T in the rope is the same as the force that the man applies to the rope. Numerically,

$$T = \frac{1}{2}(70.0 \text{ kg} + 25.0 \text{ kg})(1.80 \text{ m/s}^2 + 9.80 \text{ m/s}^2) = 551 \text{ N}.$$

(b) The rope has a *downward* acceleration of 1.80 m/s^2 through his hands.

5-89: a) The only horizontal force on the two-block combination is the horizontal component of $\overrightarrow{F}$, $F\cos\alpha$. The blocks will accelerate with $a = F\cos\alpha/(m_1 + m_2)$. b) The normal force between the blocks is $m_1 g + F\sin\alpha$; for the blocks to move together, the product of this force and μ_s must be greater than the horizontal force that the lower block exerts on the upper block. That horizontal force is one of an action-reaction pair; the reaction to this force accelerates the lower block. Thus, for the blocks to stay together, $m_2 a \leq \mu_s(m_1 g + F\sin\alpha)$. Using the result of part (a),

$$m_2 \frac{F\cos\alpha}{m_1 + m_2} \leq \mu_s(m_1 g + F\sin\alpha).$$

Solving the inequality for F gives the desired result.

5-90: The banked angle of the track has the same form as that found in Example 5-23, $\tan\beta = \frac{v_0^2}{gR}$, where v_0 is the ideal speed, 20 m/s in this case. For speeds larger than v_0, a frictional force is needed to keep the car from skidding. In this case, the inward force will consist of a part due to the normal force $\mathcal{N}$ and the friction force f; $\mathcal{N} \sin \beta + f \cos \beta = ma_{\text{rad}}$. The normal and friction forces both have vertical components; since there is no vertical acceleration, $\mathcal{N} \cos \beta - f \sin \beta = mg$. Using $f = \mu_s \mathcal{N}$ and $a_{\text{rad}} = \frac{v^2}{R} = \frac{(1.5v_0)^2}{R} = 2.25g \tan \beta$, these two relations become

$$\mathcal{N} \sin \beta + \mu_s \mathcal{N} \cos \beta = 2.25 \, mg \tan \beta,$$
$$\mathcal{N} \cos \beta - \mu_s \mathcal{N} \sin \beta = mg.$$

Dividing to cancel $\mathcal{N}$ gives

$$\frac{\sin \beta + \mu_s \cos \beta}{\cos \beta - \mu_s \sin \beta} = 2.25 \tan \beta.$$

Solving for μ_s and simplifying yields

$$\mu_s = \frac{1.25 \sin \beta \cos \beta}{1 + 1.25 \sin^2 \beta}.$$

Using $\beta = \arctan\left(\frac{(20 \text{ m/s})^2}{(9.80 \text{ m/s}^2)(120 \text{ m})}\right) = 18.79°$ gives $\mu_s = 0.34$.

5-91: a) The same analysis as in Problem 5-92 applies, but with the speed v an unknown. The equations of motion become

$$\mathcal{N} \sin \beta + \mu_s \mathcal{N} \cos \beta = mv^2/R,$$
$$\mathcal{N} \cos \beta - \mu_s \mathcal{N} \sin \beta = mg.$$

Dividing to cancel $\mathcal{N}$ gives

$$\frac{\sin \beta + \mu_s \cos \beta}{\cos \beta - \mu_s \sin \beta} = \frac{v^2}{Rg}.$$

Solving for v and substituting numerical values gives $v = 20.9$ m/s (note that the value for the coefficient of static friction must be used).

b) The same analysis applies, but the friction force must be directed up the bank; this has the same algebraic effect as replacing f with $-f$, or replacing μ_s with $-\mu_s$ (although coefficients of friction may certainly never be negative). The result is

$$v^2 = (gR)\frac{\sin \theta - \mu_s \cos \theta}{\cos \theta + \mu_s \sin \theta},$$

and substitution of numerical values gives $v = 8.5$ m/s.

5-92: The analysis of this problem is the same as that of Example 5-21; solving for v in terms of β and R, $v = \sqrt{gR \tan \beta} = \sqrt{(9.80 \text{ m/s}^2)(50.0 \text{ m}) \tan 30.0°} = 16.8$ m/s, about 60.6 km/h.

5-93: The point to this problem is that the monkey and the bananas have the same weight, and the tension in the string is the same at the point where the bananas are suspended and where the monkey is pulling; in all cases, the monkey and bananas will have the same net force and hence the same acceleration, direction and magnitude. a) The bananas move up. b) The monkey and bananas always move at the same velocity, so the distance between them stays the same. c) Both the monkey and bananas are in free fall, and as they have the same initial velocity, the distance bewteen them doesn't change. d) The bananas will slow down at the same rate as the monkey; if the monkey comes to a stop, so will the bananas.

5-94: The separated equation of motion has a lower limit of $3v_t$ instead of 0; specifically,

$$\int_{3v_t}^{v} \frac{dv}{v - v_t} = \ln \frac{v_t - v}{-2v_t} = \ln \left(\frac{v}{2v_t} - \frac{1}{2} \right) = -\frac{k}{m} t, \quad \text{or}$$

$$v = 2v_t \left[\frac{1}{2} + e^{-(k/m)t} \right].$$

Note that the speed is always greater than v_t.

5-95: a) The rock is released from rest, and so there is initially no resistive force and $a_0 = (18.0 \text{ N})/(3.00 \text{ kg}) = 6.00 \text{ m/s}^2$.

b) $(18.0 \text{ N} - (2.20 \text{ N·s/m})(3.00 \text{ m/s}))/(3.00 \text{ kg}) = 3.80 \text{ m/s}^2$. c) The net force must be 1.80 N, so $kv = 16.2 \text{ N}$ and $v = (16.2 \text{ N})/(2.20 \text{ N·s/m}) = 7.36 \text{ m/s}$. d) When the net force is equal to zero, and hence the acceleration is zero, $kv_t = 18.0 \text{ N}$ and $v_t = (18.0 \text{ N})/(2.20 \text{ N·s/m}) = 8.18 \text{ m/s}$. e) From Eq. (5-12),

$$y = (8.18 \text{ m/s}) \left[(2.00 \text{ s}) - \frac{3.00 \text{ kg}}{2.20 \text{ N·s/m}} \left(1 - e^{-((2.20 \text{ N·s/m})/(3.00 \text{ kg}))(2.00 \text{ s})} \right) \right]$$

$$= +7.78 \text{ m}.$$

From Eq. (5-10),

$$v = (8.18 \text{ m/s}) \left[1 - e^{-((2.2 \text{ N·s/m})/(3.00 \text{ kg}))(2.00 \text{ s})} \right]$$

$$= 6.29 \text{ m/s}.$$

From Eq. (5-11), but with a_0 instead of g,

$$a = (6.00 \text{ m/s}^2) e^{-((2.20 \text{ N·s/m})/(3.00 \text{ kg}))(2.00 \text{ s})} = 1.38 \text{ m/s}^2.$$

f)
$$1 - \frac{v}{v_t} = 0.1 = e^{-(k/m)t}, \quad \text{so}$$

$$t = \frac{m}{k} \ln(10) = 3.14 \text{ s}.$$

5-96: Recognizing the geometry of a 3-4-5 right triangle simplifies the calculation. For instance, the radius of the circle of the mass' motion is 0.75 m.

a) Balancing the vertical force, $T_U \frac{4}{5} - T_L \frac{4}{5} = w$, so

$$T_L = T_U - \frac{5}{4}w = 80.0 \text{ N} - \frac{5}{4}(4.00 \text{ kg})(9.80 \text{ m/s}^2) = 31.0 \text{ N}.$$

b) The net inward force is $F = \frac{3}{5}T_U + \frac{3}{5}T_L = 66.6$ N. Solving $F = ma_{rad} = m\frac{4\pi^2 R}{T^2}$ for the period T,

$$T = 2\pi\sqrt{\frac{mR}{F}} = 2\pi\sqrt{\frac{(4.00 \text{ kg})(0.75 \text{ m})}{(66.6 \text{ N})}} = 1.334 \text{ s},$$

or 0.02223 min, so the system makes 45.0 rev/min.   c) When the lower string becomes slack, the system is the same as the conical pendulum considered in Example 5-21. With $\cos\beta = 0.800$, the period is $T = 2\pi\sqrt{(1.25 \text{ m})(0.800)/(9.80 \text{ m/s}^2)} = 2.007$ s, which is the same as 29.9 rev/min.   d) The system will still be the same as a conical pendulum, but the block will drop to a smaller angle.

5-97:   a) For the same rotation rate, the magnitude of the radial acceleration is proportional to the radius, and for twins of the same mass, the needed force is proportional to the radius; Jackie is twice as far away from the center, and so must hold on with twice as much force as Jena, or 120 N.

b) $\Sigma F_{\text{Jackie}} = mv^2/r$, and $\Sigma F_{\text{Jackie}}/\Sigma F_{\text{Jena}} = r_{\text{Jackie}}/r_{\text{Jena}}$. Thus, $\Sigma F_{\text{Jackie}} = 60 \text{ N}(2) = 120 \text{ N}$. $v = \sqrt{\frac{(120 \text{ N})(3.6 \text{ m})}{30 \text{ kg}}} = 3.8$ m/s.

5-98:   The force that the seat exerts on the passenger will, in general, have both a vertical and horizontal component.   a) In both cases, the normal force $\vec{\mathcal{N}}$ will have a vertical component equal to the weight, and a horizontal component equal to mv^2/R. The apparent weight is then $\mathcal{N} = \sqrt{(mg)^2 + (mv^2/R)^2} = m\sqrt{g^2 + v^4/R^2}$.   b) See Problem 1-66. The direction of the weight (vertically down) and the inward radial acceleration must form two sides of an equilateral triangle; this occurs when the passenger is at a position on the wheel which is $\pm 30°$ from straight up.

5-99:   a)

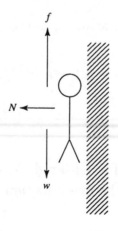

b) The upward friction force must be equal to the weight, so $\mu_s \mathcal{N} = \mu_s m(4\pi^2 R/T^2) \geq mg$ and

$$\mu_s > \frac{gT^2}{4\pi^2 R} = \frac{(9.80 \text{ m/s}^2)(1 \text{ s}/0.60 \text{ rev})^2}{4\pi^2 (2.5 \text{ m})} = 0.28.$$

c) No; both the weight and the needed normal force are proportional to the rider's mass.

5-100: a) For the tires not to lose contact, there must be a downward force on the tires. Thus, the (downward) acceleration at the top of the sphere must exceed mg, so $m\frac{v^2}{R} > mg$, and $v > \sqrt{gR} = \sqrt{(9.80 \text{ m/s}^2)(13.0 \text{ m})} = 11.3$ m/s.

b) The (upward) acceleration will then be $4g$, so the upward normal force must be $5mg = 5(110 \text{ kg})(9.80 \text{ m/s}^2) = 5390$ N.

5-101: a) What really happens (according to a nosy observer on the ground) is that you slide closer to the passenger by turning to the right. b) The analysis is the same as that of Example 5-22. In this case, the friction force should be insufficient to provide the inward radial acceleration, and so $\mu_s mg < mv^2/R$, or

$$R < \frac{v^2}{\mu_s g} = \frac{(20 \text{ m/s})^2}{(0.35)(9.80 \text{ m/s}^2)} = 120 \text{ m}$$

to two places. Why the passenger is not wearing a seat belt is another question.

5-102: The tension F in the string must be the same as the weight of the hanging block, and must also provide the resultant force necessary to keep the block on the table in uniform circular motion; $Mg = F = m\frac{v^2}{r}$, so $v = \sqrt{grM/m}$.

5-103: a) The analysis is the same as that for the conical pendulum of Example 5-21, and so

$$\beta = \arccos\left(\frac{gT^2}{4\pi^2 L}\right) = \arccos\left(\frac{(9.80 \text{ m/s}^2)(1/4.00 \text{ s})^2}{4\pi^2(0.100 \text{ m})}\right) = 81.0°.$$

b) For the bead to be at the same elevation as the center of the hoop, $\beta = 90°$ and $\cos\beta = 0$, which would mean $T = 0$; the speed of the bead would be infinite, and this is not possible. c) The expression for $\cos\beta$ gives $\cos\beta = 2.48$, which is not possible. In deriving the expression for $\cos\beta$, a factor of $\sin\beta$ was canceled, precluding the possibility that $\beta = 0$. For this situation, $\beta = 0$ is the only physical possibility.

5-104: a) Differentiating twice, $a_x = -6\beta t$ and $a_y = -2\delta$, so

$$F_x = ma_x = (2.20 \text{ kg})(-0.72 \text{ N/s})t = -(1.58 \text{ N/s})t$$
$$F_y = ma_y = (2.20 \text{ kg})(-2.00 \text{ m/s}^2) = -4.40 \text{ N}.$$

b)

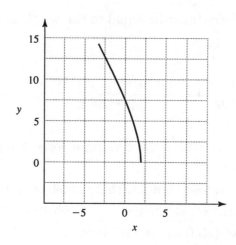

c) At $t = 3.00$ s, $F_x = -4.75$ N and $F_y = -4.40$ N, so

$$F = \sqrt{(-4.75 \text{ N})^2 + (-4.40 \text{ N})^2} = 6.48 \text{ N},$$

at an angle of arctan $\left(\frac{-4.40}{-4.75}\right) = 223°$.

5-105:

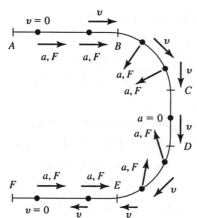

5-106: See Example 5-24.

a) $F_A = m\left(g + \frac{v^2}{R}\right) = (1.60 \text{ kg})\left(9.80 \text{ m/s}^2 + \frac{(12.0 \text{ m/s})^2}{5.00 \text{ m}}\right) = 61.8$ N.

b) $F_B = m\left(g - \frac{v^2}{R}\right) = (1.60 \text{ kg})\left(9.80 \text{ m/s}^2 - \frac{(12.0 \text{ m/s})^2}{5.00 \text{ m}}\right) = -30.4$ N, where the minus sign indicates that the track pushes *down* on the car. The magnitude of this force is 30.4 N.

5-107: The analysis is the same as for Problem 5-91; in the case of the cone, the speed is related to the period by $v = 2\pi R/T = 2\pi h \tan\beta/T$, or $T = 2\pi h \tan\beta/v$. The maximum and minimum speeds are the same as those found in Problem 5-91,

$$v_{\text{max}} = \sqrt{gh \tan\beta \frac{\cos\beta + \mu_s \sin\beta}{\sin\beta - \mu_s \cos\beta}}$$

$$v_{\text{min}} = \sqrt{gh \tan\beta \frac{\cos\beta - \mu_s \sin\beta}{\sin\beta + \mu_s \cos\beta}}.$$

The minimum and maximum values of the period T are then

$$T_{\min} = 2\pi \sqrt{\frac{h \tan \beta}{g} \frac{\sin \beta - \mu_s \cos \beta}{\cos \beta + \mu_s \sin \beta}}$$

$$T_{\max} = 2\pi \sqrt{\frac{h \tan \beta}{g} \frac{\sin \beta + \mu_s \cos \beta}{\cos \beta - \mu_s \sin \beta}}.$$

5-108: a) There are many ways to do these sorts of problems; the method presented is fairly straightforward in terms of application of Newton's laws, but involves a good deal of algebra. For both parts, take the x-direction to be horizontal and positive to the right, and the y-direction to be vertical and positive upward. The normal force between the block and the wedge is $\mathcal{N}$; the normal force between the wedge and the horizontal surface will not enter, as the wedge is presumed to have zero vertical acceleration. The horizontal acceleration of the wedge is A, and the components of acceleration of the block are a_x and a_y. The equations of motion are then

$$MA = -\mathcal{N} \sin \alpha$$

$$ma_x = \mathcal{N} \sin \alpha$$

$$ma_y = \mathcal{N} \cos \alpha - mg.$$

Note that the normal force gives the wedge a negative acceleration; the wedge is expected to move to the left. These are three equations in four unknowns, A, a_x, a_y and $\mathcal{N}$. Solution is possible with the imposition of the relation between A, a_x and a_y.

An observer on the wedge is not in an inertial frame, and should not apply Newton's laws, but the kinematic relation between the components of acceleration are not so restricted. To such an observer, the vertical acceleration of the block is a_y, but the horizontal acceleration of the block is $a_x - A$. To this observer, the block descends at an angle α, so the relation needed is

$$\frac{a_y}{a_x - A} = -\tan \alpha.$$

At this point, algebra is unavoidable. Symbolic-manipulation programs may save some solution time. A possible approach is to eliminate a_x by noting that $a_x = -\frac{M}{m}A$ (a result that anticipates conservation of momentum), using this in the kinematic constraint to eliminate a_y and then eliminating $\mathcal{N}$. The results are:

$$A = \frac{-gm}{(M+m) \tan \alpha + (M/\tan \alpha)}$$

$$a_x = \frac{gM}{(M+m) \tan \alpha + (M/\tan \alpha)}$$

$$a_y = \frac{-g(M+m) \tan \alpha}{(M+m) \tan \alpha + (M/\tan \alpha)}.$$

(b) When $M \gg m$, $A \longrightarrow 0$, as expected (the large block won't move). Also, $a_x \longrightarrow$ $\frac{g}{\tan \alpha + (1/\tan \alpha)} = g \frac{\tan \alpha}{\tan^2 \alpha + 1} = g \sin \alpha \cos \alpha$, which is the acceleration of the block ($g \sin \alpha$ in this case), with the factor of $\cos \alpha$ giving the horizontal component. Similarly, $a_y \longrightarrow$ $-g \sin^2 \alpha$.

(c) The trajectory is a spiral.

5-109: If the block is not to move vertically, the acceleration must be horizontal. The analysis is the same as that of Example 5-12, with the normal force replacing the string. The common acceleration is $a = g \tan \theta$, so the applied force must be $(M + m)a = (M + m)g \tan \theta$.

5-110: The normal force that the ramp exerts on the box will be $\mathcal{N} = w \cos \alpha - T \sin \theta$. The rope provides a force of $T \cos \theta$ up the ramp, and the component of the weight down the ramp is $w \sin \alpha$. Thus, the net force up the ramp is

$$F = T \cos \theta - w \sin \alpha - \mu_k \left(w \cos \alpha - T \sin \theta \right)$$

$$= T \left(\cos \theta + \mu_k \sin \theta \right) - w \left(\sin \alpha + \mu_k \cos \alpha \right).$$

The acceleration will be the greatest when the first term in parantheses is greatest; as in Problems 5-73 and 5-111, this occurs when $\tan \theta = \mu_k$.

5-111: a) See Exercise 5-38; $F = \mu_k w / (\cos \theta + \mu_k \sin \theta)$.

b)

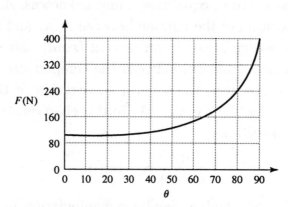

c) The expression for F is a minimum when the denominator is a maximum; the calculus is identical to that of Problem 5-71 (maximizing w for a given F gives the same result as minimizing F for a given w), and so F is minimized at $\tan \theta = \mu_k$. For $\mu_k = 0.25$, $\theta = 14.0°$, keeping an extra figure.

5-112: For convenience, take the positive direction to be down, so that for the baseball released from rest, the acceleration and velocity will be positive, and the speed of the baseball is the same as its positive component of velocity. Then the resisting force, directed against the velocity, is upward and hence negative.

a)

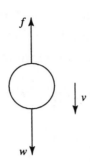

b) Newton's Second Law is then $ma = mg - Dv^2$. Initially, when $v = 0$, the acceleration is g, and the speed increases. As the speed increases, the resistive force increases and hence the acceleration decreases. This continues as the speed approaches the terminal speed. c) At terminal velocity, $a = 0$, so $v_t = \sqrt{\frac{mg}{D}}$, in agreement with Eq. (5-13). d) The equation of motion may be rewritten as $\frac{dv}{dt} = \frac{g}{v_t^2}(v_t^2 - v^2)$. This is a separable equation and may be expressed as

$$\int \frac{dv}{v_t^2 - v^2} = \frac{g}{v_t^2} \int dt, \quad \text{or}$$

$$\frac{1}{v_t}\text{arctanh}\left(\frac{v}{v_t}\right) = \frac{gt}{v_t^2},$$

so $v = v_t \tanh(gt/v_t)$.

Note: If inverse hyperbolic functions are unknown or undesirable, the integral can be done by partial fractions, in that

$$\frac{1}{v_t^2 - v^2} = \frac{1}{2v_t}\left[\frac{1}{v_t - v} + \frac{1}{v_t + v}\right],$$

and the resulting logarithms in the integrals can be solved for $v(t)$ in terms of exponentials.

5-113: Take all accelerations to be positive downward. The equations of motion are straightforward, but the kinematic relations between the accelerations, and the resultant algebra, are not immediately obvious. If the acceleration of pulley B is a_B, then $a_B = -a_3$, and a_B is the average of the accelerations of masses 1 and 2, or $a_1 + a_2 = 2a_B = -2a_3$. There can be no net force on the massless pulley B, so $T_C = 2T_A$. The five equations to be solved are then

$$m_1 g - T_A = m_1 a_1$$

$$m_2 g - T_A = m_2 a_2$$

$$m_3 g - T_C = m_3 a_3$$

$$a_1 + a_2 + 2a_3 = 0$$

$$2T_A - T_C = 0.$$

These are five equations in five unknowns, and may be solved by standard means. A symbolic-manipulation program is of great use here.

a) The accelerations a_1 and a_2 may be eliminated by using

$$2a_3 = -(a_1 + a_2) = -(2g - T_A((1/m_1) + (1/m_2))).$$

The tension T_A may be eliminated by using

$$T_A = (1/2)T_C = (1/2)m_3(g - a_3).$$

Combining and solving for a_3 gives

$$a_3 = g\frac{-4m_1m_2 + m_2m_3 + m_1m_3}{4m_1m_2 + m_2m_3 + m_1m_3}.$$

b) The acceleration of the pulley B has the same magnitude as a_3 and is in the opposite direction.

c) $$a_1 = g - \frac{T_A}{m_1} = g - \frac{T_C}{2m_1} = g - \frac{m_3}{2m_1}(g - a_3).$$

Substituting the above expression for a_3 gives

$$a_1 = g\frac{4m_1m_2 - 3m_2m_3 + m_1m_3}{4m_1m_2 + m_2m_3 + m_1m_3}.$$

d) A similar analysis (or, interchanging the labels 1 and 2) gives

$$a_2 = g\frac{4m_1m_2 - 3m_1m_3 + m_2m_3}{4m_1m_2 + m_2m_3 + m_1m_3}.$$

e) & f) Once the accelerations are known, the tensions may be found by substitution into the appropriate equation of motion, giving

$$T_A = g\frac{4m_1m_2m_3}{4m_1m_2 + m_2m_3 + m_1m_3}, \quad T_C = g\frac{8m_1m_2m_3}{4m_1m_2 + m_2m_3 + m_1m_3}.$$

g) If $m_1 = m_2 = m$ and $m_3 = 2m$, all of the accelerations are zero, $T_C = 2mg$ and $T_A = mg$. All masses and pulleys are in equilibrium, and the tensions are equal to the weights they support, which is what is expected.

5-114: In all cases, the tension in the string will be half of F.

a) $F/2 = 62$ N, which is insufficient to raise either block; $a_1 = a_2 = 0$.

b) $F/2 = 147$ N. The larger block (of weight 196 N) will not move, so $a_1 = 0$, but the smaller block, of weight 98 N, has a net upward force of 49 N applied to it, and so will accelerate upwards with $a_2 = \frac{49 \text{ N}}{10.0 \text{ kg}} = 4.9$ m/s^2.

c) $F/2 = 212$ N, so the net upward force on block A is 16 N and that on block B is 114 N, so $a_1 = \frac{16 \text{ N}}{20.0 \text{ kg}} = 0.8$ m/s^2 and $a_2 = \frac{114 \text{ N}}{10.0 \text{ kg}} = 11.4$ m/s^2.

5-115: Before the horizontal string is cut, the ball is in equilibrium, and the vertical component of the tension force must balance the weight, so $T_A \cos \beta = w$, or $T_A = w/\cos \beta$. At point B, the ball is not in equilibrium; its speed is instantaneously 0, so there is no radial acceleration, and the tension force must balance the radial component of the weight, so $T_B = w \cos \beta$, and the ratio $(T_B/T_A) = \cos^2 \beta$.

Chapter 6 Work and Kinetic Energy

6-1: a) $(2.40 \text{ N})(1.5 \text{ m}) = (3.60 \text{ J})$ b) $(-0.600 \text{ N})(1.50 \text{ m}) = -0.900 \text{ J}$
c) $3.60 \text{ J} - 0.720 \text{ J} = 2.70 \text{ J}$.

6-2: a) "Pulling slowly" can be taken to mean that the bucket rises at constant speed, so the tension in the rope may be taken to be the bucket's weight. In pulling a given length of rope, from Eq. (6-1),

$$W = Fs = mgs = (6.75 \text{ kg})(9.80 \text{ m/s}^2)(4.00 \text{ m}) = 264.6 \text{ J}.$$

b) Gravity is directed opposite to the direction of the bucket's motion, so Eq. (6-2) gives the negative of the result of part (a), or -265 J. c) The net work done on the bucket is zero.

6-3: $(25.0 \text{ N})(12.0 \text{ m}) = 300 \text{ J}$.

6-4: a) The friction force to be overcome is

$$f = \mu_k \mathcal{N} = \mu_k mg = (0.25)(30.0 \text{ kg})(9.80 \text{ m/s}^2) = 73.5 \text{ N},$$

or 74 N to two figures.

b) From Eq. (6-1), $Fs = (73.5 \text{ N})(4.5 \text{ m}) = 331 \text{ J}$. The work is positive, since the worker is pushing in the same direction as the crate's motion.

c) Since f and s are oppositely directed, Eq. (6-2) gives

$$-fs = -(73.5 \text{ N})(4.5 \text{ m}) = -331 \text{ J}.$$

d) Both the normal force and gravity act perpendicular to the direction of motion, so neither force does work. e) The net work done is zero.

6-5: a) See Exercise 5-37. The needed force is

$$F = \frac{\mu_k mg}{\cos\phi - \mu_k \sin\phi} = \frac{(0.25)(30 \text{ kg})(9.80 \text{ m/s}^2)}{\cos 30° - (0.25)\sin 30°} = 99.2 \text{ N},$$

keeping extra figures.   b) $Fs \cos\phi = (99.2 \text{ N})(4.50 \text{ m})\cos 30° = 386.5 \text{ J}$, again keeping an extra figure.   c) The normal force is $mg + F\sin\phi$, and so the work done by friction is $-(4.50 \text{ m})(0.25)((30 \text{ kg})(9.80 \text{ m/s}^2) + (99.2 \text{ N})\sin 30°) = -386.5 \text{ J}$.   d) Both the normal force and gravity act perpendicular to the direction of motion, so neither force does work. e) The net work done is zero.

6-6: From Eq. (6-2),

$$Fs \cos\phi = (180 \text{ N})(300 \text{ m})\cos 15.0° = 5.22 \times 10^4 \text{ J}.$$

6-7: $2Fs \cos\phi = 2(1.80 \times 10^6 \text{ N})(0.75 \times 10^3 \text{ m})\cos 14° = 2.62 \times 10^9 \text{ J}$, or 2.6×10^9 J to two places.

6-8: a) From Eq. (6-6),

$$K = \frac{1}{2}(1600 \text{ kg}) \left((50.0 \text{ km/h}) \left(\frac{1}{3.6} \frac{\text{m/s}}{\text{km/h}} \right) \right)^2 = 1.54 \times 10^5 \text{ J.}$$

b) Equation (6-5) gives the explicit dependence of kinetic energy on speed; doubling the speed of any object increases the kinetic energy by a factor of four.

6-9: For the T-Rex, $K = \frac{1}{2}(7000 \text{ kg})((4 \text{ km/hr})\frac{1 \text{ m/s}}{3.6 \text{ km/hr}})^2 = 4.32 \times 10^3$ J. The person's velocity would be $v = \sqrt{2(4.32 \times 10^3 \text{ J})/70 \text{ kg}} = 11.1$ m/s, or about 24.8 mi/hr.

6-10: Doubling the speed increases the kinetic energy, and hence the magnitude of the work done by friction, by a factor of four. With the stopping force given as being independent of speed, the distance must also increase by a factor of four.

6-11: Barring a balk, the initial kinetic energy of the ball is zero, and so

$$W = (1/2)mv^2 = (1/2)(0.145 \text{ kg})(32.0 \text{ m/s})^2 = 74.2 \text{ J.}$$

6-12: As the example explains, the boats have the same kinetic energy K at the finish line, so $(1/2)m_A v_A^2 = (1/2)m_B v_B^2$, or, with $m_B = 2m_A$, $v_A^2 = 2v_B^2$. a) Solving for the ratio of the speeds, $v_A/v_B = \sqrt{2}$. b) The boats are said to start from rest, so the elapsed time is the distance divided by the average speed. The ratio of the average speeds is the same as the ratio of the final speeds, so the ratio of the elapsed times is $t_B/t_A = v_A/v_B = \sqrt{2}$.

6-13: a) From Eq. (6-5), $K_2 = K_1/16$, and from Eq. (6-6), $W = -(15/16)K_1$. b) No; kinetic energies depend on the magnitudes of velocities only.

6-14: From Equations (6-1), (6-5) and (6-6), and solving for F,

$$F = \frac{\Delta K}{s} = \frac{\frac{1}{2}m(v_2^2 - v_1^2)}{s} = \frac{\frac{1}{2}(8.00 \text{ kg})((6.00 \text{ m/s})^2 - (4.00 \text{ m/s})^2)}{(2.50 \text{ m})} = 32.0 \text{ N.}$$

6-15:
$$s = \frac{\Delta K}{F} = \frac{\frac{1}{2}(0.420 \text{ kg})((6.00 \text{ m/s})^2 - (2.00 \text{ m/s})^2)}{(40.0 \text{ N})} = 16.8 \text{ cm}$$

6-16: a) If there is no work done by friction, the final kinetic energy is the work done by the applied force, and solving for the speed,

$$v = \sqrt{\frac{2W}{m}} = \sqrt{\frac{2Fs}{m}} = \sqrt{\frac{2(36.0 \text{ N})(1.20 \text{ m})}{(4.30 \text{ kg})}} = 4.48 \text{ m/s.}$$

b) The net work is $Fs - f_k s = (F - \mu_k mg)s$, so

$$v = \sqrt{\frac{2(F - \mu_k mg)s}{m}}$$

$$= \sqrt{\frac{2(36.0 \text{ N} - (0.30)(4.30 \text{ kg})(9.80 \text{ m/s}^2))(1.20 \text{ m})}{(4.30 \text{ kg})}}$$

$$= 3.61 \text{ m/s.}$$

(Note that even though the coefficient of friction is known to only two places, the difference of the forces is still known to three places.)

6-17: a) On the way up, gravity is opposed to the direction of motion, and so $W = -mgs = -(0.145 \text{ kg})(9.80 \text{ m/s}^2)(20.0 \text{ m}) = -28.4 \text{ J}$.

b)
$$v_2 = \sqrt{v_1^2 + 2\frac{W}{m}} = \sqrt{(25.0 \text{ m/s})^2 + \frac{2(-28.4 \text{ J})}{(0.145 \text{ kg})}} = 15.26 \text{ m/s}.$$

c) No; in the absence of air resistance, the ball will have the same speed on the way down as on the way up. On the way down, gravity will have done both negative and positive work on the ball, but the net work will be the same.

6-18: a) Gravity acts in the same direction as the watermelon's motion, so Eq. (6-1) gives

$$W = Fs = mgs = (4.80 \text{ kg})(9.80 \text{ m/s}^2)(25.0 \text{ m}) = 1176 \text{ J}.$$

b) Since the melon is released from rest, $K_1 = 0$, and Eq. (6-6) gives

$$K = K_2 = W = 1176 \text{ J}.$$

6-19: a) Combining Equations (6-5) and (6-6) and solving for v_2 algebraically,

$$v_2 = \sqrt{v_1^2 + 2\frac{W_{\text{tot}}}{m}} = \sqrt{(4.00 \text{ m/s})^2 + \frac{2(10.0 \text{ N})(3.0 \text{ m})}{(7.00 \text{ kg})}} = 4.96 \text{ m/s}.$$

Keeping extra figures in the intermediate calculations, the acceleration is $a = (10.0 \text{ kg·m/s}^2)/(7.00 \text{ kg}) = 1.429 \text{ m/s}^2$. From Eq. (2-13), with appropriate change in notation,

$$v_2^2 = v_1^2 + 2as = (4.00 \text{ m/s})^2 + 2(1.429 \text{ m/s}^2)(3.0 \text{ m}),$$

giving the same result.

6-20: The normal force does no work. The work-energy theorem, along with Eq. (6-5), gives

$$v = \sqrt{\frac{2K}{m}} = \sqrt{\frac{2W}{m}} = \sqrt{2gh} = \sqrt{2gL\sin\theta},$$

where $h = L\sin\theta$ is the vertical distance the block has dropped, and θ is the angle the plane makes with the horizontal. Using the given numbers,

$$v = \sqrt{2(9.80 \text{ m/s}^2)(0.75 \text{ m})\sin 36.9°} = 2.97 \text{ m/s}.$$

6-21: a) The friction force is $\mu_k mg$, which is directed against the car's motion, so the net work done is $-\mu_k mgs$. The change in kinetic energy is $\Delta K = -K_1 = -(1/2)mv_0^2$, and so $s = v_0^2/2\mu_k g$. b) From the result of part (a), the stopping distance is proportional to the square of the initial speed, and so for an initial speed of 60 km/h,

$s = (91.2 \text{ m})(60.0/80.0)^2 = 51.3$ m. (This method avoids the intermediate calculation of μ_k, which in this case is about 0.279.)

6-22: The intermediate calculation of the spring constant may be avoided by using Eq. (6-9) to see that the work is proportional to the square of the extension; the work needed to compress the spring 4.00 cm is $(12.0 \text{ J}) \left(\frac{4.00 \text{ cm}}{3.00 \text{ cm}}\right)^2 = 21.3$ J.

6-23: a) The magnitude of the force is proportional to the magnitude of the extension or compression;

$$(160 \text{ N})(0.015 \text{ m}/0.050 \text{ m}) = 48 \text{ N}, \qquad (160 \text{ N})(0.020 \text{ m}/0.050 \text{ m}) = 64 \text{ N}.$$

b) There are many equivalent ways to do the necessary algebra. One way is to note that to stretch the spring the original 0.050 m requires $\frac{1}{2} \left(\frac{160 \text{ N}}{0.050 \text{ m}}\right) (0.050 \text{ m})^2 = 4$ J, so that stretching 0.015 m requires $(4 \text{ J})(0.015/0.050)^2 = 0.360$ J and compressing 0.020 m requires $(4 \text{ J})(0.020/0.050)^2 = 0.64$ J. Another is to find the spring constant $k = (160 \text{ N}) \div (0.050 \text{ m}) = 3.20 \times 10^3$ N/m, from which $(1/2)(3.20 \times 10^3 \text{ N/m})(0.015 \text{ m})^2 = 0.360$ J and $(1/2)(3.20 \times 10^3 \text{ N/m})(0.020 \text{ m})^2 = 0.64$ J.

6-24: The work can be found by finding the area under the graph, being careful of the sign of the force. The area under each triangle is 1/2 base $\times$ height.

  a) 1/2 (8 m)(10 N) = 40 J.
  b) 1/2 (4 m)(10 N) = +20 J.
  c) 1/2 (12 m)(10 N) = 60 J.

6-25: Use the Work–Energy Theorem and the results of problem 24.

  a) $v = \sqrt{\dfrac{(2)(40 \text{ J})}{10 \text{ kg}}} = 2.83$ m/s.
  b) At $x = 12$ m, the 40 Joules of kinetic energy will have been increased by 20 J, so

$$v = \sqrt{\frac{(2)(60 \text{ J})}{10 \text{ kg}}} = 3.46 \text{ m/s}.$$

6-26: a) The average force is $(80.0 \text{ J})/(0.200 \text{ m}) = 400$ N, and the force needed to hold the platform in place is twice this, or 800 N. b) From Eq. (6-9), doubling the distance quadruples the work so an extra 240 J of work must be done. The maximum force is quadrupled, 1600 N.

Both parts may of course be done by solving for the spring constant $k = 2(80.0 \text{ J}) \div (0.200 \text{ m})^2 = 4.00 \times 10^3$ N/m, giving the same results.

6-27: a) The static friction force would need to be equal in magnitude to the spring force, $\mu_s mg = kd$ or $\mu_s = \frac{(20.0 \text{ N/m})(0.086 \text{ m})}{(0.100 \text{ kg})(9.80 \text{ m/s}^2)} = 1.76$, which is quite large. (Keeping extra figures in the intermediate calculation for d gives a different answer.) b) In Example 6-8, the relation

$$\mu_k mgd + \frac{1}{2}kd^2 = \frac{1}{2}mv_1^2$$

was obtained, and d was found in terms of the known initial speed v_1. In this case, the condition on d is that the static friction force at maximum extension just balances the spring force, or $kd = \mu_s mg$. Solving for v_1^2 and substituting,

$$
\begin{aligned}
v_1^2 &= \frac{k}{m}d^2 + 2gd\mu_k d \\
&= \frac{k}{m}\left(\frac{\mu_s mg}{k}\right)^2 + 2\mu_k g\left(\frac{\mu_s mg}{k}\right) \\
&= \frac{mg^2}{k}\left(\mu_s^2 + 2\mu_s\mu_k\right) \\
&= \left(\frac{(0.10 \text{ kg})(9.80 \text{ m/s}^2)^2}{(20.0 \text{ N/m})}\right)((0.60)^2 + 2(0.60)(0.47)),
\end{aligned}
$$

from which $v_1 = 0.67$ m/s.

6-28: a) The spring is pushing on the block in its direction of motion, so the work is positive, and equal to the work done in compressing the spring. From either Eq. (6-9) or Eq. (6-10), $W = \frac{1}{2}kx^2 = \frac{1}{2}(200 \text{ N/m})(0.025 \text{ m})^2 = 0.06$ J.

b) The work-energy theorem gives

$$
v = \sqrt{\frac{2W}{m}} = \sqrt{\frac{2(0.06 \text{ J})}{(4.0 \text{ kg})}} = 0.18 \text{ m/s}.
$$

6-29: The work done in any interval is the area under the curve, easily calculated when the areas are unions of triangles and rectangles. a) The area under the trapezoid is 4.0 N·m = 4.0 J. b) No force is applied in this interval, so the work done is zero. c) The area of the triangle is 1.0 N·m = 1.0 J, and since the curve is *below* the axis ($F_x < 0$), the work is negative, or -1.0 J. d) The net work is the sum of the results of parts (a), (b) and (c), 3.0 J. e) $+1.0$ J $- 2.0$ J $= -1.0$ J.

6-30: a) $K = 4.0$ J, so $v = \sqrt{2K/m} = \sqrt{2(4.0 \text{ J})/(2.0 \text{ kg})} = 2.00$ m/s. b) No work is done between $x = 3.0$ m and $x = 4.0$ m, so the speed is the same, 2.00 m/s. c) $K = 3.0$ J, so $v = \sqrt{2K/m} = \sqrt{2(3.0 \text{ J})/(2.0 \text{ kg})} = 1.73$ m/s.

6-31: a) The spring does positive work on the sled and rider; $(1/2)kx^2 = (1/2)mv^2$, or $v = x\sqrt{k/m} = (0.375 \text{ m})\sqrt{(4000 \text{ N/m})/(70 \text{ kg})} = 2.83$ m/s. b) The net work done by the spring is $(1/2)k(x_1^2 - x_2^2)$, so the final speed is

$$
v = \sqrt{\frac{k}{m}(x_1^2 - x_2^2)} = \sqrt{\frac{(4000 \text{ N/m})}{(70 \text{ kg})}((0.375 \text{ m})^2 - (0.200 \text{ m})^2)} = 2.40 \text{ m/s}.
$$

6-32: a) From Eq. (6-14), with $dl = R\,d\phi$,

$$
W = \int_{P_1}^{P_2} F\cos\phi\,dl = 2wR\int_0^{\theta_0}\cos\phi\,d\phi = 2wR\sin\theta_0.
$$

In an equivalent geometric treatment, when $\overrightarrow{F}$ is horizontal, $\overrightarrow{F}\cdot d\overrightarrow{l} = F\,dx$, and the total work is $F = 2w$ times the horizontal distance, in this case (see Fig. 6-18(a)) $R\sin\theta_0$, giving the same result.   b) The ratio of the forces is $\frac{2w}{w\tan\theta_0} = 2\cot\theta_0$.

c)
$$\frac{2wR\sin\theta_0}{wR(1-\cos\theta_0)} = 2\frac{\sin\theta_0}{(1-\cos\theta_0)} = 2\cot\frac{\theta_0}{2}.$$

6-33:   a) The initial and final (at the maximum distance) kinetic energy is zero, so the positive work done by the spring, $(1/2)kx^2$, must be the opposite of the negative work done by gravity, $-mgL\sin\theta$, or

$$x = \sqrt{\frac{2mgL\sin\theta}{k}} = \sqrt{\frac{2(0.0900\text{ kg})(9.80\text{ m/s}^2)(1.80\text{ m})\sin 40.0°}{(640\text{ N/m})}} = 5.7\text{ cm}.$$

b) The intermediate calculation of the initial compression can be avoided by considering that between the point 0.80 m from the launch to the maximum distance, gravity does a negative amount of work given by $-(0.0900\text{ kg})(9.80\text{ m/s}^2)(1.80\text{ m} - 0.80\text{ m})\sin 40.0° = -0.567$ J, and so the kinetic energy of the glider at this point is 0.567 J.

6-34:   The initial and final kinetic energies of the brick are both zero, so the net work done on the brick by the spring and gravity is zero, so $(1/2)kd^2 - mgh = 0$, or $d = \sqrt{2mgh/k} = \sqrt{2(1.80\text{ kg})(9.80\text{ m/s}^2)(3.6\text{ m})/(450\text{ N/m})} = 0.53$ m. The spring will provide an upward force while the spring and the brick are in contact. When this force goes to zero, the spring is at its uncompressed length.

6-35:   The total power is $(165\text{ N})(9.00\text{ m/s}) = 1.485 \times 10^3$ W, so the power per rider is 742.5 W, or about 1.0 hp (which is a very large output, and cannot be sustained for long periods).

6-36:   a)
$$\frac{(1.0 \times 10^{19}\text{ J/yr})}{(3.16 \times 10^7\text{ s/yr})} = 3.2 \times 10^{11}\text{ W}.$$

b)
$$\frac{3.2 \times 10^{11}\text{ W}}{2.6 \times 10^8\text{ folks}} = 1.2\text{ kW/person}.$$

c)
$$\frac{3.2 \times 10^{11}\text{ W}}{(0.40)1.0 \times 10^3\text{ W/m}^2} = 8.0 \times 10^8\text{ m}^2 = 800\text{ km}^2.$$

6-37:   The power is $P = F \cdot v$. F is the weight, mg, so $P = (700\text{ kg})(9.8\text{ m/s}^2)(2.5\text{ m/s}) = 17.15$ kW. 17.15 kW/75 kW. $= .23$, or about 23% of the engine power is used in climbing.

6-38:   a) The number per minute would be the average power divided by the work (mgh) required to lift one box,

$$\frac{(0.50\text{ hp})(746\text{ W/hp})}{(30\text{ kg})(9.80\text{ m/s}^2)(0.90\text{ m})} = 1.41\text{ /s},$$

or 84.6 /min.   b) Similarly,

$$\frac{(100\text{ W})}{(30\text{ kg})(9.80\text{ m/s}^2)(0.90\text{ m})} = 0.378\text{ /s},$$

or 22.7 /min.

6-39:   The total mass that can be raised is

$$\frac{(40.0\text{ hp})(746\text{ W/hp})(16.0\text{ s})}{(9.80\text{ m/s}^2)(20.0\text{ m})} = 2436\text{ kg},$$

so the maximum number of passengers is $\frac{1836\text{ kg}}{65.0\text{ kg}} = 28.$

6-40:   From any of Equations (6-15), (6-16), (6-18) or (6-19),

$$P = \frac{Wh}{t} = \frac{(3800\text{ N})(2.80\text{ m})}{(4.00\text{ s})} = 2.66 \times 10^3\text{ W} = 3.57\text{ hp}.$$

6-41: $F = \dfrac{(0.70)P_{\text{ave}}}{v} = \dfrac{(0.70)(280{,}000\text{ hp})(746\text{ W/hp})}{(65\text{ km/h})((1\text{ km/h})/(3.6\text{ m/s}))} = 8.1 \times 10^6\text{ N}.$

6-42:   Here, Eq. (6-19) is the most direct. Gravity is doing negative work, so the rope must do positive work to lift the skiers. The force $\vec{F}$ is gravity, and $F = Nmg$, where N is the number of skiers on the rope. The power is then

$$
\begin{aligned}
P &= (Nmg)(v)\cos\phi \\
&= (50)(70\text{ kg})(9.80\text{ m/s}^2)(12.0\text{ km/h})\left(\frac{1\text{ m/s}}{3.6\text{ km/h}}\right)\cos(90.0° - 15.0°) \\
&= 2.96 \times 10^4\text{ W}.
\end{aligned}
$$

Note that Eq. (1-18) uses ϕ as the angle between the force and velocity vectors; in this case, the force is vertical, but the angle 15.0° is measured from the horizontal, so $\phi = 90.0° - 15.0°$ is used.

6-43:   a) In terms of the acceleration a and the time t since the force was applied, the speed is $v = at$ and the force is ma, so the power is $P = Fv = (ma)(at) = ma^2t$.   b) The power at a given time is proportional to the square of the acceleration, tripling the acceleration would mean increasing the power by a factor of nine.   c) If the magnitude of the net force is the same, the acceleration will be the same, and the needed power is proportional to the time. At $t = 15.0$ s, the needed power is three times that at 5.0 s, or 108 W.

6-44:

$$
\begin{aligned}
\frac{dK}{dt} &= \frac{d}{dt}\left(\frac{1}{2}mv^2\right) \\
&= mv\frac{dv}{dt} \\
&= mva = mav \\
&= Fv = P.
\end{aligned}
$$

6-45: a) $(540 \text{ N})(30 \text{ m/s}) = 1.6 \times 10^4 \text{ W}.$

b)
$$\frac{(540 \text{ N})(30 \text{ m/s})(3600 \text{ s})}{(0.15)(3.5 \times 10^7 \text{ J})} = 11 \text{ L}$$

c) Keeping extra figures in the intermediate calculation,

$$\frac{11.11 \text{ L}}{(30 \text{ m/s})(3600 \text{ s})} = 0.103 \text{ L/km}, \qquad \text{and}$$

$$0.103 \text{ L/km} \times \frac{1 \text{ gal}}{3.788 \text{ L}} \times \frac{1.609 \text{ km}}{1 \text{ mi}} = 0.044 \text{ gal/mi}.$$

6-46: a) $\quad F = \dfrac{P}{v} = \dfrac{28.0 \times 10^3 \text{ W}}{(60.0 \text{ km/h})((1 \text{ m/s})/(3.6 \text{ km/h}))} = 1.68 \times 10^3 \text{ N}.$

b) The speed is lowered by a factor of one-half, and the resisting force is lowered by a factor of $(0.65 + 0.35/4)$, and so the power at the lower speed is

$$(28.0 \text{ kW})(0.50)(0.65 + 0.35/4) = 10.3 \text{ kW} = 13.8 \text{ hp}.$$

c) Similarly, at the higher speed,

$$(28.0 \text{ kW})(2.0)(0.65 + 0.35 \times 4) = 114.8 \text{ kW} = 154 \text{ hp}.$$

6-47: a)
$$\frac{(8.00 \text{ hp})(746 \text{ W/hp})}{(60.0 \text{ km/h})((1 \text{ m/s})/(3.6 \text{ km/h}))} = 358 \text{ N}$$

b) The extra power needed is

$$mgv_{\parallel} = (1800 \text{ kg})(9.80 \text{ m/s}^2)\frac{60.0 \text{ km/h}}{3.6\frac{\text{km/h}}{\text{m/s}}} \sin(\arctan(1/10)) = 29.3 \text{ kW} = 39.2 \text{ hp},$$

so the total power is 47.2 hp. (Note: If the sine of the angle is approximated by the tangent, the third place will be different.) c) Similarly,

$$mgv_{\parallel} = (1800 \text{ kg})(9.80 \text{ m/s}^2)\frac{60.0 \text{ km/h}}{3.6\frac{\text{km/h}}{\text{m/s}}} \sin(\arctan(0.010)) = 2.94 \text{ kW} = 3.94 \text{ hp}.$$

This is the rate at which work is done on the car by gravity. The engine must do work on the car at a rate of 4.06 hp. d) In this case, approximating the sine of the slope by the tangent is appropriate, and the grade is

$$\frac{(8.00 \text{ hp})(746 \text{ W/hp})}{(1800 \text{ kg})(9.80 \text{ m/s}^2)(60.0 \text{ km/h})((1 \text{ m/s})/(3.6 \text{ km/h}))} = 0.0203,$$

very close to a 2% grade.

6-48: The mass changes the magnitude of the rolling friction proportionate to the mass, but the air friction remains unchanged. The percent changes in the power are then a) $\frac{180 \text{ N}(0.060)}{220 \text{ N}} = 4.9\%$ and b) $\frac{180 \text{ N}(0.060)}{540 \text{ N}} = 2.0\%$.

6-49: a) $(140 \text{ N})(3.80 \text{ m}) = 532 \text{ J}$ b) $(20.0 \text{ kg})(9.80 \text{ m/s}^2)(3.80 \text{ m})(-\sin 25°) = -315 \text{ J}$

c) The normal force does no work.

d)
$$W_f = -f_k s = -\mu_k \mathcal{N} s = -\mu_k mgs \cos\theta$$
$$= -(0.30)(20.0 \text{ kg})(9.80 \text{ m/s}^2)(3.80 \text{ m}) \cos 25° = -203 \text{ J}$$

e) $532 \text{ J} - 315 \text{ J} - 203 \text{ J} = 15 \text{ J}$ (14.7 J to three figures).

f) The result of part (e) is the kinetic energy at the top of the ramp, so the speed is $v = \sqrt{2K/m} = \sqrt{2(14.7 \text{ J})/(20.0 \text{ kg})} = 1.21 \text{ m/s}$.

6-50: The work per unit mass is $(W/m) = gh$.

a) The man does work, $(9.8 \text{ N/kg})(0.4 \text{ m}) = 3.92 \text{ J/kg}$.

b) $(3.92 \text{ J/kg})/(70 \text{ J/kg}) \times 100 = 5.6\%$.

c) The child does work, $(9.8 \text{ N/kg})(0.2 \text{ m}) = 1.96 \text{ J/kg}$. $(1.96 \text{ J/kg})/(70 \text{ J/kg}) \times 100 = 2.8\%$.

d) If both the man and the child can do work at the rate of 70 J/kg, and if the child only needs to use 1.96 J/kg instead of 3.92 J/kg, the child should be able to do more pull ups.

6-51: a) Moving a distance L along the ramp, $s_{in} = L$, $s_{out} = L \sin\alpha$, so $IMA = \frac{1}{\sin\alpha}$.

b) If $AMA = IMA$, $(F_{out}/F_{in}) = (s_{in}/s_{out})$ and so $(F_{out})(s_{out}) = (F_{in})(s_{in})$, or $W_{out} = W_{in}$.

c)

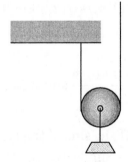

d)
$$E = \frac{W_{out}}{W_{in}} = \frac{(F_{out})(s_{out})}{(F_{in})(s_{in})} = \frac{F_{out}/F_{in}}{s_{in}/s_{out}} = \frac{AMA}{IMA}.$$

6-52: a) $m = \dfrac{w}{g} = \dfrac{-W_g/s}{g} = \dfrac{(7.35 \times 10^3 \text{ J})}{(9.80 \text{ m/s}^2)(18.0 \text{ m})} = 41.7 \text{ kg}$.

b) $\mathcal{N} = \dfrac{W_N}{s} = \dfrac{8.25 \times 10^3 \text{ J}}{18.0 \text{ m}} = 458 \text{ N}$.

c) The weight is $mg = \dfrac{W_g}{s} = 408 \text{ N}$, so the acceleration is the net force divided by the mass, $\dfrac{458 \text{ N} - 408 \text{ N}}{41.7 \text{ kg}} = 1.2 \text{ m/s}^2$.

6-53: a)

$$\frac{1}{2}mv^2 = \frac{1}{2}m\left(\frac{2\pi R}{T}\right)^2 = \frac{1}{2}(86,400 \text{ kg})\left(\frac{2\pi(6.66 \times 10^6 \text{ m})}{(90.1 \text{ min})(60 \text{ s/min})}\right)^2 = 2.59 \times 10^{12} \text{ J}.$$

b) $(1/2)mv^2 = (1/2)(86,400 \text{ kg})((1.00 \text{ m})/(3.00 \text{ s}))^2 = 4.80 \times 10^3 \text{ J}.$

6-54: a) $\quad W_f = -f_k s = -\mu_k mg \cos\theta\, s$

$$= -(0.31)(5.00 \text{ kg})(9.80 \text{ m/s}^2)\cos 12.0°(1.50 \text{ m}) = -22.3 \text{ J}$$

(keeping an extra figure)   b) $(5.00 \text{ kg})(9.80 \text{ m/s}^2)\sin 12.0°(1.50 \text{ m}) = 15.3 \text{ J}.$ c) The normal force does no work.   d) $15.3 \text{ J} - 22.3 \text{ J} = -7.0 \text{ J}.$ e) $K_2 = K_1 + W = (1/2)(5.00 \text{ kg})(2.2 \text{ m/s})^2 - 7.0 \text{ J} = 5.1 \text{ J}$, and so $v_2 = \sqrt{2(5.1 \text{ J})/(5.00 \text{ kg})} = 1.4 \text{ m/s}.$

6-55: See Problem 6-54: The work done is negative, and is proportional to the distance s that the package slides along the ramp, $W = mg(\sin\theta - \mu_k \cos\theta)s$. Setting this equal to the (negative) change in kinetic energy and solving for s gives

$$s = -\frac{(1/2)mv_1^2}{mg(\sin\theta - \mu_k \cos\theta)} = -\frac{v_1^2}{2g(\sin\theta - \mu_k \cos\theta)}$$

$$= -\frac{(2.2 \text{ m/s})^2}{2(9.80 \text{ m/s}^2)(\sin 12° - (0.31)\cos 12°)} = 2.6 \text{ m}.$$

As a check of the result of Problem 6-54, $(2.2 \text{ m/s})\sqrt{1 - (1.5 \text{ m})/(2.6 \text{ m})} = 1.4 \text{ m/s}.$

6-56: a) From Eq. (6-7),

$$W = \int_{x_1}^{x_2} F_x\, dx = -k\int_{x_1}^{x_2} \frac{dx}{x^2} = -k\left[-\frac{1}{x}\right]_{x_1}^{x_2} = k\left(\frac{1}{x_2} - \frac{1}{x_1}\right).$$

The force is given to be attractive, so $F_x < 0$, and k must be positive. If $x_2 > x_1$, $\frac{1}{x_2} < \frac{1}{x_1}$, and $W < 0$.   b) Taking "slowly" to be constant speed, the net force on the object is zero, so the force applied by the hand is opposite F_x, and the work done is negative of that found in part (a), or $k\left(\frac{1}{x_1} - \frac{1}{x_2}\right)$, which is positive if $x_2 > x_1$.   c) The answers have the same magnitude but opposite signs; this is to be expected, in that the net work done is zero.

6-57: a) $\alpha x_a^3 = (4.00 \text{ N/m}^3)(1.00 \text{ m})^3 = 4.00 \text{ N}.$

b) $\alpha x_b^3 = (4.00 \text{ N/m}^3)(2.00 \text{ m})^3 = 32.0 \text{ N}.$ c) Equation 6-7 gives the force needed to move an object against the force; the work done by the force is the negative of this,

$$-\int_{x_1}^{x_2} \alpha x^3\, dx = -\frac{\alpha}{4}\left(x_2^4 - x_1^4\right).$$

With $x_1 = x_a = 1.00 \text{ m}$ and $x_2 = x_b = 2.00 \text{ m}$, $W = -15.0 \text{ J}$, this work is negative.

6-58: From Eq. (6-7), with $x_1 = 0$,

$$W = \int_0^{x_2} F\, dx = \int_0^{x_2} \left(kx - bx^2 + cx^3\right) dx = \frac{k}{2}x_2^2 - \frac{b}{3}x_2^3 + \frac{c}{4}x_2^4$$

$$= (50.0 \text{ N/m})x_2^2 - (233 \text{ N/m}^2)x_2^3 + (3000 \text{ N/m}^3)x_2^4.$$

a) When $x_2 = 0.050$ m, $W = 0.115$ J, or 0.12 J to two figures.   b) When $x_2 = -0.050$ m, $W = 0.173$ J, or 0.17 J to two figures.   c) It's easier to stretch the spring; the quadratic $-bx^2$ term is always in the $-x$-direction, and so the needed force, and hence the needed work, will be less when $x_2 > 0$.

6-59:   a) $T = ma_{rad} = m\frac{v^2}{R} = (0.120 \text{ kg})\frac{(0.70 \text{ m/s})^2}{(0.40 \text{ m})} = 0.147$ N, or 0.15 N to two figures.

b) At the later radius and speed, the tension is $(0.120 \text{ kg})\frac{(2.80 \text{ m/s})^2}{(0.10 \text{ m})} = 9.41$ N, or 9.4 N to two figures.   c) The surface is frictionless and horizontal, so the net work is the work done by the string. For a massless and frictionless string, this is the same as the work done by the person, and is equal to the change in the block's kinetic energy, $K_2 - K_1 = (1/2)m\,(v_2^2 - v_1^2) = (1/2)(0.120 \text{ kg})((2.80 \text{ m/s})^2 - (0.70 \text{ m/s})^2) = 0.441$ J. Note that in this case, the tension cannot be perpendicular to the block's velocity at all times; the string is in the radial direction, and for the radius to change, the block must have some non-zero component of velocity in the radial direction.

6-60:   a) This is similar to Problem 6-56, but here $\alpha > 0$ (the force is repulsive), and $x_2 < x_1$, so the work done is again negative;

$$W = \alpha \left(\frac{1}{x_1} - \frac{1}{x_2} \right) = (2.12 \times 10^{-26} \text{ N·m}^2((0.200 \text{ m}^{-1}) - (1.25 \times 10^9 \text{ m}^{-1}))$$

$$= -2.65 \times 10^{-17} \text{ J}.$$

Note that x_1 is so large compared to x_2 that the term $\frac{1}{x_1}$ is negligible. Then, using Eq. (6-13)) and solving for v_2,

$$v_2 = \sqrt{v_1^2 + \frac{2W}{m}} = \sqrt{(3.00 \times 10^5 \text{ m/s})^2 + \frac{2(-2.65 \times 10^{-17} \text{ J})}{(1.67 \times 10^{-27} \text{ kg})}} = 2.41 \times 10^5 \text{ m/s}.$$

b) With $K_2 = 0$, $W = -K_1$. Using $W = -\frac{\alpha}{x_2}$,

$$x_2 = \frac{\alpha}{K_1} = \frac{2\alpha}{mv_1^2} = \frac{2(2.12 \times 10^{-26} \text{ N·m}^2)}{(1.67 \times 10^{-27} \text{ kg})(3.00 \times 10^5 \text{ m/s})^2} = 2.82 \times 10^{-10} \text{ m}.$$

c) The repulsive force has done no net work, so the kinetic energy and hence the speed of the proton have their original values, and the speed is 3.00×10^5 m/s.

6-61:   The velocity and acceleration as functions of time are

$$v(t) = \frac{dx}{dt} = 2\alpha t + 3\beta t^2, \qquad a(t) = 2\alpha + 6\beta t$$

a) $v(t = 4.00 \text{ s}) = 2(0.20 \text{ m/s}^2)(4.00 \text{ s}) + 3(0.02 \text{ m/s}^3)(4.00 \text{ s})^2 = 2.56$ m/s.

b) $ma = (6.00 \text{ kg})(2(0.20 \text{ m/s}^2) + 6(0.02 \text{ m/s}^3)(4.00 \text{ s}) = 5.28$ N.

c) $W = K_2 - K_1 = K_2 = (1/2)(6.00 \text{ kg})(2.56 \text{ m/s})^2 = 19.7$ J.

6-62: In Eq. (6-14), $dl = dx$ and $\phi = 31.0°$ is constant, and so

$$W = \int_{P_1}^{P_2} F \cos \phi \, dl = \int_{x_1}^{x_2} F \cos \phi \, dx$$

$$= (5.00 \text{ N/m}^2) \cos 31.0° \int_{1.00 \text{ m}}^{1.50 \text{ m}} x^2 \, dx = 3.39 \text{ J}.$$

The final speed of the object is then

$$v_2 = \sqrt{v_1^2 + \frac{2W}{m}} = \sqrt{(4.00 \text{ m/s})^2 + \frac{2(3.39 \text{ J})}{(0.250 \text{ kg})}} = 6.57 \text{ m/s}.$$

6-63: a) $K_2 - K_1 = (1/2)m \left(v_2^2 - v_1^2 \right)$

$$= (1/2)(80.0 \text{ kg})((1.50 \text{ m/s})^2 - (5.00 \text{ m/s})^2) = -910 \text{ J}.$$

b) The work done by gravity is $-mgh = -(80.0 \text{ kg})(9.80 \text{ m/s}^2)(5.20 \text{ m}) = -4.08 \times 10^3 \text{ J}$, so the work done by the rider is $-910 \text{ J} - (-4.08 \times 10^3 \text{ J}) = 3.17 \times 10^3 \text{ J}$.

6-64: a) $W = \int_{x_0}^{\infty} \frac{b}{x^n} \, dx = \frac{b}{(-(n-1))x^{n-1}} \Big|_{x_0}^{\infty} = \frac{b}{(n-1)x_0^{n-1}}.$

Note that for this part, for $n > 1$, $x^{1-n} \to 0$ as $x \to \infty$. **b)** When $0 < n < 1$, the improper integral must be used,

$$W = \lim_{x_2 \to \infty} \left[\frac{b}{(n-1)} \left(x_2^{n-1} - x_0^{n-1} \right) \right],$$

and because the exponent on the x_2^{n-1} is positive, the limit does not exist, and the integral diverges. This is interpreted as the force F doing an infinite amount of work, even though $F \to 0$ as $x_2 \to \infty$.

6-65: Setting the (negative) work done by the spring to the needed (negative) change in kinetic energy, $\frac{1}{2}kx^2 = \frac{1}{2}mv_0^2$, and solving for the spring constant,

$$k = \frac{mv_0^2}{x^2} = \frac{(1200 \text{ kg})(0.65 \text{ m/s})^2}{(0.070 \text{ m})^2} = 1.03 \times 10^5 \text{ N/m}.$$

6-66: a) Equating the work done by the spring to the gain in kinetic energy, $\frac{1}{2}kx_0^2 = \frac{1}{2}mv^2$, so

$$v = \sqrt{\frac{k}{m}}x_0 = \sqrt{\frac{400 \text{ N/m}}{0.0300 \text{ kg}}}(0.060 \text{ m}) = 6.93 \text{ m/s}.$$

b) W_{tot} must now include friction, so $\frac{1}{2}mv^2 = W_{\text{tot}} = \frac{1}{2}kx_0^2 - fx_0$, where f is the magnitude of the friction force. Then,

$$v = \sqrt{\frac{k}{m}x_0^2 - \frac{2f}{m}x_0}$$

$$= \sqrt{\frac{400 \text{ N/m}}{0.0300 \text{ kg}}(0.06 \text{ m})^2 - \frac{2(6.00 \text{ N})}{(0.0300 \text{ kg})}(0.06 \text{ m})} = 4.90 \text{ m/s}.$$

c) The greatest speed occurs when the acceleration (and the net force) are zero, or $kx = f$, $x = \frac{f}{k} = \frac{6.00\ \text{N}}{400\ \text{N/m}} = 0.0150$ m. To find the speed, the net work is $W_{\text{tot}} = \frac{1}{2}k\left(x_0^2 - x^2\right) - f\left(x_0 - x\right)$, so the maximum speed is

$$v_{\max} = \sqrt{\frac{k}{m}\left(x_0^2 - x^2\right) - \frac{2f}{m}\left(x_0 - x\right)}$$

$$= \sqrt{\frac{400\ \text{N/m}}{(0.0300\ \text{kg})}\left((0.060\ \text{m})^2 - (0.0150\ \text{m})^2\right) - \frac{2(6.00\ \text{N})}{(0.0300\ \text{kg})}(0.060\ \text{m} - 0.0150\ \text{m})}$$

$$= 5.20\ \text{m/s},$$

which is larger than the result of part (b) but smaller than the result of part (a).

6-67: Denote the initial compression of the spring by x and the distance from the initial position by L. Then, the work done by the spring is $\frac{1}{2}kx^2$ and the work done by friction is $-\mu_k mg(x + L)$; this form takes into account the fact that while the spring is compressed, the frictional force is still present (see Problem 6-66). The initial and final kinetic energies are both zero, so the net work done is zero, and $\frac{1}{2}kx^2 = \mu_k mg(x + L)$. Solving for L,

$$L = \frac{(1/2)kx^2}{\mu_k mg} - x = \frac{(1/2)(250\ \text{N/m})(0.250\ \text{m})^2}{(0.30)(2.50\ \text{kg})(9.80\ \text{m/s}^2)} - (0.250\ \text{m}) = 0.813\ \text{m},$$

or 0.81 m to two figures. Thus the book moves .81 m + .25 m = 1.06 m, or about 1.1 m.

6-68: The work done by gravity is $W_g = -mgL\sin\theta$ (negative since the cat is moving up), and the work done by the applied force is FL, where F is the magnitude of the applied force. The total work is

$$W_{\text{tot}} = (100\ \text{N})(2.00\ \text{m}) - (7.00\ \text{kg})(9.80\ \text{m/s}^2)(2.00\ \text{m})\sin 30° = 131.4\ \text{J}.$$

The cat's initial kinetic energy is $\frac{1}{2}mv_1^2 = \frac{1}{2}(7.00\ \text{kg})(2.40\ \text{m/s})^2 = 20.2\ \text{J}$, and

$$v_2 = \sqrt{\frac{2(K_1 + W)}{m}} = \sqrt{\frac{2(20.2\ \text{J} + 131.4\ \text{J})}{(7.00\ \text{kg})}} = 6.58\ \text{m/s}.$$

6-69: In terms of the bumper compression x and the initial speed v_0, the necessary relations are

$$\frac{1}{2}kx^2 = \frac{1}{2}mv_0^2, \quad kx < 5mg.$$

Combining to eliminate k and then x, the two inequalties are

$$x > \frac{v^2}{5g} \quad \text{and} \quad k < 25\frac{mg^2}{v^2}.$$

a) Using the given numbers,

$$x > \frac{(20.0 \text{ m/s})^2}{5(9.80 \text{ m/s}^2)} = 8.16 \text{ m},$$

$$k < 25\frac{(1700 \text{ kg})(9.80 \text{ m/s}^2)^2}{(20.0 \text{ m/s})^2} = 1.02 \times 10^4 \text{ N/m}.$$

b) A distance of 8 m is not commonly available as space in which to stop a car.

6-70: The students do positive work, and the force that they exert makes an angle of 30.0° with the direction of motion. Gravity does negative work, and is at an angle of 60.0° with the chair's motion, so the total work done is $W_{\text{tot}} = ((600 \text{ N}) \cos 30.0° - (85.0 \text{ kg})(9.80 \text{ m/s}^2) \cos 60.0°)(2.50 \text{ m}) = 257.8 \text{ J}$, and so the speed at the top of the ramp is

$$v_2 = \sqrt{v_1^2 + \frac{2W_{\text{tot}}}{m}} = \sqrt{(2.00 \text{ m/s})^2 + \frac{2(257.8 \text{ J})}{(85.0 \text{ kg})}} = 3.17 \text{ m/s}.$$

Note that extra figures were kept in the intermediate calculation to avoid roundoff error.

6-71: a) At maximum compression, the spring (and hence the block) is not moving, so the block has no kinetic energy. Therefore, the work done *by* the block is equal to its initial kinetic energy, and the maximum compression is found from $\frac{1}{2}kX^2 = \frac{1}{2}mv^2$, or

$$X = \sqrt{\frac{m}{k}}v = \sqrt{\frac{5.00 \text{ kg}}{500 \text{ N/m}}}(6.00 \text{ m/s}) = 0.600 \text{ m}.$$

b) Solving for v in terms of a known X,

$$v = \sqrt{\frac{k}{m}}X = \sqrt{\frac{500 \text{ N/m}}{5.00 \text{ kg}}}(0.150 \text{ m}) = 1.50 \text{ m/s}.$$

6-72: The total work done is the sum of that done by gravity (on the hanging block) and that done by friction (on the block on the table). The work done by gravity is $(6.00 \text{ kg})gh$ and the work done by friction is $-\mu_k(8.00 \text{ kg})gh$, so

$$W_{\text{tot}} = (6.00 \text{ kg} - (0.25)8.00 \text{ kg})(9.80 \text{ m/s}^2)(1.50 \text{ m}) = 58.8 \text{ J}.$$

This work increases the kinetic energy of both blocks;

$$W_{\text{tot}} = \frac{1}{2}(m_1 + m_2)v^2,$$

so

$$v = \sqrt{\frac{2(58.8 \text{ J})}{(14.00 \text{ kg})}} = 2.90 \text{ m/s}.$$

6-73: See Problem 6-72. Gravity does positive work, while friction does negative work. Setting the net (negative) work equal to the (negative) change in kinetic energy,

$$(m_1 - \mu_k m_2)gh = -\frac{1}{2}(m_1 + m_2)v^2,$$

and solving for μ_k gives

$$\mu_k = \frac{m_1 + (1/2)(m_1 + m_2)v^2/gh}{m_2}$$

$$= \frac{(6.00 \text{ kg}) + (1/2)(14.00 \text{ kg})(0.900 \text{ m/s})^2/((9.80 \text{ m/s}^2)(2.00 \text{ m}))}{(8.00 \text{ kg})}$$

$$= 0.79.$$

6-74: The arrow will acquire the energy that was used in drawing the bow (*i.e.*, the work done by the archer), which will be the area under the curve that represents the force as a function of distance. One possible way of estimating this work is to approximate the F *vs.* x curve as a parabola which goes to zero at $x = 0$ and $x = x_0$, and has a maximum of F_0 at $x = \frac{x_0}{2}$, so that $F(x) = \frac{4F_0}{x_0^2}x(x_0 - x)$. This may seem like a crude approximation to the figure, but it has the ultimate advantage of being easy to integrate;

$$\int_0^{x_0} F\,dx = \frac{4F_0}{x_0^2}\int_0^{x_0}(x_0 x - x^2)\,dx = \frac{4F_0}{x_0^2}\left(x_0\frac{x_0^2}{2} - \frac{x_0^3}{3}\right) = \frac{2}{3}F_0 x_0.$$

With $F_0 = 200$ N and $x_0 = 0.75$ m, $W = 100$ J. The speed of the arrow is then $\sqrt{\frac{2W}{m}} = \sqrt{\frac{2(100 \text{ J})}{(0.025 \text{ kg})}} = 89$ m/s. Other ways of finding the area under the curve in Fig. (6-27) should give similar results.

6-75: a) $(800 \text{ kg})(9.80 \text{ m/s}^2)(14.0 \text{ m}) = 1.098 \times 10^5$ J, or 1.10×10^5 J to three figures.

b) $\qquad\qquad (1/2)(800 \text{ kg})(18.0 \text{ m/s})^2 = 1.30 \times 10^5$ J.

c)
$$\frac{1.10 \times 10^5 \text{ J} + 1.30 \times 10^5 \text{ J}}{60 \text{ s}} = 3.99 \text{ kW}.$$

6-76:
$$P = Fv = mav$$
$$= m(2\alpha + 6\beta t)(2\alpha t + 3\beta t^2)$$
$$= m(4\alpha^2 t + 18\alpha\beta t^2 + 18\beta^2 t^3)$$
$$= (0.96 \text{ N/s})t + (0.43 \text{ N/s}^2)t^2 + (0.043 \text{ N/s}^3)t^3.$$

At $t = 4.00$ s, the power output is 13.5 W.

6-77: Let t equal the number of seconds she walks every day. Then, $(280 \text{ J/s})t + (100 \text{ J/s})(86400 \text{ s} - t) = 1.1 \times 10^7$ J. Solving for t, $t = 13,111$ s $= 3.6$ hours.

6-78: a) The hummingbird produces energy at a rate of 0.7 J/s to 1.75 J/s. At 10 beats/s, the bird must expend between 0.07 J/beat and 0.175 J/beat.

b) The steady output of the athlete is 500 W/70 kg = 7 W/kg, which is below the 10 W/kg necessary to stay aloft. Though the athlete can expend 1400 W/70 kg = 20 W/kg for short periods of time, no human-powered aircraft could stay aloft for very long. Movies of early attempts at human-powered flight bear out this observation.

6-79: From the chain rule, $P = \frac{d}{dt}W = \frac{d}{dt}(mgh) = \frac{dm}{dt}gh$, for ideal efficiency. Expressing the mass rate in terms of the volume rate and solving gives

$$\frac{(2000 \times 10^6 \text{ W})}{(0.92)(9.80 \text{ m/s}^2)(170 \text{ m})(1000 \text{ kg/m}^3)} = 1.30 \times 10^3 \frac{\text{m}^3}{\text{s}}.$$

6-80: a) The power P is related to the speed by $Pt = K = \frac{1}{2}mv^2$, so $v = \sqrt{\frac{2Pt}{m}}$.

b)
$$a = \frac{dv}{dt} = \frac{d}{dt}\sqrt{\frac{2Pt}{m}} = \sqrt{\frac{2P}{m}}\frac{d}{dt}\sqrt{t} = \sqrt{\frac{2P}{m}}\frac{1}{2\sqrt{t}} = \sqrt{\frac{P}{2mt}}.$$

c)
$$x - x_0 = \int v\,dt = \sqrt{\frac{2P}{m}}\int t^{\frac{1}{2}}\,dt = \sqrt{\frac{2P}{m}}\frac{2}{3}t^{\frac{3}{2}} = \sqrt{\frac{8P}{9m}}t^{\frac{3}{2}}.$$

6-81: a) $(7500 \times 10^{-3} \text{ kg}^3)(1.05 \times 10^3 \text{ kg/m}^3)(9.80 \text{ m/s}^2)(1.63 \text{ m}) = 1.26 \times 10^5 \text{ J}.$
   b) $(1.26 \times 10^5 \text{ J})/(86,400 \text{ s}) = 1.46 \text{ W}.$

6-82: a) The number of cars is the total power available divided by the power needed per car,

$$\frac{13.4 \times 10^6 \text{ W}}{(2.8 \times 10^3 \text{ N})(27 \text{ m/s})} = 177,$$

rounding down to the nearest integer.

b) To accelerate a total mass M at an acceleration a and speed v, the extra power needed is Mav. To climb a hill of angle α, the extra power needed is $Mg\sin\alpha\,v$. These will be nearly the same if $a \sim g\sin\alpha$; if $g\sin\alpha \sim g\tan\alpha \sim 0.10 \text{ m/s}^2$, the power is about the same as that needed to accelerate at 0.10 m/s^2.

c) $(1.10 \times 10^6 \text{ kg})(9.80 \text{ m/s}^2)(0.010)(27 \text{ m/s}) = 2.9 \text{ MW}.$ d) The power per car needed is that used in part (a), plus that found in part (c) with M being the mass of a single car. The total number of cars is then

$$\frac{13.4 \times 10^6 \text{ W} - 2.9 \times 10^6 \text{ W}}{(2.8 \times 10^3 \text{ N} + (8.2 \times 10^4 \text{ kg})(9.80 \text{ m/s}^2)(0.010))(27 \text{ m/s})} = 36,$$

rounding to the nearest integer.

6-83: a) $P_0 = Fv = (53 \times 10^3 \text{ N})(45 \text{ m/s}) = 2.4 \text{ MW}.$
   b) $P_1 = mav = (9.1 \times 10^5 \text{ kg})(1.5 \text{ m/s}^2)(45 \text{ m/s}) = 61 \text{ MW}.$
   c) Approximating $\sin\alpha$ by $\tan\alpha$, and using the component of gravity down the incline as $mg\sin\alpha$, $P_2 = (mg\sin\alpha)v = (9.1 \times 10^5 \text{ kg})(9.80 \text{ m/s}^2)(0.015)(45 \text{ m/s}) = 6.0 \text{ MW}.$

6-84: a) Along this path, y is constant, and the displacement is parallel to the force, so $W = \alpha y \int x\,dx = (2.50 \text{ N/m}^2)(3.00 \text{ m})\frac{(2.00 \text{ m})^2}{2} = 15.0 \text{ J}.$
   b) Since the force has no y-component, no work is done moving in the y-direction.

c) Along this path, y varies with position along the path, given by $y = 1.5x$, so $F_x = \alpha(1.5x)x = 1.5\alpha x^2$, and

$$W = \int F_x\, dx = 1.5\alpha \int x^2\, dx = 1.5(2.50\ \text{N/m}^2)\frac{(2.00\ \text{m})^3}{3} = 10.0\ \text{J}.$$

6-85: a) $P = Fv = (F_{\text{roll}} + F_{\text{air}})v$

$$= ((0.0045)(62.0\ \text{kg})(9.80\ \text{m/s}^2)$$
$$+ (1/2)(1.00)(0.463\ \text{m}^2)(1.2\ \text{kg/m}^3)(12.0\ \text{m/s})^2)(12.0\ \text{m/s})$$
$$= 513\ \text{W}.$$

b)
$$((0.0030)(59.0\ \text{kg})(9.80\ \text{m/s}^2)$$
$$+ (1/2)(0.88)(0.366\ \text{m}^2)(1.2\ \text{kg/m}^3)(12.0\ \text{m/s})^2)(12.0\ \text{m/s})$$
$$= 355\ \text{W}.$$

c)
$$((0.0030)(59.0\ \text{kg})(9.80\ \text{m/s}^2)$$
$$+ (1/2)(0.88)(0.366\ \text{m}^2)(1.2\ \text{kg/m}^3)(6.0\ \text{m/s})^2)(6.0\ \text{m/s})$$
$$= 52\ \text{W}.$$

6-86: Use the Work–Energy Theorem, $W = \Delta KE$, and integrate to find the work.

$$\Delta KE = 0 - \frac{1}{2}mv_0^2 \quad \text{and} \quad W = \int_0^x (-mg\sin\alpha - \mu mg\cos\alpha)dx.$$

Then,

$$W = -mg\int_0^x (\sin\alpha + Ax\cos\alpha)dx,\quad W = -mg\left[\sin\alpha x + \frac{Ax^2}{2}\cos\alpha\right].$$

Set $W = \Delta KE$.

$$-\frac{1}{2}mv_0^2 = -mg\left[\sin\alpha x + \frac{Ax^2}{2}\cos\alpha\right].$$

To eliminate x, note that the box comes to a rest when the force of static friction balances the component of the weight directed down the plane. So, $mg\sin\alpha = Ax\, mg\cos\alpha$; solve this for x and substitute into the previous equation.

$$x = \frac{\sin\alpha}{A\cos\alpha}.$$

Then,

$$\frac{1}{2}v_0^2 = +g\left[\sin\alpha\frac{\sin\alpha}{A\cos\alpha} + \frac{A\left(\frac{\sin\alpha}{A\cos\alpha}\right)^2}{2}\cos\alpha\right],$$

and upon canceling factors and collecting terms, $v_0^2 = \dfrac{3g \sin^2 \alpha}{A \cos \alpha}$. Or the box will remain stationary whenever $v_0^2 \geq \dfrac{3g \sin^2 \alpha}{A \cos \alpha}$.

6-87: a) Denote the position of a piece of the spring by l; $l = 0$ is the fixed point and $l = L$ is the moving end of the spring. Then the velocity of the point corresponding to l, denoted u, is $u(l) = v\dfrac{l}{L}$ (when the spring is moving, l will be a function of time, and so u is an implicit function of time). The mass of a piece of length dl is $dm = \dfrac{M}{L}\,dl$, and so

$$dK = \frac{1}{2}\,dm\,u^2 = \frac{1}{2}\frac{Mv^2}{L^3}l^2\,dl,$$

and

$$K = \int dK = \frac{Mv^2}{2L^3}\int_0^L l^2\,dl = \frac{Mv^2}{6}.$$

b) $\frac{1}{2}kx^2 = \frac{1}{2}mv^2$, so $v = \sqrt{(k/m)}x = \sqrt{(3200 \text{ N/m})/(0.053 \text{ kg})}(2.50 \times 10^{-2} \text{ m}) = 6.1 \text{ m/s}$.
c) With the mass of the spring included, the work that the spring does goes into the kinetic energies of both the ball and the spring, so $\frac{1}{2}kx^2 = \frac{1}{2}mv^2 + \frac{1}{6}Mv^2$. Solving for v,

$$v = \sqrt{\frac{k}{m + M/3}}x = \sqrt{\frac{(3200 \text{ N/m})}{(0.053 \text{ kg}) + (0.243 \text{ kg})/3}}(2.50 \times 10^{-2} \text{ m}) = 3.9 \text{ m/s}.$$

d) Algebraically,

$$\frac{1}{2}mv^2 = \frac{(1/2)kx^2}{(1 + M/3m)} = 0.40 \text{ J} \qquad \text{and}$$

$$\frac{1}{6}Mv^2 = \frac{(1/2)kx^2}{(1 + 3m/M)} = 0.60 \text{ J}.$$

6-88: In both cases, a given amount of fuel represents a given amount of work W_0 that the engine does in moving the plane forward against the resisting force. In terms of the range R and the (presumed) constant speed v,

$$W_0 = RF = R\left(\alpha v^2 + \frac{\beta}{v^2}\right).$$

In terms of the time of flight T, $R = vT$, so

$$W_0 = vTF = T\left(\alpha v^3 + \frac{\beta}{v}\right).$$

a) Rather than solve for R as a function of v, differentiate the first of these relations with respect to v, setting $\frac{dW_0}{dv} = 0$ to obtain $\frac{dR}{dv}F + R\frac{dF}{dv} = 0$. For the maximum range, $\frac{dR}{dv} = 0$, so $\frac{dF}{dv} = 0$. Performing the differentiation, $\frac{dF}{dv} = 2\alpha v - 2\beta/v^3 = 0$, which is solved for

$$v = \left(\frac{\beta}{\alpha}\right)^{1/4} = \left(\frac{3.5 \times 10^5 \text{ N·m}^2/\text{s}^2}{0.30 \text{ N·s}^2/\text{m}^2}\right)^{1/4} = 32.9 \text{ m/s} = 118 \text{ km/h}.$$

b) Similarly, the maximum time is found by setting $\frac{d}{dv}(Fv) = 0$; performing the differentiation, $3\alpha v^2 - \beta/v^2 = 0$, which is solved for

$$v = \left(\frac{\beta}{3\alpha}\right)^{1/4} = \left(\frac{3.5 \times 10^5 \text{ N·m}^2/\text{s}^2}{3(0.30 \text{ N·s}^2/\text{m}^2)}\right)^{1/4} = 25 \text{ m/s} = 90 \text{ km/h}.$$

6-89: a) The walk will take one-fifth of an hour, 12 min. From the graph, the oxygen consumption rate appears to be about 12 cm^3/kg·min, and so the total energy is

$$(12 \text{ cm}^3/\text{kg·min})(70 \text{ kg})(12 \text{ min})(20 \text{ J/cm}^3) = 2.0 \times 10^5 \text{ J}.$$

b) The run will take 6 min. Using an estimation of the rate from the graph of about 33 cm^3/kg·min gives an energy consumption of about 2.8×10^5 J.   c) The run takes 4 min, and with an estimated rate of about 50 cm^3/kg·min, the energy used is about 2.8×10^5 J.   d) Walking is the most efficient way to go. In general, the point where the slope of the line from the origin to the point on the graph is the smallest is the most efficient speed.

6-90: From $\vec{F} = m\vec{a}$, $F_x = ma_x$, $F_y = ma_y$ and $F_z = ma_z$. The generalization of Eq. (6-11) is then

$$a_x = v_x \frac{dv_x}{dx}, \qquad a_y = v_y \frac{dv_y}{dy}, \qquad a_z = v_z \frac{dv_z}{dz}.$$

The total work is then

$$W_{\text{tot}} = \int_{(x_1, y_1, z_1)}^{(x_2, y_2, z_2)} F_x \, dx + F_y \, dy + F_z \, dz$$

$$= m \left(\int_{x_1}^{x_2} v_x \frac{dv_x}{dx} \, dx + \int_{y_1}^{y_2} v_y \frac{dv_y}{dy} \, dy + \int_{z_1}^{z_2} v_z \frac{dv_z}{dz} \, dz \right)$$

$$= m \left(\int_{v_{x1}}^{v_{x2}} v_x \, dv_x + \int_{v_{y1}}^{v_{y2}} v_y \, dv_y + \int_{v_{z1}}^{v_{z2}} v_z \, dv_z \right)$$

$$= \frac{1}{2} m \left(v_{x2}^2 - v_{x1}^2 + v_{y2}^2 - v_{y1}^2 + v_{z2}^2 - v_{z1}^2 \right)$$

$$= \frac{1}{2} m v_2^2 - \frac{1}{2} m v_1^2.$$

Chapter 7 Potential Energy and Energy Conservation

7-1: From Eq. (7-2),

$$mgy = (800 \text{ kg})(9.80 \text{ m/s}^2)(440 \text{ m}) = 3.45 \times 10^6 \text{ J} = 3.45 \text{ MJ}.$$

7-2: a) For constant speed, the net force is zero, so the required force is the sack's weight, $(5.00 \text{ kg})(9.80 \text{ m/s}^2) = 49 \text{ N}$. b) The lifting force acts in the same direction as the sack's motion, so the work is equal to the weight times the distance, $(49.00 \text{ N})(15.0 \text{ m}) = 735 \text{ J}$; this work becomes potential energy. Note that the result is independent of the speed, and that an extra figure was kept in part (b) to avoid roundoff error.

7-3: In Eq. (7-7), taking $K_1 = 0$ (as in Example 6-5) and $U_2 = 0$, $K_2 = U_1 + W_{\text{other}}$. Friction does negative work $-fy$, so $K_2 = mgy - fy$; solving for the speed v_2,

$$v_2 = \sqrt{\frac{2(mg - f)y}{m}} = \sqrt{\frac{2((200 \text{ kg})(9.80 \text{ m/s}^2) - 60 \text{ N})(3.00 \text{ m})}{(200 \text{ kg})}} = 7.55 \text{ m/s}.$$

7-4: a) The rope makes an angle of $\arcsin(\frac{3.0 \text{ m}}{6.0 \text{ m}}) = 30°$ with the vertical. The needed horizontal force is then $w \tan\theta = (120 \text{ kg})(9.80 \text{ m/s}^2)\tan 30° = 679 \text{ N}$, or $6.8 \times 10^2 \text{ N}$ to two figures. b) In moving the bag, the rope does no work, so the worker does an amount of work equal to the change in potential energy, $(120 \text{ kg})(9.80 \text{ m/s}^2)(6.0 \text{ m})(1 - \cos 30°) = 0.95 \times 10^3 \text{ J}$. Note that this is not the product of the result of part (a) and the horizontal displacement; the force needed to keep the bag in equilibrium varies as the angle is changed.

7-5: a) In the absence of air resistance, Eq. (7-5) is applicable. With $y_1 - y_2 = 22.0 \text{ m}$, solving for v_2 gives

$$v_2 = \sqrt{v_1^2 + 2g(y_2 - y_1)} = \sqrt{(12.0 \text{ m/s})^2 + 2(9.80 \text{ m/s}^2)(22.0 \text{ m})} = 24.0 \text{ m/s}.$$

b) The result of part (a), and any application of Eq. (7-5), depends only on the magnitude of the velocities, not the directions, so the speed is again 24.0 m/s. c) The ball thrown upward would be in the air for a longer time and would be slowed more by air resistance.

7-6: a) (Denote the top of the ramp as point 2.) In Eq. (7-7), $K_2 = 0$, $W_{\text{other}} = -(35 \text{ N}) \times (2.5 \text{ m}) = -87.5 \text{ J}$, and taking $U_1 = 0$ and $U_2 = mgy_2 = (12 \text{ kg})(9.80 \text{ m/s}^2)(2.5 \text{ m} \sin 30°) = 147 \text{ J}$, $v_1 = \sqrt{\frac{2(147 \text{ J} + 87.5 \text{ J})}{12 \text{ kg}}} = 6.25 \text{ m/s}$, or 6.3 m/s to two figures. Or, the work done by friction and the change in potential energy are both proportional to the distance the crate moves up the ramp, and so the initial speed is proportional to the square root of the distance up the ramp; $(5.0 \text{ m/s})\sqrt{\frac{2.5 \text{ m}}{1.6 \text{ m}}} = 6.25 \text{ m/s}$.

b) In part a), we calculated W_{other} and U_2. Using Eq. (7-7), $K_2 = \frac{1}{2}(12\,\text{kg})(11.0\,\text{m/s})^2 - 87.5\,\text{J} - 147\,\text{J} = 491.5\,\text{J}$

$$v_2 = \sqrt{\frac{2K_2}{m}} = \sqrt{\frac{2(491.5\,\text{J})}{(12\,\text{kg})}} = 9.05\,\text{m/s}.$$

7-7: As in Example 7-7, $K_2 = 0$, $U_2 = 94\,\text{J}$, and $U_3 = 0$. The work done by friction is $-(35\,\text{N})(1.6\,\text{m}) = -56\,\text{J}$, and so $K_3 = 38\,\text{J}$, and $v_3 = \sqrt{\frac{2(38\,\text{J})}{12\,\text{kg}}} = 2.5\,\text{m/s}.$

7-8: The speed is v and the kinetic energy is $4K$. The work done by friction is proportional to the normal force, and hence the mass, and so each term in Eq. (7-7) is proportional to the total mass of the crate, and the speed at the bottom is the same for any mass. The kinetic energy is proportional to the mass, and for the same speed but four times the mass, the kinetic energy is quadrupled.

7-9: In Eq. (7-7), $K_1 = 0$, W_{other} is given as $-0.22\,\text{J}$, and taking $U_2 = 0$, $K_2 = mgR - 0.22\,\text{J}$, so

$$v_2 = \sqrt{2\left((9.80\,\text{m/s}^2)(0.50\,\text{m}) - \frac{0.22\,\text{J}}{0.20\,\text{kg}}\right)} = 2.8\,\text{m/s}.$$

7-10: Tarzan is lower than his original height by a distance $l(\cos 30 - \cos 45)$, so his speed is

$$v = \sqrt{2gl(\cos 30° - \cos 45°)} = 7.9\,\text{m/s},$$

a bit quick for conversation.

7-11: a) The force is applied parallel to the ramp, and hence parallel to the oven's motion, and so $W = Fs = (110\,\text{N})(8.0\,\text{m}) = 880\,\text{J}$. b) Because the applied force $\vec{F}$ is parallel to the ramp, the normal force is just that needed to balance the component of the weight perpendicular to the ramp, $N = w\cos\alpha$, and so the friction force is $f_k = \mu_k mg\cos\alpha$ and the work done by friction is

$$W_f = -\mu_k mg\cos\alpha\, s = -(0.25)(10.0\,\text{kg})(9.80\,\text{m/s}^2)\cos 37°(8.0\,\text{m}) = -157\,\text{J},$$

keeping an extra figure. c) $mgs\sin\alpha = (10.0\,\text{kg})(9.80\,\text{m/s}^2)(8.0\,\text{m})\sin 37° = 472\,\text{J}$, again keeping an extra figure. d) $880\,\text{J} - 472\,\text{J} - 157\,\text{J} = 251\,\text{J}$. e) In the direction up the ramp, the net force is

$$F - mg\sin\alpha - \mu_k mg\cos\alpha$$
$$= 110\,\text{N} - (10.0\,\text{kg})(9.80\,\text{m/s}^2)(\sin 37° + (0.25)\cos 37°)$$
$$= 31.46\,\text{N},$$

so the acceleration is $(31.46\,\text{N})/(10.0\,\text{kg}) = 3.15\,\text{m/s}^2$. The speed after moving up the ramp is $v = \sqrt{2as} = \sqrt{2(3.15\,\text{m/s}^2)(8.0\,\text{m})} = 7.09\,\text{m/s}$, and the kinetic energy is $(1/2)mv^2 = 252\,\text{J}$. (In the above, numerical results of specific parts may differ in the third place if extra figures are not kept in the intermediate calculations.)

7-12: a) At the top of the swing, when the kinetic energy is zero, the potential energy (with respect to the bottom of the circlular arc) is $mgl(1 - \cos\theta)$, where l is the length of the string and θ is the angle the string makes with the vertical. At the bottom of the swing, this potential energy has become kinetic energy, so $mgl(1 - \cos\theta) = \frac{1}{2}mv^2$, or $v = \sqrt{2gl(1 - \cos\theta)} = \sqrt{2(9.80 \text{ m/s}^2)(0.80 \text{ m})(1 - \cos 45°)} = 2.1 \text{ m/s}$. b) At $45°$ from the vertical, the speed is zero, and there is no radial acceleration; the tension is equal to the radial component of the weight, or $mg\cos\theta = (0.12 \text{ kg})(9.80 \text{ m/s}^2)\cos 45° = 0.83$ N. c) At the bottom of the circle, the tension is the sum of the weight and the radial acceleration,

$$mg + mv_2^2/l = mg(1 + 2(1 - \cos 45°)) = 1.86 \text{ N},$$

or 1.9 N to two figures. Note that this method does not use the intermediate calculation of v.

7-13: Of the many ways to find energy in a spring in terms of the force and the distance, one way (which avoids the intermediate calculation of the spring constant) is to note that the energy is the product of the average force and the distance compressed or extended. a) $(1/2)(800 \text{ N})(0.200 \text{ m}) = 80.0$ J. b) The potential energy is proportional to the square of the compression or extension; $(80.0 \text{ J})(0.050 \text{ m}/0.200 \text{ m})^2 = 5.0$ J.

7-14: $U = \frac{1}{2}ky^2$, where y is the vertical distance the spring is stretched when the weight $w = mg$ is suspended. $y = \frac{mg}{k}$, and $k = \frac{F}{x}$, where x and F are the quantities that "calibrate" the spring. Combining,

$$U = \frac{1}{2}\frac{(mg)^2}{F/x} = \frac{1}{2}\frac{((60.0 \text{ kg})(9.80 \text{ m/s}^2))^2}{(720 \text{ N}/0.150 \text{ m})} = 36.0 \text{ J}.$$

7-15: a) Solving Eq. (7-9) for x, $x = \sqrt{\frac{2U}{k}} = \sqrt{\frac{2(3.20 \text{ J})}{(1600 \text{ N/m})}} = 0.063$ m.

b) Denote the initial height of the book as h and the maximum compression of the spring by x. The final and initial kinetic energies are zero, and the book is initially a height $x + h$ above the point where the spring is maximally compressed. Equating initial and final potential energies, $\frac{1}{2}kx^2 = mg(x + h)$. This is a quadratic in x, the solution to which is

$$x = \frac{mg}{k}\left[1 \pm \sqrt{1 + \frac{2kh}{mg}}\right]$$

$$= \frac{(1.20 \text{ kg})(9.80 \text{ m/s}^2)}{(1600 \text{ N/m})}\left[1 \pm \sqrt{1 + \frac{2(1600 \text{ N/m})(0.80 \text{ m})}{(1.20 \text{ kg})(9.80 \text{ m/s}^2)}}\right]$$

$$= 0.116 \text{ m}, \quad -0.101 \text{ m}.$$

The second (negative) root is not unphysical, but represents an extension rather than a compression of the spring. To two figures, the compression is 0.12 m.

7-16: a) In going from rest in the slingshot's pocket to rest at the maximum height, the potential energy stored in the rubber band is converted to gravitational potential energy; $U = mgy = (10 \times 10^{-3} \text{ kg})(9.80 \text{ m/s}^2)(22.0 \text{ m}) = 2.16$ J.

b) Because gravitational potential energy is proportional to mass, the larger pebble rises only 8.8 m.

c) The lack of air resistance and no deformation of the rubber band are two possible assumptions.

7-17: The initial kinetic energy and the kinetic energy of the brick at its greatest height are both zero. Equating initial and final potential energies, $\frac{1}{2}kx^2 = mgh$, where h is the greatest height. Solving for h,

$$h = \frac{kx^2}{2mg} = \frac{(1800 \text{ N/m})(0.15 \text{ m})^2}{2(1.20 \text{ kg})(9.80 \text{ m/s}^2)} = 1.7 \text{ m}.$$

7-18: As in Example 7-8, $K_1 = 0$ and $U_1 = 0.0250$ J. For $v_2 = 0.20$ m/s, $K_2 = 0.0040$ J, so $U_2 = 0.0210$ J $= \frac{1}{2}kx^2$, so $x = \pm\sqrt{\frac{2(0.0210 \text{ J})}{5.00 \text{ N/m}}} = \pm 0.092$ m. In the absence of friction, the glider will go through the equilibrium position and pass through $x = -0.092$ m with the same speed, on the opposite side of the equilibrium position.

7-19: a) In this situation, $U_2 = 0$ when $x = 0$, so $K_2 = 0.0250$ J and $v_2 = \sqrt{\frac{2(0.0250 \text{ J})}{0.200 \text{ kg}}} =$ 0.500 m/s. b) If $v_2 = 2.50$ m/s, $K_2 = (1/2)(0.200 \text{ kg})(2.50 \text{ m/s})^2 = 0.625$ J $= U_1$, so $x_1 = \sqrt{\frac{2(0.625 \text{ J})}{5.00 \text{ N/m}}} = 0.500$ m. Or, because the speed is 5 times that of part (a), the kinetic energy is 25 times that of part (a), and the initial extension is 5×0.100 m $= 0.500$ m.

7-20: a) The work done by friction is

$$W_{\text{other}} = -\mu_k mg\Delta x = -(0.05)(0.200 \text{ kg})(9.80 \text{ m/s}^2)(0.020 \text{ m}) = -0.00196 \text{ J},$$

so $K_2 = 0.00704$ J and $v_2 = \sqrt{\frac{2(0.00704 \text{ J})}{0.200 \text{ kg}}} = 0.27$ m/s. b) In this case $W_{\text{other}} = -0.0098$ J, so $K_2 = 0.0250$ J $- 0.0098$ J $= 0.0152$ J, and $v_2 = \sqrt{\frac{2(0.0152 \text{ J})}{0.200 \text{ kg}}} = 0.39$ m/s.

c) In this case, $K_2 = 0$, $U_2 = 0$, so $U_1 + W_{\text{other}} = 0 = 0.0250$ J $- \mu_k(0.200$ kg$)$ $(9.80 \text{ m/s}^2) \times (0.100 \text{ m})$, or $\mu_k = 0.13$.

7-21: a) In this case, $K_1 = 625,000$ J as before, $W_{\text{other}} = -17,000$ J and

$$\begin{aligned}
U_2 &= (1/2)ky_2^2 + mgy_2 \\
&= (1/2)(1.41 \times 10^5 \text{ N/m})(-1.00 \text{ m})^2 + (2000 \text{ kg})(9.80 \text{ m/s}^2)(-1.00 \text{ m}) \\
&= 50,900 \text{ J}.
\end{aligned}$$

The kinetic energy is then $K_2 = 625,000$ J $-50,900$ J $-17,000$ J $= 557,100$ J, corresponding to a speed $v_2 = 23.6$ m/s. b) The elevator is moving down, so the friction force is up (tending to stop the elevator, which is the idea). The net upward force is then $-mg + f - kx = -(2000 \text{ kg})(9.80 \text{ m/s}^2) + 17,000 \text{ N} - (1.41 \times 10^5 \text{ N/m})(-1.00 \text{ m}) = 138,400 \text{ N}$, for an upward acceleration of 69.2 m/s^2.

7-22: From $\frac{1}{2}kx^2 = \frac{1}{2}mv^2$, the relations between m, v, k and x are

$$kx^2 = mv^2, \quad kx = 5mg.$$

Dividing the first by the second gives $x = \frac{v^2}{5g}$, and substituting this into the second gives $k = 25\frac{mg^2}{v^2}$, so a) & b),

$$x = \frac{(2.50 \text{ m/s})^2}{5(9.80 \text{ m/s}^2)} = 0.128 \text{ m},$$

$$k = 25\frac{(1160 \text{ kg})(9.80 \text{ m/s}^2)^2}{(2.50 \text{ m/s})^2} = 4.46 \times 10^5 \text{ N/m}.$$

7-23: a) Gravity does negative work, $-(0.75 \text{ kg})(9.80 \text{ m/s}^2)(16 \text{ m}) = -118 \text{ J}$. b) Gravity does 118 J of positive work. c) Zero d) Conservative; gravity does no net work on any complete round trip.

7-24: a) & b) $-(0.050 \text{ kg})(9.80 \text{ m/s}^2)(5.0 \text{ m}) = -2.5 \text{ J}$.

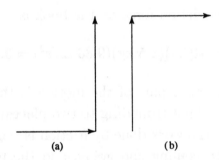

(a)   (b)

c) Gravity is conservative, as the work done to go from one point to another is path-independent.

7-25: a) The displacement is in the y-direction, and since $\overrightarrow{F}$ has no y-component, the work is zero.

b) $$\int_{P_1}^{P_2} \overrightarrow{F} \cdot d\overrightarrow{l} = -12 \int_{x_1}^{x_2} x^2 \, dx = -\frac{12}{3}\left(x_2^3 - x_1^3\right) = -0.104 \text{ m}^3.$$

c) The negative of the answer to part (b), 0.104 m^3 d) The work is independent of path, and the force is conservative. The corresponding potential energy is $U = \frac{12x^3}{3} = 4x^3$.

7-26: a) From $(0, 0)$ to $(0, L)$, $x = 0$ and so $\overrightarrow{F} = 0$, and the work is zero. From $(0, L)$ to (L, L), $\overrightarrow{F}$ and $d\overrightarrow{l}$ are perpendicular, so $\overrightarrow{F} \cdot d\overrightarrow{l} = 0$, and the net work along this path is zero. b) From $(0, 0)$ to $(L, 0)$, $\overrightarrow{F} \cdot d\overrightarrow{l} = 0$. From $(L, 0)$ to (L, L), the work is that found in the example, $W_2 = CL^2$, so the total work along the path is CL^2. c) Along the diagonal path, $x = y$, and so $\overrightarrow{F} \cdot d\overrightarrow{l} = Cy \, dy$; integrating from 0 to L gives $\frac{CL^2}{2}$. (It is not a coincidence that this is the average to the answers to parts (a) and (b).) d) The work depends on path, and the field is not conservative.

7-27: a) When the book moves to the left, the friction force is to the right, and the work is $-(1.2 \text{ N})(3.0 \text{ m}) = -3.6 \text{ J}$. b) The friction force is now to the left, and the work is again -3.6 J.   c) -7.2 J.   d) The net work done by friction for the round trip is not zero, and friction is not a conservative force.

7-28:   The friction force has magnitude $\mu_k mg = (0.20)(30.0 \text{ kg})(9.80 \text{ m/s}^2) = 58.8$ N. a) For each part of the move, friction does $-(58.8 \text{ N})(10.6 \text{ m}) = -623$ J, so the total work done by friction is -1.2 kN.   b) $-(58.8 \text{ N})(15.0 \text{ m}) = -882$ N.

   c)

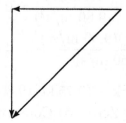

The net work done by friction depends on the path, so friction is not a conservative force.

7-29:   The magnitude of the friction force on the book is

$$\mu_k mg = (0.25)(1.5 \text{ kg})(9.80 \text{ m/s}^2) = 3.68 \text{ N}.$$

a) The work done during each part of the motion is the same, and the total work done is $-2(3.68 \text{ N})(8.0 \text{ m}) = -59$ J (rounding to two places).   b) The magnitude of the displacement is $\sqrt{2}(8.0 \text{ m})$, so the work done by friction is $-\sqrt{2}(8.0 \text{ m})(3.68 \text{ N}) = -42$ N. c) The work is the same both coming and going, and the total work done is the same as in part (a), -59 J.   d) The work required to go from one point to another is not path independent, and the work required for a round trip is not zero, so friction is not a conservative force.

7-30:   a) $\frac{1}{2}k(x_1^2 - x_2^2)$ b) $-\frac{1}{2}k(x_1^2 - x_2^2)$. The total work is zero; the spring force is conservative   c) From x_1 to x_3, $W = -\frac{1}{2}k(x_3^2 - x_1^2)$. From x_3 to x_2, $W = -\frac{1}{2}k(x_2^2 - x_3^2)$. The net work is $-\frac{1}{2}k(x_2^2 - x_1^2)$. This is the same as the result of part (a).

7-31:   From Eq. (7-17), the force is

$$F_x = -\frac{dU}{dx} = C_6 \frac{d}{dx}\left(\frac{1}{x^6}\right) = -\frac{6C_6}{x^7}.$$

The minus sign means that the force is attractive.

7-32:   From Eq. (7-15), $F_x = -\frac{dU}{dx} = -4\alpha x^3 = -(4.8 \text{ J/m}^4)x^3$, and so

$$F_x(-0.800 \text{ m}) = -(4.8 \text{ J/m}^4)(-0.80 \text{ m})^3 = 2.46 \text{ N}.$$

7-33: $\frac{\partial U}{\partial x} = 2kx + k'y$, $\frac{\partial U}{\partial y} = 2ky + k'x$ and $\frac{\partial U}{\partial z} = 0$, so from Eq. (7-19),

$$\vec{F} = -(2kx + k'y)\,\hat{i} - (2ky + k'x)\,\hat{j}.$$

7-34: From Eq. (7-19), $\vec{F} = -\frac{\partial U}{\partial x}\hat{i} - \frac{\partial U}{\partial y}\hat{j}$, since U has no z-dependence. $\frac{\partial U}{\partial x} = \frac{-2\alpha}{x^3}$ and $\frac{\partial U}{\partial y} = \frac{-2\alpha}{y^3}$, so

$$\vec{F} = -\alpha\left(\frac{-2}{x^3}\hat{i} + \frac{-2}{y^3}\hat{j}\right).$$

7-35: a) $F_r = -\frac{\partial U}{\partial r} = 12\frac{a}{r^{13}} - 6\frac{b}{r^7}.$

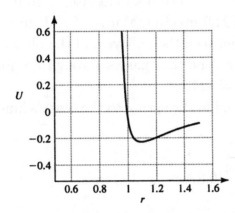
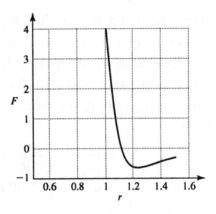

b) Setting $F_r = 0$ and solving for r gives $r_{\min} = (2a/b)^{1/6}$. This is the minimum of potential energy, so the equilibrium is stable.

c)
$$U(r_{\min}) = \frac{a}{r_{\min}^{12}} - \frac{b}{r_{\min}^6}$$

$$= \frac{a}{((2a/b)^{1/6})^{12}} - \frac{b}{((2a/b)^{1/6})^6}$$

$$= \frac{ab^2}{4a^2} - \frac{b^2}{2a} = -\frac{b^2}{4a}.$$

To separate the particles means to remove them to zero potential energy, and requires the negative of this, or $E_0 = b^2/4a$. d) The expressions for E_0 and $r_{\min}$ in terms of a and b are

$$E_0 = \frac{b^2}{4a} \qquad r_{\min}^6 = \frac{2a}{b}.$$

Mulitplying the first by the second and solving for b gives $b = 2E_0 r_{\min}^6$, and substituting this into the first and solving for a gives $a = E_0 r_{\min}^{12}$. Using the given numbers,

$$a = (1.54 \times 10^{-18}\ \text{J})(1.13 \times 10^{-10}\ \text{m})^{12} = 6.68 \times 10^{-138}\ \text{J·m}^{12}$$
$$b = 2(1.54 \times 10^{-18}\ \text{J})(1.13 \times 10^{-10}\ \text{m})^6 = 6.41 \times 10^{-78}\ \text{J·m}^6.$$

(Note: the numerical value for a might not be within the range of standard calculators, and the powers of ten may have to be handled seperately.)

7-36: a) Considering only forces in the x-direction, $F_x = -\frac{dU}{dx}$, and so the force is zero when the slope of the U vs x graph is zero, at points b and d. b) Point b is at a potential minimum; to move it away from b would require an input of energy, so this point is stable. c) Moving away from point d involves a decrease of potential energy, hence an increase in kinetic energy, and the marble tends to move further away, and so d is an unstable point.

7-37: a) At constant speed, the upward force of the three ropes must balance the force, so the tension in each is one-third of the man's weight. The tension in the rope is the force he exerts, or $(70.0 \text{ kg})(9.80 \text{ m/s}^2)/3 = 229$ N. b) The man has risen 1.20 m, and so the increase in his potential energy is $(70.0 \text{ kg})(9.80 \text{ m/s}^2)(1.20 \text{ m}) = 823$ J. In moving up a given distance, the total length of the rope between the pulleys and the platform changes by three times this distance, so the length of rope that passes through the man's hands is $3 \times 1.20 \text{ m} = 3.60 \text{ m}$, and $(229 \text{ N})(3.6 \text{ m}) = 824$ J.

7-38: a) Equating the potential energy stored in the spring to the block's kinetic energy, $\frac{1}{2}kx^2 = \frac{1}{2}mv^2$, or

$$v = \sqrt{\frac{k}{m}}x = \sqrt{\frac{400 \text{ N/m}}{2.00 \text{ kg}}}(0.220 \text{ m}) = 3.11 \text{ m/s}.$$

b) Using energy methods directly, the initial potential energy of the spring is the final gravitational potential energy, $\frac{1}{2}kx^2 = mgL \sin\theta$, or

$$L = \frac{\frac{1}{2}kx^2}{mg \sin\theta} = \frac{\frac{1}{2}(400 \text{ N/m})(0.220 \text{ m})^2}{(2.00 \text{ kg})(9.80 \text{ m/s}^2)\sin 37.0°} = 0.821 \text{ m}.$$

7-39: The initial and final kinetic energies are both zero, so the work done by the spring is the negative of the work done by friction, or $\frac{1}{2}kx^2 = \mu_k mgl$, where l is the distance the block moves. Solving for μ_k,

$$\mu_k = \frac{(1/2)kx^2}{mgl} = \frac{(1/2)(100 \text{ N/m})(0.20 \text{ m})^2}{(0.50 \text{ kg})(9.80 \text{ m/s}^2)(1.00 \text{ m})} = 0.41.$$

7-40: a) $U_A - U_B = mg(h - 2R) = \frac{1}{2}mv_A^2$. From previous considerations, the speed at the top must be at least $\sqrt{gR}$. Thus,

$$mg(h - 2R) > \frac{1}{2}mgR, \quad \text{or} \quad h > \frac{5}{2}R.$$

b) $U_A - U_C = (2.50)Rmg = K_C$, so

$$v_C = \sqrt{(5.00)gR} = \sqrt{(5.00)(9.80 \text{ m/s}^2)(20.0 \text{ m})} = 31.3 \text{ m/s}.$$

The radial acceleration is $a_{\text{rad}} = \frac{v_C^2}{R} = 49.0 \text{ m/s}^2$. The tangential direction is down, the normal force at point C is horizontal, there is no friction, so the only downward force is gravity, and $a_{\text{tan}} = g = 9.80 \text{ m/s}^2$.

7-41: The net work done during the trip down the barrel is the sum of the energy stored in the spring, the (negative) work done by friction and the (negative) work done by gravity. Using $\frac{1}{2}kx^2 = \frac{1}{2}(F^2/k)$, the performer's kinetic energy at the top of the barrel is

$$K = \frac{1}{2}\frac{(4400 \text{ N})^2}{1100 \text{ N/m}} - (40 \text{ N})(4.0 \text{ m}) - (60 \text{ kg})(9.80 \text{ m/s}^2)(2.5 \text{ m}) = 7.17 \times 10^3 \text{ J},$$

and his speed is $\sqrt{\frac{2(7.17 \times 10^3 \text{ J})}{60 \text{ kg}}} = 15.5$ m/s.

7-42: To be at equilibrium at the bottom, with the spring compressed a distance x_0, the spring force must balance the component of the weight down the ramp plus the largest value of the static friction, or $kx_0 > w\sin\theta + f$. The work-energy theorem requires that the energy stored in the spring is equal to the sum of the work done by friction, the work done by gravity and the initial kinetic energy, or

$$\frac{1}{2}kx_0^2 = (w\sin\theta - f)L + \frac{1}{2}mv^2,$$

where L is the total length traveled down the ramp and v is the speed at the top of the ramp. With the given parameters, $\frac{1}{2}kx_0^2 = 248$ J and $kx_0 = 1.10 \times 10^3$ N. Squaring the second of these expressions and dividing by the first and solving for k gives $k = 2440$ N/m.

7-43: The potential energy has decreased by $(12.0 \text{ kg})(9.80 \text{ m/s}^2)(2.00 \text{ m}) - (4.0 \text{ kg}) \times (9.80 \text{ m/s}^2)(2.00 \text{ m}) = 156.8$ J. The kinetic energy of the masses is then $\frac{1}{2}(m_1 + m_2)v^2 = (8.0 \text{ kg})v^2 = 156.8$ J, so the common speed is $v = \sqrt{\frac{(156.8 \text{ J})}{8.0 \text{ kg}}} = 4.43$ m/s, or 4.4 m/s to two figures.

7-44: a) The energy stored may be found directly from

$$\frac{1}{2}ky_2^2 = K_1 + W_{\text{other}} - mgy_2 = 625{,}000 \text{ J} - 51{,}000 \text{ J} - (-58{,}800 \text{ J}) = 6.33 \times 10^5 \text{ J}.$$

b) Denote the upward distance from point 2 by h. The kinetic energy at point 2 and at the height h are both zero, so the energy found in part (a) is equal to the negative of the work done by gravity and friction, $-(mg + f)h = -((2000 \text{ kg})(9.80 \text{ m/s}^2) + 17{,}000 \text{ N})h = (36{,}600 \text{ N})h$, so $h = \frac{6.33 \times 10^5 \text{ J}}{3.66 \times 10^4 \text{ J}} = 17.3$ m. c) The net work done on the elevator between the highest point of the rebound and the point where it next reaches the spring is $(mg - f)(h - 3.00 \text{ m}) = 3.72 \times 10^4$ J. Note that on the way down, friction does negative work. The speed of the elevator is then $\sqrt{\frac{2(3.72 \times 10^4 \text{ J})}{2000 \text{ kg}}} = 6.10$ m/s. d) When the elevator next comes to rest, the total work done by the spring, friction, and gravity must be the negative of the kinetic energy K_3 found in part (c), or

$$K_3 = 3.72 \times 10^4 \text{ J} = -(mg - f)x_3 + \frac{1}{2}kx_3^2 = -(2{,}600 \text{ N})x_3 + (7.03 \times 10^4 \text{ N/m})x_3^2.$$

(In this calculation, the value of k was recalculated to obtain better precision.) This is a quadratic in x_3, the positive solution to which is

$$x_3 = \frac{1}{2(7.03 \times 10^4 \text{ N/m})}$$
$$\times \left[2.60 \times 10^3 \text{ N} + \sqrt{(2.60 \times 10^3 \text{ N})^2 + 4(7.03 \times 10^4 \text{ N/m})(3.72 \times 10^4 \text{ J})} \right]$$
$$= 0.746 \text{ m},$$

corresponding to a force of 1.05×10^5 N and a stored energy of 3.91×10^4 J. It should be noted that different ways of rounding the numbers in the intermediate calculations may give different answers.

7-45: The two design conditions are expressed algebraically as $ky = f + mg = 3.66 \times 10^4$ N (the condition that the elevator remains at rest when the spring is compressed a distance y; y will be taken as positive) and $\frac{1}{2}mv^2 + mgy - fy = \frac{1}{2}kx^2$ (the condition that the change in energy is the work $W_{\text{other}} = -fy$). Eliminating y in favor of k by $y = \frac{3.66 \times 10^4 \text{ N}}{k}$ leads to

$$\frac{1}{2} \frac{(3.66 \times 10^4 \text{ N})^2}{k} + \frac{(1.70 \times 10^4 \text{ N})(3.66 \times 10^4 \text{ N})}{k}$$
$$= 62.5 \times 10^4 \text{ J} + \frac{(1.96 \times 10^4 \text{ N})(3.66 \times 10^4 \text{ N})}{k}.$$

This is actually not hard to solve for $k = 919$ N/m, and the corresponding x is 39.8 m. This is a very weak spring constant, and would require a space below the operating range of the elevator about four floors deep, which is not reasonable.   **b)** At the lowest point, the spring exerts an upward force of magnitude $f + mg$. Just before the elevator stops, however, the friction force is also directed upward, so the net force is $(f + mg) + f - mg = 2f$, and the upward acceleration is $\frac{2f}{m} = 17.0$ m/s^2.

7-46: One mass rises while the other falls, so the net loss of potential energy is

$$(0.5000 \text{ kg} - 0.2000 \text{ kg})(9.80 \text{ m/s}^2)(0.400 \text{ m}) = 1.176 \text{ J}.$$

This is the sum of the kinetic energies of the animals. If the animals are equidistant from the center, they have the same speed, so the kinetic energy of the combination is $\frac{1}{2}m_{\text{tot}}v^2$, and

$$v = \sqrt{\frac{2(1.176 \text{ J})}{(0.7000 \text{ kg})}} = 1.83 \text{ m/s}.$$

7-47: **a)** The kinetic energy of the potato is the work done by gravity (or the potential energy lost), $\frac{1}{2}mv^2 = mgl$, or $v = \sqrt{2gl} = \sqrt{2(9.80 \text{ m/s}^2)(2.50 \text{ m})} = 7.00$ m/s.

   b)
$$T - mg = m\frac{v^2}{l} = 2mg,$$

so $T = 3mg = 3(0.100 \text{ kg})(9.80 \text{ m/s}^2) = 2.94$ N.

7-48: a) The change in total energy is the work done by the air,

$$(K_2 + U_2) - (K_1 + U_1) = m \left(\frac{1}{2} \left(v_2^2 - v_1^2 \right) + g y_2 \right)$$

$$= (0.145 \text{ kg}) \left(\begin{array}{c} (1/2)((18.6 \text{ m/s})^2 - (30.0 \text{ m/s})^2 \\ -(40.0 \text{ m/s})^2) + (9.80 \text{ m/s}^2)(53.6 \text{ m}) \end{array} \right)$$

$$= -80.0 \text{ J}.$$

b) Similarly,

$$(K_3 + U_3) - (K_2 + U_2) = (0.145 \text{ kg}) \left(\begin{array}{c} (1/2)((11.9 \text{ m/s})^2 + (-28.7 \text{ m/s})^2 \\ -(18.6 \text{ m/s})^2) - (9.80 \text{ m/s}^2)(53.6 \text{ m}) \end{array} \right)$$

$$= -31.3 \text{ J}.$$

c) The ball is moving slower on the way down, and does not go as far (in the x-direction), and so the work done by the air is smaller in magnitude.

7-49: a) For a friction force f, the total work done sliding down the pole is $mgd - fd$. This is given as being equal to mgh, and solving for f gives

$$f = mg \frac{(d-h)}{d} = mg \left(1 - \frac{h}{d} \right).$$

When $h = d$, $f = 0$, as expected, and when $h = 0$, $f = mg$; there is no net force on the fireman. b) $(75 \text{ kg})(9.80 \text{ m/s}^2)(1 - \frac{1.0 \text{ m}}{2.5 \text{ m}}) = 441$ N. c) The net work done is $(mg - f)(d - y)$, and this must be equal to $\frac{1}{2}mv^2$. Using the above expression for f,

$$\frac{1}{2}mv^2 = (mg - f)(d - y)$$

$$= mg \left(\frac{h}{d} \right)(d - y)$$

$$= mgh \left(1 - \frac{y}{d} \right),$$

from which $v = \sqrt{2gh(1 - y/d)}$. When $y = 0$, $v = \sqrt{2gh}$, which is the original condition. When $y = d$, $v = 0$; the fireman is at the top of the pole.

7-50: a) The skier's kinetic energy at the bottom can be found from the potential energy at the top minus the work done by friction, $K_1 = mgh - W_F = (60.0 \text{ kg})(9.8 \text{ N/kg})(65.0 \text{ m}) -$ 10,500 J, or $K_1 = 38,200 \text{ J} - 10,500 \text{ J} = 27,720 \text{ J}$. Then $v_1 = \sqrt{\frac{2K}{m}} = \sqrt{\frac{2(27,720 \text{ J})}{60 \text{ kg}}} = 30.4 \text{ m/s}$.

b) $K_2 = K_1 - (W_F + W_A) = 27{,}720\,\text{J} - (\mu_k mgd + f_{air}d)$, $K_2 = 27{,}720\,\text{J} - [(.2)(588\,\text{N}) \times (82\,\text{m}) + (160\,\text{N})(82\,\text{m})]$, or $K_2 = 27{,}720\,\text{J} - 22{,}763\,\text{J} = 4957\,\text{J}$. Then,

$$v_2 = \sqrt{\frac{2K}{m}} = \sqrt{\frac{2(4957\,\text{J})}{60\,\text{kg}}} = 12.85\,\text{m/s} \approx 12.9\,\text{m/s}.$$

c) Use the Work-Energy Theorem to find the force. $W = \Delta KE$, $F = KE/d = (4957\,\text{J})/(2.5\,\text{m}) = 1983\,\text{N} \approx 2000\,\text{N}$.

7-51: The skier is subject to both gravity and a normal force; it is the normal force that causes her to go in a circle, and when she leaves the hill, the normal force vanishes. The vanishing of the normal force is the condition that determines when she will leave the hill. As the normal force approaches zero, the necessary (inward) radial force is the radial component of gravity, or $mv^2/R = mg\cos\alpha$, where R is the radius of the snowball. The speed is found from conservation of energy; at an angle α, she has descended a vertical distance $R(1 - \cos\alpha)$, so $\frac{1}{2}mv^2 = mgR(1 - \cos\alpha)$, or $v^2 = 2gR(1 - \cos\alpha)$. Using this in the previous relation gives $2(1 - \cos\alpha) = \cos\alpha$, or $\alpha = \arccos(\frac{2}{3}) = 48.2°$. This result does not depend on the skier's mass, the radius of the snowball, or g.

7-52: If the speed of the rock at the top is v_t, then conservation of energy gives the speed v_b from $\frac{1}{2}mv_b^2 = \frac{1}{2}mv_t^2 + mg(2R)$, R being the radius of the circle, and so $v_b^2 = v_t^2 + 4gR$. The tension at the top and bottom are found from $T_t + mg = \frac{mv_t^2}{R}$ and $T_b - mg = \frac{mv_b^2}{R}$, so $T_b - T_t = \frac{m}{R}(v_b^2 - v_t^2) + 2mg = 6mg = 6w$.

7-53: a) The magnitude of the work done by friction is the kinetic energy of the package at point B, or $\mu_k mgL = \frac{1}{2}mv_B^2$, or

$$\mu_k = \frac{(1/2)v_B^2}{gL} = \frac{(1/2)(4.80\,\text{m/s})^2}{(9.80\,\text{m/s}^2)(3.00\,\text{m})} = 0.392.$$

b)
$$\begin{aligned}
W_{other} &= K_B - U_A \\
&= \frac{1}{2}(0.200\,\text{kg})(4.80\,\text{m/s})^2 - (0.200\,\text{kg})(9.80\,\text{m/s}^2)(1.60\,\text{m}) \\
&= -0.832\,\text{J}.
\end{aligned}$$

Equivalently, since $K_A = K_B = 0$, $U_A + W_{AB} + W_{BC} = 0$, or

$$W_{AB} = -U_A - W_{BC} = mg(-(1.60\,\text{m}) - (0.300)(-3.00\,\text{m})) = -0.832\,\text{J}.$$

7-54: Denote the distance the truck moves up the ramp by x. $K_1 = \frac{1}{2}mv_0^2$, $U_1 = mgL\sin\alpha$, $K_2 = 0$, $U_2 = mgx\sin\beta$ and $W_{other} = -\mu_r mgx\cos\beta$. From $W_{other} = (K_2 + U_2) - (K_1 + U_1)$, and solving for x,

$$x = \frac{K_1 + mgL\sin\alpha}{mg(\sin\beta + \mu_r\cos\beta)} = \frac{(v_0^2/2g) + L\sin\alpha}{\sin\beta + \mu_r\cos\beta}.$$

7-55: a) Taking $U(0) = 0$,

$$U(x) = -\int_0^x F_x \, dx = \frac{\alpha}{2}x^2 + \frac{\beta}{3}x^3 = (30.0 \text{ N/m})x^2 + (6.00 \text{ N/m}^2)x^3.$$

b) $\qquad K_2 = U_1 - U_2$

$$= \left((30.0 \text{ N/m})(1.00 \text{ m})^2 + (6.00 \text{ N/m}^2)(1.00 \text{ m})^3\right)$$
$$- \left((30.0 \text{ N/m})(0.50 \text{ m})^2 + (6.00 \text{ N/m}^2)(0.50 \text{ m})^3\right)$$
$$= 27.75 \text{ J},$$

and so $v_2 = \sqrt{\frac{2(27.75 \text{ J})}{0.900 \text{ kg}}} = 7.85 \text{ m/s}$.

7-56: The force increases both the gravitational potential energy of the block and the potential energy of the spring. If the block is moved slowly, the kinetic energy can be taken as constant, so the work done by the force is the increase in potential energy, $\Delta U = mga \sin\theta + \frac{1}{2}k(a\theta)^2$.

7-57: With $U_2 = 0$, $K_1 = 0$, $K_2 = \frac{1}{2}mv_2^2 = U_1 = \frac{1}{2}kx^2 + mgh$, and solving for v_2,

$$v_2 = \sqrt{\frac{kx^2}{m} + 2gh} = \sqrt{\frac{(1900 \text{ N/m})(0.045 \text{ m})^2}{(0.150 \text{ kg})} + 2(9.80 \text{ m/s}^2)(1.20 \text{ m})} = 7.01 \text{ m/s}.$$

7-58: a) In this problem, use of algebra avoids the intermediate calculation of the spring constant k. If the original height is h and the maximum compression of the spring is d, then $mg(h + d) = \frac{1}{2}kd^2$. The speed needed is when the spring is compressed $\frac{d}{2}$, and from conservation of energy, $mg(h + d/2) - \frac{1}{2}k(d/2)^2 = \frac{1}{2}mv^2$. Substituting for k in terms of $h + d$,

$$mg\left(h + \frac{d}{2}\right) - \frac{mg(h + d)}{4} = \frac{1}{2}mv^2,$$

which simplifies to

$$v^2 = 2g\left(\frac{3}{4}h + \frac{1}{4}d\right).$$

Insertion of numerical values gives $v = 6.14 \text{ m/s}$. b) If the spring is compressed a distance x, $\frac{1}{2}kx^2 = mgx$, or $x = \frac{2mg}{k}$. Using the expression from part (a) that gives k in terms of h and d,

$$x = (2mg)\frac{d^2}{2mg(h + d)} = \frac{d^2}{h + d} = 0.0210 \text{ m}.$$

7-59: The first condition, that the maximum height above the release point is h, is expressed as $\frac{1}{2}kx^2 = mgh$. The magnitude of the acceleration is largest when the spring is compressed to a distance x; at this point the net upward force is $kx - mg = ma$, so the second condition is expressed as $x = (m/k)(g+a)$. a) Substituting the second expression into the first gives

$$\frac{1}{2}k\left(\frac{m}{k}\right)^2(g+a)^2 = mgh, \quad \text{or} \quad k = \frac{m(g+a)^2}{2gh}.$$

b) Substituting this into the expression for x gives $x = \frac{2gh}{g+a}$.

7-60: Following the hint, the force constant k is found from $w = mg = kd$, or $k = \frac{mg}{d}$. When the fish falls from rest, its gravitational potential energy decreases by mgy; this becomes the potential energy of the spring, which is $\frac{1}{2}ky^2 = \frac{1}{2}\frac{mg}{d}y^2$. Equating these,

$$\frac{1}{2}\frac{mg}{d}y^2 = mgy, \quad \text{or} \quad y = 2d.$$

7-61: With $K_1 = 0$,

$$\begin{aligned}
U_1 &= K_2 + U_2 - W_{\text{other}} \\
&= \frac{1}{2}mv_2^2 + mgL\sin\alpha - (-mgL\mu_k\cos\alpha) \\
&= m\left(v_2^2/2 + gL(\sin\alpha + \mu_k\cos\alpha)\right) \\
&= (1.50\text{ kg})\left((7.00\text{ m/s})^2/2 + (9.80\text{ m/s}^2)(6.00\text{ m})(\sin 30.0° + (0.50)\cos 30.0°)\right) \\
&= 119\text{ J}.
\end{aligned}$$

7-62: a) From either energy or force considerations, the speed before the block hits the spring is

$$\begin{aligned}
v &= \sqrt{2gL(\sin\theta - \mu_k\cos\theta)} \\
&= \sqrt{2(9.80\text{ m/s}^2)(4.00\text{m})(\sin 53.1° - (0.20)\cos 53.1°)} \\
&= 7.30\text{m/s}.
\end{aligned}$$

b) This does require energy considerations; the combined work done by gravity and friction is $mg(L+d)(\sin\theta - \mu_k\cos\theta)$, and the potential energy of the spring is $\frac{1}{2}kd^2$, where d is the maximum compression of the spring. This is a quadratic in d, which can be written as

$$d^2\frac{k}{2mg(\sin\theta - \mu_k\cos\theta)} - d - L = 0.$$

The factor multiplying d^2 is $4.504\,\text{m}^{-1}$, and use of the quadratic formula gives $d = 1.06\,\text{m}$.

c) The easy thing to do here is to recognize that the presence of the spring determines d, but at the end of the motion the spring has no potential energy, and the distance below the starting point is determined solely by how much energy has been lost to friction. If the block ends up a distance y below the starting point, then the block has moved a distance $L + d$ down the incline and $L + d - y$ up the incline. The magnitude of the friction force is the same in both directions, $\mu_k mg\cos\theta$, and so the work done by friction is $-\mu_k(2L + 2d - y)mg\cos\theta$. This must be equal to the change in gravitational potential energy, which is $-mgy \sin\theta$. Equating these and solving for y gives

$$y = (L+d)\frac{2\mu_k \cos\theta}{\sin\theta + \mu_k \cos\theta} = (L+d)\frac{2\mu_k}{\tan\theta + \mu_k}.$$

Using the value of d found in part (b) and the given values for μ_k and θ gives $y = 1.32\,\text{m}$.

7-63: a)
$$K_B = W_{\text{other}} - U_B = (20.0\,\text{N})(0.25\,\text{m}) - (1/2)(40.0\,\text{N/m})(.25\,\text{m})^2 = 3.75\,\text{J},$$

so $v_B = \sqrt{\frac{2(3.75\,\text{J})}{0.500\,\text{kg}}} = 3.87\,\text{m/s}$, or $3.9\,\text{m/s}$ to two figures.   b) At this point (point C),

$K_C = 0$, and so $U_C = W_{\text{other}}$ and $x_c = -\sqrt{\frac{2(5.00\,\text{J})}{40.0\,\text{N/m}}} = -0.50\,\text{m}$ (the minus sign denotes a displacement to the left in Fig. (7-65)), which is $0.10\,\text{m}$ from the wall.

7-64: The kinetic energy K' after moving up the ramp the distance s will be the energy initially stored in the spring, plus the (negative) work done by gravity and friction, or

$$K' = \frac{1}{2}kx^2 - mg(\sin\alpha + \mu_k \cos\alpha)s.$$

Minimizing the speed is equivalent to minimizing K', and differentiating the above expression with respect to α and setting $\frac{dK'}{d\alpha} = 0$ gives

$$0 = -mgs(\cos\alpha - \mu_k \sin\alpha),$$

or $\tan\alpha = \frac{1}{\mu_k}$, $\alpha = \arctan\left(\frac{1}{\mu_k}\right)$. Pushing the box straight up ($\alpha = 90°$) maximizes the vertical displacement h, but not $s = h/\sin\alpha$.

7-65: Let $x_1 = 0.18\,\text{m}$, $x_2 = 0.71\,\text{m}$. The spring constants (assumed identical) are then known in terms of the unknown weight w, $4kx_1 = w$. The speed of the brother at a given height h above the point of maximum compression is then found from

$$\frac{1}{2}(4k)x_2^2 = \frac{1}{2}\left(\frac{w}{g}\right)v^2 + mgh,$$

or

$$v^2 = \frac{(4k)g}{w}x_2^2 - 2gh = g\left(\frac{x_2^2}{x_1} - 2h\right),$$

so $v = \sqrt{(9.80\,\text{m/s}^2)((0.71\,\text{m})^2/(0.18\,\text{m}) - 2(0.90\,\text{m}))} = 3.13\,\text{m/s}$, or $3.1\,\text{m/s}$ to two figures.   b) Setting $v = 0$ and solving for h,

$$h = \frac{2kx_2^2}{mg} = \frac{x_2^2}{2x_1} = 1.40\,\text{m},$$

or $1.4\,\text{m}$ to two figures.   c) No; the distance x_1 will be different, and the ratio $\frac{x_2^2}{x_1} = \frac{(x_1 + 0.53\,\text{m})^2}{x_1} = x_1(1 + \frac{0.53\,\text{m}}{x_1})^2$ will be different. Note that on a small planet, with lower g, x_1 will be smaller and h will be larger.

7-66: a) Yes; rather than considering arbitrary paths, consider that

$$\vec{F} = -\left[\frac{\partial}{\partial y}\left(-\frac{Cy^3}{3}\right)\right]\hat{j}.$$

b) No; consider the same path as in Example 7-13 (the field is not the same). For this force, $\vec{F} = 0$ along Leg 1, $\vec{F}\cdot d\vec{l} = 0$ along legs 2 and 4, but $\vec{F}\cdot d\vec{l} \neq 0$ along Leg 3.

7-67: a) Along this line, $x = y$, so $\vec{F}\cdot d\vec{l} = -\alpha y^3\,dy$, and

$$\int_{y_1}^{y_2} F_y\,dy = -\frac{\alpha}{4}\left(y_2^4 - y_1^4\right) = -50.6\text{ J}.$$

b) Along the first leg, $dy = 0$ and so $\vec{F}\cdot d\vec{l} = 0$. Along the second leg, $x = 3.00$ m, so $F_y = -(7.50\text{ N/m}^2)y^2$, and

$$\int_{y_1}^{y_2} F_y\,dy = -(7.5/3\text{ N/m}^2)\left(y_2^3 - y_1^3\right) = -67.5\text{ J}.$$

c) The work done depends on the path, and the force is not conservative.

7-68: a)

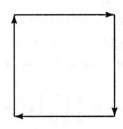

b) (1): $x = 0$ along this leg, so $\vec{F} = 0$ and $W = 0$.   (2): Along this leg, $y = 1.50$ m, so $\vec{F}\cdot d\vec{l} = (3.00\text{ N/m})x\,dx$, and $W = (1.50\text{ N/m})((1.50\text{ m})^2 - 0) = 3.8$ J   (3) $\vec{F}\cdot d\vec{l} = 0$, so $W = 0$ (4) $y = 0$, so $\vec{F} = 0$ and $W = 0$. The work done in moving around the closed path is 3.8 J.   c) The work done in moving around a closed path is not zero, and the force is not conservative.

7-69: a) For the given proposed potential $U(x)$, $-\frac{dU}{dx} = -kx + F$, so this is a possible potential function. For this potential, $U(0) = -F^2/2k$, not zero. Setting the zero of potential is equivalent to adding a constant to the potential; any additive constant will not change the derivative, and will correspond to the same force.   b) At equilibrium, the force is zero; solving $-kx + F = 0$ for x gives $x_0 = F/k$. $U(x_0) = -F^2/k$, and this is a minimum of U, and hence a stable point.

c)

d) No; $F_{\text{tot}} = 0$ at only one point, and this is a stable point.   e) The extreme values of x correspond to zero velocity, hence zero kinetic energy, so $U(x_\pm) = E$, where $x_\pm$ are the extreme points of the motion. Rather than solve a quadratic, note that $U(x) = \frac{1}{2}k(x - F/k)^2 - F^2/k$, so $U(x_\pm) = E$ becomes

$$\frac{1}{2}k\left(x_\pm - \frac{F}{k}\right)^2 - F^2/k = \frac{F^2}{k}$$

$$x_\pm - \frac{F}{k} = \pm 2\frac{F}{k},$$

$$x_+ = 3\frac{F}{k} \qquad x_- = -\frac{F}{k}.$$

f) The maximum kinetic energy occurs when $U(x)$ is a minimum, the point $x_0 = F/k$ found in part (b). At this point $K = E - U = (F^2/k) - (-F^2/k) = 2F^2/k$, so $v = 2F/\sqrt{mk}$.

7-70:   a) The slope of the U vs. x curve is negative at point A, so F_x is positive (Eq. (7-17)). b) The slope of the curve at point B is positive, so the force is negative. c) The kinetic energy is a maximum when the potential energy is a minimum, and that figures to be at around $0.75\,\text{m}$. d) The curve at point C looks pretty close to flat, so the force is zero. e) The object had zero kinetic energy at point A, and in order to reach a point with more potential energy than $U(A)$, the kinetic energy would need to be negative. Kinetic energy is never negative, so the object can never be at any point where the potential energy is larger than $U(A)$. On the graph, that looks to be at about $2.2\,\text{m}$. f) The point of minimum potential (found in part (c)) is a stable point, as is the relative minimum near $1.9\,\text{m}$. g) The only potential maximum, and hence the only point of unstable equilibrium, is at point C.

7-71:   a) Eliminating β in favor of α and x_0 ($\beta = \alpha/x_0$),

$$U(x) = \frac{\alpha}{x^2} - \frac{\beta}{x} = \frac{\alpha\, x_0^2}{x_0^2\, x^2} - \frac{\alpha}{x_0 x} = \frac{\alpha}{x_0^2}\left[\left(\frac{x_0}{x}\right)^2 - \left(\frac{x_0}{x}\right)\right].$$

$U(x_0) = \frac{\alpha}{x_0^2}(1 - 1) = 0$. $U(x)$ is positive for $x < x_0$ and negative for $x > x_0$ (α and β must be taken as positive).

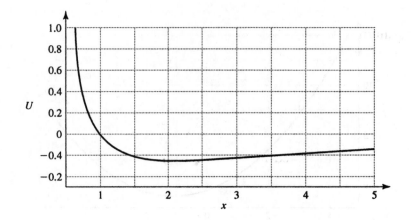

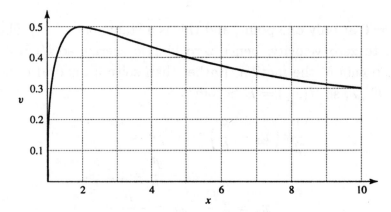

b) $$v(x) = \sqrt{-\frac{2}{m}U} = \sqrt{\left(\frac{2\alpha}{mx_0^2}\right)\left(\left(\frac{x_0}{x}\right)^2 - \left(\frac{x_0}{x}\right)\right)}.$$

The proton moves in the positive x-direction, speeding up until it reaches a maximum speed (see part (c)), and then slows down, although it never stops. The minus sign in the square root in the expression for $v(x)$ indicates that the particle will be found only in the region where $U < 0$, that is, $x > x_0$.

c) The maximum speed corresponds to the maximum kinetic energy, and hence the minimum potential energy. This minimum occurs when $\frac{dU}{dx} = 0$, or

$$\frac{dU}{dx} = \frac{\alpha}{x_0}3\left[-2\left(\frac{x_0}{x}\right)^3 + \left(\frac{x_0}{x}\right)^2\right] = 0,$$

which has the solution $x = 2x_0$. $U(2x_0) = -\frac{\alpha}{4x_0^2}$, so $v = \sqrt{\frac{\alpha}{2mx_0^2}}$. d) The maximum speed occurs at a point where $\frac{dU}{dx} = 0$, and from Eq. (7-15), the force at this point is zero. e) $x_1 = 3x_0$, and $U(3x_0) = -\frac{2}{9}\frac{\alpha}{x_0^2}$; $v(x) = \sqrt{\frac{2}{m}(U(x_1) - U(x))} = \sqrt{\frac{2}{m}\left[\left(\frac{-2}{9}\frac{\alpha}{x_0^2}\right) - \frac{\alpha}{x_0^2}\left(\left(\frac{x_0}{x}\right)^2 - \frac{x_0}{x}\right)\right]} = \sqrt{\frac{2\alpha}{mx_0^2}\left(\left(\frac{x_0}{x}\right) - \left(\frac{x_0}{x}\right)^2 - 2/9\right)}$. The particle is confined to the region where $U(x) < U(x_1)$.

The maximum speed still occurs at $x = 2x_0$, but now the particle will oscillate between x_1 and some minimum value (see part (f)). f) Note that $U(x) - U(x_1)$ can be written as

$$\frac{\alpha}{x_0^2}\left[\left(\frac{x_0}{x}\right)^2 - \left(\frac{x_0}{x}\right) + \left(\frac{2}{9}\right)\right] = \frac{\alpha}{x_0^2}\left[\left(\frac{x_0}{x}\right) - \frac{1}{3}\right]\left[\left(\frac{x_0}{x}\right) - \frac{2}{3}\right],$$

which is zero (and hence the kinetic energy is zero) at $x = 3x_0 = x_1$ and $x = \frac{3}{2}x_0$. Thus, when the particle is released from x_0, it goes on to infinity, and doesn't reach any maximum distance. When released from x_1, it oscillates between $\frac{3}{2}x_0$ and $3x_0$.

Chapter 8 Momentum, Impulse, and Collisions

8-1: a) $(10{,}000 \text{ kg})(12.0 \text{ m/s}) = 1.20 \times 10^5 \text{ kg·m/s}$.

   b) (i) Five times the speed, 60.0 m/s.   (ii) $\sqrt{5}(12.0 \text{ m/s}) = 26.8 \text{ m/s}$.

8-2: See Exercise 8-3(a); the iceboats have the same kinetic energy, so the boat with the larger mass has the larger magnitude of momentum by a factor of $\sqrt{(2m)/(m)} = \sqrt{2}$.

8-3: a)
$$K = \frac{1}{2}mv^2 = \frac{1}{2}\frac{m^2v^2}{m} = \frac{1}{2}\frac{p^2}{m}.$$

b) From the result of part (a), for the same kinetic energy, $\frac{p_1^2}{m_1} = \frac{p_2^2}{m_2}$, so the larger mass baseball has the greater momentum; $(p_{\text{bird}}/p_{\text{ball}}) = \sqrt{0.040/0.145} = 0.525$. From the result of part (b), for the same momentum $K_1m_1 = K_2m_2$, so $K_1w_1 = K_2w_2$; the woman, with the smaller weight, has the larger kinetic energy. $(K_{\text{man}}/K_{\text{woman}}) = 450/700 = 0.643$.

8-4: From Eq. (8-2),

$$p_x = mv_x = (0.420 \text{ kg})(4.50 \text{ m/s})\cos 20.0° = 1.78 \text{ kg·m/s}$$

$$p_y = mv_y = (0.420 \text{ kg})(4.50 \text{ m/s})\sin 20.0° = 0.646 \text{ kg·m/s}.$$

8-5: The y-component of the total momentum is

$$(0.145 \text{ kg})(1.30 \text{ m/s}) + (0.0570 \text{ kg})(-7.80 \text{ m/s}) = -0.256 \text{ kg·m/s}.$$

This quantity is negative, so the total momentum of the system is in the $-y$-direction.

8-6: From Eq. (8-2), $p_y = -(0.145 \text{ kg})(7.00 \text{ m/s}) = -1.015 \text{ kg·m/s}$, and $p_x = (0.045 \text{ kg})(9.00 \text{ m/s}) = 0.405 \text{ kg·m/s}$, so the total momentum has magnitude

$$p = \sqrt{p_x^2 + p_y^2} = \sqrt{(-0.405 \text{ kg·m/s})^2 + (-1.015 \text{ kg·m/s})^2} = 1.09 \text{ kg·m/s},$$

and is at an angle $\arctan(\frac{-1.015}{+.405}) = -68°$, using the value of the arctangent function in the fourth quadrant ($p_x > 0$, $p_y < 0$).

8-7: $\frac{\Delta p}{\Delta t} = \frac{(0.0450 \text{ kg})(25.0 \text{ m/s})}{2.00 \times 10^{-3} \text{ s}} = 563 \text{ N}$. The weight of the ball is less than half a newton, so the weight is not significant while the ball and club are in contact.

8-8: a) The magnitude of the velocity has changed by

$$(45.0 \text{ m/s}) - (-55.0 \text{ m/s}) = 100.0 \text{ m/s},$$

and so the magnitude of the change of momentum is $(0.145 \text{ kg})(100.0 \text{ m/s}) = 14.500 \text{ kg·m/s}$, or 14.5 kg·m/s to three figures. This is also the magnitude of the impulse. b) From Eq. (8-8), the magnitude of the average applied force is $\frac{14.500 \text{ kg·m/s}}{2.00 \times 10^{-3} \text{ s}} = 7.25 \times 10^3 \text{ N}$.

8-9: a) Considering the $+x$-components, $p_2 = p_1 + J = (0.16 \text{ kg})(3.00 \text{ m/s}) + (25.0 \text{ N}) \times (0.05 \text{ s}) = 1.73 \text{ kg·m/s}$, and the velocity is 10.8 m/s in the $+x$-direction. b) $p_2 = 0.48 \text{ kg·m/s} + (-12.0 \text{ N})(0.05 \text{ s}) = -0.12 \text{ kg·m/s}$, and the velocity is $+0.75 \text{ m/s}$ in the $-x$-direction.

8-10: a) $\vec{F}\,t = (1.04 \times 10^5 \text{ kg·m/s})\hat{j}$. b) $(1.04 \times 10^5 \text{ kg·m/s})\hat{j}$.

c) $\frac{(1.04 \times 10^5 \text{ kg·m/s})}{(95{,}000 \text{ kg})}\hat{j} = (1.10 \text{ m/s})\hat{j}$. d) The initial velocity of the shuttle is not known; the change in the square of the speed is not the square of the change of the speed.

8-11: a) With $t_1 = 0$,

$$J_x = \int_0^{t_2} F_x\, dt = (0.80 \times 10^7 \text{ N/s})t_2^2 - (2.00 \times 10^9 \text{ N/s}^2)t_2^3,$$

which is 18.8 kg·m/s, and so the impulse delivered between $t = 0$ and $t_2 = 2.50 \times 10^{-3}$ s is $(18.8 \text{ kg·m/s})\hat{i}$. b) $J_y = -(0.145 \text{ kg})(9.80 \text{ m/s}^2)(2.50 \times 10^{-3} \text{ s})$, and the impulse is $(-3.55 \times 10^{-3} \text{ kg·m/s})\hat{j}$. c) $\frac{J_x}{t_2} = 7.52 \times 10^3$ N, so the average force is $(7.52 \times 10^3 \text{ N})\hat{i}$.

d) $$\vec{P}_2 = \vec{P}_1 + \vec{J}$$

$$= -(0.145 \text{ kg})(40.0\,\hat{i} + 5.0\,\hat{j}) \text{ m/s} + (18.8\,\hat{i} - 3.55 \times 10^{-3}\,\hat{j})$$

$$= (13.0 \text{ kg·m/s})\hat{i} - (0.73 \text{ kg·m/s})\hat{j}.$$

The velocity is the momentum divided by the mass, or $(89.7 \text{ m/s})\hat{i} - (5.0 \text{ m/s})\hat{j}$.

8-12: The change in the ball's momentum in the x-direction (taken to be positive to the right) is $(0.145 \text{ kg})(-(65.0 \text{ m/s})\cos 30° - 50.0 \text{ m/s}) = -15.41$ kg·m/s, so the x-component of the average force is

$$\frac{-15.41 \text{ kg·m/s}}{1.75 \times 10^{-3} \text{ s}} = -8.81 \times 10^3 \text{ N},$$

and the y-component of the force is

$$\frac{(0.145 \text{ kg})(65.0 \text{ m/s})\sin 30°}{(1.75 \times 10^{-3} \text{ s})} = 2.7 \times 10^3 \text{ N}.$$

8-13: a) $$J = \int_{t_1}^{t_2} F\, dt = A(t_2 - t_1) + \frac{B}{3}\left(t_2^3 - t_1^3\right),$$

or $J = At_2 + (B/3)t_2^3$ if $t_1 = 0$. b) $v = \frac{p}{m} = \frac{J}{m} = \frac{A}{m}t_2 + \frac{B}{3m}t_2^3$.

8-14: The impulse imparted to the player is opposite in direction but of the same magnitude as that imparted to the puck, so the player's speed is $\frac{(0.160 \text{ kg})(20.0 \text{ m/s})}{(75.0 \text{ kg})} = 4.27$ cm/s, in the direction opposite to the puck's.

8-15: a) You and the snowball now share the momentum of the snowball when thrown so your speed is $\frac{(0.400 \text{ kg})(10.0 \text{ m/s})}{(70.0 \text{ kg} + 0.400 \text{ kg})} = 5.68$ cm/s. b) The change in the snowball's momentum is $(0.400 \text{ kg})(18.0 \text{ m/s}) = 7.20$ kg·m/s), so your speed is $\frac{7.20 \text{ kg·m/s}}{70.0 \text{ kg}} = 10.3$ cm/s.

8-16: a) The final momentum is

$$(0.250 \text{ kg})(-0.120 \text{ m/s}) + (0.350 \text{ kg})(0.650 \text{ m/s}) = 0.1975 \text{ kg·m/s},$$

taking positive directions to the right. a) Before the collision, puck B was at rest, so all of the momentum is due to puck A's motion, and

$$v_{A1} = \frac{p}{m_A} = \frac{0.1975 \text{ kg·m/s}}{0.250 \text{ kg}} = 0.790 \text{ m/s}.$$

b) $$\Delta K = K_2 - K_1 = \frac{1}{2}m_A v_{A2}^2 + \frac{1}{2}m_B v_{B2}^2 - \frac{1}{2}m_a v_{A1}^2$$

$$= \frac{1}{2}(0.250 \text{ kg})(-0.120 \text{ m/s})^2 + \frac{1}{2}(0.350 \text{ kg})(0.650 \text{ m/s})^2$$

$$- \frac{1}{2}(0.250 \text{ kg})(0.7900 \text{ m/s})^2$$

$$= -0.0023 \text{ J}.$$

Note that an extra figure was kept in the intermediate calculation to avoid roundoff error.

8-17: The change in velocity is the negative of the change in Gretzky's momentum, divided by the defender's mass, or

$$v_{B2} = v_{B1} - \frac{m_A}{m_B}(v_{A2} - v_{A1})$$

$$= -5.00 \text{ m/s} - \frac{756 \text{ N}}{900 \text{ N}}(1.50 \text{ m/s} - 13.0 \text{ m/s})$$

$$= 4.66 \text{ m/s}.$$

Positive velocities are in Gretzky's original direction of motion, so the defender has changed direction.

b) $$K_2 - K_1 = \frac{1}{2}m_A\left(v_{A2}^2 - v_{A1}^2\right) + \frac{1}{2}m_B\left(v_{B_2}^2 - v_{B_1}^2\right)$$

$$= \frac{1}{2(9.80 \text{ m/s}^2)}\left[\begin{array}{l}(756 \text{ N})\left((1.50 \text{ m/s})^2 - (13.0 \text{ m/s})^2\right) \\ +(900 \text{ N})\left((4.66 \text{ m/s})^2 - (-5.00 \text{ m/s})^2\right)\end{array}\right]$$

$$= -6.58 \text{ kJ}.$$

8-18: Take the direction of the bullet's motion to be the positive direction. The total momentum of the bullet, rifle, and gas must be zero, so

$$(0.00720 \text{ kg})(601 \text{ m/s} - 1.85 \text{ m/s}) + (2.80 \text{ kg})(-1.85 \text{ m/s}) + p_{gas} = 0,$$

and $p_{gas} = 0.866$ kg·m/s. Note that the speed of the bullet is found by subtracting the speed of the rifle from the speed of the bullet relative to the rifle.

8-19: a) See Exercise 8-21; $v_A = \left(\frac{3.00 \text{ kg}}{1.00 \text{ kg}}\right)(0.800 \text{ m/s}) = 3.60 \text{ m/s}.$

b) $(1/2)(1.00 \text{ kg})(3.60 \text{ m/s})^2 + (1/2)(3.00 \text{ kg})(1.200 \text{ m/s})^2 = 8.64$ J.

8-20: In the absence of friction, the horizontal component of the hat-plus-adversary system is conserved, and the recoil speed is

$$\frac{(4.50 \text{ kg})(22.0 \text{ m/s})\cos 36.9°}{(120 \text{ kg})} = 0.66 \text{ m/s}.$$

8-21: a) Taking v_A and v_B to be magnitudes, conservation of momentum is expressed as $m_A v_A = m_B v_B$, so $v_B = \frac{m_A}{m_B} v_A$.

b)
$$\frac{K_A}{K_B} = \frac{(1/2)m_A v_A^2}{(1/2)m_B v_B^2} = \frac{m_A v_A^2}{m_B((m_A/m_B)v_A)^2} = \frac{m_B}{m_A}.$$

(This result may be obtained using the result of Exercise 8-3.)

8-22: Let Rebecca's original direction of motion be the x-direction. a) From conservation of the x-component of momentum,

$$(45.0 \text{ kg})(13.0 \text{ m/s}) = (45.0 \text{ kg})(8.0 \text{ m/s}) \cos 53.1° + (65.0 \text{ kg})v_x,$$

so $v_x = 5.67$ m/s. If Rebecca's final motion is taken to have a positive y-component, then

$$v_y = -\frac{(45.0 \text{ kg})(8.0 \text{ m/s}) \sin 53.1°}{(65.0 \text{ kg})} = -4.43 \text{ m/s}.$$

Daniel's final speed is

$$\sqrt{v_x^2 + v_y^2} = \sqrt{(5.67 \text{ m/s})^2 + (-4.43 \text{ m/s})^2} = 7.20 \text{ m/s},$$

and his direction is $\arctan(\frac{-4.43}{5.67}) = -38°$ from the x-axis, which is $91.1°$ from the direction of Rebecca's final motion.

b) $\Delta K = \frac{1}{2}(45.0 \text{ kg})(8.0 \text{ m/s})^2 + \frac{1}{2}(65.0 \text{ kg})(7.195 \text{ m/s})^2 - \frac{1}{2}(45.0 \text{ kg})(13.0 \text{ m/s})^2$
$$= -680 \text{ J}.$$

Note that an extra figure was kept in the intermediate calculation.

8-23: $(m_{\text{Kim}} + m_{\text{Ken}})(3.00 \text{ m/s}) = m_{\text{Kim}}(4.00 \text{ m/s}) + m_{\text{Ken}}(2.25 \text{ m/s})$, so
$$\frac{m_{\text{Kim}}}{m_{\text{Ken}}} = \frac{(3.00 \text{ m/s}) - (2.25 \text{ m/s})}{(4.00 \text{ m/s}) - (3.00 \text{ m/s})} = 0.750,$$

and Kim weighs $(0.750)(700 \text{ N}) = 525$ N.

8-24: The original momentum is $(24,000 \text{ kg})(4.00 \text{ m/s}) = 9.60 \times 10^4$ kg·m/s, the final mass is $24,000 \text{ kg} + 3000 \text{ kg} = 27,000$ kg, and so the final speed is

$$\frac{9.60 \times 10^4 \text{ kg·m/s}}{2.70 \times 10^4 \text{ kg}} = 3.56 \text{ m/s}.$$

8-25: Denote the final speeds as v_A and v_B and the initial speed of puck A as v_0, and omit the common mass. Then, the condition for conservation of momentum is

$$v_0 = v_A \cos 30.0° + v_B \cos 45.0°$$
$$0 = v_A \sin 30.0° - v_B \sin 45.0°.$$

The 45.0° angle simplifies the algebra, in that $\sin 45.0° = \cos 45.0°$, and so the v_B terms cancel when the equations are added, giving

$$v_A = \frac{v_0}{\cos 30.0° + \sin 30.0°} = 29.3 \text{ m/s}.$$

From the second equation, $v_B = \frac{v_A}{\sqrt{2}} = 20.7$ m/s.   b) Again neglecting the common mass,

$$\frac{K_2}{K_1} = \frac{(1/2)(v_A^2 + v_B^2)}{(1/2)v_0^2} = \frac{(29.3 \text{ m/s})^2 + (20.7 \text{ m/s})^2}{(40.0 \text{ m/s})^2} = 0.804,$$

so 19.6% of the original energy is dissipated.

8-26:   a) From $m_1v_1 + m_2v_2 = m_1v + m_2v = (m_1 + m_2)v$, $v = \frac{m_1v_1 + m_2v_2}{m_1 + m_2}$. Taking positive velocities to the right, $v_1 = -3.00$ m/s and $v_2 = 1.20$ m/s, so $v = -1.60$ m/s.

b)
$$\Delta K = \frac{1}{2}(0.500 \text{ kg} + 0.250 \text{ kg})(-1.60 \text{ m/s})^2$$
$$-\frac{1}{2}(0.500 \text{ kg})(-3.00 \text{ m/s})^2 - \frac{1}{2}(0.250 \text{ kg})(1.20 \text{ m/s})^2$$
$$= -1.47 \text{ J}.$$

8-27:   For the truck, $M = 6320$ kg, and $V = 10$ m/s, for the car, $m = 1050$ kg and $v = -15$ m/s (the negative sign indicates a westbound direction).

a) Conservation of momentum requires $(M + m)v' = MV + mv$, or

$$v' = \frac{(6320 \text{ kg})(10 \text{ m/s}) + (1050 \text{ kg})(-15 \text{m/s})}{(6320 \text{ kg} + 1050 \text{ kg})} = 6.4 \text{ m/s eastbound}.$$

b) $V = \frac{-mv}{M} = \frac{-(1050 \text{ kg})(-15 \text{ m/s})}{6320 \text{ kg}} = 2.5$ m/s.

c) $\Delta KE = -281$kJ for part (a) and $\Delta KE = -138$ kJ for part (b).

8-28:   Take north to be the x-direction and east to be the y-direction (these choices are arbitrary). Then, the final momentum is the same as the initial momentum (for a sufficiently muddy field), and the velocity components are

$$v_x = \frac{(110 \text{ kg})(8.8 \text{ m/s})}{(195 \text{ kg})} = 5.0 \text{ m/s}$$

$$v_y = \frac{(85 \text{ kg})(7.2 \text{ m/s})}{(195 \text{ kg})} = 3.1 \text{ m/s}.$$

The magnitude of the velocity is then $\sqrt{(5.0 \text{ m/s})^2 + (3.1 \text{ m/s})^2} = 5.9$ m/s, at an angle of $\arctan\left(\frac{3.1}{5.0}\right) = 32°$ east of north.

8-29: Take east to be the x-direction and north to be the y-direction (again, these choices are arbitrary). The components of the common velocity after the collision are

$$v_x = \frac{(1400 \text{ kg})(-35.0 \text{ km/h})}{(4200 \text{ kg})} = -11.67 \text{ km/h}$$

$$v_y = \frac{(2800 \text{ kg})(-50.0 \text{ km/h})}{(4200 \text{ kg})} = -33.33 \text{ km/h}.$$

The velocity has magnitude $\sqrt{(-11.67 \text{ km/h})^2 + (-33.33 \text{ km/h})^2} = 35.3 \text{ km/h}$ and is at a direction $\arctan\left(\frac{-33.33}{-11.67}\right) = 70.7°$ south of west.

8-30: The initial momentum of the car must be the x-component of the final momentum as the truck had no initial x-component of momentum, so

$$v_{\text{car}} = \frac{p_x}{m_{\text{car}}} = \frac{(m_{\text{car}} + m_{\text{truck}})v \cos\theta}{m_{\text{car}}}$$

$$= \frac{2850 \text{ kg}}{950 \text{ kg}}(16.0 \text{ m/s}) \cos(90° - 24°)$$

$$= 19.5 \text{ m/s}.$$

Similarly, $v_{\text{truck}} = \frac{2850}{1900}(16.0 \text{ m/s}) \sin 66° = 21.9 \text{ m/s}.$

8-31: The speed of the block immediately after being struck by the bullet may be found from either force or energy considerations. Either way, the distance s is related to the speed v_{block} by $v^2 = 2\mu_k g s$. The speed of the bullet is then

$$v_{\text{bullet}} = \frac{m_{\text{block}} + m_{\text{bullet}}}{m_{\text{bullet}}}\sqrt{2\mu_k g s}$$

$$= \frac{1.205 \text{ kg}}{5.00 \times 10^{-3} \text{ kg}}\sqrt{2(0.20)(9.80 \text{ m/s}^2)(0.230 \text{ m})}$$

$$= 229 \text{ m/s},$$

or 2.3×10^2 m/s to two places.

8-32: a) The final speed of the bullet-block combination is

$$V = \frac{12.0 \times 10^{-3} \text{ kg}}{6.012 \text{ kg}}(380 \text{ m/s}) = 0.758 \text{ m/s}.$$

Energy is conserved after the collision, so $(m + M)gy = \frac{1}{2}(m + M)V^2$, and

$$y = \frac{1}{2}\frac{V^2}{g} = \frac{1}{2}\frac{(0.758 \text{ m/s})^2}{(9.80 \text{ m/s}^2)} = 0.0293 \text{ m} = 2.93 \text{ cm}.$$

b) $K_1 = \frac{1}{2}mv^2 = \frac{1}{2}(12.0 \times 10^{-3} \text{ kg})(380 \text{ m/s})^2 = 866$ J.
c) From part a), $K_2 = \frac{1}{2}(6.012 \text{ kg})(0.758 \text{ m/s})^2 = 1.73$ J.

8-33: In the notation of Example 8-10, with the smaller glider denoted as A, conservation of momentum gives $(1.50)v_{A2} + (3.00)v_{B2} = -5.40$ m/s. The relative velocity has switched direction, so $v_{A2} - v_{B2} = -3.00$ m/s. Multiplying the second of these relations by (3.00) and adding to the first gives $(4.50)v_{A2} = -14.4$ m/s, or $v_{A2} = -3.20$ m/s, with the minus sign indicating a velocity to the left. This may be substituted into either relation to obtain $v_{B2} = -0.20$ m/s; or, multiplying the second relation by (1.50) and subtracting from the first gives $(4.50)v_{B_2} = -0.90$ m/s, which is the same result.

8-34:   a) In the notation of Example 8-10, with the large marble (originally moving to the right) denoted as A, $(3.00)v_{A2} + (1.00)v_{B2} = 0.200$ m/s. The relative velocity has switched direction, so $v_{A2} - v_{B2} = -0.600$ m/s. Adding these eliminates v_{B2} to give $(4.00)v_{A2} = -0.400$ m/s, or $v_{A2} = -0.100$ m/s, with the minus sign indicating a final velocity to the left. This may be substituted into either of the two relations to obtain $v_{B2} = 0.500$ m/s; or, the second of the above relations may be multiplied by 3.00 and subtracted from the first to give $(4.00)v_{B2} = 2.00$ m/s, the same result.

   b) $\Delta P_A = -0.009$ kg $\cdot$ m/s, $\Delta P_B = 0.009$ kg $\cdot$ m/s

   c) $\Delta K_A = -4.5 \times 10^{-4}, \Delta K_B = 4.5 \times 10^{-4}$. Because the collision is elastic, the numbers have the same magnitude.

8-35: Algebraically, $v_{B2} = \sqrt{20}$ m/s. This substitution and the cancellation of common factors and units allow the equations in α and β to be reduced to

$$2 = \cos\alpha + \sqrt{1.8}\cos\beta$$
$$0 = \sin\alpha - \sqrt{1.8}\sin\beta.$$

Solving for $\cos\alpha$ and $\sin\alpha$, squaring and adding gives

$$\left(2 - \sqrt{1.8}\cos\beta\right)^2 + \left(\sqrt{1.8}\sin\beta\right)^2 = 1.$$

Minor algebra leads to $\cos\beta = \frac{1.2}{\sqrt{1.8}}$, or $\beta = 26.57°$. Substitution of this result into the first of the above relations gives $\cos\alpha = \frac{4}{5}$, and $\alpha = 36.87°$.

8-36:   a) Using Eq. (8-24), $\frac{v_A}{v} = \frac{1}{1}\frac{u-2\,u}{u+2\,u} = \frac{1}{3}$.   b) The kinetic energy is proportional to the square of the speed, so $\frac{K_A}{K} = \frac{1}{9}$ c) The magnitude of the speed is reduced by a factor of $\frac{1}{3}$ after each collision, so after N collisions, the speed is $\left(\frac{1}{3}\right)^N$ of its original value. To find N, consider

$$\left(\frac{1}{3}\right)^N = \frac{1}{59{,}000}, \quad \text{or}$$
$$3^N = 59{,}000$$
$$N\ln(3) = \ln(59{,}000)$$
$$N = \frac{\ln(59{,}000)}{\ln(3)} = 10.$$

to the nearest integer. Of course, using the logarithm in any base gives the same result.

8-37: a) In Eq. (8-24), let $m_A = m$ and $m_B = M$. Solving for M gives

$$M = m\,\frac{v - v_A}{v + v_A}.$$

In this case, $v = 1.50 \times 10^7$ m/s, and $v_A = -1.20 \times 10^7$ m/s, with the minus sign indicating a rebound. Then, $M = m\,\frac{1.50 + 1.20}{1.50 + (-1.20)} = 9m$. Either Eq. (8-25) may be used to find $V_B = \frac{v}{5} = 3.00 \times 10^6$ m/s, or Eq. (8-23), which gives $v_B = (1.50 \times 10^7 \text{ m/s}) + (-1.20 \times 10^7 \text{ m/s})$, the same result.

8-38: From Eq. (8-28),

$$x_{\text{cm}} = \frac{(0.30 \text{ kg})(0.20 \text{ m}) + (0.40 \text{ kg})(0.10 \text{ m}) + (0.20 \text{ kg})(-0.30 \text{ m})}{(0.90 \text{ kg})} = +0.044 \text{ m},$$

$$y_{\text{cm}} = \frac{(0.30 \text{ kg})(0.30 \text{ m}) + (0.40 \text{ kg})(-0.40 \text{ m}) + (0.20 \text{ kg})(0.60 \text{ m})}{(0.90 \text{ kg})} = 0.056 \text{ m}.$$

8-39: Measured from the center of the sun,

$$\frac{(1.99 \times 10^{30} \text{ kg})(0) + (1.90 \times 10^{27} \text{ kg})(7.78 \times 10^{11} \text{ m})}{1.99 \times 10^{30} \text{ kg} + 1.90 \times 10^{27} \text{ kg}} = 7.42 \times 10^8 \text{ m}.$$

The center of mass of the system lies outside the sun.

8-40: a) Measured from the rear car, the position of the center of mass is, from Eq. (8-28), $\frac{(1800 \text{ kg})(40.0 \text{ m})}{(1200 \text{ kg} + 1800 \text{ kg})} = 24.0$ m, which is 16.0 m behind the leading car.

b) $(1200 \text{ kg})(12.0 \text{ m/s}) + (1800 \text{ kg})(20.0 \text{ m/s}) = 5.04 \times 10^4$ kg·m/s.

c) From Eq. (8-30),

$$v_{\text{cm}} = \frac{(1200 \text{ kg})(12.0 \text{ m/s}) + (1800 \text{ kg})(20.0 \text{ m/s})}{(1200 \text{ kg} + 1800 \text{ kg})} = 16.8 \text{ m/s}.$$

d) $(1200 \text{ kg} + 1800 \text{ kg})(16.8 \text{ m/s}) = 5.04 \times 10^4$ kg·m/s.

8-41: a) With $x_1 = 0$ in Eq. (8-28),

$$m_1 = m_2\left((x_2/x_{\text{cm}}) - 1\right) = (0.10 \text{ kg})((8.0 \text{ m})/(2.0 \text{ m}) - 1) = 0.30 \text{ kg}.$$

b) $\vec{P} = M\vec{v}_{\text{cm}} = (0.40 \text{ kg})(5.0 \text{ m/s})\hat{i} = (2.0 \text{ kg·m/s})\hat{i}$.   c) In Eq. (8-32), $\vec{v}_2 = 0$, so $\vec{v}_1 = \vec{P}/(0.30 \text{ kg}) = (6.7 \text{ m/s})\hat{i}$.

8-42: As in Example 8-15, the center of mass remains at rest, so there is zero net momentum, and the magnitudes of the speeds are related by $m_1 v_1 = m_2 v_2$, or $v_2 = (m_1/m_2)v_1 = (60.0 \text{ kg}/90.0 \text{ kg})(0.70 \text{ m/s}) = 0.47$ m/s.

8-43: See Exercise 8-41(a); with $y_1 = 0$, Eq. (8-28) gives $m_1 = m_2((y_2/y_{\text{cm}}) - 1) = (0.50 \text{ kg})((6.0 \text{ m})/(2.4 \text{ m}) - 1) = 0.75$ kg, so the total mass of the system is 1.25 kg.

b) $\vec{a}_{cm} = \frac{d}{dt}\vec{v}_{cm} = (1.50 \text{ m/s}^3)t\hat{i}.$

c) $\vec{F} = m\vec{a}_{cm} = (1.25 \text{ kg})(1.50 \text{ m/s}^3)(3.0 \text{ s})\hat{i} = (5.63 \text{ N})\hat{i}.$

8-44:   a) $p_z = 0$, so $F_z = 0$. The x-component of force is

$$F_x = \frac{dp_x}{dt} = (-1.50 \text{ N/s})t.$$

$$F_y = \frac{dp_y}{dt} = 0.25 \text{ N}$$

b) Setting $F_x = 0$ and solving for t gives $t = 0$ s.

8-45:   a) From Eq. (8-38), $F = (1600 \text{ m/s})(0.0500 \text{ kg/s}) = 80.0$ N.   b) The absence of atmosphere would not prevent the rocket from operating. The rocket could be steered by ejecting the fuel in a direction with a component perpendicular to the rocket's velocity, and braked by ejecting in a direction parallel (as opposed to antiparallel) to the rocket's velocity.

8-46:   It turns out to be more convenient to do part (b) first; the thrust is the force that accelerates the astronaut and MMU, $F = ma = (70 \text{ kg} + 110 \text{ kg})(0.029 \text{ m/s}^2) = 5.22$ N. a) Solving Eq. (8-38) for $|dm|$,

$$|dm| = \frac{F\,dt}{v_{ex}} = \frac{(5.22 \text{ N})(5.0 \text{ s})}{(490 \text{ m/s})} = 53 \text{ gm}.$$

8-47:   Solving for the magnitude of dm in Eq. (8-39),

$$|dm| = \frac{ma}{v_{ex}}dt = \frac{(6000 \text{ kg})(25.0 \text{ m/s}^2)}{(2000 \text{ m/s})}(1 \text{ s}) = 75.0 \text{ kg}.$$

8-48:   Solving Eq. (8-34) for v_{ex} and taking the magnitude to find the exhaust speed, $v_{ex} = a\frac{m}{dm/dt} = (15.0 \text{ m/s}^2)(160 \text{ s}) = 2.4$ km/s. In this form, the quantity $\frac{m}{dm/dt}$ is approximated by $\frac{m}{\Delta m/\Delta t} = \frac{m}{\Delta m}\Delta t = 160$ s.

8-49:   a) The average thrust is the impulse divided by the time, so the ratio of the average thrust to the maximum thrust is $\frac{(10.0 \text{ N·s})}{(13.3 \text{ N})(1.70 \text{ s})} = 0.442$.   b) Using the average force in Eq. (8-38), $v_{ex} = \frac{F\,dt}{|dm|} = \frac{10.0 \text{ N·s}}{0.0125 \text{ kg}} = 800$ m/s.   c) Using the result of part (b) in Eq. (8-40), $v = (800 \text{ m/s})\ln(0.0258/0.0133) = 530$ m/s.

8-50:   Solving Eq. (8-4) for the ratio $\frac{m_0}{m}$, with $v_0 = 0$,

$$\frac{m_0}{m} = \exp\left(\frac{v}{v_{ex}}\right) = \exp\left(\frac{8.00 \text{ km/s}}{2.10 \text{ km/s}}\right) = 45.1.$$

8-51:   Solving Eq. (8-40) for $\frac{m}{m_0}$, the fraction of the original rocket mass that is not fuel,

$$\frac{m}{m_0} = \exp\left(-\frac{v}{v_{ex}}\right).$$

a) For $v = 1.00 \times 10^{-3}c = 3.00 \times 10^5$ m/s, $\exp(-(3.00 \times 10^5 \text{ m/s})/(2000 \text{ m/s})) = 7.2 \times 10^{-66}$.   b) For $v = 3000$ m/s, $\exp(-(3000 \text{ m/s})/(2000 \text{ m/s})) = 0.22$.

8-52: The ratios that appear in Eq. (8-42) are $\frac{0.0176}{1.0176}$ and $\frac{1}{1.0176}$, so the kinetic energies are
a) $\frac{0.0176}{1.0176}(6.54 \times 10^{-13}$ J$) = 1.13 \times 10^{-14}$ J and b) $\frac{1}{1.0176}(6.54 \times 10^{-13}$ J$) = 6.43 \times 10^{-13}$ J.
Note that the energies do not add to 6.54×10^{-13} J exactly, due to roundoff.

8-53: Solving Eq. (8-42) for m_2,

$$m_2 = m_1 \frac{K_1}{Q - K_1} = (6.65 \times 10^{-27} \text{ kg}) \frac{(9.650 \times 10^{-13} \text{ J})}{(0.020 \times 10^{-12} \text{ J})} = 3.2 \times 10^{-25} \text{ kg}.$$

Note that the difference in energies, and hence the nuclear mass, is known to only two figures.

8-54: The "missing momentum" is

$$5.60 \times 10^{-22} \text{ kg·m/s} - (3.50 \times 10^{-25} \text{ kg})(1.14 \times 10^3 \text{ m/s}) = 1.61 \times 10^{-22} \text{ kg·m/s}.$$

Since the electron has momentum to the right, the neutrino's momentum must be to the left.

8-55: a) The magnitude of the recoil momentum is the magnitude of the vector sum of the electron and antineutrino momenta; these are given to be at right angles, so the magnitude of the sum is

$$\sqrt{(3.60 \times 10^{-22} \text{ kg·m/s})^2 + (5.20 \times 10^{-22} \text{ kg·m/s})^2} = 6.32 \times 10^{-22} \text{ kg·m/s}.$$

b) $p/m \ll c$, so the kinetic energy is calculated nonrelativistically, $K = p^2/2m = (6.32 \times 10^{-22} \text{ kg·m/s})^2/2(3.5 \times 10^{-25} \text{ kg}) = 5.71 \times 10^{-19}$ J.

8-56: a) The speed of the ball before and after the collision with the plate are found from the heights. The impulse is the mass times the sum of the speeds, $J = m(v_1 + v_2) = m(\sqrt{2gy_1} + \sqrt{2gy_2}) = (0.040 \text{ kg})\sqrt{2(9.80 \text{ m/s}^2)}(\sqrt{2.00 \text{ m}} + \sqrt{1.60 \text{ m}}) = 0.47 \text{ N·s}$.
b) $\frac{J}{\Delta t} = (0.47 \text{ N·s}/2.00 \times 10^{-3} \text{ s}) = 237 \text{ N}$.

8-57:

$$\vec{p} = \int \vec{F} \, dt = (\alpha t^3/3)\hat{i} + (\beta t + \gamma t^2/2)\hat{j} = (8.33 \text{ N/s}^2 t^3)\hat{i} + (30.0 \text{ N} t + 2.5 \text{ N/s} t^2)\hat{j}.$$

After 0.500 s, $\vec{p} = (1.04 \text{ kg·m/s})\hat{i} + (15.63 \text{ kg·m/s})\hat{j}$, and the velocity is

$$\vec{v} = \vec{p}/m = (0.52 \text{ m/s})\hat{i} + (7.82 \text{ m/s})\hat{j}.$$

8-58: a)
$$J_x = F_x t = (-380 \text{ N})(3.00 \times 10^{-3} \text{ s}) = -1.14 \text{ N·s}$$
$$J_y = F_y t = (110 \text{ N})(3.00 \times 10^{-3} \text{ s}) = 0.33 \text{ N·s}.$$

b)
$$v_{2x} = v_{1x} + J_x/m = (20.0 \text{ m/s}) + \frac{(-1.14 \text{ N·s})}{((0.560 \text{ N})/(9.80 \text{ m/s}^2))} = 0.05 \text{ m/s}$$

$$v_{2y} = v_{1y} + J_y/m = (-4.0 \text{ m/s}) + \frac{(0.33 \text{ N·s})}{((0.560 \text{ N})/(9.80 \text{ m/s}^2))} = 1.78 \text{ m/s}.$$

8-59: The total momentum of the final combination is the same as the initial momentum; for the speed to be one-fifth of the original speed, the mass must be five times the original mass, or 15 cars.

8-60: The momentum of the convertible must be the south component of the total momentum, so

$$v_{\text{con}} = \frac{(8000 \text{ kg·m/s}) \cos 60.0°}{(1500 \text{ kg})} = 2.67 \text{ m/s}.$$

Similarly, the speed of the station wagon is

$$v_{\text{sw}} = \frac{(8000 \text{ kg·m/s}) \sin 60.0°}{(2000 \text{ kg})} = 3.46 \text{ m/s}.$$

8-61: The total momentum must be zero, and the velocity vectors must be three vectors of the same magnitude that sum to zero, and hence must form the sides of an equilateral triangle. One puck will move 60° north of east and the other will move 60° south of east.

8-62: a) $m_A v_{Ax} + m_B v_{Bx} + m_C v_{Cx} = m_{tot} v_x$, therefore

$$v_{Cx} = \frac{(0.100 \text{ kg})(0.50 \text{ m/s}) - (0.020 \text{ kg})(-1.50 \text{ m/s}) - (0.030 \text{ kg})(-0.50 \text{ m/s}) \cos 60°}{0.050 \text{ kg}}$$

$$v_{Cx} = 1.75 \text{ m/s}$$

Similarly,

$$v_{Cy} = \frac{(0.100 \text{ kg})(0 \text{ m/s}) - (0.020 \text{ kg})(0 \text{ m/s}) - (0.030 \text{ kg})(-0.50 \text{ m/s}) \sin 60°}{0.050 \text{ kg}}$$

$$v_{Cy} = 0.26 \text{ m/s}$$

b) $\Delta K = \frac{1}{2}(0.100 \text{ kg})(0.5 \text{ m/s})^2 - \frac{1}{2}(0.020 \text{ kg})(1.50 \text{ m/s})^2 - \frac{1}{2}(0.030 \text{ kg})(0.50 \text{ m/s})^2 - \frac{1}{2}(0.050 \text{ kg}) \times [(1.75 \text{ m/s})^2 + (0.26 \text{ m/s})^2] = -0.092 \text{ J}$

8-63: a) To throw the mass sideways, a sideways force must be exerted on the mass, and hence a sideways force is exerted on the car. The car is given to remain on track, so some other force (the tracks on the car) act to give a net horizontal force of zero on the car, which continues at 5.00 m/s east.

b) If the mass is thrown with backward with a speed of 5.00 m/s relative to the initial motion of the car, the mass is at rest relative to the ground, and has zero momentum. The speed of the car is then $(5.00 \text{ m/s})\frac{(200 \text{ kg})}{(175 \text{ kg})} = 5.71 \text{ m/s}$, and the car is still moving east.

c) The combined momentum of the mass and car must be the same before and after the mass hits the car, so the speed is $\frac{(200 \text{ kg})(5.00 \text{ m/s}) + (25.0 \text{ kg})(-6.00 \text{ m/s})}{(225 \text{ kg})} = 3.78 \text{ m/s}$, with the car still moving east.

8-64: The total mass of the car is changing, but the speed of the sand as it leaves the car is the same as the speed of the car, so there is no change in the velocity of either the car or the sand (the sand acquires a downward velocity after it leaves the car, and is stopped on the tracks *after* it leaves the car). Another way of regarding the situation is that v_{ex} in Equations (8-37), (8-38) and (8-39) is zero, and the car does not accelerate. In any event, the speed of the car remains constant at 15.0 m/s. In Exercise 8-24, the rain is given as falling vertically, so its velocity relative to the car as it hits the car is not zero.

8-65: a) The ratio of the kinetic energy of the Nash to that of the Packard is $\frac{m_N v_N^2}{m_P v_P^2} = \frac{(840 \text{ kg})(9 \text{ m/s})^2}{(1620 \text{ kg})(5 \text{ m/s})^2} = 1.68$. b) The ratio of the momentum of the Nash to that of the Packard is $\frac{m_N v_N}{m_P v_P} = \frac{(840 \text{ kg})(9 \text{ m/s})}{(1620 \text{ kg})(5 \text{ m/s})} = 0.933$, therefore the Packard has the greater magnitude of momentum. c) The force necessary to stop an object with momentum P in time t is $F = -P/t$. Since the Packard has the greater momentum, it will require the greater force to stop it. The ratio is the same since the time is the same, therefore $F_N/F_P = 0.933$. d) By the work-kinetic energy theorem, $F = \frac{\Delta K}{d}$. Therefore, since the Nash has the greater kinetic energy, it will require the greater force to stop it in a given distance. Since the distance is the same, the ratio of the forces is the same as that of the kinetic energies, $F_N/F_P = 1.68$.

8-66: The recoil force is the momentum delivered to each bullet times the rate at which the bullets are fired,

$$F_{\text{ave}} = (7.45 \times 10^{-3} \text{ kg})(293 \text{ m/s}) \left(\frac{1000 \text{ bullets/min}}{60 \text{ s/min}} \right) = 36.4 \text{ N}.$$

8-67: (This problem involves solving a quadratic. The method presented here formulates the answer in terms of the parameters, and avoids intermediate calculations, including that of the spring constant.)

Let the mass of the frame be M and the mass of the putty be m. Denote the distance that the frame stretches the spring by x_0, the height above the frame from which the putty is dropped as h, and the maximum distance the frame moves from its initial position (with the frame attached) as d.

The collision between the putty and the frame is completely inelastic, and the common speed after the collision is $v_0 = \sqrt{2gh} \frac{m}{m+M}$. After the collision, energy is conserved, so that

$$\frac{1}{2}(m + M)v_0^2 + (m + M)gd = \frac{1}{2}k\left((d + x_0)^2 - x_0^2\right), \quad \text{or}$$

$$\frac{1}{2}\frac{m^2}{m + M}(2gh) + (m + M)gd = \frac{1}{2}\frac{mg}{x_0}\left((d + x_0)^2 - x_0^2\right),$$

where the above expression for v_0, and $k = mg/x_0$ have been used. In this form, it is seen that a factor of g cancels from all terms. After performing the algebra, the quadratic for d becomes

$$d^2 - d\left(2x_0 \frac{m}{M}\right) - 2hx_0 \frac{m^2}{m + M} = 0,$$

which has as its positive root

$$d = x_0\left[\left(\frac{m}{M}\right) + \sqrt{\left(\frac{m}{M}\right)^2 + 2\frac{h}{x_0}\left(\frac{m^2}{M(m + M)}\right)}\right].$$

For this situation, $m = 4/3M$ and $h/x_0 = 6$, so

$$d = 0.232 \text{ m}.$$

8-68: a) After impact, the block-bullet combination has a total mass of 1.00 kg, and the speed V of the block is found from $\frac{1}{2}M_{\text{total}}V^2 = \frac{1}{2}kX^2$, or $V = \sqrt{\frac{k}{m}}X$. The spring constant k is determined from the calibration; $k = \frac{0.75 \text{ N}}{2.50 \times 10^{-3} \text{ m}} = 300 \text{ N/m}$. Combining,

$$V = \sqrt{\frac{300 \text{ N/m}}{1.00 \text{ kg}}} (15.0 \times 10^{-2} \text{ m}) = 2.60 \text{ m/s}.$$

b) Although this is not a pendulum, the analysis of the inelastic collision is the same;

$$v = \frac{M_{\text{total}}}{m} V = \frac{1.00 \text{ kg}}{8.0 \times 10^{-3} \text{ kg}} (2.60 \text{ m/s}) = 325 \text{ m/s}.$$

8-69: a) Take the original direction of the bullet's motion to be the x-direction, and the direction of recoil to be the y-direction. The components of the stone's velocity after impact are then

$$v_x = \left(\frac{6.00 \times 10^{-3} \text{ kg}}{0.100 \text{ kg}}\right)(350 \text{ m/s}) = 21.0 \text{ m/s},$$

$$v_y = -\left(\frac{6.00 \times 10^{-3} \text{ kg}}{0.100 \text{ kg}}\right)(250 \text{ m/s}) = 15.0 \text{ m/s},$$

and the stone's speed is $\sqrt{(21.0 \text{ m/s})^2 + (15.0 \text{ m/s})^2} = 25.8 \text{ m/s}$, at an angle of arctan $\left(\frac{15.0}{21.0}\right) = 35.5°$. b) $K_1 = \frac{1}{2}(6.00 \times 10^{-3} \text{ kg})(350 \text{ m/s})^2 = 368 \text{ J}$, $K_2 = \frac{1}{2}(6.00 \times 10^{-3} \text{ kg})(250 \text{ m/s})^2 + \frac{1}{2}(0.100 \text{ kg})(25.8 \text{ m}^2/\text{s}^2) = 221 \text{ J}$, so the collision is not perfectly elastic.

8-70: a) The stuntman's speed before the collision is $v_{0\text{s}} = \sqrt{2gy} = 9.9 \text{ m/s}$. The speed after the collision is

$$v = \frac{m_{\text{s}}}{m_{\text{s}} + m_{\text{v}}} v_{0\text{s}} = \frac{80.0 \text{ kg}}{80.0 \text{ kg} + 70.0 \text{ kg}} (9.9 \text{ m/s}) = 5.3 \text{ m/s}.$$

b) Momentum is not conserved during the slide. From the work-energy theorem, the distance x is found from $\frac{1}{2}m_{\text{total}}v^2 = \mu_{\text{k}}m_{\text{total}}gx$, or

$$x = \frac{v^2}{2\mu_{\text{k}}g} = \frac{(5.28 \text{ m/s})^2}{2(0.25)(9.80 \text{ m/s}^2)} = 5.7 \text{ m}.$$

Note that an extra figure was needed for V in part (b) to avoid roundoff error.

8-71:   a) The coefficient of friction, from either force or energy considerations, is $\mu_k = v^2/2gs$, where v is the speed of the block after the bullet passes through. The speed of the block is determined from the momentum lost by the bullet, $(4.00 \times 10^{-3}$ kg$)(280$ m/s$) = 1.12$ kg·m/s, and so the coefficient of kinetic friction is

$$\mu_k = \frac{((1.12 \text{ kg·m/s})/(0.80 \text{ kg}))^2}{2(9.80 \text{ m/s}^2)(0.45 \text{ m})} = 0.22.$$

b) $\frac{1}{2}(4.00 \times 10^{-3}$ kg$)\left((400 \text{ m/s})^2 - (120 \text{ m/s})^2\right) = 291$ J.   c) From the calculation of the momentum in part (a), the block's initial kinetic energy was $\frac{p^2}{2m} = \frac{(1.12 \text{ kg·m/s})^2}{2(0.80 \text{ kg})} = 0.784$ J.
8-72:   The speed of the block after the bullet has passed through (but before the block has begun to rise; this assumes a large force applied over a short time, a situation characteristic of bullets) is

$$V = \sqrt{2gy} = \sqrt{2(9.80 \text{ m/s}^2)(0.45 \times 10^{-2} \text{ m})} = 0.297 \text{ m/s}.$$

The final speed v of the bullet is then

$$v = \frac{p}{m} = \frac{mv_0 - MV}{m} = v_0 - \frac{M}{m}V$$

$$= 450 \text{ m/s} - \frac{1.00 \text{ kg}}{5.00 \times 10^{-3} \text{ kg}}(0.297 \text{ m/s}) = 390.6 \text{ m/s},$$

or 390 m/s to two figures.
8-73:   a) Using the notation of Eq. (8-24),

$$K_0 - K_2 = \frac{1}{2}mv^2 - \frac{1}{2}mv_A^2$$

$$= \frac{1}{2}mv^2 \left(1 - \left(\frac{m-M}{m+M}\right)^2\right)$$

$$= K_0 \left(\frac{(m+M)^2 - (m-M)^2}{(m+M)^2}\right)$$

$$= K_0 \left(\frac{4mM}{(m+M)^2}\right).$$

b) Of the many ways to do this calculation, the most direct way is to differentiate the expression of part (a) with respect to M and set equal to zero;

$$0 = (4mK_0) \frac{d}{dM} \left(\frac{M}{(m+M)^2}\right), \quad \text{or}$$

$$0 = \frac{1}{(m+M)^2} - \frac{2M}{(m+M)^3}$$

$$0 = (m+M) - 2M$$

$$m = M.$$

c) From Eq. (8-24), with $m_A = m_B = m$, $v_A = 0$; the neutron has lost all of its kinetic energy.

8-74: Even though one of the masses is not known, the analysis of Section (8-5) leading to Eq. (8-26) is still valid, and $v_{\text{red}} = 0.200 \text{ m/s} + 0.050 \text{ m/s} = 0.250 \text{ m/s}$. b) The mass m_{red} may be found from either energy or momentum considerations. From momentum conservation,

$$m_{\text{red}} = \frac{(0.040 \text{ kg})(0.200 \text{ m/s} - 0.050 \text{ m/s})}{(0.250 \text{ m/s})} = 0.024 \text{ kg}.$$

As a check, note that

$$K_1 = \tfrac{1}{2}(0.040 \text{ kg})(0.200 \text{ m/s})^2 = 8.0 \times 10^{-4} \text{ J}, \quad \text{and}$$

$$K_2 = \tfrac{1}{2}(0.040 \text{ kg})(0.050 \text{ m/s})^2 + \tfrac{1}{2}(0.024 \text{ kg})(0.250 \text{ m/s})^2 = 8.0 \times 10^{-4} \text{ J},$$

so $K_1 = K_2$, as it must for a perfectly elastic collision.

8-75: a) In terms of the primed coordinates,

$$v_A^2 = \left(\overrightarrow{v}'_A + \overrightarrow{v}_{\text{cm}}\right) \cdot \left(\overrightarrow{v}'_A + \overrightarrow{v}_{\text{cm}}\right)$$

$$= \overrightarrow{v}'_A \cdot \overrightarrow{v}'_A + \overrightarrow{v}_{\text{cm}} \cdot \overrightarrow{v}_{\text{cm}} + 2\overrightarrow{v}'_A \cdot \overrightarrow{v}_{\text{cm}}$$

$$= v_A'^2 + v_{\text{cm}}^2 + 2\overrightarrow{v}'_A \cdot \overrightarrow{v}_{\text{cm}},$$

with a similar expression for v_B^2. The total kinetic energy is then

$$K = \frac{1}{2} m_A v_A^2 + \frac{1}{2} m_B v_B^2$$

$$= \frac{1}{2} m_A \left(v_A'^2 + v_{\text{cm}}^2 + 2\overrightarrow{v}'_A \cdot \overrightarrow{v}_{\text{cm}}\right) + \frac{1}{2} m_B \left(v_B'^2 + v_{\text{cm}}^2 + 2\overrightarrow{v}'_B \cdot \overrightarrow{v}_{\text{cm}}\right)$$

$$= \frac{1}{2}\left(m_A + m_B\right) v_{\text{cm}}^2 + \frac{1}{2}\left(m_A v_A'^2 + m_B v_B'^2\right)$$

$$+ 2\left[m_A \overrightarrow{v}'_A \cdot \overrightarrow{v}_{\text{cm}} + m_B \overrightarrow{v}'_B \cdot \overrightarrow{v}_{\text{cm}}\right].$$

The last term in brackets can be expressed as

$$2\left(m_A \overrightarrow{v}'_A + m_B \overrightarrow{v}'_B\right) \cdot \overrightarrow{v}_{\text{cm}},$$

and the term

$$m_A \overrightarrow{v}'_A + m_B \overrightarrow{v}'_B = m_A \overrightarrow{v}_A + m_B \overrightarrow{v}_B - \left(m_A + m_B\right) \overrightarrow{v}_{\text{cm}}$$

$$= \mathbf{0},$$

and so the term in square brackets in the expression for the kinetic energy vanishes, showing the desired result.   b) In any collision for which other forces may be neglected, the velocity of the center of mass does not change, and the $\frac{1}{2}Mv_{cm}^2$ in the kinetic energy will not change. The other terms can be zero (for a perfectly inelastic collision, which is not likely), but never negative, so the minimum possible kinetic energy is $\frac{1}{2}Mv_{cm}^2$.

8-76: a) The relative speed of approach before the collision is the relative speed at which the balls separate after the collision. Before the collision, they are approaching with relative speed $2v$, and so after the collision they are receding with speed $2v$. In the limit that the larger ball has the much larger mass, its speed after the collision will be unchanged (the limit as $m_A \gg m_B$ in Eq. (8-24)), and so the small ball will move upward with speed $3v$. b) With three times the speed, the ball will rebound to a height nine times greater than the initial height.

8-77: a) If the crate has final speed v, J&J have speed 4.00 m/s$-v$ relative to the ice, and so $(15.0 \text{ kg})v = (120.0 \text{ kg})(4.00 \text{ m/s} - v)$. Solving for v, $v = \frac{(120.0 \text{ kg})(4.00 \text{ m/s})}{(135.0 \text{ kg})} = 3.56$ m/s. b) After Jack jumps, the speed of the crate is $\frac{(75.0 \text{ kg})}{(135.0 \text{ kg})}(4.00 \text{ m/s}) = 2.222$ m/s, and the momentum of Jill and the crate is 133.3 kg·m/s. After Jill jumps, the crate has a speed v and Jill has speed 4.00 m/s$-v$, and so 133.3 kg·m/s $= (15.0 \text{ kg})v - (45.0 \text{ kg})(4.00 \text{ m/s} - v)$, and solving for v gives $v = 5.22$ m/s. c) Repeating the calculation for part (b) with Jill jumping first gives a final speed of 4.67 m/s.

8-78: a) For the x- and y-directions, respectively, and m as the common mass of a proton,

$$mv_{A1} = mv_{A2}\cos\alpha + mv_2\cos\beta$$
$$0 = mv_{A2}\sin\alpha - mv_{B2}\sin\beta$$

or

$$v_{A1} = v_{A2}\cos\alpha + v_{B2}\cos\beta$$
$$0 = v_{A2}\sin\alpha - v_{B2}\sin\beta.$$

b) After minor algebra,

$$v_{A1}^2 = v_{A2}^2 + v_{B2}^2 + 2v_{A2}v_{B2}(\cos\alpha\cos\beta - \sin\alpha\sin\beta)$$
$$= v_{A2}^2 + v_{B2}^2 + 2v_{A2}v_{B2}\cos(\alpha + \beta).$$

c) For a perfectly elastic collision,

$$\frac{1}{2}mv_{A1}^2 = \frac{1}{2}mv_{A2}^2 + \frac{1}{2}mv_{B2}^2 \quad \text{or} \quad v_{A1}^2 = v_{A2}^2 + v_{B2}^2.$$

Substitution into the above result gives $\cos(\alpha + \beta) = 0$. d) The only positive angle with zero cosine is $\frac{\pi}{2}$ (90°).

8-79: See Problem 8-78. Puck B moves at an angle 65.0° (i.e. 90° $-$ 25° $=$ 65°) from the original direction of puck A's motion, and from conservation of momentum in the

y-direction, $v_{B2} = 0.466v_{A2}$. Substituting this into the expression for conservation of momentum in the x-direction, $v_{A2} = v_{A1}/(\cos 25.0° + 0.466\cos 65°) = 13.6$ m/s, and so $v_{B2} = 6.34$ m/s.

As an alternative, a coordinate system may be used with axes along the final directions of motion (from Problem 8-78, these directions are known to be perpendicular). The initial direction of the puck's motion is 25.0° from the final direction, so $v_{A2} = v_{A1}\cos 25.0°$ and $v_{B2} = v_{A1}\cos 65.0$, giving the same results.

8-80: Since mass is proportional to weight, the given weights may be used in determining velocities from conservation of momentum. Taking the positive direction to the left,

$$v = \frac{(800 \text{ N})(5.00 \text{ m/s})\cos 30.0° - (600 \text{ N})(7.00 \text{ m/s})\cos 36.9°}{1000 \text{ N}} = 0.105 \text{ m/s},$$

8-81: a) From symmetry, the center of mass is on the vertical axis, a distance $(L/2)\cos(\alpha/2)$ from the apex. b) The center of mass is on the (vertical) axis of symmetry, a distance $2(L/2)/3 = L/3$ from the center of the bottom of the $\sqcup$. c) Using the wire frame as a coordinate system, the coordinates of the center of mass are equal, and each is equal to $(L/2)/2 = L/4$. The distance of this point from the corner is $(1/\sqrt{8})L = (0.353)L$. This may also be found from consideration of the situation of part (a), with $\alpha = 45°$. d) By symmetry, the center of mass is in the center of the equilateral triangle, a distance $(L/2)(\tan 60°) = L/\sqrt{12} = (0.289)L$ above the center of the base.

8-82: The trick here is to notice that the final configuration is the same as if the canoe (assumed symmetrical) has been rotated about its center of mass. Initially, the center of mass is a distance $\frac{(45.0 \text{ kg})(1.5 \text{ m})}{(105 \text{ kg})} = 0.643$ m from the center of the canoe, so in rotating about this point the center of the canoe would move 2×0.643 m $= 1.29$ m.

8-83: Neglecting friction, the total momentum is zero, and your speed will be one-fifth of the slab's speed, or 0.40 m/s.

8-84: The trick here is to realize that the center of mass will continue to move in the original parabolic trajectory, "landing" at the position of the original range of the projectile. Since the explosion takes place at the highest point of the trajectory, and one fragment is given to have zero speed after the explosion, neither fragment has a vertical component of velocity immediately after the explosion, and the second fragment has *twice* the velocity the projectile had before the explosion. a) The fragments land at positions symmetric about the original target point. Since one lands at $\frac{1}{2}R$, the other lands at

$$\frac{3}{2}R = \frac{3}{2}\frac{v_0^2}{g}\sin 2\alpha_0 = \frac{3}{2}\frac{(80 \text{ m/s})^2}{(9.80 \text{ m/s}^2)}\sin 120° = 848 \text{ m}.$$

b) In terms of the mass m of the original fragment and the speed v before the explosion, $K_1 = \frac{1}{2}mv^2$ and $K_2 = \frac{1}{2}\frac{m}{2}(2v)^2 = mv^2$, so $\Delta K = mv^2 - \frac{1}{2}mv^2 = \frac{1}{2}mv^2$. The speed v is related to v_0 by $v = v_0 \cos \alpha_0$, so

$$\Delta K = \frac{1}{2} m v_0^2 \cos^2 \alpha_0 = \frac{1}{2}(20.0 \text{ kg})((80 \text{ m/s}) \cos 60.0°)^2 = 1.60 \times 10^4 \text{ J}.$$

8-85: The information is not sufficient to use conservation of energy. Denote the emitted neutron that moves in the $+y$-direction by the subscript 1 and the emitted neutron that moves in the $-y$-direction by the subscript 2. Using conservation of momentum in the x- and y-directions, and neglecting the common factor of the mass of a neutron,

$$v_0 = (2v_0/3) \cos 10° + v_1 \cos 45° + v_2 \cos 30°$$

$$0 = (2v_0/3) \sin 10° + v_1 \sin 45° - v_2 \sin 30°.$$

With $\sin 45° = \cos 45°$, these two relations may be subtracted to eliminate v_1, and rearrangement gives

$$v_0(1 - (2/3) \cos 10° + (2/3) \sin 10°) = v_2(\cos 30° + \sin 30°),$$

from which $v_2 = 1.01 \times 10^3$ m/s or 1.0×10^3 m/s to two figures. Substitution of this into either of the momentum relations gives $v_1 = 221$ m/s.   b) All that is known is that there is no z-component of momentum, and so only the ratio of the speeds can be determined. The ratio is the inverse of the ratio of the masses, so $v_{Kr} = (1.5)v_{Ba}$.

8-86:   a) With block B initially at rest, $v_{cm} = \frac{m_A}{m_A+m_B} v_{A1}$. b) Since there is no net external force, the center of mass moves with constant velocity, and so a frame that moves with the center of mass is an inertial reference frame. c) The velocities have only x-components, and the x-components are $u_{A1} = v_{A1} - v_{cm} = \frac{m_B}{m_A+m_B} v_{A1}$, $u_{B1} = -v_{cm} = -\frac{m_A}{m_A+m_B} v_{A1}$. Then, $P_{cm} = m_A u_{A1} + m_B u_{B1} = 0$. d) Since there is zero momentum in the center-of-mass frame before the collision, there can be no momentum after the collision; the momentum of each block after the collision must be reversed in direction. The only way to conserve kinetic energy is if the momentum of each has the same magnitude, so in the center-of-mass frame, the blocks change direction but keep the same speeds. Symbolically, $u_{A2} = -u_{A1}$, $u_{B2} = -u_{B1}$. e) The velocities all have only x-components; these components are $u_{A1} = \frac{0.200}{0.600} 6.00$ m/s $= 2.00$ m/s, $u_{B1} = -\frac{0.400}{0.600} 6.00$ m/s $= -4.00$ m/s, $u_{A2} = -2.00$ m/s, $u_{B2} = 4.00$ m/s, and $v_{A2} = +2.00$ m/s, $v_{B2} = 8.00$ m/s. Equation (8-24) predicts $v_{A2} = +\frac{1}{3}v_{A1}$ and Eq. (8-25) predicts $v_{B2} = \frac{4}{3}v_{A1}$, which are in agreement with the above.

8-87:   a) If the objects stick together, their relative speed is zero and $\epsilon = 0$.   b) From Eq. (8-27), the relative speeds are the same, and $\epsilon = 1$.   c) Neglecting air resistance, the speeds before and after the collision are $\sqrt{2gh}$ and $\sqrt{2gH_1}$, and $\epsilon = \frac{\sqrt{2gH_1}}{\sqrt{2gh}} = \sqrt{H_1/h}$. d) From part (c), $H_1 = \epsilon^2 h = (0.85)^2(1.2 \text{ m}) = 0.87$ m.   e) $H_{k+1} = H_k \epsilon^2$, and by induction $H_n = \epsilon^{2n} h$.   f) $(1.2 \text{ m})(0.85)^{16} = 8.9$ cm.

8-88:   a) The decrease in potential energy $(-\Delta < 0)$ means that the kinetic energy increases. In the center of mass frame of two hydrogen atoms, the net momentum is necessarily zero and after the atoms combine and have a common velocity, that velocity must have zero magnitude, a situation precluded by the necessarily positive kinetic energy. b) The initial momentum is zero before the collision, and must be zero after the collision. Denote the common initial speed as v_0, the final speed of the hydrogen atom as v, the final speed of the hydrogen molecule as V, the common mass of the hydrogen atoms as m and the mass of the hydrogen molecule as $2m$. After the collision, the two particles must be moving in opposite directions, and so to conserve momentum, $v = 2V$. From conservation of energy,

$$\frac{1}{2}(2m)V^2 - \Delta + \frac{1}{2}mv^2 = 3\tfrac{1}{2}\,mv_0^2$$
$$mV^2 - \Delta + 2mV^2 = \tfrac{3}{2}mv_0^2$$
$$V^2 = \frac{v_0^2}{2} + \frac{\Delta}{3m},$$

from which $V = 1.203 \times 10^4$ m/s, or 1.20×10^4 m/s to two figures and the hydrogen atom speed is $v = 2.41 \times 10^4$ m/s.

8-89: a) The wagon, after coming down the hill, will have speed $\sqrt{2gL\sin\alpha} = 10$ m/s. After the "collision", the speed is $\left(\frac{300 \text{ kg}}{435 \text{ kg}}\right)(10 \text{ m/s}) = 6.9$ m/s, and in the 5.0 s, the wagon will not reach the edge. b) The "collision" is completely inelastic, and kinetic energy is not conserved. The change in kinetic energy is $\frac{1}{2}(435 \text{ kg})(6.9 \text{ m/s})^2 - \frac{1}{2}(300 \text{ kg})(10 \text{ m/s})^2 = -4769$ J, so about 4800 J is lost.

8-90: a) Including the extra force, Eq. (8-37) becomes

$$m\frac{dv}{dt} = -v_{\text{ex}}\frac{dm}{dt} - mg,$$

where the positive direction is taken upwards (usually a sign of good planning).   b) Diving by a factor of the mass m,

$$a = \frac{dv}{dt} = -\frac{v_{\text{ex}}}{m}\frac{dm}{dt} - g.$$

c) $20 \text{ m/s}^2 - 9.80 \text{ m/s}^2 = 10.2 \text{ m/s}^2$. d) $3327 \text{ m/s} - (9.80 \text{ m/s}^2)(90 \text{ s}) = 2.45$ km/s, which is about three-fourths the speed found in Example 8-17.

8-91: a) From Eq. (8-40), $v = v_{\text{ex}}\ln\left(\frac{13,000 \text{ kg}}{3,300 \text{ kg}}\right) = (1.37)v_{\text{ex}}$.

 b) $v_{\text{ex}}\ln(13,000/4,000) = (1.18)v_{\text{ex}}$.

 c) $(1.18)v_{\text{ex}} + v_{\text{ex}}\ln(1000/300) = (2.38)v_{\text{ex}}$.   d) Setting the result of part (c) equal to 7.00 km/s and solving for v_{ex} gives $v_{\text{ex}} = 2.94$ km/s.

8-92: a) There are two contribution to F_{net}, $F_{\text{net}} = v_{\text{ex}}|dm/dt| - v|dm/dt|$, or $F_{\text{net}} = (v_{\text{ex}} - v)|dm/dt|$.

b) $F_{net}/|dm/dt| = (1300 \text{ N})/(150 \text{ kg/s}) = 8.66 \text{ m/s} = 31 \text{ km/h}$. This is equal to $v_{ex} - v$.

8-93: a) For $t < 0$ the rocket is at rest. For $0 \le t \le 90$ s, Eq. (8-40) is valid, and $v(t) = (2400 \text{ m/s}) \ln(1/(1 - (t/120 \text{ s})))$. At $t = 90$ s, this speed is 3.33 km/s, and this is also the speed for $t > 90$ s.

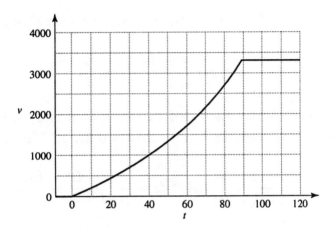

b) The acceleration is zero for $t < 0$ and $t > 90$ s. For $0 \le t \le 90$ s, Eq. (8-39) gives, with $\frac{dm}{dt} = -m_0/120$ s, $a = \frac{20 \text{ m/s}^2}{(1 - (t/120 \text{ s}))}$.

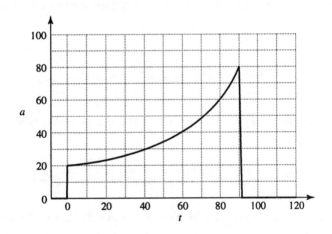

c) The maximum acceleration occurs at the latest time of firing, $t = 90$ s, at which time the acceleration is, from the result of part (a), $\frac{20 \text{ m/s}^2}{(1 - 90/120)} = 80 \text{ m/s}^2$, and so the astronaut is subject to a force of 6.0 kN, about eight times her weight on earth.

8-94: The impulse applied to the cake is $J = \mu_{k1}mgt = mv$, where m is the mass of the cake and v is its speed after the impulse is applied. The distance d that the cake moves during this time is then $d = \frac{1}{2}\mu_{k1}gt^2$. While sliding on the table, the cake must lose its kinetic energy to friction, or $\mu_{k2}mg(r - d) = \frac{1}{2}mv^2$. Simplification and substitution for v

gives $r - d = \frac{1}{2}g\frac{\mu_{k1}^2}{\mu_{k2}}t^2$, and substituting for d in terms of t^2 gives

$$r = \frac{1}{2}gt^2\left(\mu_{k1} + \frac{\mu_{k1}^2}{\mu_{k2}}\right) = \frac{1}{2}gt^2\frac{\mu_{k1}}{\mu_{k2}}(\mu_{k1} + \mu_{k2}),$$

which gives $t = 0.59$ s.

8-95: a) Noting that $dm = \frac{M}{L}dx$ avoids the intermediate variable ρ. Then,

$$x_{cm} = \frac{1}{M}\int_0^L x\frac{M}{L}dx = \frac{L}{2}.$$

b) In this case, the mass M may be found in terms of ρ and L, specifically by using $dm = \rho A\, dx = \alpha Ax\, dx$ to find that $M = \alpha A\int x\, dx = \alpha AL^2/2$. Then,

$$x_{cm} = \frac{2}{\alpha AL^2}\int_0^L \alpha Ax^2\, dx = \frac{2}{\alpha AL^2}\frac{L^3}{3} = \frac{2L}{3}.$$

8-96: By symmetry, $x_{cm} = 0$. Using plane polar coordinates leads to an easier integration, and using the Theorem of Pappus $(2\pi y_{cm}(\frac{\pi a^2}{2}) = \frac{4}{3}\pi a^3)$ is easiest of all, but the method of Problem 8-95 involves Cartesian coordinates.

For the x-coordinate, $dm = \rho t\sqrt{a^2 - x^2}\, dx$, which is an even function of x, so $\int x\, dx = 0$. For the y-coordinate, $dm = \rho t 2\sqrt{a^2 - y^2}\, dy$, and the range of integration is from 0 to a, so

$$y_{cm} = \frac{2\rho t}{M}\int_0^a y\sqrt{a^2 - y^2},\, dy.$$

Making the substitutions $M = \frac{1}{2}\rho\pi a^2 t$, $u = a^2 - y^2$, $du = -2y$, and

$$y_{cm} = \frac{-2}{\pi a^2}\int_{a^2}^0 u^{\frac{1}{2}}\, du = \frac{-4}{3\pi a^2}\left[u^{\frac{3}{2}}\right]_{a^2}^0 = \frac{4a}{3\pi}.$$

8-97: a) The tension in the rope at the point where it is suspended from the table is $T = (\lambda x)g$, where x is the length of rope over the edge, hanging vertically. In raising the rope a distance $-dx$, the work done is $(\lambda g)x(-dx)$ (dx is negative). The total work done is then

$$-\int_{l/4}^0 (\lambda g)\, x\, dx = (\lambda g)\frac{x^2}{2}\Big|_0^{l/4} = \frac{\lambda gl^2}{32}.$$

b) The center of mass of the hanging piece is initially a distance $l/8$ below the top of the table, and the hanging weight is $(\lambda g)(l/4)$, so the work required to raise the rope is $(\lambda g)(l/4)(l/8) = \lambda gl^2/32$, as before.

8-98:   a) For constant acceleration a, the downward velocity is $v = at$ and the distance x that the drop has fallen is $x = \frac{1}{2}at^2$. Substitution into the differential equation gives

$$\frac{1}{2}at^2 g = \frac{1}{2}at^2 a + (at)^2 = \frac{3}{2}a^2t^2,$$

the non-zero solution of which is $a = \frac{g}{3}$.

   b) $$\frac{1}{2}at^2 = \frac{1}{2}\left(\frac{9.80 \text{ m/s}^2}{3}\right)(3.00 \text{ s})^2 = 14.7 \text{ m}.$$

c) $kx = (2.00 \text{ g/m})(14.7 \text{ m}) = 29.4 \text{ g}.$

Chapter 9   Rotation of Rigid Bodies

9-1:   a)
$$\frac{1.50 \text{ m}}{2.50 \text{ m}} = 0.60 \text{ rad} = 34.4°.$$

b)
$$\frac{(14.0 \text{ cm})}{(128°)(\pi \text{ rad}/180°)} = 6.27 \text{ cm}.$$

c)
$$(1.50 \text{ m})(0.70 \text{ rad}) = 1.05 \text{ m}.$$

9-2:   a)
$$\left(1900 \frac{\text{rev}}{\text{min}}\right) \times \left(\frac{2\pi \text{ rad}}{\text{rev}}\right)\left(\frac{1 \text{ min}}{60 \text{ s}}\right) = 199 \text{ rad/s}.$$

b)
$$(35° \times \pi \text{ rad}/180°)/(199 \text{ rad/s}) = 3.07 \times 10^{-3} \text{ s}.$$

9-3:   a) $\alpha = \frac{d\omega}{dt} = (12.0 \text{ rad/s}^3)t$, so at $t = 3.5$ s, $\alpha = 42$ rad/s^2. The angular acceleration is proportional to the time, so the average angular acceleration between any two times is the arithmetic average of the angular accelerations.   b) $\omega = (6.0 \text{ rad/s}^3)t^2$, so at $t = 3.5$ s, $\omega = 73.5$ rad/s. The angular velocity is not a linear function of time, so the average angular velocity is not the arithmetic average or the angular velocity at the midpoint of the interval.

9-4:   a) $\alpha(t) = \frac{d\omega}{dt} = -2\beta t = (-1.60 \text{ rad/s}^3)t.$

b) $\alpha(3.0 \text{ s}) = (-1.60 \text{ rad/s}^3)(3.0 \text{ s}) = -4.80 \text{ rad/s}^2.$

$$\alpha_{\text{ave}} = \frac{\omega(3.0 \text{ s}) - \omega(0)}{3.0 \text{ s}} = \frac{-2.20 \text{ rad/s} - 5.00 \text{ rad/s}}{3.0 \text{ s}} = -2.40 \text{ rad/s}^2,$$

which is half as large (in magnitude) as the acceleration at $t = 3.0$ s.

9-5:   a) $\omega = \gamma + 3\beta t^2 = (0.400 \text{ rad/s}) + (0.036 \text{ rad/s}^3)t^2$ b) At $t = 0$, $\omega = \gamma = 0.400$ rad/s. c) At $t = 5.00$ s, $\omega = 1.3$ rad/s, $\theta = 3.50$ rad, so $\omega_{\text{ave}} = \frac{3.50 \text{ rad}}{5.00 \text{ s}} = 0.70$ rad/s. The acceleration is not constant, but increasing, so the angular velocity is larger than the average angular velocity.

9-6: $\omega = (250 \text{ rad/s}) - (40.0 \text{ rad/s}^2)t - (4.50 \text{ rad/s}^3)t^2$, $\alpha = -(40.0 \text{ rad/s}^2) - (9.00 \text{ rad/s}^3)t$.   a) Setting $\omega = 0$ results in a quadratic in t; the only positive time at which $\omega = 0$ is $t = 4.23$ s.   b) At $t = 4.23$ s, $\alpha = -78.1$ rad/s^2.   c) At $t = 4.23$ s, $\theta = 586$ rad $= 93.3$ rev.   d) At $t = 0$, $\omega = 250$ rad/s.   e) $\omega_{\text{ave}} = \frac{586 \text{ rad}}{4.23 \text{ s}} = 138$ rad/s.

9-7:   a) $\omega = \frac{d\theta}{dt} = 2bt - 3ct^2$ and $\alpha = \frac{d\omega}{dt} = 2b - 6ct$. b) Setting $\alpha = 0$, $t = \frac{b}{3c}$.

9-8:   a) $\omega = \omega_0 + \alpha t = 1.50 \text{ rad/s} + (0.300 \text{ rad/s}^2)(2.50 \text{ s}) = 2.25$ rad/s.

b) $\theta = \omega_0 t + 1/2\alpha t^2 = (1.50 \text{ rad/s})(2.50 \text{ s}) + \frac{1}{2}(0.300 \text{ rad/s}^2)(2.50 \text{ s})^2 = 4.69$ rad.

9-9: a) $$\frac{(500 \text{ rev/min} - 200 \text{ rev/min}) \times \left(\frac{1 \text{min}}{60 \text{ s}}\right)}{(4.00 \text{ s})} = 1.25 \frac{\text{rev}}{\text{s}^2}.$$

The number of revolutions is the average angular velocity, 350 rev/min, times the time interval of 0.067 min, or 23.33 rev. b) The angular velocity will decrease by another 200 rev/min in a time $\frac{200 \text{ rev/min}}{60 \text{ s/min}} \cdot \frac{1}{1.25 \text{ rev/s}^2} = 2.67$ s.

9-10: a) Solving Eq. (9-7) for t gives $t = \frac{\omega - \omega_0}{\alpha}$.
Rewriting Eq. (9-11) as $\theta - \theta_0 = t(\omega_0 + \frac{1}{2}\alpha t)$ and substituting for t gives

$$\theta - \theta_0 = \left(\frac{\omega - \omega_0}{\alpha}\right)\left(\omega_0 + \frac{1}{2}(\omega - \omega_0)\right)$$

$$= \frac{1}{\alpha}(\omega - \omega_0)\left(\frac{\omega + \omega_0}{2}\right)$$

$$= \frac{1}{2\alpha}\left(\omega^2 - \omega_0^2\right),$$

which when rearranged gives Eq. (9-12).

b) $\alpha = (1/2)(1/\Delta\theta)(\omega^2 - \omega_0^2) = (1/2)(1/(7.00 \text{ rad}))((16.0 \text{ rad/s})^2 - (12.0 \text{ rad/s})^2) = 8 \text{ rad/s}^2$.

9-11: a) From Eq. (9-7), with $\omega_0 = 0$, $t = \frac{\omega}{\alpha} = \frac{36.0 \text{ rad/s}}{1.50 \text{ rad/s}^2} = 24.0$ s.

b) From Eq. (9-12), with $\omega_0 = 0$, $\theta - \theta_0 = \frac{(36.0 \text{ rad/s})^2}{2(1.50 \text{ rad/s}^2)} = 432 \text{ rad} = 68.8$ rev.

9-12: a) The average angular velocity is $\frac{162 \text{ rad}}{4.00 \text{ s}} = 40.5 \text{ rad/s}$, and so the initial angular velocity is $2\omega_{\text{ave}} - \omega_2 = \omega_0$, $\omega_0 = -27 \text{ rad/s}$.

b) $$\alpha = \frac{\Delta\omega}{\Delta t} = \frac{108 \text{ rad/s} - (-27 \text{ rad/s})}{4.00 \text{ s}} = 33.8 \text{ rad/s}^2.$$

9-13: From Eq. (9-11),

$$\omega_0 = \frac{\theta - \theta_0}{t} - \frac{\alpha t}{2} = \frac{60.0 \text{ rad}}{4.00 \text{ s}} - \frac{(2.25 \text{ rad/s}^2)(4.00 \text{ s})}{2} = 10.5 \text{ rad/s}.$$

9-14: From Eq. (9-7), with $\omega_0 = 0$, $\alpha = \frac{\omega}{t} = \frac{140 \text{ rad/s}}{6.00 \text{ s}} = 23.33 \text{ rad/s}^2$. The angle is most easily found from $\theta = \omega_{\text{ave}}t = (70 \text{ rad/s})(6.00 \text{ s}) = 420 \text{ rad}$.

9-15: From Eq. (9-12), with $\omega = 0$, the number of revolutions is proportional to the square of the initial angular velocity, so tripling the initial angular velocity increases the number of revolutions by 9, to 9.0 rev.

9-16: The following table gives the revolutions and the angle θ through which the wheel has rotated for each instant in time and each of the three situations:

t	(a) rev's	(a) θ	(b) rev's	(b) θ	(c) rev's	(c) θ
0.05	0.50	180	0.03	11.3	0.44	158
0.10	1.00	360	0.13	45	0.75	270
0.15	1.50	540	0.28	101	0.94	338
0.20	2.00	720	0.50	180	1.00	360

The θ and ω graphs are as follows:

a)

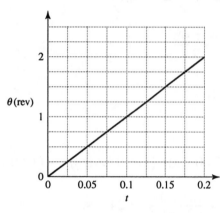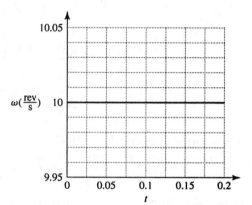

b)

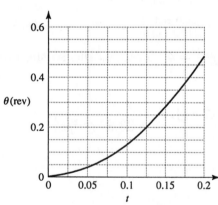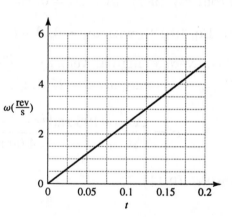

c)

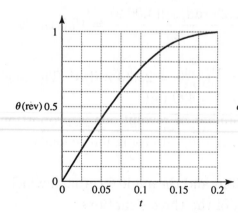

9-17:   a) Before the circuit breaker trips, the angle through which the wheel turned was $(24.0 \text{ rad/s})(2.00 \text{ s}) + (30.0 \text{ rad/s}^2)(2.00 \text{ s})^2/2 = 108 \text{ rad}$, so the total angle is $108 \text{ rad} + 432 \text{ rad} = 540 \text{ rad}$. b) The angular velocity when the circuit breaker trips is $(24.0 \text{ rad/s}) + (30.0 \text{ rad/s}^2)(2.00 \text{ s}) = 84 \text{ rad/s}$, so the average angular velocity while the wheel is slowing is 42.0 rad/s, and the time to slow to a stop is $\frac{432 \text{ rad}}{42.0 \text{ rad/s}} = 10.3 \text{ s}$, so the time when the wheel stops is 12.3 s.   c) Of the many ways to find the angular acceleration, the most direct is to use the intermediate calculation of part (b) to find that while slowing down $\Delta\omega = -84 \text{ rad/s}$ so $\alpha = \frac{-84 \text{ rad/s}}{10.3 \text{ s}} = -8.17 \text{ rad/s}^2$.

9-18:   a) Equation (9-7) is solved for $\omega_0 = \omega - \alpha t$, which gives $\omega_{\text{ave}} = \omega - \frac{\alpha}{2}t$, or $\theta - \theta_0 = \omega t - \frac{1}{2}\alpha t^2$.   b) $2\left(\frac{\omega}{t} - \frac{\Delta\theta}{t^2}\right) = -0.125 \text{ rad/s}^2$.   c) $\omega - \alpha t = 5.5 \text{ rad/s}$.

9-19:   The horizontal component of velocity is $r\omega$, so the magnitude of the velocity is
a) 47.1 m/s

b) $\sqrt{\left((5.0 \text{ m})(90 \text{ rev/min})\left(\frac{\pi}{30}\frac{\text{rad/s}}{\text{rev/min}}\right)\right)^2 + (4.0 \text{ m/s})^2} = 47.3 \text{ m/s}.$

9-20:   a) $\frac{1.25 \text{ m/s}}{25.0 \times 10^{-3} \text{ m}} = 50.0 \text{ rad/s}$, $\frac{1.25 \text{ m}}{58.0 \times 10^{-3} \text{ m}} = 21.55 \text{ rad/s}$, or 21.6 rad/s to three figures.

b) $(1.25 \text{ m/s})(74.0 \text{ min})(60 \text{ s/min}) = 5.55 \text{ km}.$

c) $\alpha = \frac{50.0 \text{ rad/s} - 21.55 \text{ rad/s}}{(74.0 \text{ min})(60 \text{ s/min})} = 6.41 \times 10^{-3} \text{ rad/s}^2.$

9-21:   a) $\omega^2 r = (6.00 \text{ rad/s})^2(0.500 \text{ m}) = 18 \text{ m/s}^2.$

b) $v = \omega r = (6.00 \text{ rad/s})(0.500 \text{ m}) = 3.00 \text{ m/s}$, and $\frac{v^2}{r} = \frac{(3.00 \text{ m/s})^2}{(0.500 \text{ m})} = 18 \text{ m/s}^2.$

9-22:   From $a_{\text{rad}} = \omega^2 r$,

$$\omega = \sqrt{\frac{a}{r}} = \sqrt{\frac{400{,}000 \times 9.80 \text{ m/s}^2}{2.50 \times 10^{-2} \text{ m}}} = 1.25 \times 10^4 \text{ rad/s},$$

which is $(1.25 \times 10^4 \text{ rad/s})\left(\frac{1 \text{ rev}/2\pi \text{ rad}}{1 \text{ min}/60 \text{ s}}\right) = 1.20 \times 10^5 \text{ rev/min}.$

9-23:   a) $a_{\text{rad}} = 0$, $a_{\text{tan}} = \alpha r = (0.600 \text{ rad/s}^2)(0.300 \text{ m}) = 0.180 \text{ m/s}^2$ and so $a = 0.180 \text{ m/s}^2$.   b) $\theta = \frac{\pi}{3}$ rad, so

$$a_{\text{rad}} = \omega^2 r = 2(0.600 \text{ rad/s}^2)(\pi/3 \text{ rad})(0.300 \text{ m}) = 0.377 \text{ m/s}^2.$$

The tangential acceleration is still 0.180 m/s^2, and so

$$a = \sqrt{(0.180 \text{ m/s}^2)^2 + (0.377 \text{ m/s}^2)^2} = 0.418 \text{ m/s}^2.$$

c) For an angle of 120°, $a_{\text{rad}} = 0.754 \text{ m/s}^2$, and $a = 0.775 \text{ m/s}^2$, since a_{tan} is still 0.180 m/s^2.

9-24:   a) $\omega = \omega_0 + \alpha t = 0.250 \text{ rev/s} + (0.900 \text{ rev/s}^2)(0.200 \text{ s}) = 0.430 \text{ rev/s}$ (note that since ω_0 and α are given in terms of revolutions, it's not necessary to convert to radians).
b) $\omega_{\text{ave}}\Delta t = (0.340 \text{ rev/s})(0.2 \text{ s}) = 0.068 \text{ rev}.$ c) Here, the conversion to radians must be

made to use Eq. (9-13), and

$$v = r\omega = \left(\frac{0.750 \text{ m}}{2}\right)(0.430 \text{ rev/s} \times 2\pi \text{ rad/rev}) = 1.01 \text{ m/s}.$$

d) Combining Equations (9-14) and (9-15),

$$a = \sqrt{a_{\text{rad}}^2 + a_{\text{tan}}^2} = \sqrt{(\omega^2 r)^2 + (\alpha r)^2}$$

$$= \left[((0.430 \text{ rev/s} \times 2\pi \text{ rad/rev})^4 (0.375 \text{ m}))^2 \right.$$

$$\left. + ((0.900 \text{ rev/s}^2 \times 2\pi \text{ rad/rev})(0.375 \text{ m}))^2 \right]^{\frac{1}{2}}$$

$$= 3.46 \text{ m/s}^2.$$

9-25:
$$r = \frac{a_{\text{rad}}}{\omega^2} = \frac{(3000)(9.80 \text{ m/s}^2)}{\left((5000 \text{ rev/min}) \left(\frac{\pi}{30} \frac{\text{rad/s}}{\text{rev/min}}\right)\right)^2} = 10.7 \text{ cm},$$

so the diameter is more than 12.7 cm, contrary to the claim.

9-26: a) Combining Equations (9-13) and (9-15),

$$a_{\text{rad}} = \omega^2 r = \omega^2 \left(\frac{v}{\omega}\right) = \omega v.$$

b) From the result of part (a), $\omega = \frac{a_{\text{rad}}}{v} = \frac{0.500 \text{ m/s}^2}{2.00 \text{ m/s}} = 0.250 \text{ rad/s}$.

9-27: a) $\omega r = (1250 \text{ rev/min}) \left(\frac{\pi}{30} \frac{\text{rad/s}}{\text{rev/min}}\right) \left(\frac{12.7 \times 10^{-3} \text{ m}}{2}\right) = 0.831 \text{ m/s}$.

b) $\frac{v^2}{r} = \frac{(0.831 \text{ m/s})^2}{(12.7 \times 10^{-3} \text{ m})/2} = 109 \text{ m/s}^2$.

9-28: a) $\alpha = \frac{a_{\text{tan}}}{r} = \frac{-10.0 \text{ m/s}^2}{0.200 \text{ m}} = -50.0 \text{ rad/s}^2$ b) At $t = 3.00$ s, $v = 50.0$ m/s and $\omega = \frac{v}{r} = \frac{50.0 \text{ m/s}}{0.200} = 250$ rad/s, and at $t = 0$, $v = 50.0$ m/s $+ (-10.0 \text{ m/s}^2)(0 - 3.00 \text{ s}) = 80.0$ m/s, so $\omega = 400$ rad/s. c) $\omega_{\text{ave}} t = (325 \text{ rad/s})(3.00 \text{ s}) = 975$ rad $= 155$ rev. d) $v = \sqrt{a_{\text{rad}} r} = \sqrt{(9.80 \text{ m/s}^2)(0.200 \text{ m})} = 1.40$ m/s. This speed will be reached at time $\frac{50.0 \text{ m/s} - 1.40 \text{ m/s}}{10.0 \text{ m/s}^2} = 4.86$ s after $t = 3.00$ s, or at $t = 7.86$ s. (There are many equivalent ways to do this calculation.)

9-29: a) For a given radius and mass, the force is proportional to the square of the angular velocity; $\left(\frac{640 \text{ rev/min}}{423 \text{ rev/min}}\right)^2 = 2.29$ (note that conversion to rad/s is not necessary for this part). b) For a given radius, the tangential speed is proportional to the angular velocity; $\frac{640}{423} = 1.51$ (again conversion of the units of angular speed is not necessary). c) $(640 \text{ rev/min}) \left(\frac{\pi}{30} \frac{\text{rad/s}}{\text{rev/min}}\right) \left(\frac{0.470 \text{ m}}{2}\right) = 15.75$ m/s, or 15.7 m/s to three figures, and $a_{\text{rad}} = \frac{v^2}{r} = \frac{(15.75 \text{ m/s})^2}{(0.470 \text{ m})/2} = 1.06 \times 10^3 \text{ m/s}^2 = 108g$.

9-30: The distances of the masses from the axis are $\frac{L}{4}$, $\frac{L}{4}$ and $\frac{3L}{4}$, and so from Eq. (9-16), the moment of inertia is

$$I = m\left(\frac{L}{4}\right)^2 + m\left(\frac{L}{4}\right)^2 + m\left(\frac{3L}{4}\right)^2 = \frac{11}{16}mL^2.$$

9-31: The moment of inertia of the cylinder is $M\frac{L^2}{12}$ and that of each cap is $m\frac{L^2}{4}$, so the moment of inertia of the combination is $\left(\frac{M}{12} + \frac{m}{2}\right) L^2$.

9-32: Since the rod is 500 times as long as it is wide, it can be considered slender.

a) From Table (9-2(a)),

$$I = \frac{1}{12}ML^2 = \frac{1}{12}(0.042 \text{ kg})(1.50 \text{ m})^2 = 7.88 \times 10^{-3} \text{ kg·m}^2.$$

b) From Table (9-2(b)),

$$I = \frac{1}{3}ML^2 = \frac{1}{3}(0.042 \text{ kg})(1.50 \text{ m})^2 = 3.15 \times 10^{-2} \text{ kg·m}^2.$$

c) For this slender rod, the moment of inertia about the axis is obtained by considering it as a solid cylinder, and from Table (9-2(f)),

$$I = \frac{1}{2}MR^2 = \frac{1}{2}(0.042 \text{ kg})(1.5 \times 10^{-3} \text{ m})^2 = 4.73 \times 10^{-8} \text{ kg·m}^2.$$

9-33: a) For each mass, the square of the distance from the axis is $2(0.200 \text{ m})^2 = 8.00 \times 10^{-2} \text{ m}^2$, and the moment of inertia is $4(0.200 \text{ kg})(0.800 \times 10^{-2} \text{ m}^2) = 6.40 \times 10^{-2} \text{ kg·m}^2$. b) Each sphere is 0.200 m from the axis, so the moment of inertia is $4(0.200 \text{ kg})(0.200 \text{ m})^2 = 3.20 \times 10^{-2} \text{ kg·m}^2$. c) The two masses through which the axis passes do not contribute to the moment of inertia. The moment for *each* mass is $mr^2 \cdot I = 2(0.2 \text{ kg})(0.2\sqrt{2})^2 = 0.032 \text{ kg} \cdot \text{m}^2$.

9-34: a) In the expression of Eq. (9-16), each term will have the mass multiplied by f^3 and the distance multiplied by f, and so the moment of inertia is multiplied by $f^3(f)^2 = f^5$. b) $(2.5)(48)^5 = 6.37 \times 10^8$.

9-35: Each of the eight spokes may be treated as a slender rod about an axis through an end, so the moment of inertia of the combination is

$$I = m_{\text{rim}}R^2 + 8\left(\frac{m_{\text{spoke}}}{3}\right) R^2$$

$$= \left[(1.40 \text{ kg}) + \frac{8}{3}(0.280 \text{ kg})\right] (0.300 \text{ m})^2$$

$$= 0.193 \text{ kg·m}^2.$$

9-36: a) From Eq. (9-17),with I from Table (9-2(f)),

$$K = \frac{1}{2}\frac{1}{12}mL^2\omega^2 = \frac{1}{24}(117 \text{ kg})(2.08 \text{ m})^2\left(2400 \ \frac{\text{rev}}{\text{min}} \times \frac{2\pi \text{ rad/rev}}{60 \text{ s/min}}\right)^2 = 1.3 \times 10^6 \text{ J.}$$

b) From $mgy = K$,

$$y = \frac{K}{mg} = \frac{(1.3 \times 10^6 \text{ J})}{(117 \text{ kg})(9.80 \text{ m/s}^2)} = 1.16 \times 10^3 \text{ m} = 1.16 \text{ km}.$$

9-37: a) The units of moment of inertia are $[\text{kg}]\,[\text{m}^2]$ and the units of ω are equivalent to $[\text{s}^{-1}]$ and so the product $\frac{1}{2}I\omega^2$ has units equivalent to $[\text{kg·m·s}^{-2}] = [\text{kg·(m/s)}^2]$, which are the units of Joules. A radian is a ratio of distances and is therefore unitless.

b) $K = \pi^2 Iw^2/1800$, when ω is in rev/min.

9-38: Solving Eq. (9-17) for I,

$$I = \frac{2K}{\omega^2} = \frac{2(0.025 \text{ J})}{\left(45 \text{ rev/min} \times \frac{2\pi}{60} \frac{\text{rad/s}}{\text{rev/min}}\right)^2} = 2.25 \times 10^{-3} \text{ kg·m}^2.$$

9-39: From Eq. (9-17), $K_2 - K_1 = \frac{1}{2}I\left(\omega_2^2 - \omega_1^2\right)$, and solving for I,

$$I = 2\frac{(K_2 - K_1)}{(\omega_2^2 - \omega_1^2)}$$

$$= 2\frac{(-500 \text{ J})}{((520 \text{ rev/min})^2 - (650 \text{ rev/min})^2)\left(\frac{\pi}{30} \frac{\text{rad/s}}{\text{rev/min}}\right)^2}$$

$$= 0.600 \text{ kg·m}^2.$$

9-40: The work done on the cylinder is PL, where L is the length of the rope. Combining Equations (9-17), (9-13) and the expression for I from Table (9-2(g)),

$$PL = \frac{1}{2}\frac{w}{g}v^2, \quad \text{or} \quad P = \frac{1}{2}\frac{w}{g}\frac{v^2}{L} = \frac{(40.0 \text{ N})(6.00 \text{ m/s})^2}{2(9.80 \text{ m/s}^2)(5.00 \text{ m})} = 14.7 \text{ N}.$$

9-41: Expressing ω in terms of a_{rad}, $\omega^2 = \frac{a_{\text{rad}}}{R}$. Combining with $I = \frac{1}{2}MR^2$, Eq. (9-17) becomes

$$K = \frac{1}{2}\frac{1}{2}MRa_{\text{rad}} = \frac{(70.0 \text{ kg})(1.20 \text{ m})(3500 \text{ m/s}^2)}{4} = 7.35 \times 10^4 \text{ J}.$$

9-42: a) With $I = MR^2$, the expression for v is

$$v = \sqrt{\frac{2gh}{1 + M/m}}.$$

b) This expression is smaller than that for the solid cylinder; more of the cylinder's mass is concentrated at its edge, so for a given speed, the kinetic energy of the cylinder is larger. A larger fraction of the potential energy is converted to the kinetic energy of the cylinder, and so less is available for the falling mass.

9-43: a) $\omega = \frac{2\pi}{T}$, so Eq. (9-17) becomes $K = 2\pi^2 I/T^2$.

b) Differentiating the expression found in part (a) with respect to T, $\frac{dK}{dt} = (-4\pi^2 I/T^3)\frac{dT}{dt}$.

c) $2\pi^2(8.0 \text{ kg·m}^2)/(1.5 \text{ s})^2 = 70.2$ J, or 70 J to two figures.

d) $(-4\pi^2(8.0 \text{ kg·m}^2)/(1.5 \text{ s})^3)(0.0060) = -0.56$ W.

9-44: The center of mass has fallen half of the length of the rope, so the change in gravitational potential energy is

$$-\frac{1}{2}mgL = -\frac{1}{2}(3.00 \text{ kg})(9.80 \text{ m/s}^2)(10.0 \text{ m}) = -147 \text{ J}.$$

9-45: $(120 \text{ kg})(9.80 \text{ m/s}^2)(0.700 \text{ m}) = 823$ J.

9-46: In Eq. (9-19), $I_{cm} = MR^2$ and $d = R^2$, so $I_P = 2MR^2$.

9-47: $\frac{2}{3}MR^2 = \frac{2}{5}MR^2 + Md^2$, so $d^2 = \frac{4}{15}R^2$, and the axis comes nearest to the center of the sphere at a distance $d = \left(2/\sqrt{15}\right)R = (0.516)R$.

9-48: Using the parallel-axis theorem to find the moment of inertia of a thin rod about an axis through its end and perpendicular to the rod,

$$I_P = I_{cm} + Md^2 = \frac{M}{12}L^2 + M\left(\frac{L}{2}\right)^2 = \frac{M}{3}L^2.$$

9-49: $I_P = I_{cm} + md^2$, so $I = \frac{1}{12}m(a^2 + b^2) + m\left(\left(\frac{a}{2}\right)^2 + \left(\frac{b}{2}\right)^2\right)$, which gives

$$I = \frac{1}{12}m(a^2 + b^2) + \frac{1}{4}m(a^2 + b^2), \text{ or } I = \frac{1}{3}m(a^2 + b^2).$$

9-50: a) $I = \frac{1}{12}Ma^2$ b) $I = \frac{1}{12}Mb^2$

9-51: In Eq. (9-19), $I_{cm} = \frac{M}{12}L^2$ and $d = (L/2 - h)$, so

$$I_P = M\left[\frac{1}{12}L^2 + \left(\frac{L}{2} - h\right)^2\right]$$

$$= M\left[\frac{1}{12}L^2 + \frac{1}{4}L^2 - Lh + h^2\right]$$

$$= M\left[\frac{1}{3}L^2 - Lh + h^2\right],$$

which is the same as found in Example 9-12.

9-52: The analysis is identical to that of Example 9-13, with the lower limit in the integral being zero and the upper limit being R, and the mass $M = \pi L\rho R^2$. The result is $I = \frac{1}{2}MR^2$, as given in Table (9-2(f)).

9-53: With $dm = \frac{M}{L}dx$,

$$I = \int_0^L x^2 \frac{M}{L} dx = \frac{M}{L}\frac{x^3}{3}\Big|_0^L = \frac{M}{3}L^2.$$

9-54: For this case, $dm = \gamma\, dx$.

a) $$M = \int dm = \int_0^L \gamma x\, dx = \gamma \frac{x^2}{2}\Big|_0^L = \frac{\gamma L^2}{2}.$$

b)
$$I = \int_0^L x^2(\gamma x)\,dx = \gamma\left.\frac{x^4}{4}\right|_0^L = \frac{\gamma L^4}{4} = \frac{M}{2}L^2.$$

This is larger than the moment of inertia of a uniform rod of the same mass and length, since the mass density is greater further away from the axis than nearer the axis.

c)
$$I = \int_0^L (L-x)^2 \gamma x\,dx$$
$$= \gamma\int_0^L \left(L^2 x - 2Lx^2 + x^3\right)\,dx$$
$$= \gamma\left(L^2\frac{x^2}{2} - 2L\frac{x^3}{3} + \frac{x^4}{4}\right)\Big|_0^L$$
$$= \gamma\frac{L^4}{12}$$
$$= \frac{M}{6}L^2.$$

This is a third of the result of part (b), reflecting the fact that more of the mass is concentrated at the right end.

9-55: a) For a clockwise rotation, $\vec{\omega}$ will be out of the page. b) The upward direction crossed into the radial direction is, by the right-hand rule, counterclockwise. $\vec{\omega}$ and $\vec{r}$ are perpendicular, so the magnitude of $\vec{\omega}\times\vec{r}$ is $\omega r = v$. c) Geometrically, $\vec{\omega}$ is perpendicular to $\vec{v}$, and so $\vec{\omega}\times\vec{v}$ has magnitude $\omega v = a_{\text{rad}}$, and from the right-hand rule, the upward direction crossed into the counterclockwise direction is inward, the direction of $\vec{a}_{\text{rad}}$. Algebraically,

$$\vec{a}_{\text{rad}} = \vec{\omega}\times\vec{v} = \vec{\omega}\times\left(\vec{\omega}\times\vec{r}\right)$$
$$= \vec{\omega}\left(\vec{\omega}\cdot\vec{r}\right) - \vec{r}\left(\vec{\omega}\cdot\vec{\omega}\right)$$
$$= -\omega^2\vec{r},$$

where the fact that $\vec{\omega}$ and $\vec{r}$ are perpendicular has been used to eliminate their dot product.

9-56: a) For constant angular acceleration $\theta = \frac{\omega^2}{2\alpha}$, and so $a_{\text{rad}} = \omega^2 r = 2\alpha\theta r$.

b) Denoting the angle that the acceleration vector makes with the radial direction as β, and using Equations (9-14) and (9-15),

$$\tan\beta = \frac{a_{\text{tan}}}{a_{\text{rad}}} = \frac{\alpha r}{\omega^2 r} = \frac{\alpha r}{2\alpha\theta r} = \frac{1}{2\theta},$$

so $\theta = \frac{1}{2\tan\beta} = \frac{1}{2\tan 36.9°} = 0.666$ rad.

9-57: a) $\omega = \dfrac{d\theta}{dt} = 2\gamma t - 3\beta t^2 = (6.40\text{ rad/s}^2)t - (1.50\text{ rad/s}^3)t^2.$

b) $\alpha = \dfrac{d\omega}{dt} = 2\gamma - 6\beta t = (6.40\text{ rad/s}^2) - (3.00\text{ rad/s}^3)t.$

c) An extreme of angular velocity occurs when $\alpha = 0$, which occurs at

$t = \frac{\gamma}{3\beta} = \frac{3.20 \text{ rad/s}^2}{1.50 \text{ rad/s}^3} = 2.13$ s, and at this time

$$\omega = (2\gamma)(\gamma/3\beta) - (3\beta)(\gamma/3\beta)^2 = \gamma^2/3\beta = \frac{(3.20 \text{ rad/s}^2)^2}{3(0.500 \text{ rad/s}^3)} = 6.83 \text{ rad/s}.$$

9-58: a) By successively integrating Equations (9-5) and (9-3),

$$\omega = \gamma t - \frac{\beta}{2}t^2 = (1.80 \text{ rad/s}^2)t - (0.125 \text{ rad/s}^3)t^2,$$

$$\theta = \frac{\gamma}{2}t^2 - \frac{\beta}{6}t^3 = (0.90 \text{ rad/s}^2)t^2 - (0.042 \text{ rad/s}^3)t^3.$$

b) The maximum positive angular velocity occurs when $\alpha = 0$, or $t = \frac{\gamma}{\beta}$; the angular velocity at this time is

$$\omega = \gamma\left(\frac{\gamma}{\beta}\right) - \frac{\beta}{2}\left(\frac{\gamma}{\beta}\right)^2 = \frac{1}{2}\frac{\gamma^2}{\beta} = \frac{1}{2}\frac{(1.80 \text{ rad/s}^2)^2}{(0.25 \text{ rad/s}^3)} = 6.48 \text{ rad/s}.$$

The maximum angular displacement occurs when $\omega = 0$, at time $t = \frac{2\gamma}{\beta}$ ($t = 0$ is an inflection point, and $\theta(0)$ is not a maximum) and the angular displacement at this time is

$$\theta = \frac{\gamma}{2}\left(\frac{2\gamma}{\beta}\right)^2 - \frac{\beta}{6}\left(\frac{2\gamma}{\beta}\right)^3 = \frac{2}{3}\frac{\gamma^3}{\beta^2} = \frac{2}{3}\frac{(1.80 \text{ rad/s}^2)^3}{(0.25 \text{ rad/s}^3)^2} = 62.2 \text{ rad}.$$

9-59: a) The scale factor is 20.0, so the actual speed of the car would be 35 km/h = 9.72 m/s   b) $(1/2)mv^2 = 8.51$ J.   c) $\omega = \sqrt{\frac{2K}{I}} = 652$ rad/s.

9-60: a) $\alpha = \frac{a_{\tan}}{r} = \frac{3.00 \text{ m/s}^2}{60.0 \text{ m}} = 0.050 \text{ rad/s}^2$.   b) $\alpha t = (0.05 \text{ rad/s}^2)(6.00 \text{ s}) = 0.300$ rad/s.

c) $a_{\text{rad}} = \omega^2 r = (0.300 \text{ rad/s})^2(60.0 \text{ m}) = 5.40 \text{ m/s}^2$.

d)

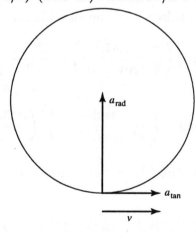

e) $a = \sqrt{a_{\text{rad}}^2 + a_{\tan}^2} = \sqrt{(5.40 \text{ m/s}^2)^2 + (3.00 \text{ m/s}^2)^2} = 6.18 \text{ m/s}^2$,

and the magnitude of the force is $F = ma = (1240 \text{ kg})(6.18 \text{ m/s}^2) = 7.66$ kN.

f) $\arctan\left(\frac{a_{\text{rad}}}{a_{\tan}}\right) = \arctan\left(\frac{5.40}{3.00}\right) = 60.9°$.

9-61: a) Expressing angular frequencies in units of revolutions per minute may be ac-
comodated by changing the units of the dynamic quantities; specifically,

$$\omega_2 = \sqrt{\omega_1^2 + \frac{2W}{I}}$$

$$= \sqrt{(300 \text{ rev/min})^2 + \left(\frac{2(-4000 \text{ J})}{16.0 \text{ kg}\cdot\text{m}^2}\right) \Big/ \left(\frac{\pi}{30} \frac{\text{rad/s}}{\text{rev/min}}\right)^2}$$

$$= 211 \text{ rev/min}.$$

b) At the initial speed, the 4000 J will be recovered; if this is to be done in 5.00 s, the
power must be $\frac{4000 \text{ J}}{5.00 \text{ s}} = 800$ W.

9-62: a) The angular acceleration will be zero when the speed is a maximum, which
is at the bottom of the circle. The speed, from energy considerations, is $v = \sqrt{2gh} = \sqrt{2gR(1 - \cos\beta)}$, where β is the angle from the vertical at release, and

$$\omega = \frac{v}{R} = \sqrt{\frac{2g}{R}(1 - \cos\beta)} = \sqrt{\frac{2(9.80 \text{ m/s}^2)}{(2.50 \text{ m})}(1 - \cos 36.9°)} = 1.25 \text{ rad/s}.$$

b) α will again be 0 when the meatball again passes through the lowest point.

c) a_{rad} is directed toward the center, and $a_{\text{rad}} = \omega^2 R$, $a_{\text{rad}} = (1.25 \text{ rad/s})^2(2.50 \text{ m}) = 3.93 \text{ m/s}^2$.

d) $a_{\text{rad}} = \omega^2 R = (2g/R)(1 - \cos\beta)R = (2g)(1 - \cos\beta)$, independent of R.

9-63: a) $(60.0 \text{ rev/s})(2\pi \text{ rad/rev})(0.45 \times 10^{-2} \text{ m}) = 1.696 \text{ m/s}$.

b) $\omega = \frac{v}{r} = \frac{1.696 \text{ m/s}}{2.00 \times 10^{-2} \text{ m}} = 84.8 \text{ rad/s}$.

9-64: The second pulley, with half the diameter of the first, must have twice the angular
velocity, and this is the angular velocity of the saw blade.

a) $(2(3450 \text{ rev/min}))\left(\frac{\pi}{30} \frac{\text{rad/s}}{\text{rev/min}}\right)\left(\frac{0.208 \text{ m}}{2}\right) = 75.1 \text{ m/s}.$

b) $a_{\text{rad}} = \omega^2 r = \left(2(3450 \text{ rev/min})\left(\frac{\pi}{30} \frac{\text{rad/s}}{\text{rev/min}}\right)\right)^2 \left(\frac{0.208 \text{ m}}{2}\right) = 5.43 \times 10^4 \text{ m/s}^2,$

so the force holding sawdust on the blade would have to be about 5500 times as strong
as gravity.

9-65: a)

$$\Delta a_{\text{rad}} = \omega^2 r - \omega_0^2 r = \left(\omega^2 - \omega_0^2\right) r$$

$$= [\omega - \omega_0][\omega + \omega_0] r$$

$$= \left[\frac{\omega - \omega_0}{t}\right] [(\omega + \omega_0) t] r$$

$$= [\alpha][2(\theta - \theta_0)] r.$$

b) From the above,

$$\alpha r = \frac{\Delta a_{\text{rad}}}{2\Delta\theta} = \frac{(85.0 \text{ m/s}^2 - 25.0 \text{ m/s}^2)}{2(15.0 \text{ rad})} = 2.00 \text{ m/s}^2.$$

c) Similar to the derivation of part (a),

$$\Delta K = \frac{1}{2}\omega^2 I - \frac{1}{2}\omega_0^2 I = \frac{1}{2}[\alpha][2\Delta\theta]I = I\alpha\Delta\theta.$$

d) Using the result of part (c),

$$I = \frac{\Delta K}{\alpha\,\Delta\theta} = \frac{(45.0 \text{ J} - 20.0 \text{ J})}{((2.00 \text{ m/s}^2)/(0.250 \text{ m}))(15.0 \text{ rad})} = 0.208 \text{ kg·m}^2.$$

9-66: Quantitatively, from Table (9-2), $I_A = \frac{1}{2}MR^2$, $I_B = MR^2$ and $I_C = \frac{2}{3}MR^2$.
a) Object A has the smallest moment of inertia because, of the three objects, its mass is the most concentrated near its axis. b) Conversely, object B's mass is concentrated the farthest from its axis. c) Because $I_{\text{sphere}} = 2/5MR^2$, the sphere would replace the disk as having the smallest moment of inertia.

9-67: a) See Exercise 9-44.

$$K = \frac{2\pi^2 I}{T^2} = \frac{2\pi^2(0.3308)(5.97 \times 10^{24} \text{ kg})(6.38 \times 10^6 \text{ m})^2}{(86,164 \text{ s})^2} = 2.14 \times 10^{29} \text{ J}.$$

b) $\dfrac{1}{2}M\left(\dfrac{2\pi R}{T}\right)^2 = \dfrac{2\pi^2(5.97 \times 10^{24} \text{ kg})(1.50 \times 10^{11} \text{ m})^2}{(3.156 \times 10^7 \text{ s})^2} = 2.66 \times 10^{33} \text{ J}.$

c) Since the Earth's moment of inertia is less than that of a uniform sphere, more of the Earth's mass must be concentrated near its center.

9-68: Using energy considerations, the system gains as kinetic energy the lost potential energy, mgR. The kinetic energy is

$$K = \frac{1}{2}I\omega^2 + \frac{1}{2}mv^2 = \frac{1}{2}I\omega^2 + \frac{1}{2}m(\omega R)^2 = \frac{1}{2}(I + mR^2)\omega^2.$$

Using $I = \frac{1}{2}mR^2$ and solving for ω,

$$\omega^2 = \frac{4}{3}\frac{g}{R}, \quad \text{and} \quad \omega = \sqrt{\frac{4}{3}\frac{g}{R}}.$$

9-69: a) $(0.160 \text{ kg})(-0.500 \text{ m})(9.80 \text{ m/s}^2) = -0.784 \text{ J}.$ b) The kinetic energy of the stick is 0.784 J, and so the angular velocity is

$$\omega = \sqrt{\frac{2K}{I}} = \sqrt{\frac{2K}{ML^2/3}} = \sqrt{\frac{2(0.784 \text{ J})}{(0.160 \text{ kg})(1.00 \text{ m})^2/3}} = 5.42 \text{ rad/s}.$$

This result may also be found by using the algebraic form for the kinetic energy, $K = MgL/2$, from which $\omega = \sqrt{3g/L}$, giving the same result. Note that ω is independent of the mass.

c) $v = \omega L = (5.42 \text{ rad/s})(1.00 \text{ m}) = 5.42 \text{ m/s}.$

d) $\sqrt{2gL} = 4.43$ m/s; This is $\sqrt{2/3}$ of the result of part (c).

9-70: Taking the zero of gravitational potential energy to be at the axle, the initial potential energy is zero (the rope is wrapped in a circle with center on the axle). When the rope has unwound, its center of mass is a distance πR below the axle, since the length of the rope is $2\pi R$ and half this distance is the position of the center of mass. Initially, every part of the rope is moving with speed $\omega_0 R$, and when the rope has unwound, and the cylinder has angular speed ω, the speed of the rope is ωR (the upper end of the rope has the same tangential speed as the edge of the cylinder). From conservation of energy, using $I = (1/2)MR^2$ for a uniform cylinder,

$$\left(\frac{M}{4} + \frac{m}{2}\right) R^2 \omega_0^2 = \left(\frac{M}{4} + \frac{m}{2}\right) R^2 \omega^2 - mg\pi R.$$

Solving for ω gives

$$\omega = \sqrt{\omega_0^2 + \frac{(4\pi mg/R)}{(M+2m)}},$$

and the speed of any part of the rope is $v = \omega R$.

9-71: In descending a distance d, gravity has done work $m_B gd$ and friction has done work $-\mu_k m_A gd$, and so the total kinetic energy of the system is $gd(m_B - \mu_k m_A)$. In terms of the speed v of the blocks, the kinetic energy is

$$K = \frac{1}{2}(m_A + m_B)v^2 + \frac{1}{2}I\omega^2 = \frac{1}{2}\left(m_A + m_B + I/R^2\right),$$

where $\omega = v/R$, and condition that the rope not slip, have been used. Seting the kinetic energy equal to the work done and solving for the speed v,

$$v = \sqrt{\frac{2gd\,(m_B - \mu_k m_A)}{(m_A + m_B + I/R^2)}}.$$

9-72: The gravitational potential energy which has become kinetic energy is $K = (4.00 \text{ kg} - 2.00 \text{ kg})(9.80 \text{ m/s}^2)(5.00 \text{ m}) = 98.0$ J. In terms of the common speed v of the blocks, the kinetic energy of the system is

$$K = \frac{1}{2}(m_1 + m_2)v^2 + \frac{1}{2}I\left(\frac{v}{R}\right)^2$$
$$= v^2\frac{1}{2}\left(4.00 \text{ kg} + 2.00 \text{ kg} + \frac{(0.480 \text{ kg·m}^2)}{(0.160 \text{ m})^2}\right) = v^2(12.4 \text{ kg}).$$

Solving for v gives $v = \sqrt{\frac{98.0 \text{ J}}{12.4 \text{ kg}}} = 2.81$ m/s.

9-73: The moment of inertia of the hoop about the nail is $2MR^2$ (see Exercise 9-46), and the initial potential energy with respect to the center of the loop when its center is directly below the nail is $gR(1 - \cos\beta)$. From the work-energy theorem,

$$K = \frac{1}{2}I\omega^2 = M\omega^2 R^2 = MgR(1 - \cos\beta),$$

from which $\omega = \sqrt{(g/R)(1 - \cos\beta)}$.

9-74: a) $K = \frac{1}{2}I\omega^2$

$$= \frac{1}{2}\left(\frac{1}{2}(1000 \text{ kg})(0.90 \text{ m})^2\right)\left(3000 \text{ rev/min} \times \frac{2\pi}{60}\frac{\text{rad/s}}{\text{rev/min}}\right)^2$$

$$= 2.00 \times 10^7 \text{ J}.$$

b)

$$\frac{K}{P_{\text{ave}}} = \frac{2.00 \times 10^7 \text{ J}}{1.86 \times 10^4 \text{ W}} = 1075 \text{ s},$$

which is about 18 min.

9-75: a)

$$\frac{1}{2}M_1 R_1^2 + \frac{1}{2}M_2 R_2^2 = \frac{1}{2}((0.80 \text{ kg})(2.50 \times 10^{-2} \text{ m})^2 + (1.60 \text{ kg})(5.00 \times 10^{-2} \text{ m})^2)$$
$$= 2.25 \times 10^{-3} \text{ kg·m}^2.$$

b) See Example 9-9. In this case, $\omega = v/R_1$, and so the expression for v becomes

$$v = \sqrt{\frac{2gh}{1 + (I/mR^2)}}$$

$$= \sqrt{\frac{2(9.80 \text{ m/s}^2)(2.00 \text{ m})}{(1 + ((2.25 \times 10^{-3} \text{ kg·m}^2)/(1.50 \text{ kg})(0.025 \text{ m})^2))}} = 3.40 \text{ m/s}.$$

c) The same calculation, with R_2 instead of R_1, gives $v = 4.95$ m/s. This does make sense, because for a given total energy, the disk combination will have a larger fraction of the kinetic energy with the string on the larger radius, and with this larger fraction, the disk combination must be moving faster.

9-76: a) In the case that no energy is lost, the rebound height h' is related to the speed v by $h' = \frac{v^2}{2g}$, and with the form for h given in Example 9-9, $h' = \frac{h}{1 + M/2m}$. b) Considering the system as a whole, some of the initial potential energy of the mass went into the kinetic energy of the cylinder. Considering the mass alone, the tension in the string did work on the mass, so its total energy is not conserved.

9-77: a) The initial moment of inertia is $I_0 = \frac{1}{2}MR^2$. The piece punched has a mass of $\frac{M}{16}$ and a moment of inertia with respect to the axis of the original disk of

$$\frac{M}{16}\left[\frac{1}{2}\left(\frac{R}{4}\right)^2 + \left(\frac{R}{2}\right)^2\right] = \frac{9}{512}MR^2.$$

The moment of inertia of the remaining piece is then

$$I = \frac{1}{2} MR^2 - \frac{9}{512} MR^2 = \frac{247}{512} MR^2.$$

b) $I = \frac{1}{2}MR^2 + M(R/2)^2 - \frac{1}{2}(M/16)(R/4)^2 = \frac{383}{512}MR^2$.

9-78: a) From the parallel-axis theorem, the moment of inertia is
$I_P = (2/5)MR^2 + ML^2$, and

$$\frac{I_P}{ML^2} = \left(1 + \left(\frac{2}{5}\right)\left(\frac{R}{L}\right)^2\right).$$

If $R = (0.05)L$, the difference is $(2/5)(0.05)^2 = 0.001$. b) $(I_{\text{rod}}/ML^2) = (m_{\text{rod}}/3M)$, which is 0.33% when $m_{\text{rod}} = (0.01)M$.

9-79: a) With respect to O, each element r_i^2 in Eq. (9-17) is $x_i^2 + y_i^2$, and so

$$I_O = \sum_i m_i r_i^2 = \sum_i m_i (x_i^2 + y_i^2) = \sum_i m_i x_i^2 + \sum_i m_i y_i^2 = I_x + I_y.$$

b) Two perpendicular axes, both perpendicular to the washer's axis, will have the same moment of inertia about those axes, and the perpendicular-axis theorem predicts that they will sum to the moment of inertia about the washer axis, which is $\frac{M}{2}(R_1^2 + R_2^2)$, and so $I_x = I_y = \frac{M}{4}(R_1^2 + R_2^2)$.

c) From Table (9-2), $I = \frac{1}{12}m(L^2 + L^2) = \frac{1}{6}mL^2$. Since $I_0 = I_x + I_y$, and $I_x = I_y$, both I_x and I_y must be $\frac{1}{12}mL^2$.

9-80: Each side has length a and mass $\frac{M}{4}$, and the moment of inertia of each side about an axis perpendicular to the side and through its center is $\frac{1}{12}\frac{M}{4}a^2 = \frac{Ma^2}{48}$. The moment of inertia of each side about the axis through the center of the square is, from the perpendicular axis theorem, $\frac{Ma^2}{48} + \frac{M}{4}\left(\frac{a}{2}\right)^2 = \frac{Ma^2}{12}$. The total moment of inertia is the sum of the contributions from the four sides, or $4 \times \frac{Ma^2}{12} = \frac{Ma^2}{3}$.

9-81: Introduce the auxiliary variable L, the length of the cylinder, and consider thin cylindrical shells of thickness dr and radius r; the cross-sectional area of such a shell is $2\pi r\, dr$, and the mass of the shell is $dm = 2\pi r L\rho\, dr = 2\pi\alpha L r^2\, dr$. The total mass of the cylinder is then

$$M = \int dm = 2\pi L\alpha \int_0^R r^2\, dr = 2\pi L\alpha \frac{R^3}{3}$$

and the moment of inertia is

$$I = \int r^2\, dm = 2\pi L\alpha \int_0^R r^4\, dr = 2\pi L\alpha \frac{R^5}{5} = \frac{3}{5} MR^2.$$

b) This is less than the moment of inertia if all the mass were concentrated at the edge, as with a thin shell with $I = MR^2$, and is greater than that for a uniform cylinder with $I = \frac{1}{2}MR^2$, as expected.

9-82: a) From Exercise 9-43, the rate of energy loss is $\frac{4\pi^2 I}{T^3}\frac{dT}{dt}$; solving for the moment of inertia I in terms of the power P,

$$I = \frac{PT^3}{4\pi^2}\frac{1}{dT/dt} = \frac{(5 \times 10^{31}\text{ W})(0.0331\text{ s})^3}{4\pi^2}\frac{1\text{ s}}{4.22 \times 10^{-13}\text{ s}} = 1.09 \times 10^{38}\text{ kg·m}^2.$$

b) $R = \sqrt{\dfrac{5I}{2M}} = \sqrt{\dfrac{5(1.08 \times 10^{38}\text{ kg·m}^2)}{2(1.4)(1.99 \times 10^{30}\text{ kg})}} = 9.9 \times 10^3\text{ m}$, about 10 km.

c)
$$\frac{2\pi R}{T} = \frac{2\pi(9.9 \times 10^3\text{ m})}{(0.0331\text{ s})} = 1.9 \times 10^6\text{ m/s} = 6.3 \times 10^{-3}c.$$

d)
$$\frac{M}{V} = \frac{M}{(4\pi/3)R^3} = 6.9 \times 10^{17}\text{ kg/m}^3,$$

which is much higher than the density of ordinary rock by 14 orders of magnitude, and is comparable to nuclear mass densities.

9-83: a) Following the hint, the moment of inertia of a uniform sphere in terms of the mass density is $I = \frac{2}{5}MR^2 = \frac{8\pi}{15}\rho R^5$, and so the difference in the moments of inertia of two spheres with the same density ρ but different radii R_2 and R_1 is $I = \rho(8\pi/15)(R_2^5 - R_1^5)$.

b) A rather tedious calculation, summing the product of the densities times the difference in the cubes of the radii that bound the regions and multiplying by $4\pi/3$, gives $M = 5.97 \times 10^{24}$ kg. c) A similar calculation, summing the product of the densities times the difference in the fifth powers of the radii that bound the regions and multiplying by $8\pi/15$, gives $I = 8.02 \times 10^{22}$ kg·m$^2 = 0.334MR^2$.

9-84: Following the procedure used in Example 9-14 (and using z as the coordinate along the vertical axis), $r(z) = z\frac{R}{h}$, $dm = \pi\rho\frac{R^2}{h^2}z^2\,dz$ and $dI = \frac{\pi\rho}{2}\frac{R^4}{h^4}z^4\,dz$. Then,

$$I = \int dI = \frac{\pi\rho}{2}\frac{R^4}{h^4}\int_0^h z^4\,dz = \frac{\pi\rho}{10}\frac{R^4}{h^4}\left[z^5\right]_0^h = \frac{1}{10}\pi\rho R^4 h.$$

The volume of a right circular cone is $V = \frac{1}{3}\pi R^2 h$, the mass is $\frac{1}{3}\pi\rho R^2 h$, and so

$$I = \frac{3}{10}\left(\frac{\pi\rho R^2 h}{3}\right)R^2 = \frac{3}{10}MR^2.$$

9-85: a) $ds = r\,d\theta = r_0\,d\theta + \beta\theta\,d\theta$, so $s(\theta) = r_0\theta + \frac{\beta}{2}\theta^2$. b) Setting $s = vt = r_0\theta + \frac{\beta}{2}\theta^2$ gives a quadratic in θ. The positive solution is

$$\theta(t) = \frac{1}{\beta}\left[\sqrt{r_0^2 + 2\beta vt} - r_0\right].$$

(The negative solution would be going backwards, to values of r smaller than r_0.)

c) Differentiating,

$$\omega(t) = \frac{d\theta}{dt} = \frac{v}{\sqrt{r_0^2 + \beta vt}},$$

$$\alpha = \frac{d\omega}{dt} = -\frac{\beta v^2}{(r_0^2 + \beta vt)^{3/2}}.$$

The angular acceleration α is not constant.   d) $r_0 = 25.0$ mm; It is crucial that θ is measured in radians, so $\beta = (1.55 \ \mu m/rev)(1 \ rev/2\pi \ rad) = 0.247 \ \mu m/rad$. The total angle turned in 74.0 min = 4440 s is

$$\theta = \frac{1}{2.47 \times 10^{-7} \ m/rad} \left[\sqrt{\begin{array}{c} 2(2.47 \times 10^{-7} \ m/rad)(1.25 \ m/s)(4440 \ s) \\ + (25.0 \times 10^{-3} \ m)^2 \\ - 25.0 \times 10^{-3} \ m \end{array}} \right]$$

$$= 1.337 \times 10^5 \ rad$$

which is 2.13×10^4 rev.

e)

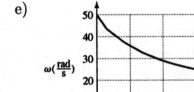
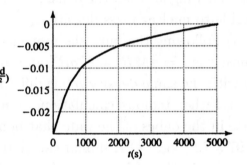

Chapter 10 Dynamics of Rotational Motion

10-1: Equation (10-2) or Eq. (10-3) is used for all parts.

    a) $(4.00 \text{ m})(10.0 \text{ N}) \sin 90° = 40.00$ N·m, out of the page.

    b) $(4.00 \text{ m})(10.0 \text{ N}) \sin 120° = 34.6$ N, out of the page.

    c) $(4.00 \text{ m})(10.0 \text{ N}) \sin 30° = 20.0$ N·m, out of the page.

    d) $(2.00 \text{ m})(10.00 \text{ N}) \sin 60° = 17.3$ N·m, into the page.

    e) The force is applied at the origin, so $\tau = 0$.

    f) $(4.00 \text{ m})(10.0 \text{ N}) \sin 180° = 0$.

10-2:
$$\tau_1 = -(8.00 \text{ N})(5.00 \text{ m}) = -40.0 \text{ N·m},$$

$$\tau_2 = (12.0 \text{ N})(2.00 \text{ m}) \sin 30° = 12.0 \text{ N·m},$$

where positive torques are taken counterclockwise, so the net torque is -28.0 N·m, with the minus sign indicating a clockwise torque, or a torque into the page.

10-3: Taking positive torques to be counterclockwise (out of the page), $\tau_1 = -(0.090 \text{ m}) \times (18.0 \text{ N}) = -1.62$ N·m, $\tau_2 = (0.09 \text{ m})(26.0 \text{ N}) = 2.34$ N·m, $\tau_3 = (\sqrt{2})(0.09 \text{ m})(14.0 \text{ N}) = 1.78$ N·m, so the net torque is 2.50 N·m, with the direction counterclockwise (out of the page). Note that for τ_3, the applied force is perpendicular to the lever arm.

10-4:
$$\tau_1 + \tau_2 = -F_1 R + F_2 R = (F_2 - F_1)R$$

$$= (5.30 \text{ N} - 7.50 \text{ N})(0.330 \text{ m}) = -0.726 \text{ N·m}.$$

10-5: a)

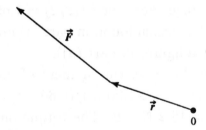

    b) Into the plane of the page.

    c) $\vec{r} \times \vec{F} = \left[(-0.450 \text{ m})\hat{i} + (0.150 \text{ m})\hat{j} \right] \times \left[(-5.00 \text{ N})\hat{i} + (4.00 \text{ N})\hat{j} \right]$

$$= \left[(-0.450 \text{ m})(4.00 \text{ N}) - (0.150 \text{ m})(-5.00 \text{ N}) \right] \hat{k}$$

$$= (-1.05 \text{ N·m})\hat{k}$$

10-6: a) $\tau = I\alpha = I\dfrac{\Delta\omega}{\Delta t} = (2.50 \text{ kg·m}^2)\dfrac{\left(400 \text{ rev/min} \times \frac{2\pi}{60} \frac{\text{rad/s}}{\text{rev/min}} \right)}{(8.00 \text{ s})} = 13.1 \text{ N·m}.$

    b) $\dfrac{1}{2}I\omega^2 = \dfrac{1}{2}(2.50 \text{ kg·m}^2)\left(400 \text{ rev/min} \times \dfrac{2\pi}{60} \dfrac{\text{rad/s}}{\text{rev/min}} \right)^2 = 2.19 \times 10^3 \text{ J}.$

10-7: $v = \sqrt{2as} = \sqrt{2(0.36 \text{ m/s}^2)(2.0 \text{ m})} = 1.2$ m/s, the same as that found in Example 9-8.

10-8:
$$\alpha = \frac{\tau}{I} = \frac{FR}{I} = \frac{(40.0 \text{ N})(0.250 \text{ m})}{(5.0 \text{ kg·m}^2)} = 2.00 \text{ rad/s}^2.$$

10-9: a)
$$\mathcal{N} = Mg + T = g\left[M + \frac{m}{1 + 2m/M}\right] = g\left[\frac{M + 3m}{1 + 2m/M}\right]$$

b) This is less than the total weight; the suspended mass is accelerating down, so the tension is less than mg. c) As long as the cable remains taut, the velocity of the mass does not affect the acceleration, and the tension and normal force are unchanged.

10-10: a) The cylinder does not move, so the net force must be zero. The cable exerts a horizontal force to the right, and gravity exerts a downward force, so the normal force must exert a force up and to the left, as shown in Fig. (10-8). b) $\mathcal{N} = \sqrt{(9.0 \text{ N})^2 + ((50 \text{ kg})(9.80 \text{ m/s}^2))^2} = 490$ N, at an angle of arctan $\left(\frac{9.0}{490}\right) = 1.1°$ from the vertical (the weight is much larger than the applied force F).

10-11:
$$\mu_k = \frac{f}{\mathcal{N}} = \frac{\tau/R}{\mathcal{N}} = \frac{I\alpha}{R\mathcal{N}} = \frac{MR(\omega_0/t)}{2\mathcal{N}}$$

$$= \frac{(50.0 \text{ kg})(0.260 \text{ m})(850 \text{ rev/min})\left(\frac{\pi}{30}\frac{\text{rad/s}}{\text{rev/min}}\right)}{2(7.50 \text{ s})(160 \text{ N})} = 0.482.$$

10-12: This is the same situation as in Example 10-3. a) $T = mg/(1 + 2m/M) = 42.0$ N. b) $v = \sqrt{2gh/(1 + M/2m)} = 11.8$ m/s c) There are many ways to find the time of fall. Rather than make the intermediate calculation of the acceleration, the time is the distance divided by the average speed, or $h/(v/2) = 1.69$ s. d) The normal force in Fig. (10-1-9(b)) is the sum of the tension found in part (a) and the weight of the windlass, a total of 159.6 N (keeping extra figures in part (a)).

10-13: See Example 10-4. In this case, the moment of inertia I is unknown, so $a_1 = (m_2 g)/(m_1 + m_2 + (I/R^2))$. a) $a_1 = 2(1.20 \text{ m})/(0.80 \text{ s})^2 = 3.75 \text{ m/s}^2$, so $T_1 = m_1 a_1 = 7.50$ N and $T_2 = m_2(g - a_1) = 18.2$ N. b) The torque on the pulley is $(T_2 - T_1)R = 0.803$ N·m, and the angular acceleration is $\alpha = a_1/R = 50 \text{ rad/s}^2$, so $I = \tau/\alpha = 0.016$ kg·m^2.

10-14:
$$\alpha = \frac{\tau}{I} = \frac{Fl}{\frac{1}{3}Ml^2} = \frac{3F}{Ml}.$$

10-15: The acceleration of the mass is related to the tension by $Ma_{cm} = Mg - T$, and the angular acceleration is related to the torque by $I\alpha = \tau = TR$, or $a_{cm} = T/M$, where $\alpha = a_{cm}/R$ and $I = MR^2$ have been used. a) Solving these for T gives $T = Mg/2 = 0.882$ N. b) Substituting the expression for T into either of the above relations gives $a_{cm} = g/2$, from which $t = \sqrt{2h/a_{cm}} = \sqrt{4h/g} = 0.553$ s. c) $\omega = v_{cm}/R = a_{cm}t/R = 33.9$ rad/s.

10-16: See Example 10-6 and Exercise 10-17. In this case, $K_2 = Mv_{cm}^2$ and $v_{cm} = \sqrt{gh}$, $\omega = v_{cm}/R = 33.9$ rad/s.

10-17: From Eq. (10-11), the fraction of the total kinetic energy that is rotational is

$$\frac{(1/2)I_{cm}\omega^2}{(1/2)Mv_{cm}^2 + (1/2)I_{cm}\omega^2} = \frac{1}{1 + (M/I_{cm})/(v_{cm}^2/\omega^2)} = \frac{1}{1 + \frac{MR^2}{I_{cm}}},$$

where $v_{cm} = R\omega$ for an object that is rolling without slipping has been used. a) $I_{cm} = (1/2)MR^2$, so the above ratio is 1/3. b) $I = (2/5)MR^2$, so the above ratio is 2/7. c) $I = 2/3\ MR^2$, so the ratio is 2/5. d) $I = 5/8\ MR^2$, so the ratio is 5/13.

10-18: a) The acceleration down the slope is $a = g\sin\theta - \frac{f}{m}$, the torque about the center of the shell is

$$\tau = Rf = I\alpha = I\frac{a}{R} = \frac{2}{3}MR^2\frac{a}{R} = \frac{2}{3}MRa,$$

so $\frac{f}{M} = \frac{2}{3}a$. Solving these relations for a and f simultaneously gives $\frac{5}{3}a = g\sin\theta$, or

$$a = \frac{3}{5}g\sin\theta = \frac{3}{5}(9.80\ \text{m/s}^2)\sin 38.0° = 3.62\ \text{m/s}^2,$$

$$f = \frac{2}{3}Ma = \frac{2}{3}(2.00\ \text{kg})(3.62\ \text{m/s}^2) = 4.83\ \text{N}.$$

The normal force is $Mg\cos\theta$, and since $f \leq \mu_s N$,

$$\mu_s \geq \frac{f}{N} = \frac{\frac{2}{3}Ma}{Mg\cos\theta} = \frac{2}{3}\frac{a}{g\cos\theta} = \frac{2}{3}\frac{\frac{3}{5}g\sin\theta}{g\cos\theta} = \frac{2}{5}\tan\theta = 0.313.$$

b) $a = 3.62\ \text{m/s}^2$ since it does not depend on the mass. The frictional force, however, is twice as large, 9.65 N, since it does depend on the mass. The minimum value of μ_s also does not change.

10-19: $wh + W_f = K_1 = (1/2)I_{cm}w_0^2 + \frac{1}{2}mv_{cm}^2$

Solving for h with $v_{cm} = Rw$

$$h = \frac{\frac{1}{2}\left(\frac{w}{9.80\ \text{m/s}^2}\right)[(0.800)(0.600\ \text{m})^2(25.0\ \text{rad/s})^2 + (0.600\ \text{m})^2(25.0\ \text{rad/s})^2]}{w}$$
$$-\frac{3500\ \text{J}}{392\ \text{N}} = 11.7\ \text{m}.$$

10-20: a)

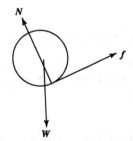

The angular speed of the ball must decrease, and so the torque is provided by a friction force that acts up the hill.

b) The friction force results in an angular acceleration, related by $I\alpha = fR$. The equation of motion is $mg\sin\beta - f = ma_{cm}$, and the acceleration and angular acceleration are related by $a_{cm} = R\alpha$ (note that positive acceleration is taken to be *down* the incline, and the relation between a_{cm} and α is correct for a friction force directed uphill). Combining,

$$mg \sin \beta = ma \left(1 + \frac{I}{mR^2}\right) = ma(7/5),$$

from which $a_{cm} = (5/7)g \sin \beta$.   c) From either of the above relations between f and a_{cm},

$$f = \frac{2}{5} ma_{cm} = \frac{2}{7} mg \sin \beta \leq \mu_s \mathcal{N} = \mu_s mg \cos \beta,$$

from which $\mu_s \geq (2/7) \tan \beta$.

10-21: a) $\omega = \alpha \Delta t = (FR/I)\Delta t = ((18.0 \text{ N})(2.40 \text{ m})/(2100 \text{ kg·m}^2))(15.0 \text{ s}) = 0.3086 \text{ rad/s}$, or 0.309 rad/s to three figures. b) $W = K_2 = (1/2)I\omega^2 = (1/2) \times (2.00 \text{ kg·m}^2)(0.3086 \text{ rad/s})^2 = 100 \text{ J}$. c) From either $P = \tau \omega_{ave}$ or $P = W/\Delta t$, $P = 6.67 \text{ W}$.

10-22: a) $$\tau = \frac{P}{\omega} = \frac{(175 \text{ hp})(746 \text{ W/hp})}{(2400 \text{ rev/min}) \left(\frac{\pi}{30} \frac{\text{rad/s}}{\text{rev/min}}\right)} = 519 \text{ N·m}.$$

b) $W = \tau \Delta \theta = (519 \text{ N·m})(2\pi) = 3261 \text{ J}$.

10-23: a) $\tau = I\alpha = I\dfrac{\Delta \omega}{\Delta t}$

$$= \frac{((1/2)(1.50 \text{ kg})(0.100 \text{ m})^2)(1200 \text{ rev/min})\left(\frac{\pi}{30} \frac{\text{rad/s}}{\text{rev/min}}\right)}{2.5 \text{ s}}$$

$$= 0.377 \text{ N·m}.$$

b) $\omega_{ave}\Delta t = \dfrac{(600 \text{ rev/min})(2.5 \text{ s})}{60 \text{ s/min}} = 25.0 \text{ rev} = 157 \text{ rad}$.   c) $\tau \Delta \theta = 59.2 \text{ J}$.

d) $K = \dfrac{1}{2}I\omega^2$

$$= \frac{1}{2}((1/2)(1.5 \text{ kg})(0.100 \text{ m})^2)\left((1200 \text{ rev/min})\left(\frac{\pi}{30} \frac{\text{rad/s}}{\text{rev/min}}\right)\right)^2$$

$$= 59.2 \text{ J},$$

the same as in part (c).

10-24: From Eq. (10-26), the power output is

$$P = \tau \omega = (4.30 \text{ N·m})\left(4800 \text{ rev/min} \times \frac{2\pi}{60} \frac{\text{rad/s}}{\text{rev/min}}\right) = 2161 \text{ W},$$

which is 2.9 hp.

10-25: a) With no load, the only torque to be overcome is friction in the bearings (neglecting air friction), and the bearing radius is small compared to the blade radius, so any frictional torque could be neglected.

b) $$F = \frac{\tau}{R} = \frac{P/\omega}{R} = \frac{(1.9 \text{ hp})(746 \text{ W/hp})}{(2400 \text{ rev/min})\left(\frac{\pi}{30} \frac{\text{rad/s}}{\text{rev/min}}\right)(0.086 \text{ m})} = 65.6 \text{ N}.$$

10-26: $I = \frac{1}{12} mL^2 = \frac{1}{12}(117 \text{ kg})(2.08 \text{ m})^2 = 42.2 \text{ kg·m}^2$

a)
$$\alpha = \frac{\tau}{I} = \frac{1950 \text{ N·m}}{42.2 \text{ kg·m}^2} = 46.2 \text{ rad/s}^2.$$

b) $\omega = \sqrt{2\alpha\theta} = \sqrt{2(46.2 \text{ rad/s}^2)(5.0 \text{ rev} \times 2\pi \text{ rad/rev})} = 53.9 \text{ rad/s}.$

c) From either $W = K = \frac{1}{2}I\omega^2$ or Eq. (10-24),

$$W = \tau\theta = (1950 \text{ N·m})(5.00 \text{ rev} \times 2\pi \text{ rad/rev}) = 6.13 \times 10^4 \text{ J}.$$

d), e)   The time may be found from the angular acceleration and the total angle, but the instantaneous power is also found from $P = \tau\omega = 105 \text{ kW}(141 \text{ hp})$. The average power is half of this, or 52.6 kW.

10-27: a) $\tau = P/\omega = (150 \times 10^3 \text{ W}) \Big/ \left((4000 \text{ rev/min}) \left(\frac{\pi}{30} \frac{\text{rad/s}}{\text{rev/min}} \right) \right) = 358 \text{ N·m}.$

b) If the tension in the rope is F, $F = w$ and so $w = \tau/R = 1.79 \times 10^3 \text{ N}$.

c) Assuming ideal efficiency, the rate at which the weight gains potential energy is the power output of the motor, or $wv = P$, so $v = P/w = 83.8 \text{ m/s}$. Equivalently, $v = \omega R$.

10-28:   As a point, the woman's moment of inertia with respect to the disk axis is mR^2, and so the total angular momentum is

$$L = L_{\text{disk}} + L_{\text{woman}} = (I_{\text{disk}} + I_{\text{woman}})\omega = \left(\frac{1}{2}M + m \right) R^2\omega$$

$$= \left(\frac{1}{2}110 \text{ kg} + 50.0 \text{ kg} \right) (4.00 \text{ m})^2 (0.500 \text{ rev/s} \times 2\pi \text{ rad/rev})$$

$$= 5.28 \times 10^3 \text{ kg·m}^2/\text{s}.$$

10-29:   a) $mvr \sin\phi = 115 \text{ kg·m}^2/\text{s}$, with a direction from the right hand rule of into the page.

b) $dL/dt = \tau = (2 \text{ kg})(9.8 \text{ N/kg}) \cdot (8 \text{ m}) \cdot \sin(90° - 36.9°) = 125 \text{ N·m} = 125 \text{ kg·m}^2/\text{s}^2$, out of the page.

10-30:   For both parts, $L = I\omega$. Also, $\omega = v/r$, so $L = I(v/r)$.

a) $L = (mr^2)(v/r) = mvr$

$L = (5.97 \times 10^{24} \text{ kg})(2.98 \times 10^4 \text{ m/s})(1.50 \times 10^{11} \text{ m}) = 2.67 \times 10^{40} \text{ kg·m}^2/\text{s}$

b) $L = (2/5 \, mr^2)(\omega)$

$L = (2/5)(5.97 \times 10^{24} \text{ kg})(6.38 \times 10^6 \text{ m})^2(2\pi \text{ rad}/(24.0 \text{ hr} \times 3600 \text{ s/hr}))$

$= 7.07 \times 10^{33} \text{ kg·m}^2/\text{s}$

10-31:   The period of a second hand is one minute, so the angular momentum is

$$L = I\omega = \frac{M}{3}l^2\frac{2\pi}{T}$$

$$= \left(\frac{6.0 \times 10^{-3} \text{ kg}}{3} \right) (15.0 \times 10^{-2} \text{ m})^2 \frac{2\pi}{60 \text{ s}} = 4.71 \times 10^{-6} \text{ kg·m}^2/\text{s}.$$

10-32: The moment of inertia is proportional to the square of the radius, and so the angular velocity will be proportional to the inverse of the square of the radius, and the final angular velocity is

$$\omega_2 = \omega_1 \left(\frac{R_1}{R_2}\right)^2 = \left(\frac{2\pi \text{ rad}}{(30 \text{ d})(86{,}400 \text{ s/d})}\right)\left(\frac{7.0 \times 10^5 \text{ km}}{16 \text{ km}}\right)^2 = 4.6 \times 10^3 \text{ rad/s}.$$

10-33: a) The net force is due to the tension in the rope, which always acts in the radial direction, so the angular momentum with respect to the hole is constant. b) $L_1 = m\omega_1 r_1^2$, $L_2 = m\omega_2 r_2^2$, and with $L_1 = L_2$, $\omega_2 = \omega_1 (r_1/r_2)^2 = 7.00$ rad/s. c) $\Delta K = (1/2)m((\omega_2 r_2)^2 - (\omega_1 r_1)^2) = 1.03 \times 10^{-2}$ J. d) No other force does work, so 1.03×10^{-2} J of work were done in pulling the cord.

10-34: The skater's initial moment of inertia is

$$I_1 = (0.400 \text{ kg·m}^2) + \frac{1}{12}(8.00 \text{ kg})(1.80 \text{ m})^2 = 2.56 \text{ kg·m}^2,$$

and her final moment of inertia is

$$I_2 = (0.400 \text{ kg·m}^2) + (8.00 \text{ kg})(25 \times 10^{-2} \text{ m})^2 = 0.9 \text{ kg·m}^2.$$

Then from Eq. (10-33),

$$\omega_2 = \omega_1 \frac{I_1}{I_2} = (0.40 \text{ rev/s})\frac{2.56 \text{ kg·m}^2}{0.9 \text{ kg·m}^2} = 1.14 \text{ rev/s}.$$

Note that conversion from rev/s to rad/s is not necessary.

10-35: If she had tucked, she would have made $(2)(3.6 \text{ kg·m}^2)/(18 \text{ kg·m}^2) = 0.40$ rev in the last 1.0 s, so she would have made $(0.40 \text{ rev})(1.5/1.0) = 0.60$ rev in the total 1.5 s.

10-36: Let

$$I_1 = I_0 = 1200 \text{ kg·m}^2,$$
$$I_2 = I_0 + mR^2 = 1200 \text{ kg·m}^2 + (40.0 \text{ kg})(2.00 \text{ m})^2 = 1360 \text{ kg·m}^2.$$

Then, from Eq. (10-33),

$$\omega_2 = \omega_1 \frac{I_1}{I_2} = \left(\frac{2\pi \text{ rad}}{6.00 \text{ s}}\right)\frac{1200 \text{ kg·m}^2}{1360 \text{ kg·m}^2} = 0.924 \text{ rad/s}.$$

10-37: a) From conservation of angular momentum,

$$\omega_2 = \omega_1 \frac{I_0}{I_0 + mR^2} = \omega_1 \frac{(1/2)MR^2}{(1/2)MR^2 + mR^2} = \omega_1 \frac{1}{1 + 2m/M}$$

$$= \frac{3.0 \text{ rad/s}}{1 + 2(70)/(120)} = 1.385 \text{ rad/s}$$

or 1.39 rad/s to three figures.

b) $K_1 = (1/2)(1/2)(120 \text{ kg})(2.00 \text{ m})^2(3.00 \text{ rad/s})^2 = 1.08 \text{ kJ}$, and $K_2 = (1/2)(I_0 + (70 \text{ kg})(2.00 \text{ m})^2)\omega_2^2 = 499 \text{ J}$. In changing the parachutist's horizontal component of velocity and slowing down the turntable, friction does negative work.

10-38: Let the width of the door be l;

$$\omega = \frac{L}{I} = \frac{mv(l/2)}{(1/3)Ml^2 + m(l/2)^2}$$

$$= \frac{(0.500 \text{ kg})(12.0 \text{ m/s})(0.500 \text{ m})}{(1/3)(40.0 \text{ kg})(1.00 \text{ m})^2 + (0.500 \text{ kg})(0.500 \text{ m})^2} = 0.223 \text{ rad/s}.$$

Ignoring the mass of the mud in the denominator of the above expression gives $\omega = 0.225 \text{ rad/s}$, so the mass of the mud in the moment of inertia does affect the third significant figure.

10-39:

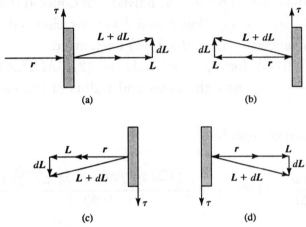

a)

b)

c)

d)

10-40: a) Since the gyroscope is precessing in a horizontal plane, there can be no net vertical force on the gyroscope, so the force that the pivot exerts must be equal in magnitude to the weight of the gyroscope, $F = w = mg = (0.140 \text{ kg})(9.80 \text{ m/s}^2) = 1.372 \text{ N}$, or 1.37 N to three figures. b) Solving Eq. (10-36) for ω,

$$\omega = \frac{wR}{I\Omega} = \frac{(1.372 \text{ N})(4.00 \times 10^{-2} \text{ m})}{(1.20 \times 10^{-4} \text{ kg·m}^2)(2\pi \text{ rad}/2.20 \text{ s})} = 160 \text{ rad/s},$$

which is $1.53 \times 10^3 \text{ rev/min}$. Note that in this and similar situations, since Ω appears in the denominator of the expression for ω, the conversion from rev/s and back to rev/min *must* be made.

c)

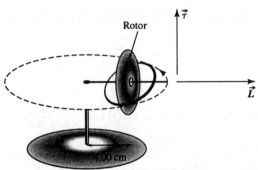

Rotor

10-41:   a) $\dfrac{K}{P} = \dfrac{(1/2)((1/2)MR^2)\omega^2}{P}$

$$= \dfrac{(1/2)((1/2)(60{,}000 \text{ kg})(2.00 \text{ m})^2)\left((500 \text{ rev/min})\left(\frac{\pi}{30}\frac{\text{rad/s}}{\text{rev/min}}\right)\right)^2}{7.46 \times 10^4 \text{ W}}$$

$= 2.21 \times 10^3$ s,

or 36.8 min.

   b) $\tau = I\Omega\omega$

$$= (1/2)(60{,}000 \text{ kg})(2.00 \text{ m})^2(500 \text{ rev/min})\left(\frac{\pi}{30}\frac{\text{rad/s}}{\text{rev/min}}\right)(1.00 \text{ °/s})\left(\frac{2\pi \text{ rad}}{360°}\right)$$

$= 1.10 \times 10^5$ N·m.

10-42:   Using Eq. (10-36) for all parts,   a) halved   b) doubled (assuming that the added weight is distributed in such a way that r and I are not changed)   c) halved (assuming that w and r are not changed)   d) doubled   e) unchanged.

10-43:   Solving Eq. (10-36) for τ, $\tau = I\omega\Omega = (2/5)MR^2\omega\Omega$. Using $\omega = \frac{2\pi \text{ rad}}{86{,}400 \text{ s}}$ and $\Omega = \frac{2\pi}{(26{,}000 \text{ y})(3.175 \times 10^7 \text{ s/y})}$ and the mass and radius of the earth from Appendix F, $\tau \sim 5.4 \times 10^{22}$ N·m.

10-44:   a) The net torque must be

$$\tau = I\alpha = I\frac{\Delta\omega}{\Delta t} = (1.86 \text{ kg·m}^2)\frac{\left(120 \text{ rev/min} \times \frac{2\pi}{60}\frac{\text{rad/s}}{\text{rev/min}}\right)}{(9.00 \text{ s})} = 2.60 \text{ N·m.}$$

This torque must be the sum of the applied force FR and the opposing frictional torques τ_f at the axle and $fr = \mu_k \mathcal{N}r$ due to the knife. Combining,

$$F = \frac{1}{R}(\tau + \tau_f + \mu_k \mathcal{N}r)$$

$$= \frac{1}{0.500 \text{ m}}((2.60 \text{ N·m}) + (6.50 \text{ N·m}) + (0.60)(160 \text{ N})(0.260 \text{ m}))$$

$$= 68.1 \text{ N.}$$

b) To maintain a constant angular velocity, the net torque τ is zero, and the force F' is $F' = \frac{1}{0.500 \text{ m}}(6.50 \text{ N·m} + 24.96 \text{ N·m}) = 62.9 \text{ N.}$ c) The time t needed to come to a stop is found by taking the magnitudes in Eq. (10-27), with $\tau = \tau_f$ constant;

$$t = \frac{L}{\tau_f} = \frac{\omega I}{\tau_f} = \frac{\left(120 \text{ rev/min} \times \frac{2\pi}{60}\frac{\text{rad/s}}{\text{rev/min}}\right)(1.86 \text{ kg·m}^2)}{(6.50 \text{ N·m})} = 3.6 \text{ s.}$$

Note that this time can also be found as $t = (9.00 \text{ s})\frac{2.60 \text{ N·m}}{6.50 \text{ N·m}}$.

10-45:   a) $I = \dfrac{\tau}{\alpha} = \dfrac{\tau\Delta t}{\Delta\omega} = \dfrac{(5.0 \text{ N·m})(2.0 \text{ s})}{(100 \text{ rev/min})\left(\frac{\pi}{30}\frac{\text{rad/s}}{\text{rev/min}}\right)} = 0.955 \text{ kg·m}^2.$

b) Rather than use the result of part (a), the magnitude of the torque is proportional to α and hence inversely proportional to $|\Delta t|$; equivalently, the magnitude of the change in angular momentum is the same and so the magnitude of the torque is again proportional to $1/|\Delta t|$. Either way, $\tau_f = (5.0 \text{ N·m}) \frac{2 \text{ s}}{125 \text{ s}} = 0.080 \text{ N·m}$.

c) $\omega_{\text{ave}} \Delta t = (50.0 \text{ rev/min})(125 \text{ s})(1 \text{ min}/60 \text{ s}) = 104.2 \text{ rev}$.

10-46: a) The moment of inertia is not given, so the angular acceleration must be found from kinematics;

$$\alpha = \frac{2\theta}{t^2} = \frac{2s}{rt^2} = \frac{2(5.00 \text{ m})}{(0.30 \text{ m})(2.00 \text{ s})^2} = 8.33 \text{ rad/s}^2.$$

b) $\alpha t = (8.33 \text{ rad/s}^2)(2.00 \text{ s}) = 16.67 \text{ rad/s}$.

c) The work done by the rope on the flywheel will be the final kinetic energy; $K = W = Fs = (40.0 \text{ N})(5.0 \text{ m}) = 200 \text{ J}$.

d)
$$I = \frac{2K}{\omega^2} = \frac{2(200 \text{ J})}{(16.67 \text{ rad/s})^2} = 1.44 \text{ kg·m}^2.$$

10-47: a)
$$P = \tau\omega = \tau\alpha t = \tau\left(\frac{\tau}{I}\right)t = \tau^2\left(\frac{t}{I}\right).$$

b) From the result of part (a), the power is $(500 \text{ W})\left(\frac{60.0}{20.0}\right)^2 = 4.50 \text{ kW}$.

c)
$$P = \tau\omega = \tau\sqrt{2\alpha\theta} = \tau\sqrt{2(\tau/I)\theta} = \tau^{3/2}\sqrt{2\theta/I}.$$

d) From the result of part (c), the power is $(500 \text{ W})\left(\frac{60.0}{20.0}\right)^{3/2} = 2.6 \text{ kW}$. e) No; the power is proportional to the time t or proportional to the square root of the angle.

10-48: a) From the right-hand rule, the direction of the torque is $\hat{i} \times \hat{j} = \hat{k}$, the $+z$ direction.

b), c)

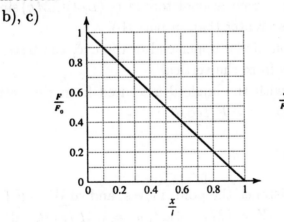
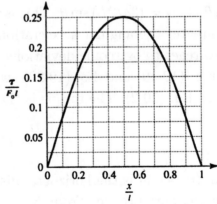

d) The magnitude of the torque is $F_0(x - x^2/L)$, which has its maximum at $L/2$.

e) The torque at $x = L/2$ is $F_0L/4$.

10-49:
$$t^2 = \frac{2\theta}{\alpha} = \frac{2\theta}{(\tau/I)} = \frac{2\theta I}{\tau}.$$

The angle in radians is $\pi/2$, the moment of inertia is

$$(1/3)((750 \text{ N})/(9.80 \text{ m/s}^2)(1.25 \text{ m}))^3 = 39.9 \text{ kg·m}^2$$

and the torque is $(220 \text{ N})(1.25 \text{ m}) = 275$ N·m. Using these in the above expression gives $t^2 = 0.455 \text{ s}^2$, so $t = 0.675$ s.

10-50: a) From geometric consideration, the lever arm and the sine of the angle between $\overrightarrow{F}$ and $\overrightarrow{r}$ are both maximum if the string is attached at the end of the rod. b) In terms of the distance x where the string is attached, the magnitude of the torque is $Fxh/\sqrt{x^2 + h^2}$. This function attains its maximum at the boundary, where $x = h$, so the string should be attached at the right end of the rod. c) As a function of x, l and h, the torque has magnitude

$$\tau = F\frac{xh}{\sqrt{(x - l/2)^2 + h^2}}.$$

This form shows that there are two aspects to increasing the torque; maximizing the lever arm l and maximizing $\sin\phi$. Differentiating τ with respect to x and setting equal to zero gives $x_{\max} = (l/2)(1 + (2h/l)^2)$. This will be the point at which to attach the string unless $2h > l$, in which case the string should be attached at the furthest point to the right, $x = l$.

10-51: a) A distance $L/4$ from the end with the clay.

b) In this case $I = (4/3)ML^2$ and the gravitational torque is $(3L/4)(2Mg)\sin\theta = (3MgL/2)\sin\theta$, so $\alpha = (9g/8L)\sin\theta$.

c) In this case $I = (1/3)ML^2$ and the gravitational torque is $(L/4)(2Mg)\sin\theta = (MgL/2)\sin\theta$, so $\alpha = (3g/2L)\sin\theta$. This is greater than in part (b).

d) The greater the angular acceleration of the upper end of the cue, the faster you would have to react to overcome deviations from the vertical.

10-52: In Fig. (10-19) and Eq. (10-22), with the angle θ measured from the vertical, $\sin\phi = \cos\theta$ in Eq. (10-2). The torque is then $\tau = FR\cos\theta$.

a)
$$W = \int_0^{\pi/2} FR\cos\theta\, d\theta = FR.$$

b) In Eq. (6-14), dl is the horizontal distance the point moves, and so $W = F\int dl = FR$, the same as part (a). c) From $K_2 = W = (MR^2/4)\omega^2$, $\omega = \sqrt{4F/MR}$. d) The torque, and hence the angular acceleration, is greatest when $\theta = 0$, at which point $\alpha = (\tau/I) = 2F/MR$, and so the maximum tangential acceleration is $2F/M$. e) Using the value for ω found in part (c), $a_{\text{rad}} = \omega^2 R = 4F/M$.

10-53: The tension in the rope must be $m(g + a) = 530$ N. The angular acceleration of the cylinder is $a/R = 3.2$ rad/s^2, and so the net torque on the cylinder must be 9.28 N·m. Thus, the torque supplied by the crank is $(530 \text{ N})(0.25 \text{ m}) + (9.28 \text{ N·m}) = 141.8$ N·m, and the force applied to the crank handle is $\frac{141.8 \text{ N·m}}{0.12 \text{ m}} = 1.2$ kN to two figures.

10-54: At the point of contact, the wall exerts a friction force f directed downward and a normal force $\mathcal{N}$ directed to the right. This is a situation where the net force on the roll is zero, but the net torque is *not* zero, so balancing torques would not be correct. Balancing vertical forces, $F_{\text{rod}} \cos \theta = f + w + F$, and balancing horizontal forces, $F_{\text{rod}} \sin \theta = \mathcal{N}$. With $f = \mu_k \mathcal{N}$, these equations become

$$F_{\text{rod}} \cos \theta = \mu_k \mathcal{N} + F + w,$$
$$F_{\text{rod}} \sin \theta = \mathcal{N}.$$

a) Eliminating $\mathcal{N}$ and solving for F_{rod} gives

$$F_{\text{rod}} = \frac{w + F}{\cos \theta - \mu_k \sin \theta} = \frac{(16.0 \text{ kg})(9.80 \text{ m/s}^2) + (40.0 \text{ N})}{\cos 30° - (0.25) \sin 30°} = 266 \text{ N}.$$

b) With respect to the center of the roll, the rod and the normal force exert zero torque. The magnitude of the net torque is $(F - f)R$, and $f = \mu_k \mathcal{N}$ may be found by insertion of the value found for F_{rod} into either of the above relations; *i.e.*, $f = \mu_k F_{\text{rod}} \sin \theta = 33.2$ N. Then,

$$\alpha = \frac{\tau}{I} = \frac{(40.0 \text{ N} - 31.54 \text{ N})(18.0 \times 10^{-2} \text{ m})}{(0.260 \text{ kg·m}^2)} = 4.71 \text{ rad/s}^2.$$

10-55: The net torque on the pulley is TR, where T is the tension in the string, and $\alpha = TR/I$. The net force on the block down the ramp is $mg(\sin \beta - \mu_k \cos \beta) - T = ma$. The acceleration of the block and the angular acceleration of the pulley are related by $a = \alpha R$. a) Multiplying the first of these relations by I/R and eliminating α in terms of a, and then adding to the second to eliminate T gives

$$a = mg \frac{(\sin \beta - \mu_k \cos \beta)}{m + I/R^2} = \frac{g(\sin \beta - \mu_k \cos \beta)}{(1 + I/mR^2)},$$

and substitution of numerical values gives 1.12 m/s^2. b) Substitution of this result into either of the above expressions involving the tension gives $T = 14.0$ N.

10-56: For a tension T in the string, $mg - T = ma$ and $TR = I\alpha = I\frac{a}{R}$. Eliminating T and solving for a gives

$$a = g \frac{m}{m + I/R^2} = \frac{g}{1 + I/mR^2},$$

where m is the mass of the hanging weight, I is the moment of inertia of the disk combination ($I = 2.25 \times 10^{-3}$ kg·m^2 from Problem 9-75) and R is the radius of the disk to which the string is attached.

a) With $m = 1.50$ kg, $R = 2.50 \times 10^{-2}$ m, $a = 2.88$ m/s^2.

b) With $m = 1.50$ kg, $R = 5.00 \times 10^{-2}$ m, $a = 6.13$ m/s^2.

The acceleration is larger in case (b); with the string attached to the larger disk, the tension in the string is capable of applying a larger torque.

10-57: Taking the torque about the center of the roller, the net torque is $fR = \alpha I$, $I = MR^2$ for a hollow cylinder, and with $\alpha = a/R$, $f = Ma$ (note that this is a relation between magnitudes; the vectors $\vec{f}$ and $\vec{a}$ are in opposite directions). The net force is $F - f = Ma$, from which $F = 2Ma$ and so $a = F/2M$ and $f = F/2$.

10-58: The accelerations of blocks A and B will have the same magnitude a. Since the cord does not slip, the angular acceleration of the pulley will be $\alpha = \frac{a}{R}$. Denoting the tensions in the cord as T_A and T_B, the equations of motion are

$$m_A g - T_A = m_A a$$
$$T_B - m_B g = m_B a$$
$$T_A - T_B = \frac{I}{R^2} a,$$

where the last equation is obtained by dividing $\tau = I\alpha$ by R and substituting for α in terms of a.

Adding the three equations eliminates both tensions, with the result that

$$a = g \frac{m_A - m_B}{m_A + m_B + I/R^2}$$

Then,

$$\alpha = \frac{a}{R} = g \frac{m_A - m_B}{m_A R + m_B R + I/R}.$$

The tensions are then found from

$$T_A = m_A(g - a) = g \frac{2 m_A m_B + m_A I/R^2}{m_A + m_B + I/R^2}$$

$$T_B = m_B(g + a) = g \frac{2 m_B m_A + m_B I/R^2}{m_A + m_B + I/R^2}.$$

As a check, it can be shown that $(T_A - T_B)R = I\alpha$.

10-59: For the disk, $K = (3/4)Mv^2$ (see Example 10-6). From the work-energy theorem, $K_1 = MgL\sin\beta$, from which

$$L = \frac{3v^2}{4g\sin\beta} = \frac{3(2.50 \text{ m/s})^2}{4(9.80 \text{ m/s}^2)\sin 30.0°} = 0.957 \text{ m}.$$

This same result may be obtained by an extension of the result of Exercise 10-20; for the disk, the acceleration is $(2/3)g\sin\beta$, leading to the same result.

b) Both the translational and rotational kinetic energy depend on the mass which cancels the mass dependence of the gravitational potential energy. Also, the moment of inertia is proportional to the square of the radius, which cancels the inverse dependence of the angular speed on the radius.

10-60: The tension is related to the acceleration of the yo-yo by $(2m)g - T = (2m)a$, and to the angular acceleration by $Tb = I\alpha = I\frac{a}{b}$. Dividing the second equation by b and adding to the first to eliminate T yields

$$a = g\frac{2m}{(2m + I/b^2)} = g\frac{2}{2 + (R/b)^2}, \qquad \alpha = g\frac{2}{2b + R^2/b},$$

where $I = 2\frac{1}{2}mR^2 = mR^2$ has been used for the moment of inertia of the yo-yo. The tension is found by substitution into either of the two equations; *e.g.,*

$$T = (2m)(g - a) = (2mg)\left(1 - \frac{2}{2 + (R/b)^2}\right) = 2mg\frac{(R/b)^2}{2 + (R/b)^2} = \frac{2mg}{(2(b/R)^2 + 1)}.$$

10-61: a) The distance the marble has fallen is $y = h - (2R - r) = h + r - 2R$. The radius of the path of the center of mass of the marble is $R - r$, so the condition that the ball stay on the track is $v^2 = g(R - r)$. The speed is determined from the work-energy theorem, $mgy = (1/2)mv^2 + (1/2)I\omega^2$. At this point, it is crucial to know that even for the curved track, $\omega = v/r$; this may be seen by considering the time T to move around the circle of radius $R - r$ at constant speed V is obtained from $2\pi(R - r) = Vt$, during which time the marble rotates by an angle $2\pi\left(\frac{R}{r} - 1\right) = \omega T$, from which $\omega = V/r$. The work-energy theorem then states $mgy = (7/10)mv^2$, and combining, canceling the factors of m and g leads to $(7/10)(R - r) = h + r - 2R$, and solving for h gives $h = (27/10)R - (17/10)r$.
b) In the absence of friction, $mgy = (1/2)mv^2$, and substitution of the expressions for y and v^2 in terms of the other parameters gives $(1/2)(R - r) = h - r - 2R$, which is solved for $h = (5/2)R - (3/2)r$.

10-62: In the first case, $\overrightarrow{F}$ and the friction force act in opposite directions, and the friction force causes a larger torque to tend to rotate the yo-yo to the right. The net force to the right is the difference $F - f$, so the net force is to the right while the net torque causes a clockwise rotation. For the second case, both the torque and the friction force tend to turn the yo-yo clockwise, and the yo-yo moves to the right. In the third case, friction tends to move the yo-yo to the right, and since the applied force is vertical, the yo-yo moves to the right.

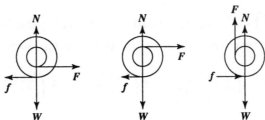

10-63: a) Because there is no vertical motion, the tension is just the weight of the hoop: $T = Mg = (0.180 \text{ kg})(9.8 \text{ N/kg}) = 1.76 \text{ N}$
b) Use $\tau = I\alpha$ to find α. The torque is RT, so $\alpha = RT/I = RT/MR^2 = T/MR = Mg/MR$, so $\alpha = g/R = (9.8 \text{ m/s}^2)/(0.08 \text{ m}) = 122.5 \text{ rad/s}^2$
c) $a = R\alpha = 9.8 \text{ m/s}^2$

d) T would be unchanged because the mass M is the same, α and a would be twice as great because I is now $\frac{1}{2}MR^2$.

10-64: a) The kinetic energy of the ball when it leaves the track (when it is still rolling without slipping) is $(7/10)mv^2$, and this must be the work done by gravity, $W = mgh$, so $v = \sqrt{10gh/7}$. The ball is in the air for a time $t = \sqrt{2y/g}$, so $x = vt = \sqrt{20hy/7}$. b) The answer does not depend on g, so the result should be the same on the moon. c) The presence of rolling friction would decrease the distance.   d) For the dollar coin, modeled as a uniform disc, $K = (3/4)mv^2$, and so $x = \sqrt{8hy/3}$.

10-65: a) $\quad v = \sqrt{\dfrac{10K}{7m}} = \sqrt{\dfrac{(10)(0.800)(1/2)(400 \text{ N/m})(0.15 \text{ m})^2}{7(0.0590 \text{ kg})}} = 9.34 \text{ m/s}.$

b) Twice the speed found in part (a), 18.7 m/s. c) If the ball is rolling without slipping, the speed of a point at the bottom of the ball is zero. d) Rather than use the intermediate calculation of the speed, the fraction of the initial energy that was converted to gravitational potential energy is $(0.800)(0.900)$, so $(0.720)(1/2)kx^2 = mgh$ and solving for h gives 5.60 m.

10-66: a)

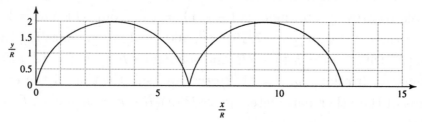

b) R is the radius of the wheel (y varies from 0 to $2R$) and T is the period of the wheel's rotation. c) Differentiating,

$$v_x = \frac{2\pi R}{T}\left[1 - \cos\left(\frac{2\pi t}{T}\right)\right] \qquad a_x = \left(\frac{2\pi}{T}\right)^2 R \sin\left(\frac{2\pi t}{T}\right)$$

$$v_y = \frac{2\pi R}{T}\sin\left(\frac{2\pi t}{T}\right) \qquad a_y = \left(\frac{2\pi}{T}\right)^2 R \cos\left(\frac{2\pi t}{T}\right).$$

d) $v_x = v_y = 0$ when $\left(\frac{2\pi t}{T}\right) = 2\pi$ or any multiple of 2π, so the times are integer multiples of the period T. The acceleration components at these times are $a_x = 0$, $a_y = \frac{4\pi^2 R}{T^2}$.

e) $\quad \sqrt{a_x^2 + a_y^2} = \left(\frac{2\pi}{T}\right)^2 R\sqrt{\cos^2\left(\frac{2\pi t}{T}\right) + \sin^2\left(\frac{2\pi t}{T}\right)} = \frac{4\pi^2 R}{T^2},$

independent of time. This is the magnitude of the radial acceleration for a point moving on a circle of radius R with constant angular velocity $\frac{2\pi}{T}$. For motion that consists of this circular motion superimposed on motion with constant velocity ($\vec{a} = 0$), the acceleration due to the circular motion will be the total acceleration.

10-67:   For rolling without slipping, the kinetic energy is $(1/2)(m+I/R^2)v^2 = (5/6)mv^2$; initially, this is 32.0 J and at the return to the bottom it is 8.0 J. Friction has done -24.0 J of work, -12.0 J each going up and down. The potential energy at the highest point was 20.0 J, so the height above the ground was $\frac{20.0 \text{ J}}{(0.600 \text{ kg})(9.80 \text{ m/s}^2)} = 3.40$ m.

10-68: Differentiating, and obtaining the answer to part (b),

$$\omega = \frac{d\theta}{dt} = 3bt^2 = 3b\left(\frac{\theta}{b}\right)^{2/3} = 3b^{1/3}\theta^{2/3},$$

$$\alpha = \frac{d\omega}{dt} = 6bt = 6b\left(\frac{\theta}{b}\right)^{1/3} = 6b^{2/3}\theta^{1/3}.$$

a)
$$W = \int I_{cm}\alpha \, d\theta = 6b^{2/3}I_{cm}\int \theta^{1/3} \, d\theta = \frac{9}{2}I_{cm}b^{2/3}\theta^{4/3}.$$

c) The kinetic energy is

$$K = \frac{1}{2}I_{cm}\omega^2 = \frac{9}{2}I_{cm}b^{2/3}\theta^{4/3},$$

in agreement with Eq. (10-25); the total work done is the change in kinetic energy.

10-69: Doing this problem using kinematics involves four unknowns (six, counting the two angular accelerations), while using energy considerations simplifies the calculations greatly. If the block and the cylinder both have speed v, the pulley has angular velocity v/R and the cylinder has angular velocity $v/2R$, the total kinetic energy is

$$K = \frac{1}{2}\left[Mv^2 + \frac{M(2R)^2}{2}(v/2R)^2 + \frac{MR^2}{2}(v/R)^2 + Mv^2\right] = \frac{3}{2}Mv^2.$$

This kinetic energy must be the work done by gravity; if the hanging mass descends a distance y, $K = Mgy$, or $v^2 = (2/3)gy$. For constant acceleration, $v^2 = 2ay$, and comparison of the two expressions gives $a = g/3$.

10-70: a) The rings and the rod exert forces on each other, but there is no net force or torque on the system, and so the angular momentum will be constant. As the rings slide toward the ends, the moment of inertia changes, and the final angular velocity is given by Eq. (10-33),

$$\omega_2 = \omega_1\frac{I_1}{I_2} = \omega_1\left[\frac{\frac{1}{12}ML^2 + 2mr_1^2}{\frac{1}{12}ML^2 + 2mr_2^2}\right] = \omega_1\frac{5.00 \times 10^{-4} \text{ kg·m}^2}{2.00 \times 10^{-3} \text{ kg·m}^2} = \frac{\omega_1}{4},$$

and so $\omega_2 = 7.5$ rev/min. Note that conversion from rev/min to rad/s is not necessary.
b) The forces and torques that the rings and the rod exert on each other will vanish, but the common angular velocity will be the same, 7.5 rev/min.

10-71: The initial angular momentum of the bullet is $(m/4)(v)(L/2)$, and the final moment of inertia of the rod and bullet is $(m/3)L^2 + (m/4)(L/2)^2 = (19/48)mL^2$. Setting the initial angular moment equal to ωI and solving for ω gives $\omega = \frac{mvL/8}{(19/48)\ mL^2} = \frac{6}{19}v/L$.

b)
$$\frac{(1/2)I\omega^2}{(1/2)(m/4)v^2} = \frac{(19/48)mL^2((6/19)(v/L))^2}{(m/4)v^2} = \frac{3}{19}.$$

10-72: Assuming the blow to be concentrated at a point (or using a suitably chosen "average" point) at a distance r from the hinge, $\Sigma\tau_{ave} = rF_{ave}$, and $\Delta L = rF_{ave}\Delta t = rJ$. The angular velocity ω is then

$$\omega = \frac{\Delta L}{I} = \frac{rF_{ave}\Delta t}{I} = \frac{(l/2)F_{ave}\Delta t}{\frac{1}{3}ml^2} = \frac{3}{2}\frac{F_{ave}\Delta t}{ml},$$

where l is the width of the door. Substitution of the given numerical values gives $\omega = 0.514$ rad/s.

10-73: a) The initial angular momentum is $L = mv(l/2)$ and the final moment of inertia is $I = I_0 + m(l/2)^2$, so

$$\omega = \frac{mv(l/2)}{(M/3)l^2 + m(l/2)^2} = 5.46 \text{ rad/s}.$$

b) $(M + m)gh = (1/2)\omega^2 I$, and after solving for h and substitution of numerical values, $h = 3.16 \times 10^{-2}$ m.   c) Rather than recalculate the needed value of ω, note that ω will be proportional to v, and hence h will be proportional to v^2; for the board to swing all the way over, $h = 0.250$ m, and so $v = (360 \text{ m/s})\sqrt{\frac{0.250 \text{ m}}{0.0316 \text{ m}}} = 1012$ m/s.

10-74: Angular momentum is conserved, so $I_0\omega_0 = I_2\omega_2$, or, using the fact that for a common mass the moment of inertia is proportional to the square of the radius, $R_0^2\omega_0 = R_2^2\omega_2$, or

$$R_0^2\omega_0 = (R_0 + \Delta R)^2 (\omega_0 + \Delta\omega) \sim R_0^2\omega_0 + 2R_0\Delta R\omega_0 + R_0^2\Delta\omega,$$

where the terms in $\Delta R\Delta\omega$ and $\Delta\omega^2$ have been omitted. Canceling the $R_0^2\omega_0$ term gives

$$\Delta R = -\frac{R_0}{2}\frac{\Delta\omega}{\omega_0} = -1.1 \text{ cm}.$$

10-75: The initial angular momentum is $L_1 = \omega_0 I_A$ and the initial kinetic energy is $K_1 = I_A\omega_0^2/2$. The final total moment of inertia is $4I_A$, so the final angular velocity is $(1/4)\omega_0$ and the final kinetic energy is $(1/2)4I_A(\omega_0/4)^2 = (1/4)K_1$. (This result may be obtained more directly from $K = L^2/I$.) Thus, $\Delta K = -(3/4)K_1$ and $K_1 = -(4/3)(-2400 \text{ J}) = 3200$ J.

10-76: The tension is related to the block's mass and speed, and the radius of the circle, by $T = m\frac{v^2}{r}$. The block's angular momentum with respect to the hole is $L = mvr$, so in terms of the angular momentum,

$$T = mv^2\frac{1}{r} = \frac{m^2v^2}{m}\frac{r^2}{r^3} = \frac{(mvr)^2}{mr^3} = \frac{L^2}{mr^3}.$$

The radius at which the string breaks can be related to the initial angular momentum by

$$r^3 = \frac{L^2}{mT_{max}} = \frac{(mv_1r_1)^2}{mT_{max}} = \frac{((0.250 \text{ kg})(4.00 \text{ m/s})(0.800 \text{ m}))^2}{(0.250 \text{ kg})(30.0 \text{ N})},$$

from which $r = 0.440$ m.

10-77: The train's speed relative to the earth is $0.600 \text{ m/s} + \omega(0.475 \text{ m})$, so the total angular momentum is

$$((0.600 \text{ m/s}) + \omega(0.475 \text{ m}))\,(1.20 \text{ kg})(0.475 \text{ m}) + \omega(1/2)(7.00 \text{ kg})\left(\frac{1.00 \text{ m}}{2}\right)^2 = 0,$$

from which $\omega = -0.298 \text{ rad/s}$, with the minus sign indicating that the turntable moves clockwise, as expected.

10-78: a), g)

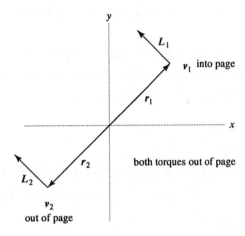

b) Using the vector product form for the angular momentum, $\overrightarrow{v}_1 = -\overrightarrow{v}_2$ and $\overrightarrow{r}_1 = -\overrightarrow{r}_2$, so

$$m\,\overrightarrow{r}_2 \times \overrightarrow{v}_2 = m\,\overrightarrow{r}_1 \times \overrightarrow{v}_1,$$

so the angular momenta are the same. c) Let $\overrightarrow{\omega} = \omega\hat{j}$. Then,

$$\overrightarrow{v}_1 = \overrightarrow{\omega} \times \overrightarrow{r}_1 = \omega\left(z\hat{i} - x\hat{k}\right), \quad \text{and}$$

$$\overrightarrow{L}_1 = m\,\overrightarrow{r}_1 \times \overrightarrow{v}_1 = m\omega\left((-xR)\hat{i} + (x^2 + y^2)\hat{j} + (xR)\hat{k}\right).$$

With $x^2 + y^2 = R^2$, the magnitude of $\overrightarrow{L}_1$ is $2m\omega R^2$, and $\overrightarrow{L}_1 \cdot \overrightarrow{\omega} = m\omega^2 R^2$, and so $\cos\theta = \frac{m\omega^2 R^2}{(2m\omega R^2)(\omega)} = \frac{1}{2}$, and $\theta = \frac{\pi}{6}$. This is true for $\overrightarrow{L}_2$ as well, so the total angular momentum makes an angle of $\frac{\pi}{6}$ with the $+y$-axis. d) From the intermediate calculation of part (c), $L_{y1} = m\omega R^2 = mvR$, so the total y-component of angular momentum is $L_y = 2mvR$. e) L_y is constant, so the net y-component of torque is zero. f) Each particle moves in a circle of radius R with speed v, and so is subject to an inward force of magnitude mv^2/R. The lever arm of this force is R, so the torque on each has magnitude mv^2. These forces are directed in opposite directions for the two particles, and the position vectors are opposite each other, so the torques have the same magnitude and direction, and the net torque has magnitude $2mv^2$.

10-79: a) The initial angular momentum with respect to the pivot is mvr, and the final total moment of inertia is $I + mr^2$, so the final angular velocity is $\omega = mvr/(mr^2 + I)$.

b) The kinetic energy after the collision is

$$K = \frac{1}{2}\omega^2\left(mr^2 + I\right) = (M + m)gh, \quad \text{or}$$

$$\omega = \sqrt{\frac{2(M + m)gh}{(mr^2 + I)}}.$$

c) Substitution of $I = Mr^2$ into either of the result of part (a) gives $\omega = \left(\frac{m}{m+M}\right)(v/r)$, and into the result of part (b), $\omega = \sqrt{2gh}(1/r)$, which are consistent with the forms for v.

10-80: The initial angular momentum is $I\omega_1 - mRv_1$, with the minus sign indicating that runner's motion is opposite the motion of the part of the turntable under his feet. The final angular momentum is $\omega_2(I + mR^2)$, so

$$\omega_2 = \frac{I\omega_1 - mRv_1}{I + mR^2}$$

$$= \frac{(80 \text{ kg·m}^2)(0.200 \text{ rad/s}) - (55.0 \text{ kg})(3.00 \text{ m})(2.8 \text{ m/s})}{(80 \text{ kg·m}^2) + (55.0 \text{ kg})(3.00 \text{ m})^2}$$

$$= -0.776 \text{ rad/s},$$

where the minus sign indicates that the turntable has reversed its direction of motion (*i.e.*, the man had the larger magnitude of angular momentum initially).

10-81: From Eq. (10-36),

$$\Omega = \frac{wr}{I\omega} = \frac{(50.0 \text{ kg})(9.80 \text{ m/s}^2)(0.040 \text{ m})}{(0.085 \text{ kg·m}^2)((6.0 \text{ m/s})/(0.33 \text{ m}))} = 12.7 \text{ rad/s},$$

or 13 rad/s to two figures, which is quite large.

10-82: The velocity of the center of mass will change by $\Delta v_{cm} = \frac{J}{m}$, and the angular velocity will change by $\Delta\omega = \frac{J(x - x_{cm})}{I}$. The change in velocity of the end of the bat will then be

$$\Delta v_{end} = \Delta v_{cm} - \Delta\omega x_{cm} = \frac{J}{m} - \frac{J(x - x_{cm})x_{cm}}{I}.$$

Setting $\Delta v_{end} = 0$ allows cancellation of J, and gives $I = (x - x_{cm})x_{cm}m$, which when solved for x is

$$x = \frac{I}{x_{cm}m} + x_{cm} = \frac{(5.30 \times 10^{-2} \text{ kg·m}^2)}{(0.600 \text{ m})(0.800 \text{ kg})} + (0.600 \text{ m}) = 0.710 \text{ m}.$$

10-83: In Fig. (10-30(a)), if the vector $\vec{r}$, and hence the vector $\vec{L}$ are not horizontal but make an angle β with the horizontal, the torque will still be horizontal (the torque must be perpendicular to the vertical weight). The magnitude of the torque will be $wr\cos\beta$, and this torque will change the direction of the horizontal component of the angular momentum, which has magnitude $L \cos \beta$. Thus, the situation of Fig. (10-32) is reproduced, but with $\vec{L}_{\text{horiz}}$ instead of $\vec{L}$. Then, the expression found in Eq. (10-36) becomes

$$\Omega = \frac{d\phi}{dt} = \frac{\left|d\vec{L}\right| \Big/ \left|\vec{L}_{\text{horiz}}\right|}{dt} = \frac{\tau}{\left|\vec{L}_{\text{horiz}}\right|} = \frac{mgr \cos \beta}{L \cos \beta} = \frac{wr}{I\omega}.$$

10-84: a)

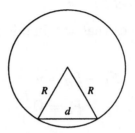

The distance from the center of the ball to the midpoint of the line joining the points where the ball is in contact with the rail is $\sqrt{R^2 - (d/2)^2}$, so $v_{\text{cm}} = \omega \sqrt{R^2 - d^2/4}$. When $d = 0$, this reduces to $v_{\text{cm}} = \omega R$, the same as rolling on a flat surface. When $d = 2R$, the rolling radius approaches zero, and $v_{\text{cm}} \to 0$ for any ω.

b)
$$K = \frac{1}{2}mv^2 + \frac{1}{2}I\omega^2$$

$$= \frac{1}{2}\left[mv_{\text{cm}}^2 + (2/5)mR^2 \left(\frac{v_{\text{cm}}}{\sqrt{R^2 - (d^2/4)}}\right)^2\right]$$

$$= \frac{mv_{\text{cm}}^2}{10}\left[5 + \frac{2}{(1 - d^2/4R^2)}\right].$$

Setting this equal to mgh and solving for v_{cm} gives the desired result.   c) The denominator in the square root in the expression for v_{cm} is larger than for the case $d = 0$, so v_{cm} is smaller. For a given speed, ω is larger than the $d = 0$ case, so a larger fraction of the kinetic energy is rotational, and the translational kinetic energy, and hence v_{cm}, is smaller. d) Setting the expression in part (b) equal to 0.95 of that of the $d = 0$ case and solving for the ratio d/R gives $d/R = 1.05$. Setting the ratio equal to 0.995 gives $d/R = 0.37$.

10-85: a)

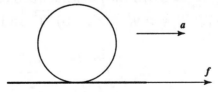

The friction force is $f = \mu_k \mathcal{N} = \mu_k Mg$, so $a = \mu_k g$. The magnitude of the angular acceleration is $\frac{fR}{I} = \frac{\mu_k MgR}{(1/2)MR^2} = \frac{2\mu_k g}{R}$.   b) Setting $v = at = \omega R = (\omega_0 - \alpha t)R$ and solving for t gives

$$t = \frac{R\omega_0}{a + R\alpha} = \frac{R\omega_0}{\mu_k g + 2\mu_k g} = \frac{R\omega_0}{3\mu_k g},$$

and

$$d = \frac{1}{2}at^2 = \frac{1}{2}(\mu_k g)\left(\frac{R\omega_0}{3\mu_k g}\right)^2 = \frac{R^2\omega_0^2}{18\mu_k g}.$$

c) The final kinetic energy is $(3/4)Mv^2 = (3/4)M(at)^2$, so the change in kinetic energy is

$$\frac{3}{4}M\left(\mu_k g\frac{R\omega_0}{3\mu_k g}\right)^2 - \frac{1}{4}MR^2\omega_0^2 = -\frac{1}{6}MR^2\omega_0^2.$$

10-86: Denoting the upward forces that the hands exert as F_L and F_R, the conditions that F_L and F_R must satisfy are

$$F_L + F_R = w$$
$$F_L - F_R = \Omega\frac{I\omega}{r},$$

where the second equation is $\tau = \Omega L$, divided by r. These two equations can be solved for the forces by first adding and then subtracting, yielding

$$F_L = \frac{1}{2}\left(w + \Omega\frac{I\omega}{r}\right)$$

$$F_R = \frac{1}{2}\left(w - \Omega\frac{I\omega}{r}\right).$$

Using the values $w = mg = (8.00 \text{ kg})(9.80 \text{ m/s}^2) = 78.4$ N and

$$\frac{I\omega}{r} = \frac{(8.00 \text{ kg})(0.325 \text{ m})^2(5.00 \text{ rev/s} \times 2\pi \text{ rad/rev})}{(0.200 \text{ m})} = 132.7 \text{ kg·m/s}$$

gives

$$F_L = 39.2 \text{ N} + \Omega(66.4 \text{ N·s}), \quad F_R = 39.2 \text{ N} - \Omega(66.4 \text{ N·s}).$$

a) $\Omega = 0$, $F_L = F_R = 39.2$ N.

b) $\Omega = 0.05 \text{ rev/s} = 0.314 \text{ rad/s}$, $F_L = 60.0$ N, $F_R = 18.4$ N.

c) $\Omega = 0.3 \text{ rev/s} = 1.89 \text{ rad/s}$, $F_L = 165$ N, $F_R = -86.2$ N, with the minus sign indicating a downward force.

d) $F_R = 0$ gives $\Omega = \frac{39.2 \text{ N}}{66.4 \text{ N·s}} = 0.575 \text{ rad/s}$, which is 0.0916 rev/s.

10-87: a) See Problem 10-76; $T = mv_1^2 r_1^2/r^3$. b) $\overrightarrow{T}$ and $d\overrightarrow{r}$ are always antiparallel, so

$$W = -\int_{r_1}^{r_2} T\, dr = mv_1^2 r_1^2 \int_{r_2}^{r_1} \frac{dr}{r^3} = \frac{mv_1^2}{2}r_1^2\left[\frac{1}{r_2^2} - \frac{1}{r_1^2}\right].$$

c) $v_2 = v_1(r_1/r_2)$, so

$$\Delta K = \frac{1}{2}m\left(v_2^2 - v_1^2\right) = \frac{mv_1^2}{2}\left[\left(\frac{r_1}{r_2}\right)^2 - 1\right],$$

which is the same as the work found in part (b).

Chapter 11   Equilibrium and Elasticity

Note: In solving static equilibrium problems, torques may be calculated about any point. The following solutions cannot possibly exhaust all of the valid ways to do any individual problem. In many instances, an origin has been chosen to simplify the calculations.

11-1:   Take the origin to be at the center of the small ball; then,

$$x_{cm} = \frac{(1.00 \text{ kg})(0) + (2.00 \text{ kg})(0.580 \text{ m})}{3.00 \text{ kg}} = 0.387 \text{ m}$$

from the center of the small ball.

11-2:   The calculation of Exercise 11-2 becomes

$$x_{cm} = \frac{(1.00 \text{ kg})(0) + (1.50 \text{ kg})(0.280 \text{ m}) + (2.00 \text{ kg})(0.580 \text{ m})}{4.50 \text{ kg}} = 0.351 \text{ m}$$

11-3:   In the notation of Example 11-1, take the origin to be the point S, and let the child's distance from this point be x. Then,

$$s_{cm} = \frac{M(-D/2) + mx}{M + m} = 0, \quad x = \frac{MD}{2m} = 1.125 \text{ m},$$

which is $(L/2 - D/2)/2$, halfway between the point S and the end of the plank.

11-4:   a) The force is applied at the center of mass, so the applied force must have the same magnitude as the weight of the door, or 300 N. In this case, the hinge exerts no force. b) With respect to the hinge, the moment arm of the applied force is twice the distance to the center of mass, so the force has half the magnitude of the weight, or 150 N. The hinge supplies an upward force of $300 \text{ N} - 150 \text{ N} = 150 \text{ N}$.

11-5: $F(8.0 \text{ m})\sin 40° = (2800 \text{ N})(10.0 \text{ m})$, so $F = 5.45 \text{ kN}$, keeping an extra figure.

11-6:   The other person lifts with a force of $160 \text{ N} - 60 \text{ N} = 100 \text{ N}$. Taking torques about the point where the 60-N force is applied,

$$(100 \text{ N})x = (160 \text{ N})(1.50 \text{ m}), \quad \text{or} \quad x = (1.50 \text{ m})\left(\frac{160 \text{ N}}{100 \text{ N}}\right) = 2.40 \text{ m}.$$

11-7:   If the board is taken to be massless, the weight of the motor is the sum of the applied forces, 1000 N. The motor is a distance $\frac{(2.00 \text{ m})(600 \text{ N})}{(1000 \text{ N})} = 1.200 \text{ m}$ from the end where the 400-N force is applied.

11-8:   The weight of the motor is $400 \text{ N} + 600 \text{ N} - 200 \text{ N} = 800 \text{ N}$. Of the myriad ways to do this problem, a sneaky way is to say that the lifters each exert 100 N to lift the board, leaving 500 N and 300 N to lift the motor. Then, the distance of the motor from the end where the 600-N force is applied is $\frac{(2.00 \text{ m})(300 \text{ N})}{(800 \text{ N})} = 0.75 \text{ m}$. The center of gravity is located at $\frac{(200 \text{ N})(1.0 \text{ m}) + (800 \text{ N})(0.75 \text{ m})}{(1000 \text{ N})} = 0.80 \text{ m}$ from the end where the 600 N force is applied.

11-9:   The torque due to T_x is $-T_x h = -\frac{Lw}{D}\cot\theta h$, and the torque due to T_y is $T_y D = Lw$. The sum of these torques is $Lw\left(1 - \frac{h}{D}\cot\theta\right)$. From Figure (11-7(b)), $h = D\tan\theta$, so the net torque due to the tension in the tendon is zero.

11-10:   a) Since the wall is frictionless, the only vertical forces are the weights of the man and the ladder, and the normal force. For the vertical forces to balance, $\mathcal{N}_2 = w_1 + w_m = 160\ \text{N} + 740\ \text{N} = 900\ \text{N}$, and the maximum frictional force is $\mu_s \mathcal{N}_2 = (0.40)(900\ \text{N}) = 360\ \text{N}$ (see Figure 11-5(b)). b) Note that the ladder makes contact with the wall at a height of 4.0 m above the ground. Balancing torques about the point of contact with the ground,

$$(4.0\ \text{m})\mathcal{N}_1 = (1.5\ \text{m})(160\ \text{N}) + (1.0\ \text{m})(3/5)(740\ \text{N}) = 684\ \text{N·m},$$

so $\mathcal{N}_1 = 171.0\ \text{N}$, keeping extra figures. This horizontal force must be balanced by the frictional force, which must then be 170 N to two figures. c) Setting the frictional force, and hence $\mathcal{N}_1$, equal to the maximum of 360 N and solving for the distance x along the ladder,

$$(4.0\ \text{m})(360\ \text{N}) = (1.50\ \text{m})(160\ \text{N}) + x(3/5)(740\ \text{N}),$$

so $x = 2.70$ m, or 2.7 m to two figures.

11-11:   Take torques about the left end of the board in Figure (11-9).   a) The force F at the support point is found from $F(1.00\ \text{m}) = +(280\ \text{N})(1.50\ \text{m}) + (500\ \text{N})(3.00\ \text{m})$, or $F = 1920\ \text{N}$.   b) The net force must be zero, so the force at the left end is $(1920\ \text{N}) - (500\ \text{N}) - (280\ \text{N}) = 1140\ \text{N}$, downward.

11-12:   a)

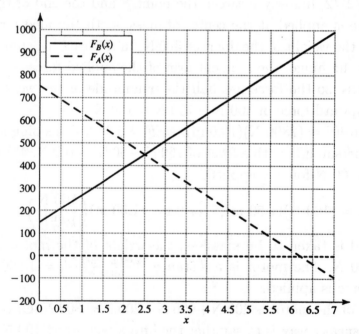

b) $x = 6.25$ m when $F_A = 0$, which is 1.25 m beyond point B.   c) Take torques about the right end. When the beam is just balanced, $F_A = 0$, so $F_B = 900\ \text{N}$. The distance that point B must be from the right end is then $\frac{(300\ \text{N})(4.50\ \text{m})}{(900\ \text{N})} = 1.50$ m.

11-13:   In both cases, the tension in the vertical cable is the weight w.   a) Denote the length of the horizontal part of the cable by L. Taking torques about the pivot point,

$TL \tan 30.0° = wL + w(L/2)$, from which $T = 2.60w$. The pivot exerts an upward vertical force of $2w$ and a horizontal force of $2.60w$, so the magnitude of this force is $3.28w$, directed $37.6°$ from the horizontal.   b) Denote the length of the strut by L, and note that the angle between the diagonal part of the cable and the strut is $15.0°$. Taking torques about the pivot point, $TL \sin 15.0° = wL \sin 45.0° + (w/2)L \sin 45°$, so $T = 4.10w$. The horizontal force exerted by the pivot on the strut is then $T \cos 30.0° = 3.55w$ and the vertical force is $(2w) + T \sin 30° = 4.05w$, for a magnitude of $5.38w$, directed $48.8°$.

11-14:   a) Taking torques about the pivot, and using the 3-4-5 geometry,

$$(4.00 \text{ m})(3/5)\text{T} = (4.00 \text{ m})(300 \text{ N}) + (2.00 \text{ m})(150 \text{ N}),$$

so $T = 625$ N. b) The horizontal force must balance the horizontal component of the force exerted by the rope, or $T(4/5) = 500$ N. The vertical force is $300 \text{ N} + 150 \text{ N} - T(3/5) = 75$ N, upwards.

11-15:   To find the horizontal force that one hinge exerts, take the torques about the other hinge; then, the vertical forces that the hinges exert have no torque. The horizontal force is found from $F_{\text{H}}(1.00 \text{ m}) = (280 \text{ N})(0.50 \text{ m})$, from which $F_{\text{H}} = 140$ N. The top hinge exerts a force away from the door, and the bottom hinge exerts a force toward the door. Note that the magnitudes of the forces must be the same, since they are the only horizontal forces.

11-16:   a) Denote the length of the boom by L, and take torques about the pivot point. The tension in the guy wire is found from

$$TL \sin 60° = (5000 \text{ N})L \cos 60.0° + (2600 \text{ N})(0.35L) \cos 60.0°,$$

so $T = 3.41$ kN. The vertical force exerted on the boom by the pivot is the sum of the wieghts, 7.60 kN and the horizontal force is the tension, 3.41 kN.   b) No; $\tan \left(\frac{F_{\text{V}}}{F_{\text{H}}} \right) \neq 0$.

11-17:   To find the tension T_{L} in the left rope, take torques about the point where the rope at the right is connected to the bar. Then, $T_{\text{L}}(3.00 \text{ m}) \sin 150° = (240 \text{ N})(1.50 \text{ m}) + (90 \text{ N})(0.50 \text{ m})$, so $T_{\text{L}} = 270$ N. The vertical component of the force that the rope at the end exerts must be $(330 \text{ N}) - (270 \text{ N}) \sin 150° = 195$ N, and the horizontal component of the force is $-(270 \text{ N}) \cos 150°$, so the tension in the rope at the right is $T_{\text{R}} = 304$ N, and $\theta = 39.9°$.

11-18:   The cable is given as perpendicular to the beam, so the tension is found by taking torques about the pivot point; $T(3.00 \text{ m}) = (1.00 \text{ kN})(2.00 \text{ m}) \cos 25.0° + (5.00 \text{ kN})(4.50 \text{ m}) \cos 25.0°$, or $T = 7.40$ kN. The vertical component of the force exerted on the beam by the pivot is the vertical component of the tension force plus the net weight, or $T \cos 25.0° + 6.00 \text{ kN} = 12.71$ kN. The horizontal force is $T \sin 25.0° = 3.13$ kN.

11-19:   a) $F_1(3.00 \text{ m}) - F_2(3.00 \text{ m} + l) = (8.00 \text{ N})(-l)$. This is given to have a magnitude of 6.40 N·m, so $l = 0.80$ m.   b) The net torque is clockwise, either by considering the figure or noting the the torque found in part (a) was negative.   c) About the point of contact of $\overrightarrow{F}_2$, the torque due to $\overrightarrow{F}_1$ is $-F_1 l$, and setting the magnitude of this torque to 6.40 N·m gives $l = 0.80$ m, and the direction is again clockwise.

11-20: From Eq. (11-10),

$$Y = F\frac{l_0}{\Delta l A} = F\frac{(0.200 \text{ m})}{(3.0 \times 10^{-2} \text{ m})(50.0 \times 10^{-4} \text{ m}^2)} = F(1333 \text{ m}^{-2}).$$

Then, $F = 25.0$ N corresponds to a Young's modulus of 3.3×10^4 Pa, and $F = 500$ N corresponds to a Young's modulus of 6.7×10^5 Pa.

11-21: $\qquad A = \dfrac{F l_0}{Y \Delta l} = \dfrac{(400 \text{ N})(2.00 \text{ m})}{(20 \times 10^{10} \text{ Pa})(0.25 \times 10^{-2} \text{ m})} = 1.60 \times 10^{-6} \text{ m}^2,$

and so $d = \sqrt{4A/\pi} = 1.43 \times 10^{-3}$ m, or 1.4 mm to two figures.

11-22: a) The strain, from Eq. (11-12), is $\frac{\Delta l}{l_0} = \frac{F}{YA}$. For steel, using Y from Table (11-1) and $A = \pi \frac{d^2}{4} = 1.77 \times 10^{-4} \text{ m}^2$,

$$\frac{\Delta l}{l_0} = \frac{(4000 \text{ N})}{(2.0 \times 10^{11} \text{ Pa})(1.77 \times 10^{-4} \text{ m}^2)} = 1.1 \times 10^{-4}.$$

Similarly, the strain for copper ($Y = 1.10 \times 10^{11}$ Pa) is 2.1×10^{-4}. b) Steel: $(1.1 \times 10^{-4}) \times (0.750 \text{ m}) = 8.5 \times 10^{-5}$ m. Copper: $(2.1 \times 10^{-4})(0.750 \text{ m}) = 1.5 \times 10^{-4}$ m.

11-23: From Eq. (11-10),

$$Y = \frac{(5000 \text{ N})(4.00 \text{ m})}{(0.50 \times 10^{-4} \text{ m}^2)(0.20 \times 10^{-2} \text{ m})} = 2.0 \times 10^{11} \text{ Pa}.$$

11-24: From Eq. (11-10),

$$Y = \frac{(65.0 \text{ kg})(9.80 \text{ m/s}^2)(45.0 \text{ m})}{(\pi(3.5 \times 10^{-3} \text{ m})^2)(1.10 \text{ m})} = 6.8 \times 10^8 \text{ Pa}.$$

11-25: a) The top wire is subject to a tension of $(16.0 \text{ kg})(9.80 \text{ m/s}^2) = 157$ N and hence a tensile strain of $\frac{(157 \text{ N})}{(20 \times 10^{10} \text{ Pa})(2.5 \times 10^{-7} \text{ m}^2)} = 3.14 \times 10^{-3}$, or 3.1×10^{-3} to two figures. The bottom wire is subject to a tension of 98.0 N, and a tensile strain of 1.96×10^{-3}, or 2.0×10^{-3} to two figures. b) $(3.14 \times 10^{-3})(0.500 \text{ m}) = 1.57$ mm, $(1.96 \times 10^{-3})(0.500 \text{ m}) = 0.98$ mm.

11-26: a) $\frac{(8000 \text{ kg})(9.80 \text{ m/s}^2)}{\pi(12.5 \times 10^{-2} \text{ m})^2} = 1.6 \times 10^6$ Pa. b) $\frac{1.6 \times 10^6 \text{ Pa}}{20 \times 10^{10} \text{ Pa}} = 0.8 \times 10^{-5}$. c) $(0.8 \times 10^{-5}) \times (2.50 \text{ m}) = 2 \times 10^{-5}$ m.

11-27: $(2.8 - 1)(1.013 \times 10^5 \text{ Pa})(50.0 \text{ m}^2) = 9.1 \times 10^6$ N.

11-28: a) The volume would increase slightly. b) The volume change would be twice as great. c) The volume change is inversely proportional to the bulk modulus for a given pressure change, so the volume change of the lead ingot would be four times that of the gold.

11-29: a) $\frac{250 \text{ N}}{0.75 \times 10^{-4} \text{ m}^2} = 3.33 \times 10^6$ Pa. b) $(3.33 \times 10^6 \text{ Pa})(2)(200 \times 10^{-4} \text{ m}^2) = 133$ kN.

11-30: a) Solving Eq. (11-14) for the volume change,

$$\Delta V = -kV\Delta P$$
$$= -(45.8 \times 10^{-11} \text{ Pa}^{-1})(1.00 \text{ m}^3)(1.16 \times 10^8 \text{ Pa} - 1.0 \times 10^5 \text{ Pa})$$
$$= -0.0531 \text{ m}^3.$$

b) The mass of this amount of water has not changed, but its volume has decreased to $1.000 \text{ m}^3 - 0.053 \text{ m}^3 = 0.947 \text{ m}^3$, and the density is now $\frac{1.03 \times 10^3 \text{ kg}}{0.947 \text{ m}^3} = 1.09 \times 10^3 \text{ kg/m}^3$.

11-31: $B = \dfrac{(600 \text{ cm}^3)(3.6 \times 10^6 \text{ Pa})}{(0.45 \text{ cm}^3)} = 4.8 \times 10^9 \text{ Pa}, \quad k = \dfrac{1}{B} = 2.1 \times 10^{-10} \text{ Pa}^{-1}.$

11-32: a) Using Equation (11-17),

$$\text{Shear strain} = \frac{F_{\parallel}}{AS} = \frac{(9 \times 10^5 \text{ N})}{[(.10 \text{ m})(.005 \text{ m})][7.5 \times 10^{10} \text{ Pa}]} = 2.4 \times 10^{-2}.$$

b) Using Equation (11-16), $x = \text{Shear strain} \cdot h = (.024)(.1 \text{ m}) = 2.4 \times 10^{-3} \text{ m}.$ $\phi = \arctan(.024) = 1.4°.$

11-33: The area A in Eq. (11-17) has increased by a factor of 9, so the shear strain for the larger object would be 1/9 that of the smaller.

11-34: Each rivet bears one-quarter of the force, so

$$\text{Shear stress} = \frac{F_{\parallel}}{A} = \frac{\frac{1}{4}(1.20 \times 10^4 \text{ N})}{\pi(.125 \times 10^{-2} \text{ m})^2} = 6.11 \times 10^8 \text{ Pa}.$$

11-35: $\frac{F}{A} = \frac{(90.8 \text{ N})}{\pi(0.92 \times 10^{-3} \text{ m})^2} = 3.41 \times 10^7 \text{ Pa}$, or 3.4×10^7 Pa to two figures.

11-36: a) $(1.6 \times 10^{-3})(20 \times 10^{10} \text{ Pa})(5 \times 10^{-6} \text{ m}^2) = 1.60 \times 10^3 \text{ N}.$ b) If this were the case, the wire would stretch 6.4 mm. c) $(6.5 \times 10^{-3})(20 \times 10^{10} \text{ Pa})(5 \times 10^{-6} \text{ m}^2) = 6.5 \times 10^3 \text{ N}.$

11-37: $a = \dfrac{F_{\text{tot}}}{m} = \dfrac{(2.40 \times 10^8 \text{ Pa})(3.00 \times 10^{-4} \text{ m}^2)/3}{(1200 \text{ kg})} - 9.80 \text{ m/s}^2 = 10.2 \text{ m/s}^2.$

11-38: $A = \frac{350 \text{ N}}{4.7 \times 10^8 \text{ Pa}} = 7.45 \times 10^{-7} \text{ m}^2$, so $d = \sqrt{4A/\pi} = 0.97 \text{ mm}.$

11-39: a) Take torques about the rear wheel, so that $fwd = wx_{\text{cm}}$, or $x_{\text{cm}} = fd.$ b) $(0.53)(2.46 \text{ m}) = 1.30 \text{ m}$ to three figures.

11-40: If Lancelot were at the end of the bridge, the tension in the cable would be (from taking torques about the hinge of the bridge) obtained from

$$T(12.0 \text{ N}) = (600 \text{ kg})(9.80 \text{ m/s}^2)(12.0 \text{ m}) + (200 \text{ kg})(9.80 \text{ m/s}^2)(6.0 \text{ m}),$$

so $T = 6860$ N. This exceeds the maximum tension that the cable can have, so Lancelot is going into the drink. To find the distance x Lancelot can ride, replace the 12.0 m multiplying Lancelot's weight by x and the tension T by $T_{\text{max}} = 5.80 \times 10^3$ N and solve for x;

$$x = \frac{(5.80 \times 10^3 \text{ N})(12.0 \text{ m}) - (200 \text{ kg})(9.80 \text{ m/s}^2)(6.0 \text{ m})}{(600 \text{ kg})(9.80 \text{ m/s}^2)} = 9.84 \text{ m}.$$

11-41: For the airplane to remain in level flight, both $\Sigma F = 0$ and $\Sigma \tau = 0$.

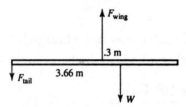

Taking the clockwise direction as positive, and taking torques about the center of mass,

$$\text{Forces: } -F_{\text{tail}} - W + F_{\text{wing}} = 0$$

$$\text{Torques: } -(3.66 \text{ m})F_{\text{tail}} + (.3 \text{ m})F_{\text{wing}} = 0$$

A shortcut method is to write a second torque equation for torques about the tail, and solve for the F_{wing}: $-(3.66 \text{ m})(6700 \text{ N}) + (3.36 \text{ m})F_{\text{wing}} = 0$. This gives $F_{\text{wing}} = 7300 \text{ N}$ (up), and $F_{\text{tail}} = 6700 \text{ N} - 7300 \text{ N} = -600 \text{ N}$ (down).

Note that the rear stabilizer provides a *downward* force, does not hold up the tail of the aircraft, but serves to counter the torque produced by the wing. Thus balance, along with weight, is a crucial factor in airplane loading.

11-42: The simplest way to do this is to consider the *changes* in the forces due to the extra weight of the box. Taking torques about the rear axle, the force on the front wheels is decreased by $3600 \text{ N}\frac{1.00 \text{ m}}{3.00 \text{ m}} = 1200 \text{ N}$, so the net force on the front wheels is $10,780 \text{ N} - 1200 \text{ N} = 9.58 \times 10^3 \text{ N}$ to three figures. The weight added to the rear wheels is then $3600 \text{ N} + 1200 \text{ N} = 4800 \text{ N}$, so the net force on the rear wheels is $8820 \text{ N} + 4800 \text{ N} = 1.36 \times 10^4 \text{ N}$, again to three figures.

b) Now we want a shift of 10,780 N away from the front axle. Therefore, $W\frac{1.00 \text{ m}}{3.00 \text{ m}} = 10,780 \text{ N}$ and so $w = 32,340 \text{ N}$.

11-43: Take torques about the pivot point, which is 2.20 m from Karen and 1.65 m from Elwood. Then $w_{\text{Elwood}}(1.65 \text{ m}) = (420 \text{ N})(2.20 \text{ m}) + (240 \text{ N})(0.20 \text{ m})$, so Elwood weighs 589 N.   b) Equilibrium is neutral.

11-44: a) Denote the weight per unit length as α, so $w_1 = \alpha(10.0 \text{ cm})$, $w_2 = \alpha(8.0 \text{ cm})$, and $w_3 = \alpha l$. The center of gravity is a distance x_{cm} to the right of point O, where

$$x_{\text{cm}} = \frac{w_1(5.0 \text{ cm}) + w_2(9.5 \text{ cm}) + w_3(10.0 \text{ cm} - l/2)}{w_1 + w_2 + w_3}$$

$$= \frac{(10.0 \text{ cm})(5.0 \text{ cm}) + (8.0 \text{ cm})(9.5 \text{ cm}) + l(10.0 \text{ cm} - l/2)}{(10.0 \text{ cm}) + (8.0 \text{ cm}) + l}.$$

Setting $x_{\text{cm}} = 0$ gives a quadratic in l, which has as its positive root $l = 28.8 \text{ cm}$.

b) Changing the material from steel to copper would have no effect on the length l since the weight of each piece would change by the same amount.

11-45: Let $\vec{r}'_i = \vec{r}_i - \vec{R}$, where $\vec{R}$ is the vector from the point O to the point P. The torque for each force with respect to point P is then $\vec{\tau}'_i = \vec{r}'_i \times \vec{F}_i$, and so the net torque is

$$\sum \vec{\tau}_i = \sum \left(\vec{r}_i - \vec{R} \right) \times \vec{F}_i$$
$$= \sum \vec{r}_i \times \vec{F}_i - \sum \vec{R} \times \vec{F}_i$$
$$= \sum \vec{r}_i \times \vec{F}_i - \vec{R} \times \sum \vec{F}_i.$$

In the last expression, the first term is the sum of the torques about point O, and the second term is given to be zero, so the net torques are the same.

11-46: From the figure (and from common sense), the force $\vec{F}_1$ is directed along the length of the nail, and so has a moment arm of $(0.0800 \text{ m}) \sin 60°$. The moment arm of $\vec{F}_2$ is 0.300 m, so

$$F_2 = F_1 \frac{(0.0800 \text{ m}) \sin 60°}{(0.300 \text{ m})} = (500 \text{ N})(0.231) = 116 \text{ N}.$$

11-47: The horizontal component must balance the applied force $\vec{F}$, and so has magnitude 120.0 N. Taking torques about point A, $(120.0 \text{ N})(4.00 \text{ m}) + F_V(3.00 \text{ m})$, so the vertical component is -160 N, with the minus sign indicating a downward component, exerting a torque in a direction opposite that of the horizontal component.

11-48: a) The tension in the string is $w_2 = 50$ N, and the horizontal force on the bar must balance the horizontal component of the force that the string exerts on the bar, and is equal to $(50 \text{ N}) \sin 37° = 30$ N, to the left in the figure. The vertical force must be $(50 \text{ N}) \cos 37° + 10 \text{ N} = 50$ N, up. b) $\arctan \left(\frac{50 \text{ N}}{30 \text{ N}} \right) = 59°$. c) $\sqrt{(30 \text{ N})^2 + (50 \text{ N})^2} = 58$ N. d) Taking torques about (and measuring the distance from) the left end, $(50 \text{ N})x = (40 \text{ N})(5.0 \text{ m})$, so $x = 4.0$ m, where only the vertical components of the forces exert torques.

11-49: a) Take torques about her hind feet. Her fore feet are 0.72 m from her hind feet, and so her fore feet together exert a force of $\frac{(190 \text{ N})(0.28 \text{ m})}{(0.72 \text{ m})} = 73.9$ N, so each foot exerts a force of 36.9 N, keeping an extra figure. Each hind foot then exerts a force of 58.1 N. b) Again taking torques about the hind feet, the force exerted by the fore feet is $\frac{(190 \text{ N})(0.28 \text{ m}) + (25 \text{ N})(0.90 \text{ m})}{0.72 \text{ m}} = 105.1$ N, so each fore foot exerts a force of 52.6 N and each hind foot exerts a force of 54.9 N.

11-50: a) Finding torques about the hinge, and using L as the length of the bridge and w_T and w_B for the weights of the truck and the raised section of the bridge,

$$TL \sin 70° = w_T \left(\tfrac{3}{4}L \right) \cos 30° + w_B \left(\tfrac{1}{2}L \right) \cos 30°, \quad \text{so}$$
$$T = \frac{\left(\tfrac{3}{4}m_T + \tfrac{1}{2}m_B \right)(9.80 \text{ m/s}^2) \cos 30°}{\sin 70°} = 2.57 \times 10^5 \text{ N}.$$

b) Horizontal: $T\cos(70°-30°) = 1.97 \times 10^5$ N. Vertical: $w_T + w_B - T\sin 40° = 2.46 \times 10^5$ N.

11-51: a) Take the torque exerted by $\overrightarrow{F}_2$ to be positive; the net torque is then $-F_1(x)\sin\phi + F_2(x+l)\sin\phi = Fl\sin\phi$, where F is the common magnitude of the forces. b) $\tau_1 = -(14.0\text{ N})(3.0\text{ m})\sin 37° = -25.3$ N·m, keeping an extra figure, and $\tau_2 = (14.0\text{ N})(4.5\text{ m})\sin 37° = 37.9$ N·m, and the net torque is 12.6 N·m. About point P, $\tau_1 = (14.0\text{ N})(3.0\text{ m})(\sin 37°) = 25.3$ N·m and $\tau_2 = (-14.0\text{ N})(1.5\text{ m})(\sin 37°) = -12.6$ N·m, and the net torque is 12.6 N·m. The result of part (a) predicts $(14.0\text{ N})(1.5\text{ m})\sin 37°$, the same result.

11-52: a) Take torques about the pivot. The force that the ground exerts on the ladder is given to be vertical, and $F_V(6.0\text{ m})\sin\theta = (250\text{ N})(4.0\text{ m})\sin\theta + (750\text{ N})(1.50\text{ m})\sin\theta$, so $F_V = 354$ N. b) There are no other horizontal forces on the ladder, so the horizontal pivot force is zero. The vertical force that the pivot exerts on the ladder must be $(750\text{ N}) + (250\text{ N}) - (354\text{ N}) = 646$ N, up, so the ladder exerts a downward force of 646 N on the pivot.

11-53: a) $V = mg + w$ and $H = T$. To find the tension, take torques about the pivot point. Then, deonting the length of the strut by L,

$$T\left(\frac{2}{3}L\right)\sin\theta = w\left(\frac{2}{3}L\right)\cos\theta + mg\left(\frac{L}{6}\right)\cos\theta, \quad \text{or}$$

$$T = \left(w + \frac{mg}{4}\right)\cot\theta.$$

b) Solving the above for w, and using the maximum tension for T,

$$w = T\tan\theta - \frac{mg}{4} = (700\text{ N})\tan 55.0° - (5.0\text{ kg})(9.80\text{ m/s}^2) = 951\text{ N}.$$

c) Solving the expression obtained in part (a) for $\tan\theta$ and letting $w \to 0$, $\tan\theta = \frac{mg}{4T} = 0.06125$, so $\theta = 4.00°$.

11-54: a) The center of mass of the beam is 1.0 m from the suspension point. Taking torques about the suspension point,

$$w(4.00\text{ m}) + (140.0\text{ N})(1.00\text{ m}) = (100\text{ N})(2.00\text{ m})$$

(note that the common factor of $\sin 30°$ has been factored out), from which $w = 15.0$ N. b) In this case, a common factor of $\sin 45°$ would be factored out, and the result would be the same.

11-55: a) Taking torques about the hinged end of the pole $(200\text{ N})(2.50\text{ m}) + (600\text{ N}) \times (5.00\text{ m}) - T_y(5.00\text{ m}) = 0$. Therefore the y-component of the tension is $T_y = 700$ N. The x-component of the tension is then $T_x = \sqrt{(1000\text{ N})^2 - (700\text{ N})^2} = 714$ N. The height above the pole that the wire must be attached is $(5.00\text{ m})\frac{700}{714} = 4.90$ m. b) The y-component of the tension remains 700 N and the x-component becomes $(714\text{ N})\frac{4.90\text{ m}}{4.40\text{ m}} = 795$ N, leading to a total tension of $\sqrt{(795\text{ N})^2 + (700\text{ N})^2} = 1059$ N, an increase of 59 N.

11-56: A and B are straightforward, the tensions being the weights suspended; $T_A = (0.0360\text{ kg})(9.80\text{ m/s}^2) = 0.353$ N, $T_B = (0.0240\text{ kg} + 0.0360\text{ kg})(9.80\text{ m/s}^2) = 0.588$ N. To find T_C and T_D, a trick making use of the right angle where the strings join is available; use a coordinate system with axes parallel to the strings. Then, $T_C = T_B \cos 36.9° = 0.470$ N, $T_D = T_B \cos 53.1° = 0.353$ N. To find T_E, take torques about the point where string F is attached;

$$T_E(1.000 \text{ m}) = T_D \sin 36.9°(0.800 \text{ m}) + T_C \sin 53.1°(0.200 \text{ m})$$
$$+ (0.120 \text{ kg})(9.80 \text{ m/s}^2)(0.500 \text{ m})$$
$$= 0.833 \text{ N·m},$$

so $T_E = 0.833$ N. T_F may be found similarly, or from the fact that $T_E + T_F$ must be the total weight of the ornament, $(0.180 \text{ kg})(9.80 \text{ m/s}^2) = 1.76$ N, from which $T_F = 0.931$ N.

11-57: a) The force will be vertical, and must support the weight of the sign, and is 300 N. Similarly, the torque must be that which balances the torque due to the sign's weight about the pivot, $(300 \text{ N})(0.75 \text{ m}) = 225$ N·m. b) The torque due to the wire must balance the torque due to the weight, again taking torques about the pivot. The minimum tension occurs when the wire is perpendicular to the lever arm, from one corner of the sign to the other. Thus, $T\sqrt{(1.50 \text{ m})^2 + (0.80 \text{ m})^2} = 225$ N·m, or $T = 132$ N. The angle that the wire makes with the horizontal is $90° - \arctan(\frac{0.80}{1.50}) = 62.0°$. Thus, the vertical component of the force that the pivot exerts is $(300 \text{ N}) - (132 \text{ N}) \sin 62.0° = 183$ N and the horizontal force is $(132 \text{ N}) \cos 62.0° = 62$ N, for a magnitude of 193 N and an angle of 71° above the horizontal.

11-58: a) $\Delta w = -\sigma(\Delta l/l)w_0 = -(0.23)(9.0 \times 10^{-4})\sqrt{4(0.30 \times 10^{-4} \text{ m}^2)/\pi} = 1.3$ μm.

b)
$$F_\perp = AY\frac{\Delta l}{l} = AY\frac{1}{\sigma}\frac{\Delta w}{w}$$
$$= \frac{(2.1 \times 10^{11} \text{ Pa})(\pi(2.0 \times 10^{-2} \text{ m})^2)}{0.42}\frac{0.10 \times 10^{-3} \text{ m}}{1.0 \times 10^{-2} \text{ m}} = 6.3 \times 10^6 \text{ N},$$

where the Young's modulus for nickel has been used.

11-59: a) The tension in the horizontal part of the wire will be 240 N. Taking torques about the center of the disk, $(240 \text{ N})(0.250 \text{ m}) - w(1.00 \text{ m})) = 0$, or $w = 60$ N. b) Balancing torques about the center of the disk in this case, $(240 \text{ N})(0.250 \text{ m}) - ((60 \text{ N})(1.00 \text{ m}) + (20 \text{ N})(2.00 \text{ m})) \cos \theta = 0$, so $\theta = 53.1°$.

11-60: a) Taking torques about the right end of the stick, the friction force is half the weight of the stick, $f = \frac{w}{2}$. Taking torques about the point where the cord is attached to the wall (the tension in the cord and the friction force exert no torque about this point), and noting that the moment arm of the normal force is $l \tan \theta$, $N \tan \theta = \frac{w}{2}$. Then, $\frac{f}{N} = \tan \theta < 0.40$, so $\theta < \arctan(0.40) = 22°$.

b) Taking torques as in part (a), and denoting the length of the meter stick as l,

$$fl = w\frac{l}{2} + w(l - x) \quad \text{and} \quad Nl \tan \theta = w\frac{l}{2} + wx.$$

In terms of the coefficient of friction μ_s,

$$\mu_s > \frac{f}{N} = \frac{\frac{l}{2} + (l-x)}{\frac{l}{2} + x} \tan\theta = \frac{3l - 2x}{l + 2x} \tan\theta.$$

Solving for x,

$$x > \frac{l}{2} \frac{3\tan\theta - \mu_s}{\mu_s + \tan\theta} = 30.2 \text{ cm.}$$

c) In the above expression, setting $x = 10$ cm and solving for μ_s gives

$$\mu_s > \frac{(3 - 20/l)\tan\theta}{1 + 20/l} = 0.625.$$

11-61: Consider torques around the point where the person on the bottom is lifting. The center of mass is displaced horizontally by a distance $(0.625 \text{ m} - 0.25 \text{ m})\sin 45°$ and the horizontal distance to the point where the upper person is lifting is $(1.25 \text{ m})\sin 45°$, and so the upper person lifts with a force of $w\frac{0.375 \sin 45°}{1.25 \sin 45°} = (0.300)w = 588$ N. The person on the bottom lifts with a force that is the difference between this force and the weight, 1.37 kN. The person above is lifting less.

11-62: a) Take torques about the upper corner of the curb. The force $\overrightarrow{F}$ acts at a perpendicular distance $R - h$ and the weight acts at a perpendicular distance $\sqrt{R^2 - (R-h)^2} = \sqrt{2Rh - h^2}$. Setting the torques equal for the minimum necessary force,

$$F = mg\frac{\sqrt{2Rh - h^2}}{R - h}.$$

b) The torque due to gravity is the same, but the force $\overrightarrow{F}$ acts at a perpendicular distance $2R - h$, so the minimum force is $(mg)\sqrt{2Rh - h^2}/(2R - h)$.   c) Less force is required when the force is applied at the top of the wheel.

11-63: a) There are several ways to find the tension. Taking torques about point B (the force of the hinge at A is given as being vertical, and exerts no torque about B), the tension acts at distance $r = \sqrt{(4.00 \text{ m})^2 + (2.00 \text{ m})^2} = 4.47$ m and at an angle of $\phi = 30° + \arctan\left(\frac{2.00}{4.00}\right) = 56.6°$. Setting $Tr\sin\phi = (500 \text{ N})(2.00 \text{ m})$ and solving for T gives $T = 268$ N.   b) The hinge at A is given as exerting no horizontal force, so taking torques about point D, the lever arm for the vertical force at point B is $(2.00 \text{ m}) + (4.00 \text{ m})\tan 30.0° = 4.31$ m, so the horizontal force at B is $\frac{(500 \text{ N})(2.00 \text{ m})}{4.31 \text{ m}} = 232$ N. Using the result of part (a), however, $(268 \text{ N})\cos 30.0° = 232$ N. In fact, finding the horizontal force at B first simplifies the calculation of the tension slightly.   c) $(500 \text{ N}) - (268 \text{ N})\sin 30.0° = 366$ N. Equivalently, the result of part (b) could be used, taking torques about point C, to get the same result.

11-64: a) The center of gravity of the top block can be as far out as the edge of the lower block. The center of gravity of this combination is then $3L/4$ from the right edge of the upper block, so the overhang is $3L/4$.   b) Take the two-block combination from part (a), and place it on the third block such that the overhang of $3L/4$ is from the right edge of the third block; that is, the center of gravity of the first two blocks is above the right edge of the third block. The center of mass of the three-block combination, measured from the right end of the bottom block, is $-L/6$ and so the largest possible overhang is $(3L/4) + (L/6) = 11L/12$. Similarly, placing this three-block combination with its center of gravity over the right edge of the fourth block allows an extra overhang of $L/8$, for a total of $25L/24$. c) As the result of part (b) shows, with only four blocks, the overhang can be larger than the length of a single block.

11-65: a) The angle at which the bale would slip is that for which $f = \mu_s \mathcal{N} = \mu_s w \cos \beta = w \sin \beta$, or $\beta = \arctan(\mu_s) = 31.0°$. The angle at which the bale would tip is that for which the center of gravity is over the lower contact point, or $\arctan(\frac{0.25 \text{ m}}{0.50 \text{ m}}) = 26.6°$, or $27°$ to two figures. b) The angle for tipping is unchanged, but the angle for slipping is $\arctan(0.40) = 21.8°$, or $22°$ to two figures.

11-66: a) $F = f = \mu_k \mathcal{N} = \mu_k mg = (0.35)(30.0 \text{ kg})(9.80 \text{ m/s}^2) = 103 \text{ N}$.

b) With respect to the forward edge of the bale, the lever arm of the weight is $\frac{0.250 \text{ m}}{2} = 0.125 \text{ m}$ and the lever arm h of the applied force is then $h = (0.125 \text{ m})\frac{mg}{F} = (0.125 \text{ m})\frac{1}{\mu_k} = \frac{0.125 \text{ m}}{0.35} = 0.36 \text{ m}$.

11-67: a) Take torques about the point where wheel B is in contact with the track. With respect to this point, the weight exerts a counterclockwise torque and the applied force and the force on wheel A both exert clockwise torques. Balancing torques, $F_A(2.00 \text{ m}) + (F)(1.60 \text{ m}) = (950 \text{ N})(1.00 \text{ m})$. Using $F = \mu_k w = 494 \text{ N}$, $F_A = 80 \text{ N}$, and $F_B = w - F_A = 870 \text{ N}$. b) Again taking torques about the point where wheel B is in contact with the track, and using $F = 494 \text{ N}$ as in part (a), $(494 \text{ N})h = (950 \text{ N})(1.00 \text{ N})$, so $h = 1.92 \text{ m}$.

11-68: a) The torque exerted by the cable about the left end is $TL \sin \theta$. For any angle θ, $\sin(180° - \theta) = \sin \theta$, so the tension T will be the same for either angle. The horizontal component of the force that the pivot exerts on the boom will be $T \cos \theta$ or $T \cos(180° - \theta) = -T \cos \theta$. b) From the result of part (a), $T \propto \frac{1}{\sin \theta}$, and this becomes infinite as $\theta \to 0$ or $\theta \to 180°$. Also, c), the tension is a minimum when $\sin \theta$ is a maximum, or $\theta = 90°$, a vertical string. d) There are no other horizontal forces, so for the boom to be in equilibrium, the pivot exerts zero horizontal force on the boom.

11-69: a) Taking torques about the contact point on the ground, $T(7.0 \text{ m}) \sin \theta = w(4.5 \text{ m}) \sin \theta$, so $T = (0.64)w = 3664 \text{ N}$. The ground exerts a vertical force on the pole, of magnitude $w - T = 2052 \text{ N}$. b) The factor of $\sin \theta$ appears in both terms of the equation representing the balancing of torques, and cancels.

11-70: a) Identifying x with Δl in Eq. (11-10), $k = YA/l_0$.

b) $(1/2)kx^2 = YAx^2/2l_0$.

11-71: a) At the bottom of the path the wire exerts a force equal in magnitude to the centripetal acceleration plus the weight,

$$F = m(((2.00 \text{ rev/s})(2\pi \text{ rad/rev}))^2(0.50 \text{ m}) + 9.80 \text{ m/s}^2) = 1.07 \times 10^3 \text{ N}.$$

From Eq. (11-10), the elongation is

$$\frac{(1.07 \times 10^3 \text{ N})(0.50 \text{ m})}{(0.7 \times 10^{11} \text{ Pa})(0.014 \times 10^{-4} \text{ m}^2)} = 5.5 \text{ mm}.$$

b) Using the same equations, at the top the force is 830 N, and the elongation is 0.0042 m.

11-72:   a)

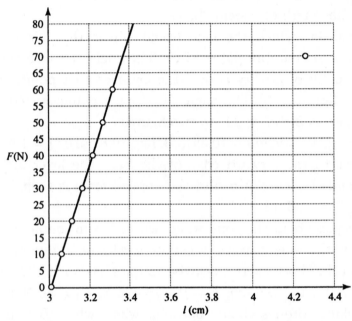

b) The ratio of the added force to the elongation, found from taking the slope of the graph, doing a least-squares fit to the linear part of the data, or from a casual glance at the data gives $\frac{F}{\Delta l} = 2.00 \times 10^4$ N/m. From Eq. (11-10),

$$Y = \frac{F}{\Delta l}\frac{l_0}{A} = (2.00 \times 10^4 \text{ N/m})\frac{(3.50 \text{ m})}{(\pi(0.35 \times 10^{-3} \text{ m})^2)} = 1.8 \times 10^{11} \text{ Pa}.$$

c) The total force at the proportional limit is 20.0 N + 60 N = 80 N, and the stress at this limit is $\frac{(80 \text{ N})}{\pi(0.35 \times 10^{-3} \text{ m})^2} = 2.1 \times 10^8$ Pa.

11-73:   a) For the same stress, the tension in wire B must be two times that in wire A, and so the weight must be suspended at a distance $(2/3)(1.05 \text{ m}) = 0.70$ m from wire A.
b) The product YA for wire B is $(4/3)$ that of wire B, so for the same strain, the tension in wire B must be $(4/3)$ that in wire A, and the weight must be 0.45 m from wire B.

11-74:   a) Solving Eq. (11-10) for Δl and using the weight for F,

$$\Delta l = \frac{Fl_0}{YA} = \frac{(1900 \text{ N})((15.0 \text{ m})}{(2.0 \times 10^{11} \text{ Pa})(8.00 \times 10^{-4} \text{ m}^2)} = 1.8 \times 10^{-4} \text{ m}.$$

b) From Example 5-21, the force that each car exerts on the cable is $F = m\omega^2 l_0 = \frac{w}{g}\omega^2 l_0$, and so

$$\Delta l = \frac{Fl_0}{YA} = \frac{w\omega^2 l_0^2}{gYA} = \frac{(1900\text{ N})(0.84\text{ rad/s})^2(15.0\text{ m})^2}{(9.80\text{ m/s}^2)(2.0\times10^{11}\text{ Pa})(8.00\times10^{-4}\text{ m}^2)} = 1.9\times10^{-4}\text{ m}.$$

11-75: Use subscripts 1 to denote the copper and 2 to denote the steel. a) From Eq. (11-10), with $\Delta l_1 = \Delta l_2$ and $F_1 = F_2$,

$$L_2 = L_1\left(\frac{A_2Y_2}{A_1Y_1}\right) = (1.40\text{ m})\left(\frac{(1.00\text{ cm}^2)(21\times10^{10}\text{ Pa})}{(2.00\text{ cm}^2)(9\times10^{10}\text{ Pa})}\right) = 1.63\text{ m}.$$

b) For nickel, $\frac{F}{A_1} = 4.00\times10^8$ Pa and for brass, $\frac{F}{A_2} = 2.00\times10^8$ Pa. c) For nickel, $\frac{4.00\times10^8\text{ Pa}}{21\times10^{10}\text{ Pa}} = 1.9\times10^{-3}$ and for brass, $\frac{2.00\times10^8\text{ Pa}}{9\times10^{10}\text{ Pa}} = 2.2\times10^{-3}$.

11-76: a) $F_{max} = YA\left(\frac{\Delta l}{l_0}\right)_{max} = (1.4\times10^{10}\text{ Pa})(3.0\times10^{-4}\text{ m}^2)(0.010) = 4.2\times10^4$ N.

b) Neglect the mass of the shins (actually the lower legs and feet) compared to the rest of the body. This allows the approximation that the compressive stress in the shin bones is uniform. The maximum height will be that for which the force exerted on each lower leg by the ground is F_{max} found in part (a), minus the person's weight. The impulse that the ground exerts is $J = (4.2\times10^4\text{ N} - (70\text{ kg})(9.80\text{ m/s}^2))(0.030\text{ s}) = 1.2\times10^3$ kg·m/s. The speed at the ground is $\sqrt{2gh}$, so $2J = m\sqrt{2gh}$ and solving for h,

$$h = \frac{1}{2g}\left(\frac{2J}{m}\right)^2 = 64\text{ m},$$

but this is not recommended.

11-77: a) Two times as much, 0.36 mm. b) One-fourth (which is $(1/2)^2$) as much, 0.045 mm. c) The Young's modulus for copper is approximately one-half that for steel, so the wire would stretch about twice as much. $(0.18\text{ mm})\frac{20\times10^{10}\text{ Pa}}{11\times10^{10}\text{ Pa}} = 0.33$ mm.

11-78: Solving Eq. (11-14) for ΔV,

$$\Delta V = -kV_0\Delta p = -kV_0\frac{mg}{A}$$
$$= -(110\times10^{-11}\text{ Pa}^{-1})(250\text{ L})\frac{(1420\text{ kg})(9.80\text{ m/s}^2)}{\pi(0.150\text{ m})^2}$$
$$= -0.0541\text{ L}.$$

The minus sign indicates that this is the volume by which the original hooch has shrunk, and is the extra volume that can be stored.

11-79: The normal component of the force is $F\cos\theta$ and the area (the intersection of the red plane and the bar in Figure (11-79)) is $A/\cos\theta$, so the normal stress is $(F/A)\cos^2\theta$. b) The tangential component of the force is $F\sin\theta$, so the shear stress is $(F/A)\sin\theta\cos\theta$. c) $\cos^2\theta$ is a maximum when $\cos\theta = 1$, or $\theta = 0$. d) The shear stress can be expressed as $(F/2A)\sin(2\theta)$, which is maximized when $\sin(2\theta) = 1$, or $\theta = \frac{90°}{2} = 45°$. Differentiation of the original expression with respect to θ and setting the derivative equal to zero gives the same result.

11-80: a) Taking torques about the pivot, the tension T in the cable is related to the weight by $T\sin\theta l_0 = mgl_0/2$, so $T = \frac{mg}{2\sin\theta}$. The horizontal component of the force that the cable exerts on the rod, and hence the horizontal component of the force that the pivot exerts on the rod, is $\frac{mg}{2}\cot\theta$ and the stress is $\frac{mg}{2A}\cot\theta$.   b)

$$\Delta l = \frac{l_0 F}{AY} = \frac{mgl_0 \cot\theta}{2AY}.$$

c) In terms of the density and length, $(m/A) = \rho l_0$, so the stress is $(\rho l_0 g/2)\cot\theta$ and the change in length is $(\rho l_0^2 g/2Y)\cot\theta$.   d) Using the numerical values, the stress is 1.4×10^5 Pa and the change in length is 2.2×10^{-6} m.   e) The stress is proportional to the length and the change in length is proportional to the square of the length, and so the quantities change by factors of 2 and 4.

11-81: a) Taking torques about the left edge of the left leg, the bookcase would tip when $F = \frac{(1500 \text{ N})(0.90 \text{ m})}{(1.80 \text{ m})} = 750$ N, and would slip when $F = (\mu_s)(1500 \text{ N}) = 600$ N, so the bookcase slides before tipping.   b) If F is vertical, there will be no net horizontal force and the bookcase could not slide. Again taking torques about the left edge of the left leg, the force necessary to tip the case is $\frac{(1500 \text{ N})(0.90 \text{ m})}{(0.10 \text{ m})} = 13.5$ kN.   c) To slide, the friction force is $f = \mu_s(w + F\cos\theta)$, and setting this equal to $F\sin\theta$ and solving for F gives

$$F = \frac{\mu_s w}{\sin\theta - \mu_s \cos\theta}.$$

To tip, the condition is that the normal force exerted by the right leg is zero, and taking torques about the left edge of the left leg, $F\sin\theta(1.80 \text{ m}) + F\cos\theta(0.10 \text{ m}) = w(0.90 \text{ m})$, and solving for F gives

$$F = \frac{w}{(1/9)\cos\theta + 2\sin\theta}.$$

Setting the expressions equal gives

$$\mu_s((1/9)\cos\theta + 2\sin\theta) = \sin\theta - \mu_s \cos\theta,$$

and solving for θ gives

$$\theta = \arctan\left(\frac{(10/9)\mu_s}{(1 - 2\mu_s)}\right) = 66°.$$

11-82: a) Taking torques about the point where the rope is fastened to the ground, the lever arm of the applied force is $\frac{h}{2}$ and the lever arm of both the weight and the normal force is $h\tan\theta$, and so $F\frac{h}{2} = (N-w)h\tan\theta$. Taking torques about the upper point (where the rope is attached to the post), $fh = F\frac{h}{2}$. Using $f \leq \mu_s \mathcal{N}$ and solving for F,

$$F \leq 2w\left(\frac{1}{\mu_s} - \frac{1}{\tan\theta}\right)^{-1} = 2(400 \text{ N})\left(\frac{1}{0.30} - \frac{1}{\tan 36.9°}\right)^{-1} = 400 \text{ N},$$

b) The above relations between F, $\mathcal{N}$ and f become

$$F\frac{3}{5}h = (\mathcal{N} - w)h\tan\theta, \quad f = \frac{2}{5}F,$$

and eliminating f and $\mathcal{N}$ and solving for F gives

$$F \leq w\left(\frac{2/5}{\mu_s} - \frac{3/5}{\tan\theta}\right)^{-1},$$

and substitution of numerical values gives 750 N to two figures. c) If the force is applied a distance y above the ground, the above relations become

$$Fy = (\mathcal{N} - w)h\tan\theta, \quad F(h - y) = fh,$$

which become, on eliminating $\mathcal{N}$ and f,

$$w \geq F\left[\frac{\left(1 - \frac{y}{h}\right)}{\mu_s} - \frac{\left(\frac{y}{h}\right)}{\tan\theta}\right].$$

As the term in square brackets approaches zero, the necessary force becomes unboundedly large. The limiting value of y is found by setting the term in square brackets equal to zero. Solving for y gives

$$\frac{y}{h} = \frac{\tan\theta}{\mu_s + \tan\theta} = \frac{\tan 36.9°}{0.30 + \tan 36.9°} = 0.71.$$

11-83: Assume that the center of gravity of the loaded girder is at $L/2$, and that the cable is attached a distance x to the right of the pivot. The sine of the angle between the lever arm and the cable is then $h/\sqrt{h^2 + ((L/2) - x)^2}$, and the tension is obtained from balancing torques about the pivot;

$$T\left[\frac{hx}{\sqrt{h^2 + ((L/2) - x)^2}}\right] = wL/2,$$

where w is the total load (the exact value of w and the position of the center of gravity do not matter for the purposes of this problem). The minimum tension will occur when the term in square brackets is a maximum; differentiating and setting the derviative equal to zero gives a maximum, and hence a minimum tension, at $x_{min} = (h^2/L) + (L/2)$. However, if $x_{min} > L$, which occurs if $h > L/\sqrt{2}$, the cable must be attached at L, the furthest point to the right.

11-84: The geometry of the 3-4-5 right triangle simplifies some of the intermediate algebra. Denote the forces on the ends of the ladders by F_L and F_R (left and right). The contact forces at the ground will be vertical, since the floor is assumed to be frictionless. a) Taking torques about the right end, $F_L(5.00 \text{ m}) = (480 \text{ N})(3.40 \text{ m}) + (360 \text{ N})(0.90 \text{ m})$, so $F_L = 391$ N. F_R may be found in a similar manner, or from $F_R = 840 \text{ N} - F_L = 449$ N. b) The tension in the rope may be found by finding the torque on each ladder, using the point A as the origin. The lever arm of the rope is 1.50 m. For the left ladder, $T(1.50 \text{ m}) = F_L(3.20 \text{ m}) - (480 \text{ N})(1.60 \text{ m})$, so $T = 322.1$ N (322 N to three figures). As a check, using the torques on the right ladder, $T(1.50 \text{ m}) = F_R(1.80 \text{ m}) - (360 \text{ N})(0.90 \text{ m})$ gives the same result. c) The horizontal component of the force at A must be equal to the tension found in part (b). The vertical force must be equal in magnitude to the difference between the weight of each ladder and the force on the bottom of each ladder, $480 \text{ N} - 391 \text{ N} = 449 \text{ N} - 360 \text{ N} = 89$ N. The magnitude of the force at A is then

$$\sqrt{(322.1 \text{ N})^2 + (89 \text{ N})^2} = 334 \text{ N}.$$

d) The easiest way to do this is to see that the added load will be distributed at the floor in such a way that $F_L' = F_L + (0.36)(800 \text{ N}) = 679$ N, and $F_R' = F_R + (0.64)(800 \text{ N}) = 961$ N. Using these forces in the form for the tension found in part (b) gives

$$T = \frac{F_L'(3.20 \text{ m}) - (480 \text{ N})(1.60 \text{ m})}{(1.50 \text{ m})} = \frac{F_R'(1.80 \text{ m}) - (360 \text{ N})(0.90 \text{ m})}{(1.50 \text{ m})} = 936.53 \text{ N},$$

which is 937 N to three figures.

11-85: The change in the volume of the oil is $= k_O V_O \Delta p$ and the change in the volume of the sodium is $= k_S V_S \Delta p$. Setting the total volume change equal to Ax (x is positive) and using $\Delta p = F/A$,

$$Ax = (k_O V_O + k_S V_S)(F/A),$$

and solving for k_s gives

$$k_S = \left(\frac{A^2 x}{F} - k_O V_O\right)\frac{1}{V_S}.$$

11-86: a) For constant temperature ($\Delta T = 0$),

$$\Delta(pV) = (\Delta p)V + p(\Delta V) = 0 \quad \text{and} \quad B = -\frac{(\Delta p)V}{(\Delta V)} = p.$$

b) In this situation,

$$(\Delta p)V^\gamma + \gamma p(\Delta V)V^{\gamma-1} = 0, \quad (\Delta p) + \gamma p\frac{\Delta V}{V} = 0,$$

and

$$B = -\frac{(\Delta p)V}{\Delta V} = \gamma p.$$

11-87: a) From Eq. (11-10), $\Delta l = \frac{(4.50 \text{ kg})(9.80 \text{ m/s}^2)(1.50 \text{ m})}{(20 \times 10^{10} \text{ Pa})(5.00 \times 10^{-7} \text{ m}^2)} = 6.62 \times 10^{-4}$ m, or 0.66 mm to two figures. b) $(4.50 \text{ kg})(9.80 \text{ m/s}^2)(0.0500 \times 10^{-2} \text{ m}) = 0.022$ J. c) The magnitude F will vary with distance; the average force is $YA(0.0250 \text{ cm}/l_0) = 16.7$ N, and so the work done by the applied force is $(16.7 \text{ N})(0.0500 \times 10^{-2} \text{ m}) = 8.35 \times 10^{-3}$ J. d) The wire is initially stretched a distance 6.62×10^{-4} m (the result of part (a)), and so the average elongation during the additional stretching is 9.12×10^{-4} m, and the average force the wire exerts is 60.8 N. The work done is negative, and equal to $-(60.8 \text{ N})(0.0500 \times 10^{-2} \text{ m}) = -3.04 \times 10^{-2}$ J. e) See Problem 11-70. The change in elastic potential energy is

$$\frac{(20 \times 10^{10} \text{ Pa})(5.00 \times 10^{-7} \text{ m}^2)}{2(1.50 \text{ m})} \left((11.62 \times 10^{-4} \text{ m})^2 - (6.62 \times 10^{-4} \text{ m})^2\right) = 3.04 \times 10^{-2} \text{ J},$$

the negative of the result of part (d). (If more figures are kept in the intermediate calculations, the agreement is exact.)

Chapter 12   Gravitation

Note: to obtain the numerical results given in this chapter, the following numerical values of certain physical quantities have been used;

$$G = 6.673 \times 10^{-11} \text{ N·m}^2/\text{kg}^2, \quad g = 9.80 \text{ m/s}^2 \quad \text{and} \quad m_E = 5.97 \times 10^{24} \text{ kg}.$$

Use of other tabulated values for these quantities may result in an answer that differs in the third significant figure.

12-1:   The ratio will be the product of the ratio of the mass of the sun to the mass of the earth and the square of the ratio of the earth-moon radius to the sun-moon radius. Using the earth-sun radius as an average for the sun-moon radius, the ratio of the forces is

$$\left(\frac{3.84 \times 10^8 \text{ m}}{1.50 \times 10^{11} \text{ m}} \right)^2 \left(\frac{1.99 \times 10^{30} \text{ kg}}{5.97 \times 10^{24} \text{ kg}} \right) = 2.18.$$

12-2:   Use of Eq. (12-1) gives

$$F_\text{g} = G\frac{m_1 m_2}{r^2} = \left(6.673 \times 10^{-11} \text{ N·m}^2/\text{kg}^2 \right) \frac{(5.97 \times 10^{24} \text{ kg})(2150 \text{ kg})}{(7.8 \times 10^5 \text{ m} + 6.38 \times 10^6)^2} = 1.67 \times 10^4 \text{ N}.$$

The ratio of this force to the satellite's weight at the surface of the earth is

$$\frac{(1.67 \times 10^4 \text{ N})}{(2150 \text{ kg})(9.80 \text{ m/s}^2)} = 0.79 = 79\%.$$

(This numerical result requires keeping one extra significant figure in the intermediate calculation.) The ratio, which is independent of the satellite mass, can be obtained directly as

$$\frac{Gm_E m/r^2}{mg} = \frac{Gm_E}{r^2 g} = \left(\frac{R_E}{r} \right)^2,$$

yielding the same result.

12-3:
$$G\frac{(nm_1)(nm_2)}{(nr_{12})^2} = G\frac{m_1 m_2}{r_{12}^2} = F_{12}.$$

12-4:   The separation of the centers of the spheres is $2R$, so the magnitude of the gravitational attraction is $GM^2/(2R)^2 = GM^2/4R^2$.

12-5:   a) Denoting the earth-sun separation as R and the distance from the earth as x, the distance for which the forces balance is obtained from

$$\frac{GM_S m}{(R - x)^2} = \frac{GM_E m}{x^2},$$

which is solved for

$$x = \frac{R}{1 + \sqrt{\frac{M_S}{M_E}}} = 2.59 \times 10^8 \text{ m}.$$

b) The ship could not be at equilibrium for long, in that the point where the forces balance is moving in a circle, and to move in that circle requires some force. The spaceship could continue toward the sun with a good navigator on board.

12-6: a) Taking force components to be positive to the right, use of Eq. (12-1) twice gives

$$F_g = \left(6.673 \times 10^{-11} \text{ N·m}^2/\text{kg}^2\right)(0.100 \text{ kg})\left[-\frac{(5.00 \text{ kg})}{(4.00 \text{ m})^2} + \frac{(10.0 \text{ kg})}{(6.00 \text{ m})^2}\right],$$

$$= -2.32 \times 10^{-13} \text{ N},$$

with the minus sign indicating a net force to the left.

b) No, the force found in part (a) is the *net* force due to the other two spheres.

12-7: $(6.673 \times 10^{-11} \text{ N·m}^2/\text{kg}^2)\dfrac{(70 \text{ kg})(7.35 \times 10^{22} \text{ kg})}{(3.78 \times 10^8 \text{ m})^2} = 2.4 \times 10^{-3} \text{ N}.$

12-8: $\dfrac{(333,000)}{(23,500)^2} = 6.03 \times 10^{-4}$

12-9: Denote the earth-sun separation as r_1 and the earth-moon separation as r_2.

a) $\qquad (Gm_M)\left[\dfrac{m_S}{(r_1 + r_2)^2} + \dfrac{m_E}{r_2^2}\right] = 6.30 \times 10^{20} \text{ N},$

toward the sun. b) The earth-moon distance is sufficiently small compared to the earth-sun distance ($r_2 \ll r_2$)that the vector from the earth to the moon can be taken to be perpendicular to the vector from the sun to the moon. The components of the gravitational force are then

$$\frac{Gm_M m_S}{r_1^2} = 4.34 \times 10^{20} \text{ N}, \qquad \frac{Gm_M m_E}{r_2^2} = 1.99 \times 10^{20} \text{ N},$$

and so the force has magnitude 4.77×10^{20} N and is directed $24.6°$ from the direction toward the sun.

c) $\qquad (Gm_M)\left[\dfrac{m_S}{(r_1 - r_2)^2} - \dfrac{m_E}{r_2^2}\right] = 2.37 \times 10^{20} \text{ N}.$

12-10: The direction of the force will be toward the larger mass, and the magnitude will be

$$\frac{Gm_2 m}{(d/2)^2} - \frac{Gm_1 m}{(d/2)^2} = \frac{4Gm(m_2 - m_1)}{d^2}.$$

12-11: For convenience of calculation, recognize that the mass of the small sphere will cancel. The acceleration is then

$$\frac{2G(0.260\text{ kg})}{(10.0 \times 10^{-2}\text{ m})^2} \times \frac{6.0}{10.0} = 2.1 \times 10^{-9}\text{ m/s}^2,$$

directed down.

12-12: Equation (12-4) gives

$$g = \frac{(6.673 \times 10^{-11}\text{ N·m}^2/\text{kg}^2)(1.5 \times 10^{22}\text{ kg})}{(1.15 \times 10^6\text{ m})^2} = 0.757\text{ m/s}^2.$$

12-13: To decrease the acceleration due to gravity by one-tenth, the distance from the earth must be increased by a factor of $\sqrt{10}$, and so the distance above the surface of the earth is

$$\left(\sqrt{10} - 1\right) R_E = 1.38 \times 10^7\text{ m}.$$

12-14: a) Using $g_E = 9.80\text{ m/s}^2$, Eq. (12-4) gives

$$g_v = \frac{Gm_v}{R_v^2} = G\left(\frac{m_v}{m_E}\right) m_E \left(\frac{R_E}{R_v}\right)^2 \left(\frac{1}{R_E^2}\right)$$
$$= \frac{Gm_E}{R_E^2}\left(\frac{m_v}{m_E}\right)\left(\frac{R_E}{R_v}\right)^2 = g_E\,(.815)\left(\frac{1}{.949}\right)^2$$
$$= (9.80\text{ m/s}^2)\,(.905)$$
$$= 8.87\ \text{m/s}^2,$$

where the subscripts v refer to the quantities pertinent to Venus. b) $(8.87\text{ m/s}^2)(5.00\text{ kg}) = $ 44.3 N.

12-15: a) See Exercise 12-14;

$$g_{\text{Titania}} = (9.80\text{ m/s}^2)\left(\frac{(8)^2}{1700}\right) = 0.369\text{ m/s}^2.$$

b) $\frac{\rho_T}{\rho_E} = \frac{m_T}{m_E} \cdot \frac{r_E^3}{r_T^3}$, or rearranging and solving for density, $\rho_T = \rho_E \cdot \frac{(1/1700)m_E}{m_E} \cdot \frac{r_E^3}{(1/8r_E)^3} = $ $(5500\text{ kg/m}^3)\left(\frac{512}{1700}\right) = 1656\text{ kg/m}^3$, or about $.3\rho_E$.

12-16: $M = \frac{gR^2}{G} = 2.44 \times 10^{21}\text{ kg}$ and $\rho = \frac{M}{(4\pi/3)R^3} = 1.30 \times 10^3\text{ kg/m}^3$.

12-17: From Eq. (12-1), $G = Fr^2/m_1m_2$, and from Eq. (12-4), $g = Gm_E/R_E^2$; combining and solving for R_E,

$$m_E = \frac{gm_1m_2R_E^2}{Fr^2} = 5.98 \times 10^{24}\text{ kg}.$$

12-18: From Example 12-4, the mass of the lander is 4000 kg. Assuming Phobos to be spherical, its mass in terms of its density ρ and radius R is $(4\pi/3)\rho R^3$, and so the gravitational force is

$$\frac{G(4\pi/3)(4000 \text{ kg})\rho R^3}{R^2} = G(4\pi/3)(4000 \text{ kg})(2000 \text{ kg/m}^3)(12 \times 10^3 \text{ m}) = 27 \text{ N}.$$

12-19: $\sqrt{2GM/R} = \sqrt{2(6.673 \times 10^{-11} \text{ N·m}^2/\text{kg}^2)(3.6 \times 10^{12} \text{ kg})/(700 \text{ m})}$
$$= 0.83 \text{ m/s}.$$

One could certainly walk that fast.

12-20: a) $F = Gm_E m/r^2$ and $|U| = Gm_E m/r$, so the altitude above the surface of the earth is $\frac{|U|}{F} - R_E = 9.36 \times 10^5$ m. b) Either of Eq. (12-1) or Eq. (12-9) can be used with the result of part (a) to find m, or noting that $U^2 = G^2 M_E^2 m^2/r^2$, $m = U^2/FGM_E = 2.55 \times 10^3$ kg.

12-21: The escape speed, from the results of Example 12-5, is $\sqrt{2GM/R}$.

a) $\sqrt{2(6.673 \times 10^{-11} \text{ N·m}^2/\text{kg}^2)(6.42 \times 10^{23} \text{ kg})/(3.40 \times 10^6 \text{ m})} = 5.02 \times 10^3$ m/s.

b) $\sqrt{2(6.673 \times 10^{-11} \text{ N·m}^2/\text{kg}^2)(1.90 \times 10^{27} \text{ kg})/(6.91 \times 10^7 \text{ m})} = 6.06 \times 10^4$ m/s.

c) Both the kinetic energy and the gravitational potential energy are proportional to the mass.

12-22: a) The kinetic energy is $K = \frac{1}{2}mv^2$, or $K = \frac{1}{2}(629 \text{ kg})(3.33 \times 10^3 \text{ m/s})^2$, or $KE = 3.49 \times 10^9$ J.

b) $U = -\frac{GMm}{r} = \frac{(6.673 \times 10^{-11} \text{ N·m}^2/\text{kg}^2)(5.97 \times 10^{24} \text{ kg})(629 \text{ kg})}{2.87 \times 10^9 \text{ m}},$

or $U = -8.73 \times 10^7$ J.

12-23: a) Eliminating the orbit radius r between Equations (12-12) and (12-14) gives

$$T = \frac{2\pi Gm_E}{v^3} = \frac{2\pi(6.673 \times 10^{-11} \text{ N·m}^2/\text{kg}^2)(5.97 \times 10^{24} \text{ kg})}{(6200 \text{ m/s})^3}$$
$$= 1.05 \times 10^4 \text{ s} = 175 \text{ min}.$$

b) $$\frac{2\pi v}{T} = 3.71 \text{ m/s}^2.$$

12-24: Substitution into Eq. (12-14) gives $T = 6.96 \times 10^3$ s, or 116 minutes.

12-25: Using Eq. (12-12),

$$v = \sqrt{\frac{(6.673 \times 10^{-11} \text{ N·m}^2/\text{kg}^2)(5.97 \times 10^{24} \text{ kg})}{(6.38 \times 10^6 \text{ m} + 7.80 \times 10^5 \text{ m})}} = 7.46 \times 10^3 \text{ m/s}.$$

12-26: Applying Kepler's third law to circular orbits, the radii of the orbits are proportional to the $\frac{2}{3}$ power of their periods. Using primes to denote the radius and period of the hypothetical planet,

$$R' = R\left(\frac{T'}{T}\right)^{2/3} = 2.1 \times 10^{10} \text{ m}.$$

12-27:   a)

$$v = \sqrt{Gm/r}$$
$$= \sqrt{(6.673 \times 10^{-11} \text{ N·m}^2/\text{kg}^2)(0.85 \times 1.99 \times 10^{30} \text{ kg})/((1.50 \times 10^{11} \text{ m})(0.11))}$$
$$= 8.27 \times 10^4 \text{ m/s}.$$

b) $2\pi r/v = 1.25 \times 10^6$ s (about two weeks).

12-28:   From either Eq. (12-14) or Eq. (12-19),

$$m_S = \frac{4\pi^2 r^3}{GT^2} = \frac{4\pi^2(1.08 \times 10^{11} \text{ m})^3}{(6.673 \times 10^{-11} \text{ N·m}^2/\text{kg}^2)((224.7 \text{ d})(8.64 \times 10^4 \text{ s/d}))^2}$$
$$= 1.98 \times 10^{30} \text{ kg}.$$

12-29:   a) 1 picture $= 1 \times 10^3$ word.   b) $(1 - 0.248)(5.92 \times 10^{12} \text{ m}) = 4.45 \times 10^{12}$ m, $(1 + 0.010)(4.50 \times 10^{12} \text{ m}) = 4.55 \times 10^{12}$ m.   c) $T = 248$ y.

12-30:   a)
$$r = \sqrt{\frac{Gm_1 m_2}{F}} = 7.07 \times 10^{10} \text{ m}.$$

b) From Eq. (12-19), using the result of part (a),

$$T = \frac{2\pi(7.07 \times 10^{10} \text{ m})^{3/2}}{\sqrt{(6.673 \times 10^{-11} \text{ N·m}^2/\text{kg}^2)(1.90 \times 10^{30} \text{ kg})}} = 1.05 \times 10^7 \text{ s} = 121 \text{ days}.$$

c) From Eq. (12-14) the radius is $(8)^{2/3} =$ four times that of the large planet's orbit, or 2.83×10^{11} m.

12-31:   a) For a circular orbit, Eq. (12-12) predicts a speed of

$$\sqrt{(6.673 \times 10^{-11} \text{ N·m}^2/\text{kg}^2)(1.99 \times 10^{30} \text{ kg})/(43 \times 10^9 \text{ m})} = 56 \text{ km/s}.$$

b) The escape speed for any object at this radius is $\sqrt{2}(56 \text{ km/s}) = 79$ km/s, so the spacecraft must be in a bound elliptical orbit.

12-32:   a) Divide the rod into differential masses dm at position l, measured from the right end of the rod. Then, $dm = dl(M/L)$, and

$$dU = -\frac{Gm\,dm}{l + x} = -\frac{GmM}{L}\frac{dl}{l + x}.$$

Integrating,

$$U = -\frac{GmM}{L}\int_0^L \frac{dl}{l + x} = -\frac{GmM}{L}\ln\left(1 + \frac{L}{x}\right).$$

For $x \gg L$, the natural logarithm is $\sim (L/x)$, and $U \to -GmM/x$. b) The x-component of the gravitational force on the sphere is

$$F_x = -\frac{\partial U}{\partial x} = \frac{GmM}{L}\frac{(-L/x^2)}{(1+(L/x))} = -\frac{GmM}{(x^2+Lx)},$$

with the minus sign indicating an attractive force. As $x \gg L$, the denominator in the above expression approaches x^2, and $F_x \to GmM/x^2$, as expected. The derivative may also be taken by expressing

$$\ln\left(1+\frac{L}{x}\right) = \ln(x+L) - \ln x$$

at the cost of a little more algebra.

12-33: a) Refer to the derivation of Eq. (12-26) and Fig. (12-16). In this case, the red ring in Fig. (12-16) has mass M and the common distance s is $\sqrt{x^2+a^2}$. Then, $U = -GMm/\sqrt{x^2+a^2}$. b) When $x \gg a$, the term in the square root approaches x^2 and $U \to -GMm/x$, as expected.

c)
$$F_x = -\frac{\partial U}{\partial x} = -\frac{GMmx}{(x^2+a^2)^{3/2}},$$

with the minus sign indicating an attractive force. d) When $x \gg a$, the term inside the parentheses in the above expression approaches x^2 and $F_x \to -GMmx/(x^2)^{3/2} = -GMm/x^2$, as expected. e) The result of part (a) indicates that $U = \frac{-GMm}{a}$ when $x = 0$. This makes sense because the mass at the center is a constant distance a from the mass in the ring. The result of part (c) indicates that $F_x = 0$ when $x = 0$. At the center of the ring, all mass elements that comprise the ring attract the particle toward the respective parts of the ring, and the net force is zero.

12-34: At the equator, the gravitational field and the radial acceleration are parallel, and taking the magnitude of the weight as given in Eq. (12-30) gives

$$w = mg_0 - ma_{\text{rad}}.$$

The difference between the measured weight and the force of gravitational attraction is the term ma_{rad}. The mass m is found by solving the first relation for m, $m = \frac{w}{g_0-a_{\text{rad}}}$. Then,

$$ma_{\text{rad}} = w\frac{a_{\text{rad}}}{g_0 - a_{\text{rad}}} = \frac{w}{(g_0/a_{\text{rad}}) - 1}.$$

Using either $g_0 = 9.80$ m/s^2 or calculating g_0 from Eq. (12-4) gives $ma_{\text{rad}} = 2.40$ N.

12-35: a) $Gm_N m/R^2 = (10.7 \text{ m/s}^2)(5.00 \text{ kg}) = 53.5$ N, or 54 N to two figures.

b) $m(g_0 - a_{\text{rad}}) = (5.00 \text{ kg})\left(10.7 \text{ m/s}^2 - \frac{4\pi^2(2.5\times10^7 \text{ m})}{[(16 \text{ h})(3600 \text{ s/h})]^2}\right) = 52.0$ N.

12-36: a)
$$\frac{GMm}{r^2} = \frac{(R_S c^2/2)m}{r^2} = \frac{mc^2 R_S}{2r^2}.$$

b)
$$\frac{(5.00 \text{ kg})(3.00 \times 10^8 \text{ m/s})^2(1.4 \times 10^{-2} \text{ m})}{2(3.00 \times 10^6 \text{ m})^2} = 350 \text{ N}.$$

c) Solving Eq. (12-32) for M,

$$M = \frac{R_S c^2}{2G} = \frac{(14.00 \times 10^{-3} \text{ m})(3.00 \times 10^8 \text{ m/s})^2}{2(6.673 \times 10^{-11} \text{ N·m}^2/\text{kg}^2)} = 9.44 \times 10^{24} \text{ kg}.$$

12-37: a) From Eq. (12-12),

$$M = \frac{Rv^2}{G} = \frac{(7.5 \text{ ly})(9.461 \times 10^{15} \text{ m/ly})(200 \times 10^3 \text{ m/s})^2}{(6.673 \times 10^{-11} \text{ N·m}^2/\text{kg}^2)}$$
$$= 4.3 \times 10^{37} \text{ kg} = 2.1 \times 10^7 \ m_S.$$

b) It would seem not.

c)
$$R_S = \frac{2GM}{c^2} = \frac{2v^2 R}{c^2} = 6.32 \times 10^{10} \text{ m},$$

which does fit.

12-38: Using the mass of the sun for M in Eq. (12-32) gives

$$R_S = \frac{2(6.673 \times 10^{-11} \text{ N·m}^2/\text{kg}^2)(1.99 \times 10^{30} \text{ kg})}{(3.00 \times 10^8 \text{ m/s})^2} = 2.95 \text{ km}.$$

That is, Eq. (12-32) may be rewritten

$$R_S = \frac{2Gm_{\text{sun}}}{c^2} \left(\frac{M}{m_{\text{sun}}}\right) = 2.95 \text{ km} \times \left(\frac{M}{m_{\text{sun}}}\right).$$

Using 3.0 km instead of 2.95 km is accurate to 1.7%.

12-39: $\dfrac{R_S}{R_E} = \dfrac{2(6.67 \times 10^{-11} \text{ Nm}^2/\text{kg}^2)(5.97 \times 10^{24} \text{ kg})}{(3 \times 10^8 \text{ m/s})^2(6.38 \times 10^6 \text{ m})} = 1.4 \times 10^{-9}.$

12-40: a) From symmetry, the net gravitational force will be in the direction 45° from the $+x$-axis (bisecting the x and y axes), with magnitude

$$(6.673 \times 10^{-11} \text{ N·m}^2/\text{kg}^2)(0.0150 \text{ kg}) \left[\frac{(2.0 \text{ kg})}{(2(0.50 \text{ m})^2)} + 2\frac{(1.0 \text{ kg})}{(0.50 \text{ m})^2} \sin 45°\right]$$
$$= 9.67 \times 10^{-12} \text{ N}.$$

b) The initial displacement is so large that the initial potential may be taken to be zero. From the work-energy theorem,

$$\frac{1}{2}mv^2 = Gm \left[\frac{(2.0 \text{ kg})}{\sqrt{2}(0.50 \text{ m})} + 2\frac{(1.0 \text{ kg})}{(0.50 \text{ m})}\right].$$

Canceling the factor of m and solving for v, and using the numerical values gives $v = 3.02 \times 10^{-5}$ m/s.

12-41: The geometry of the 3-4-5 triangle is available to simplify some of the algebra. The components of the gravitational force are

$$F_y = \frac{(6.673 \times 10^{-11} \text{ N·m}^2/\text{kg}^2)(0.500 \text{ kg})(80.0 \text{ kg})}{(5.000 \text{ m})^2} \frac{3}{5}$$

$$= 6.406 \times 10^{-11} \text{ N}$$

$$F_x = -(6.673 \times 10^{-11} \text{ N·m}^2/\text{kg}^2)(0.500 \text{ kg}) \left[\frac{(60.0 \text{ kg})}{(4.000 \text{ m})^2} + \frac{(80.0 \text{ kg})}{(5.000 \text{ m})^2} \frac{4}{5} \right]$$

$$= -2.105 \times 10^{-10} \text{ N},$$

so the magnitude is 2.20×10^{-10} N and the direction of the net gravitational force is 163° counterclockwise from the +x-axis. b) At $x = 0, y = 1.39$ m.

12-42: a) The direction from the origin to the point midway between the two large masses is $\arctan \left(\frac{0.100 \text{ m}}{0.200 \text{ m}} \right) = 26.6°$, which is not the angle (14.6°) found in the example. b) The common lever arm is 0.100 m, and the force on the upper mass is at an angle of 45° from the lever arm. The net torque is

$$(6.673 \times 10^{-11} \text{ N·m}^2/\text{kg}^2)(0.0100 \text{ kg})(0.500 \text{ kg}) \left[\frac{(0.100 \text{ m}) \sin 45°}{2(0.200 \text{ m})^2} - \frac{(0.100 \text{ m})}{(0.200 \text{ m})^2} \right]$$

$$= -5.39 \times 10^{-13} \text{ N·m},$$

with the minus sign indicating a clockwise torque. c) There can be no net torque due to gravitational fields with respect to the center of gravity, and so the center of gravity in this case is not at the center of mass.

12-43: a) The simplest way to approach this problem is to find the force between the spacecraft and the center of mass of the earth-moon system, which is 4.67×10^6 m from the center of the earth.

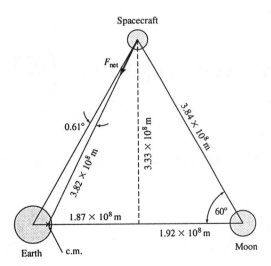

The distance from the spacecraft to the center of mass of the earth-moon system is 3.82×10^8 m. Using the Law of Gravitation, the force on the spacecraft is 3.4 N, an angle of 0.61°

from the earth-spacecraft line. This equilateral triangle arrangement of the earth, moon and spacecraft is a solution of the Lagrange Circular Restricted Three-Body Problem. The spacecraft is at one of the earth-moon system Lagrange points. The Trojan asteroids are found at the corresponding Jovian Lagrange points.

b) The work is $W = -\frac{GMm}{r} = -\frac{6.673\times10^{-11} \text{ N·m}^2/\text{kg}^2)(5.97\times10^{24} \text{ kg}+7.35\times10^{22} \text{ kg})(1250 \text{ kg})}{3.84\times10^8 \text{ m}}$, or $W = -1.31 \times 10^9$ J.

12-44: Denote the 25-kg sphere by a subscript 1 and the 100-kg sphere by a subscript 2. a) Linear momentum is conserved because we are ignoring all other forces, that is, the net external force on the system is zero. Hence, $m_1v_1 = m_2v_2$. This relationship is useful in solving part (b) of this problem. b) From the work-energy theorem,

$$Gm_1m_2\left[\frac{1}{r_f} - \frac{1}{r_i}\right] = \frac{1}{2}\left(m_1v_1^2 + m_2v_2^2\right)$$

and from conservation of momentum the speeds are related by $m_1v_1 = m_2v_2$. Using the conservation of momentum relation to eliminate v_2 in favor of v_1 and simplifying yields

$$v_1^2 = \frac{2Gm_2^2}{m_1 + m_2}\left[\frac{1}{r_f} - \frac{1}{r_i}\right],$$

with a similar expression for v_2. Substitution of numerical values gives $v_1 = 1.63 \times 10^{-5}$ m/s, $v_2 = 4.08 \times 10^{-6}$ m/s. The magnitude of the relative velocity is the sum of the speeds, 2.04×10^{-5} m/s.

c) The distances the centers of the spheres travel (x_1 and x_2) is proportional to their acceleration, and $\frac{x_1}{x_2} = \frac{a_1}{a_2} = \frac{m_2}{m_1}$, or $x_1 = 4x_2$. When the spheres finally make contact, their centers will be a distance of $2R$ apart, or $x_1 + x_2 + 2R = 40$ m, or $x_2 + 4x_2 + 2R = 40$ m. Thus, $x_2 = 8$ m $- 0.4R$, and $x_1 = 32$ m $- 1.6R$.

12-45: Solving Eq. (12-14) for r,

$$R^3 = Gm_E\left(\frac{T}{2\pi}\right)^2$$

$$= (6.673 \times 10^{-11} \text{ N·m}^2/\text{kg}^2)(5.97 \times 10^{24} \text{ kg})\left(\frac{(27.3 \text{ d})(86,400 \text{ s/d})}{2\pi}\right)^2$$

$$= 5.614 \times 10^{25} \text{ m}^3,$$

from which $r = 3.83 \times 10^8$ m.

12-46: $g = |\vec{g}| = \frac{(6.673 \times 10^{-11} \text{ N·m}^2/\text{kg}^2)(20.0 \text{ kg})}{(1.50 \text{ m})^2} = 5.93 \times 10^{-10}$ N/kg, directed toward the center of the sphere.

12-47: a) From Eq. (12-14),

$$r^3 = Gm_E\left(\frac{T}{2\pi}\right)^2 = (6.673 \times 10^{-11} \text{ N·m}^2/\text{kg}^2)(5.97 \times 10^{24} \text{ kg})\left(\frac{86,164 \text{ s}}{2\pi}\right)^2$$

$$= 7.492 \times 10^{22} \text{ m}^3,$$

and so $h = r - R_E = 3.58 \times 10^7$ m. Note that the period to use for the earth's rotation is the siderial day, not the solar day (see Section 12-8).   b) For these observers, the satellite is below the horizon.

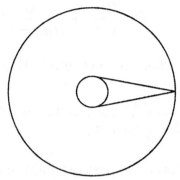

12-48: Using Eq. (12-12) with the mass and radius of Dactyl,

$$v = \sqrt{\frac{(6.673 \times 10^{-11} \text{ N·m}^2/\text{kg}^2)(3.6 \times 10^{12} \text{ kg})}{700 \text{ m}}} = 0.59 \text{ m/s}.$$

Many people can achieve this speed by walking.

12-49: In terms of the density ρ, the ratio M/R is $(4\pi/3)\rho R^2$, and so the escape speed is

$$v = \sqrt{(8\pi/3)(6.673 \times 10^{-11} \text{ N·m}^2/\text{kg}^2)(2500 \text{ kg/m}^3)(150 \times 10^3 \text{ m})^2} = 177 \text{ m/s}.$$

12-50: a) Following the hint, use as the escape velocity $v = \sqrt{2gh}$, where h is the height one can jump from the surface of the earth. Equating this to the expression for the escape speed found in Problem 12-49,

$$2gh = \frac{8\pi}{3}\rho G R^2, \qquad \text{or} \qquad R^2 = \frac{3}{4\pi}\frac{gh}{\rho G},$$

where $g = 9.80 \text{ m/s}^2$ is for the surface of the earth, not the asteroid. Using $h = 1 \text{ m}$ (variable for different people, of course), $R = 3.7 \text{ km}$. As an alternative, if one's jump speed is known, the analysis of Problem 12-49 shows that for the same density, the escape speed is proportional to the radius, and one's jump speed as a fraction of 60 m/s gives the largest radius as a fraction of 50 km.   b) With $a = v^2/R$, $\rho = \frac{3a}{4\pi GR} = 3.03 \times 10^3 \text{ kg/m}^3$.

12-51: The fractional error is

$$1 - \frac{mgh}{Gmm_\text{E}\left(\frac{1}{R_\text{E}} - \frac{1}{R_\text{E}+h}\right)} = 1 - \frac{g}{Gm_\text{E}}\,(R_\text{E} + h)(R_\text{E}).$$

At this point, it is advantageous to use the algebraic expression for g as given in Eq. (12-4) instead of numerical values to obtain the fractional difference as $1 - (R_\text{E} + h)/R_\text{E} = -h/R_\text{E}$, so if the fractional difference is -1%, $h = (0.01)R_\text{E} = 6.4 \times 10^4 \text{ m}$.

If the algebraic form for g in terms of the other parameters is not used, and the numerical values from Appendix F are used along with $g = 9.80 \text{ m/s}^2$, $h/R_\text{E} = 8.7 \times 10^{-3}$, which is qualitatively the same.

12-52:   a) The total gravitational potential energy in this model is

$$U = -Gm \left[\frac{m_E}{r} + \frac{m_M}{R_{EM} - r} \right].$$

b) See Exercise 12-5. The point where the net gravitational field vanishes is

$$r = \frac{R_{EM}}{1 + \sqrt{m_M/m_E}} = 3.46 \times 10^8 \text{ m}.$$

Using this value for r in the expression in part (a) and the work-energy theorem, including the initial potential energy of $-Gm(m_E/R_E + m_M/(R_{EM} - R_E))$ gives 11.1 km/s.   c) The final distance from the earth is not R_M, but the Earth-moon distance minus the radius of the moon, or 3.823×10^8 m. From the work-energy theorem, the rocket impacts the moon with a speed of 2.9 km/s.

12-53:   One can solve this problem using energy conservation, units of J/kg for energy, and basic concepts of orbits. $E = K + U$, or $-\frac{GM}{2a} = \frac{1}{2}v^2 - \frac{GM}{r}$, where E, K and U are the energies per unit mass, v is the circular orbital velocity of 1655 m/s at the lunicentric distance of 1.79×10^6 m. The total energy at this distance is -1.37×10^6 J/Kg. When the velocity of the spacecraft is reduced by 20 m/s, the total energy becomes

$$E = \frac{1}{2}(1655 \text{ m/s} - 20 \text{ m/s})^2 - \frac{(6.673 \times 10^{-11} \text{ N·m}^2/\text{kg}^2)(7.35 \times 10^{22} \text{ kg})}{(1.79 \times 10^6 \text{ m})},$$

or $E = -1.40 \times 10^6$ J/kg. Since $E = -\frac{GM}{2a}$, we can solve for a, $a = 1.748 \times 10^6$ m, the semi-major axis of the new elliptical orbit. The old distance of 1.79×10^6 m is now the apolune distance, and the perilune can be found from $a = \frac{r_a + r_p}{2}$, $r_p = 1.706 \times 10^6$ m. Obviously this is less than the radius of the moon, so the spacecraft crashes! At the surface, $U = -\frac{GM}{R_m}$, or, $U = -2.818 \times 10^6$ J/kg. Since the total energy at the surface is -1.40×10^6 J/kg, the kinetic energy at the surface is 1.415×10^6 J/kg. So, $\frac{1}{2}v^2 = 1.415 \times 10^6$ J/kg, or $v = 1.682 \times 10^3$ m/s = 6057 km/h.

12-54:   Combining Equations (12-13) and (3-28) and setting $a_{rad} = 9.80$ m/s² (so that $w = 0$ in Eq. (12-30)),

$$T = 2\pi \sqrt{\frac{R}{a_{rad}}} = 5.07 \times 10^3 \text{ s},$$

which is 84.5 min, or about an hour and a half.

12-55:   The change in gravitational potential energy is

$$\Delta U = \frac{Gm_E m}{(R_E + h)} - \frac{Gm_E m}{R_E} = -Gm_E m \frac{h}{R_E (R_R + h)},$$

so the speed of the hammer is, from the work-energy theorem,

$$\sqrt{\frac{2Gm_{\mathrm{E}}h}{(R_{\mathrm{E}}+h)\,R_{\mathrm{E}}}}.$$

(See Problem 12-51 for the necessity of this form.)

12-56:   a) The energy the satellite has as it sits on the surface of the Earth is $E_{\mathrm{i}} = \frac{-GmM_{\mathrm{E}}}{R_{\mathrm{E}}}$. The energy it has when it is in orbit at a radius $R \approx R_{\mathrm{E}}$ is $E_{\mathrm{f}} = \frac{-GmM_{\mathrm{E}}}{2R_{\mathrm{E}}}$. The work needed to put it in orbit is the difference between these: $W = E_{\mathrm{f}} - E_{\mathrm{i}} = \frac{GmM_{\mathrm{E}}}{2R_{\mathrm{E}}}$.

   b) The total energy of the satellite far away from the Earth is zero, so the additional work needed is $0 - \left(\frac{-GmM_{\mathrm{E}}}{2R_{\mathrm{E}}}\right) = \frac{GmM_{\mathrm{E}}}{2R_{\mathrm{E}}}$.

   c) The work needed to put the satellite into orbit was the same as the work needed to put the satellite from orbit to the edge of the universe.

12-57:   The escape speed will be

$$v = \sqrt{2G\left[\frac{m_{\mathrm{E}}}{R_{\mathrm{E}}} + \frac{m_{\mathrm{S}}}{R_{\mathrm{ES}}}\right]} = 4.35 \times 10^4 \text{ m/s}.$$

a) Making the simplifying assumption that the direction of launch is the direction of the earth's motion in its orbit, the speed relative to the earth is

$$v - \frac{2\pi R_{\mathrm{ES}}}{T} = 4.35 \times 10^4 \text{ m/s} - \frac{2\pi(1.50 \times 10^{11} \text{ m})}{(3.156 \times 10^7 \text{ s})} = 1.37 \times 10^4 \text{ m/s}.$$

b) The rotational speed at Cape Canaveral is $\frac{2\pi(6.38 \times 10^6 \text{ m})\cos 28.5°}{86164 \text{ s}} = 4.09 \times 10^2$ m/s, so the speed relative to the surface of the earth is 1.33×10^4 m/s.   c) In French Guiana, the rotational speed is 4.63×10^2 m/s, so the speed relative to the surface of the earth is 1.32×10^4 m/s.

12-58:   a) The SI units of energy are kg· m^2/s^2, so the SI units for ϕ are m^2/s^2. Also, it is known from kinetic energy considerations that the dimensions of energy, kinetic or potential, are mass × speed^2, so the dimensions of gravitational potential must be the same as speed^2.   b) $\phi = -\frac{U}{m} = -\frac{Gm_{\mathrm{E}}}{r}$.

   c)
$$\Delta\phi = Gm_{\mathrm{E}}\left[\frac{1}{r_{\mathrm{f}}} - \frac{1}{R_{\mathrm{E}}}\right] = 3.68 \times 10^6 \text{ J/kg}.$$

d) $m\Delta\phi = 5.53 \times 10^{10}$ J. (An extra figure was kept in the intermediate calculations.)

12-59:   a) The period of the asteroid is $T = \frac{2\pi a^{3/2}}{GM}$. Inserting 3×10^{11} m for a gives 2.84 y and 5×10^{11} m gives a period of 6.11 y.

   b) If the period is 5.93 y, then $a = 4.90 \times 10^{11}$ m.

   c) This happens because $0.4 = 2/5$, another ratio of integers. So once every 5 orbits of the asteroid and 2 orbits of Jupiter, the asteroid is at its perijove distance. Solving when $T = 4.74$ y, $a = 4.22 \times 10^{11}$ m.

12-60:   a) In moving to a lower orbit by whatever means, gravity does positive work, and so the speed does increase.   b) From $\frac{v^2}{r} = \frac{Gm_E}{r^2}$, $v = (Gm_E)^{1/2}r^{-1/2}$, so

$$\Delta v = (Gm_E)^{1/2} \left(\frac{-\Delta r}{2} \right) r^{-3/2} = \left(\frac{\Delta r}{2} \right) \sqrt{\frac{Gm_E}{r^3}}.$$

Note that a positive Δr is given as a decrease in radius. Similarly, the kinetic energy is $K = (1/2)mv^2 = (1/2)Gm_Em/r$, and so $\Delta K = (1/2)(Gm_Em/r^2)\Delta r$, $\Delta U = -(Gm_Em/r^2)\Delta r$ and $W = \Delta U + \Delta K = -(Gm_Em/2r^2)\Delta r$, in agreement with part (a).   c) $v = \sqrt{Gm_E/r} = 7.72 \times 10^3$ m/s, $\Delta v = (\Delta r/2)\sqrt{Gm_E/r^3} = 28.9$ m/s, $E = -Gm_Em/2r = -8.95 \times 10^{10}$ J (from Eq. (12-15)), $\Delta K = (Gm_Em/2r^2)(\Delta r) = 6.70 \times 10^8$ J, $\Delta U = -2\Delta K = -1.34 \times 10^9$ J and $W = -\Delta K = -6.70 \times 10^8$ J.   d) As the term "burns up" suggests, the energy is converted to heat or is dissipated in the collisions of the debris with the ground.

12-61:   a) The stars are separated by the diameter of the circle $d = 2R$, so the gravitational force is $\frac{GM^2}{4R^2}$.   b) The gravitational force found in part (b) is related to the radial acceleration by $F_g = Ma_{rad} = Mv^2/R$ for each star, and substituting the expression for the force from part (a) and solving for v gives $v = \sqrt{GM/4R}$. The period is $T = \frac{2\pi R}{v} = \sqrt{16\pi^2 R^3/GM} = 4\pi R^{3/2}/\sqrt{GM}$.   c) The initial gravitational potential energy is $-GM^2/2R$ and the initial kinetic energy is $2(1/2)Mv^2 = GM^2/4R$, so the total mechanical energy is $-GM^2/4R$. If the stars have zero speed when they are very far apart, the energy needed to separate them is $GM^2/4R$.

12-62:   a) The radii R_1 and R_2 are measured with respect to the center of mass, and so $M_1R_1 = M_2R_2$, and $R_1/R_2 = M_2/M_1$.   b) If the periods were different, the stars would move around the circle with respect to one another, and their separations would not be constant; the orbits would not remain circular. Employing qualitative physical principles, the forces on each star are equal in magnitude, and in terms of the periods, the product of the mass and the radial accelerations are

$$\frac{4\pi^2 M_1 R_1}{T_1^2} = \frac{4\pi^2 M_2 R_2}{T_2^2}.$$

From the result of part (a), the numerators of these expressions are equal, and so the denominators are equal, and the periods are the same. To find the period in the symmetric form desired, there are many possible routes. An elegant method, using a bit of hindsight, is to use the above expressions to relate the periods to the force $F_g = \frac{GM_1M_2}{(R_1+R_2)^2}$, so that equivalent expressions for the period are

$$M_2T^2 = \frac{4\pi^2 R_1 (R_1 + R_2)^2}{G}$$
$$M_1T^2 = \frac{4\pi^2 R_2 (R_1 + R_2)^2}{G}.$$

Adding the expressions gives

$$(M_1 + M_2)T^2 = \frac{4\pi^2 (R_1 + R_2)^3}{G} \quad \text{or} \quad T = \frac{2\pi (R_1 + R_2)^{3/2}}{\sqrt{G(M_1 + M_2)}}.$$

c) First we must find the radii of each orbit given the speed and period data. In a circular orbit, $v = \frac{2\pi R}{T}$, or $R = \frac{vT}{2\pi}$. Thus, $R_\alpha = \frac{(36 \times 10^3 \text{ m/s})(137 \text{ d} \times 86,400 \text{ s/d})}{2\pi} = 6.78 \times 10^{10}$ m, and $R_\beta = \frac{(12 \times 10^3 \text{ m/s})(137 \text{ d} \times 86,400 \text{ s/d})}{2\pi} = 2.26 \times 10^{10}$ m. Now find the sum of the masses and use $M_\alpha R_\alpha = M_\beta R_\beta$, and the fact that $R_\alpha = 3R_\beta$. $(M_\alpha + M_\beta) = \frac{4\pi^2 (R_\alpha + R_\beta)^3}{T^2 G}$, inserting the values of T, and the radii, $(M_\alpha + M_\beta) = \frac{4\pi^2 (6.78 \times 10^{10} \text{m} + 2.26 \times 10^{10} \text{ m})^3}{(137 \text{ d} \times 86,400 \text{ s/d})^2 (6.673 \times 10^{-11} \text{ N·m}^2/\text{kg}^2)}$. $M_\alpha + M_\beta = 3.12 \times 10^{30}$ kg. Since $M_\beta = M_\alpha R_\alpha / R_\beta = 3M_\alpha$, $4M_\alpha = 3.12 \times 10^{30}$ kg, or $M_\alpha = 7.80 \times 10^{29}$ kg, and $M_\beta = 2.34 \times 10^{30}$ kg.

d) Let α refer to the star and β refer to the black hole. Use the relationships derived in parts (a) and (b): $R_\beta = (M_\alpha / M_\beta)R_\alpha = (0.67/3.8)R_\alpha = (0.176)R_\alpha$, $(R_\alpha + R_\beta) = \sqrt[3]{\frac{(M_\alpha + M_\beta)T^2 G}{4\pi^2}}$, and $v = \frac{2\pi R}{T}$. For R_α, inserting the values for M and T and R_β, $(R_\alpha + 0.176 R_\alpha) = \sqrt[3]{\frac{[(0.67 + 3.8)(1.99 \times 10^{30} \text{ kg})](7.75 \text{ h} \times 3600 \text{ s/h})^2 (6.673 \times 10^{-11} \text{ N·m}^2/\text{kg}^2)}{4\pi^2}}$. Thus, for V616 Monocerotis, $R_\alpha = 1.9 \times 10^9$ m, $v_\alpha = 4.4 \times 10^2$ km/s and for the black hole: $R_\beta = 34 \times 10^8$ m, $v_\beta = 77$ km/s.

12-63: From conservation of energy, the speed at the closer distance is

$$v = \sqrt{v_0^2 + 2Gm_S \left(\frac{1}{r_f} - \frac{1}{r_i} \right)} = 6.8 \times 10^4 \text{ m/s}.$$

12-64: Using conservation of energy,

$$\frac{1}{2} m_M v_a^2 - \frac{GM_S m_M}{r_a} = \frac{1}{2} m_M v_p^2 - \frac{GM_S m_M}{r_p}, \quad \text{or}$$

$$v_p = \sqrt{v_a^2 - 2GM_S \left(\frac{1}{r_a} - \frac{1}{r_p} \right)} = 2.650 \times 10^4 \text{ m/s}.$$

The subscripts a and p denote aphelion and perihelion.

To use conservation of angular momentum, note that at the extremes of distance (periheleion and aphelion), Mars' velocity vector must be perpendicular to its radius vector, and so the magnitude of the angular momentum is $L = mrv$. Since L is a constant, the product rv must be a constant, and so

$$v_p = v_a \frac{r_a}{r_p} = (2.198 \times 10^4 \text{ m/s}) \frac{(2.492 \times 10^{11} \text{ m})}{(2.067 \times 10^{11} \text{ m})} = 2.650 \times 10^4 \text{ m/s},$$

a confirmation of Kepler's Laws.

12-65: a) The semimajor axis is the average of the perigee and apogee distances, $a = \frac{1}{2}((R_E + h_p) + (R_E + h_a)) = 8.58 \times 10^6$ m. From Eq. (12-19) with the mass of the earth, the period of the orbit is

$$T = \frac{2\pi a^{3/2}}{\sqrt{GM_E}} = 7.91 \times 10^3 \text{ s},$$

a little more than two hours.   b) See Problem 12-64; $\frac{v_p}{v_a} = \frac{r_a}{r_p} = 1.53$.   c) The equation that represents conservation of energy (apart from a common factor of the mass of the spacecraft) is

$$\frac{1}{2}v_p^2 - \frac{Gm_E}{r_p} = \frac{1}{2}v_a^2 - \frac{Gm_E}{r_a} = \frac{1}{2}\left(\frac{r_p}{r_a}\right)^2 v_p^2 - \frac{Gm_E}{r_a},$$

where conservation of angular momentum has been used to eliminate v_a in favor of v_p. Solving for v_p^2 and simplifying,

$$v_p^2 = \frac{2Gm_E r_a}{r_p(r_p + r_a)} = 7.71 \times 10^7 \text{ m}^2/\text{s}^2,$$

from which $v_p = 8.43 \times 10^3$ m/s and $v_a = 5.51 \times 10^3$ m/s.   d) The escape speed for a given distance is $v_e = \sqrt{2GM/r}$, and so the difference between escape speed and v_p is, after some algebra,

$$v_e - v_p = \sqrt{\frac{2Gm_E}{r_p}}\left[1 - 1/\sqrt{1 + (r_p/r_a)}\right].$$

Using the given values for the radii gives $v_e - v_p = 2.41 \times 10^3$ m/s. The similar calculation at apogee gives $v_e - v_a = 3.26 \times 10^3$ m/s, so it is more efficient to fire the rockets at perigee. Note that in the above, the escape speed v_e is different at the two points, $v_{pe} = 1.09 \times 10^4$ m/s and $v_{ae} = 8.77 \times 10^3$ m/s.

12-66:   a) From the value of g at the poles,

$$m_U = \frac{g_U R_U^2}{G} = \frac{(11.1 \text{ m/s}^2)(2.556 \times 10^7 \text{ m})^2}{(6.673 \times 10^{-11} \text{ N·m}^2/\text{kg}^2)} = 1.09 \times 10^{26} \text{ kg}.$$

b) $Gm_U/r^2 = g_U\,(R_U/r)^2 = 0.432$ m/s^2.   c) $Gm_M/R_M^2 = 0.080$ m/s^2.   d) No; Miranda's gravity is sufficient to retain objects released near its surface.

12-67:   Using Eq. (12-15), with the mass M_m instead of the mass of the earth, the energy needed is

$$\Delta E = \frac{Gm_m m}{2}\left[\frac{1}{r_i} - \frac{1}{r_f}\right]$$

$$= \frac{(6.673 \times 10^{-11} \text{ N·m}^2/\text{kg}^2)(6.42 \times 10^{23} \text{ kg})(3000 \text{ kg})}{2}$$

$$\times \left[\frac{1}{(2.00 \times 10^6 \text{ m} + 3.40 \times 10^6 \text{ m})} - \frac{1}{4.00 \times 10^6 \text{ m}}\right]$$

$$= 4.17 \times 10^9 \text{ J}.$$

12-68: a) The semimajor axis is 4×10^{15} m and so the period is

$$\frac{2\pi(4 \times 10^{15} \text{ m})^{3/2}}{\sqrt{(6.673 \times 10^{-11} \text{ N·m}^2/\text{kg}^2)(1.99 \times 10^{30} \text{ kg})}} = 1.38 \times 10^{14} \text{ s},$$

which is about 4 million years. b) Using the earth-sun distance as an estimate for the distance of closest approach, $v = \sqrt{2Gm_S/R_{ES}} = 4 \times 10^4$ m/s. c) $(1/2)mv^2 = Gm_Sm/R = 10^{24}$ J. This is far larger than the energy of a volcanic eruption and is comparable to the energy of burning the fossil fuel.

12-69: a) From Eq. (12-14) with the mass of the sun,

$$r = \left[\begin{array}{c} (6.673 \times 10^{-11} \text{ N·m}^2/\text{kg}^2)(1.99 \times 10^{30} \text{ kg}) \\ \times \left((3 \times 10^4 \text{ y})(3.156 \times 10^7 \text{ s/y})\right)^2 /4\pi^2 \end{array} \right]^{1/3} = 1.4 \times 10^{14} \text{ m}.$$

This is about 24 times the orbit radius of Pluto and about 1/250 of the way to Alpha Centauri.

12-70: The direct calculation of the force that the sphere exerts on the ring is slightly more involved than the calculation of the force that the ring exerts on the ball. These forces are equal in magnitude but opposite in direction, so it will suffice to do the latter calculation. By symmetry, the force on the sphere will be along the axis of the ring in Fig. (12-27), toward the ring. Each mass element dM of the ring exerts a force of magnitude $\frac{Gm\,dM}{a^2+x^2}$ on the sphere, and the x-component of this force is

$$\frac{Gm\,dM}{a^2+x^2} \frac{x}{\sqrt{a^2+x^2}} = \frac{Gm\,dM\,x}{(a^2+x^2)^{3/2}}.$$

As $x \gg a$, the denominator approaches x^3 and $F \to \frac{GMm}{x^2}$, as expected, and so the force on the sphere is $GmMx/\left(a^2+x^2\right)^{3/2}$, in the $-x$-direction. The sphere attracts the ring with a force of the same magnitude. (This is an alternative but equivalent way of obtaining the result of parts (c) and (d) of Exercise 12-33.)

12-71: Divide the rod into differential masses dm at position l, measured from the right end of the rod. Then, $dm = dl(M/L)$, and the contribution dF_x from each piece is $dF_x = -\frac{GmM\,dl}{(l+x)^2 L}$. Integrating from $l = 0$ to $l = L$ gives

$$F = -\frac{GmM}{L} \int_0^L \frac{dl}{(l+x)^2} = \frac{GmM}{L} \left[\frac{1}{x+L} - \frac{1}{x} \right] = -\frac{GmM}{x(x+L)},$$

with the negative sign indicating a force to the left. The magnitude is $F = \frac{GmM}{x(x+L)}$. As $x \gg L$, the denominator approaches x^2 and $F \to \frac{GmM}{x^2}$, as expected. (This is an alternative but equivalent way of obtaining the result of part (b) Exercise 12-33.)

12-72: a) From the result shown in Example 12-9, the force is attractive and its magnitude is proportional to the distance the object is from the center of the earth. Comparison with equations (6-8) and (7-9) show that the gravitational potential energy is given by

$$U(r) = \frac{Gm_{\mathrm{E}}m}{2R_{\mathrm{E}}^3}r^2.$$

This is also given by the integral of F_{g} from 0 to r with respect to distance. b) From part (a), the initial gravitational potential energy is $\frac{Gm_{\mathrm{E}}m}{2R_{\mathrm{E}}}$. Equating initial potential energy and final kinetic energy (initial kinetic energy and final potential energy are both zero) gives $v^2 = \frac{Gm_{\mathrm{E}}}{R_{\mathrm{E}}}$, so $v = 7.90 \times 10^3$ m/s.

12-73:   a) $T = \frac{2\pi r^{3/2}}{\sqrt{GM_{\mathrm{E}}}}$, therefore $T + \Delta T = \frac{2\pi}{\sqrt{GM_{\mathrm{E}}}}(r + \Delta r)^{3/2} = \frac{2\pi r^{3/2}}{\sqrt{GM_{\mathrm{E}}}}\left(1 + \frac{\Delta r}{r}\right)^{3/2} \approx \frac{2\pi r^{3/2}}{\sqrt{GM_{\mathrm{E}}}}\left(1 + \frac{3\Delta r}{2r}\right) = T + \frac{3\pi r^{1/2}\Delta r}{\sqrt{GM_{\mathrm{E}}}}$. Since $v = \sqrt{\frac{GM_{\mathrm{E}}}{r}}$, $\Delta T = \frac{3\pi\Delta r}{v}$. $v = \sqrt{GM_{\mathrm{E}}}\,r^{-1/2}$, therefore $v - \Delta v = \sqrt{GM_{\mathrm{E}}}(r + \Delta r)^{-1/2} = \sqrt{GM_{\mathrm{E}}}\,r^{-1/2}\left(1 + \frac{\Delta r}{r}\right)^{-1/2} \approx \sqrt{GM_{\mathrm{E}}}\,r^{-1/2}\left(1 - \frac{\Delta r}{2r}\right) = v - \frac{\sqrt{GM_{\mathrm{E}}}}{2r^{3/2}}\Delta r$. Since $T = \frac{2\pi r^{3/2}}{\sqrt{GM_{\mathrm{E}}}}$, $\Delta v = \frac{\pi\Delta r}{T}$.

b) Note: Because of the small change in r, several significant figures are needed to see the results. Starting with $T = \frac{2\pi r^{3/2}}{\sqrt{GM}}$ (Eq. (12-14)), $T = 2\pi r/v$, and $v = \sqrt{\frac{GM}{r}}$ (Eq. (12-12)) find the velocity and period of the initial orbit: $v = \sqrt{\frac{(6.673\times 10^{-11}\text{ N}\cdot\text{m}^2/\text{kg}^2)(5.97\times 10^{24}\text{ kg})}{6.776\times 10^6\text{ m}}} = 7.672 \times 10^3$ m/s, and $T = 2\pi r/v = 5549$ s $= 92.5$ min. We then can use the two derived equations to approximate the ΔT and Δv, $\Delta T = \frac{3\pi\Delta r}{v}$ and $\Delta v = \frac{\pi\Delta r}{T}$. $\Delta T = \frac{3\pi(100\text{ m})}{7.672\times 10^3\text{ m/s}} = 0.1228$ s, and $\Delta v = \frac{\pi\Delta r}{T} = \frac{\pi(100\text{ m})}{(5549\text{ s})} = .05662$ m/s. Before the cable breaks, the shuttle will have traveled a distance d, $d = \sqrt{(125\text{ m}^2) - (100\text{ m}^2)} = 75$ m. So, $(75\text{ m})/(.05662\text{ m/s}) = 1324.7$ s $= 22$ min. It will take 22 minutes for the cable to break.

c) The ISS is moving faster than the space shuttle, so the total angle it covers in an orbit must be 2π radians more than the angle that the space shuttle covers before they are once again in line. Mathematically, $\frac{vt}{r} - \frac{(v - \Delta v)t}{(r + \Delta r)} = 2\pi$. Using the binomial theorem and neglecting terms of order $\Delta v\Delta r$, $\frac{vt}{r} - \frac{(v - \Delta v)t}{r}\left(1 + \frac{\Delta r}{r}\right)^{-1} \approx t\left(\frac{\Delta v}{r} + \frac{v\Delta r}{r^2}\right) = 2\pi$. Therefore, $t = \frac{2\pi r}{(\Delta v + \frac{v\Delta r}{r})} = \frac{vT}{\frac{\pi\Delta r}{T} + \frac{v\Delta r}{r}}$. Since $2\pi r = vT$ and $\Delta r = \frac{v\Delta T}{3\pi}$, $t = \frac{vT}{\frac{\pi}{T}(\frac{v\Delta T}{3\pi}) + \frac{2\pi}{T}(\frac{v\Delta T}{3\pi})} = \frac{T^2}{\Delta T}$, as was to be shown. $t = \frac{T^2}{\Delta T} = \frac{(5549\text{ s})^2}{(0.1228\text{ s})} = 2.5 \times 10^8$ s $= 2900$ d $= 7.9$ y. It is highly doubtful the shuttle crew would survive the congressional hearings if they miss!

12-74:   a) To get from the circular orbit of the earth to the transfer orbit, the spacecraft's energy must increase, and the rockets are fired in the direction opposite that of the motion, that is, in the direction that increases the speed. Once at the orbit of Mars, the energy needs to be increased again, and so the rockets need to be fired in the direction opposite that of the motion. From Fig. (12-30), the semimajor axis of the transfer orbit is the arithmetic average of the orbit radii of the earth and Mars, and so from Eq. (12-19), the energy of spacecraft while in the transfer orbit is intermediate between the energies of the circular orbits. Returning from Mars to the earth, the procedure is reversed, and the rockets are fired against the direction of motion.   b) The time will be half the period as given in Eq. (12-19), with the semimajor axis a being the average of the orbit radii,

$a = 1.89 \times 10^{11}$ m, so

$$t = \frac{T}{2} = \frac{\pi(1.89 \times 10^{11} \text{ m})^{3/2}}{\sqrt{(6.673 \times 10^{-11} \text{ N·m}^2/\text{kg}^2)(1.99 \times 10^{30} \text{ kg})}} = 2.24 \times 10^7 \text{ s},$$

which is more than $8\frac{1}{2}$ months.   c) During this time, Mars will pass through an angle of $(360°)\frac{(2.24 \times 10^7 \text{ s})}{(687 \text{ d})(86,400 \text{ s/d})} = 135.9°$, and the spacecraft passes through an angle of $180°$, so the angle between the earth-sun line and the Mars-sun line must be $44.1°$.

12-75:   a) There are many ways of approaching this problem; two will be given here.

I) Denote the orbit radius as r and the distance from this radius to either ear as δ. Each ear, of mass m, can be modeled as subject to two forces, the gravitational force from the black hole and the tension force (actually the forces from the body tissues), denoted by F. Then, the force equations for the two ears are

$$\frac{GMm}{(r-\delta)^2} - F = m\omega^2(r - \delta)$$
$$\frac{GMm}{(r+\delta)^2} + F = m\omega^2(r + \delta),$$

where ω is the common angular frequency. The first equation reflects the fact that one ear is closer to the black hole, is subject to a larger gravitational force, has a smaller acceleration, and needs the force F to keep it in the circle of radius $r - \delta$. The second equation reflects the fact that the outer ear is further from the black hole and is moving in a circle of larger radius and needs the force F to keep in in the circle of radius $r + \delta$.

Dividing the first equation by $r - \delta$ and the second by $r + \delta$ and equating the resulting expressions eliminates ω, and after a good deal of algebra,

$$F = (3GMm\delta)\frac{(r + \delta)}{(r^2 - \delta^2)^2}.$$

At this point it is prudent to neglect δ in the sum and difference, but recognize that F is proportional to δ, and numerically $F = \frac{3GMm\delta}{r^3} = 2.1$ kN. (Using the result of Exercise 12-38 to express the graviational force in terms of the Schwartzschild radius gives the same result to two figures.)

II) Using the same notation,

$$\frac{GMm}{(r+\delta)^2} - F = m\omega^2(r + \delta),$$

where δ can be of either sign. Replace the product $m\omega^2$ with its value for $\delta = 0$, $m\omega^2 = GMm/r^3$ and solve for

$$F = (GMm)\left[\frac{r+\delta}{r^3} - \frac{1}{(r+\delta)^2}\right] = \frac{GMm}{r^3}\left[r + \delta - r\left(1 + (\delta/r)\right)^{-2}\right].$$

Using the binomial theorem to expand the term in square brackets in powers of δ/r,

$$F = \frac{GMm}{r^3}\left[r + \delta - r\left(1 - 2(\delta/r)\right)\right] = \frac{GMm}{r^3}\,(3\delta),$$

the same result as above.

Method (I) avoids using the binomial theorem or Taylor series expansions; the approximations are made only when numerical values are inserted and higher powers of δ are found to be numerically insignificant. Method (II) uses the fact that even though the center of gravity is not at the center of mass, their separation is small compared to r, and so the period of the orbit of a point mass at that position can be used to characterize the motion.

If noninertial frames are allowed, the same result may be obtained by considering the frame of the astronaut; the difference in the position of the ears contributes a difference between the gravitational forces of magnitude $\frac{2GMm\delta}{r^3}$ and the fact that the astronaut is in a frame that rotates with frequency ω as found above means that in her frame the ears need an extra force of $m\omega^2\delta$, and the tension is the same. (This method was suggested by a first-year MIT undergraduate.)

This tension is much larger than that which could be sustained by human tissue, and the astronaut is in trouble.   b) See the discussion above; the center of gravity is not the center of mass.

12-76:   As suggested in the problem, divide the disk into rings of radius r and thickness dr. Each ring has an area $dA = 2\pi r\,dr$ and mass $dM = \frac{M}{\pi a^2}\,dA = \frac{2M}{a^2}\,r\,dr$. The magnitude of the force that this small ring exerts on the mass m is then $(Gm\,dM)(x/(r^2 + x^2)^{3/2})$, the expression found in Problem 12-70, with dM instead of M and the variable r instead of a. Thus, the contribution dF to the force is

$$dF = \frac{2GMmx}{a^2}\frac{r\,dr}{(x^2 + r^2)^{3/2}}.$$

The total force F is then the integral over the range of r;

$$F = \int dF = \frac{2GMmx}{a^2}\int_0^a \frac{r}{(x^2 + r^2)^{3/2}}\,dr.$$

The integral (either by looking in a table or making the substitution $u = r^2 + a^2$) is

$$\int_0^a \frac{r}{(x^2 + r^2)^{3/2}}\,dr = \left[\frac{1}{x} - \frac{1}{\sqrt{a^2 + x^2}}\right] = \frac{1}{x}\left[1 - \frac{x}{\sqrt{a^2 + x^2}}\right].$$

Substitution yields the result

$$F = \frac{2GMm}{a^2}\left[1 - \frac{x}{\sqrt{a^2 + x^2}}\right].$$

The second term in brackets can be written as

$$\frac{1}{\sqrt{1+(a/x)^2}} = \left(1+(a/x)^2\right)^{-1/2} \approx 1 - \frac{1}{2}\left(\frac{a}{x}\right)^2$$

if $x \gg a$, where the binomial approximation (or first-order Taylor series expansion) has been used. Substitution of this into the above form gives

$$F \approx \frac{GMm}{x^2},$$

as it should.

12-77: From symmetry, the component of the gravitational force parallel to the rod is zero. To find the perpendicular component, divide the rod into segments of length dx and mass $dm = dx\,\frac{M}{2L}$, positioned at a distance x from the center of the rod. The magnitude of the gravitational force from each segment is

$$dF = \frac{Gm\,dM}{x^2+a^2} = \frac{GmM}{2L}\frac{dx}{x^2+a^2}.$$

The component of dF perpendicular to the rod is $dF\,\frac{a}{\sqrt{x^2+a^2}}$, and so the net gravitational force is

$$F = \int_{-L}^{L} dF = \frac{GmMa}{2L}\int_{-L}^{L}\frac{dx}{(x^2+a^2)^{3/2}}.$$

The integral can be found in a table, or found by making the substitution $x = a\tan\theta$. Then, $dx = a\sec^2\theta\,d\theta$, $(x^2+a^2) = a^2\sec^2\theta$, and so

$$\int \frac{dx}{(x^2+a^2)^{3/2}} = \int \frac{a\sec^2\theta\,d\theta}{a^3\sec^3\theta} = \frac{1}{a^2}\int\cos\theta\,d\theta = \frac{1}{a^2}\sin\theta = \frac{x}{a^2\sqrt{x^2+a^2}},$$

and the definite integral is

$$F = \frac{GmM}{a\sqrt{a^2+L^2}}.$$

When $a \gg L$, the term in the square root approaches a^2 and $F \to \frac{GmM}{a^2}$, as expected.

Chapter 13   Periodic Motion

13-1:   a) $T = \frac{1}{f} = 4.55 \times 10^{-3}$ s, $\omega = \frac{2\pi}{T} = 2\pi f = 1.38 \times 10^3$ rad/s.   b) $\frac{1}{4(220 \text{ Hz})} = 1.14 \times 10^{-3}$ s, $\omega = 2\pi f = 5.53 \times 10^3$ rad/s.

13-2:   a) Since the glider is released from rest, its initial displacement (0.120 m) is the amplitude. b) The glider will return to its original position after another 0.80 s, so the period is 1.60 s. c) The frequency is the reciprocal of the period (Eq. (13-2)), $f = \frac{1}{1.60 \text{ s}} = 0.625$ Hz.

13-3:   The period is $\frac{0.50 \text{ s}}{440} = 1.14 \times 10^{-3}$ and the angular frequency is $\omega = \frac{2\pi}{T} = 5.53 \times 10^3$ rad/s.

13-4:   The period will be twice the time given as being between the times at which the glider is at the equilibrium position (see Fig. (13-6));

$$k = \omega^2 m = \left(\frac{2\pi}{T}\right)^2 m = \left(\frac{2\pi}{2(2.60 \text{ s})}\right)^2 (0.200 \text{ kg}) = 0.292 \text{ N/m}.$$

13-5:   a) $T = \frac{1}{f} = 0.167$ s.   b) $\omega = 2\pi f = 37.7$ rad/s.   c) $m = \frac{k}{\omega^2} = 0.084$ kg.

13-6:   Solving Eq. (13-12) for k,

$$k = m \left(\frac{2\pi}{T}\right)^2 = (0.600 \text{ kg}) \left(\frac{2\pi}{0.150 \text{ s}}\right)^2 = 1.05 \times 10^3 \text{ N/m}.$$

13-7:   From Eq. (13-12) and Eq. (13-10), $T = 2\pi \sqrt{\frac{0.500 \text{ kg}}{140 \text{ N/m}}} = 0.375$ s, $f = \frac{1}{T} = 2.66$ Hz, $\omega = 2\pi f = 16.7$ rad/s.

13-8:   a) $a = \frac{d^2x}{dt^2} = -\omega^2 A \sin(\omega t + \beta) = -\omega^2 x$, so $x(t)$ is a solution to Eq. (13-4) if $\omega^2 = \frac{k}{m}$.   b) $a = 2A\omega$, a constant, so Eq. (13-4) is not satisfied.   c) $v = \frac{dx}{dt} = i\omega A e^{i(\omega t + \beta)}$, $a = \frac{dv}{dt} = (i\omega)^2 A e^{i(\omega t + \beta)} = -\omega^2 x$, so $x(t)$ is a solution to Eq. (13-4) if $\omega^2 = k/m$.

13-9:   a) $x = (3.0 \text{ mm}) \cos\left((2\pi)(440 \text{ Hz})t\right)$ b) $(3.0 \times 10^{-3} \text{ m})(2\pi)(440 \text{ Hz}) = 8.29$ m/s, $(3.0 \text{ mm})(2\pi)^2(440 \text{ Hz})^2 = 2.29 \times 10^4$ m/s^2.   c) $j(t) = (6.34 \times 10^7 \text{ m/s}^3) \sin((2\pi)(440 \text{ Hz})t)$, $j_{max} = 6.34 \times 10^7$ m/s^3.

13-10:   a) From Eq. (13-19), $A = \left|\frac{v_0}{\omega}\right| = \left|\frac{v_0}{\sqrt{k/m}}\right| = 0.98$ m.   b) Equation (13-18) is indeterminant, but from Eq. (13-14), $\phi = \pm \frac{\pi}{2}$, and from Eq. (13-17), $\sin\phi > 0$, so $\phi = +\frac{\pi}{2}$.
c) $\cos(\omega t + (\pi/2)) = \sin\omega t$, so $x = (-0.98 \text{ m}) \sin((12.2 \text{ rad/s})t))$.

13-11:   With the same value for ω, Eq. (13-19) gives $A = 0.383$ m and Eq. (13-18) gives

$$\phi = \arctan\left(-\frac{(-4.00 \text{ m/s})}{(0.200 \text{ m})\sqrt{300 \text{ N/m}/2.00 \text{ kg}}}\right) = 1.02 \text{ rad} = 58.5°,$$

and $x = (0.383 \text{ m}) \cos\left((12.2 \text{ rad/s})t + 1.02 \text{ rad}\right)$.

13-12:   For SHM, $a = -\omega^2 x = -(2\pi f)^2 x = -(2\pi(2.5 \text{ Hz}))^2(1.1 \times 10^{-2} \text{ m}) = -2.71$ m/s^2.
b) From Eq. (13-19) the amplitude is 1.46 cm, and from Eq. (13-18) the phase angle is

0.715 rad. The angular frequency is $2\pi f = 15.7$ rad/s, so

$$x = (1.46 \text{ cm}) \cos((15.7 \text{ rad/s})t + 0.715 \text{ rad})$$
$$v = (-22.9 \text{ cm/s}) \sin((15.7 \text{ rad/s})t + 0.715 \text{ rad})$$
$$a = (-359 \text{ cm/s}^2) \cos((15.7 \text{ rad/s})t + 0.715 \text{ rad}).$$

13-13: The equation describing the motion is $x = A \sin \omega t$; this is best found from either inspection or from Eq. (13-14) (Eq. (13-18) involves an infinite argument of the arctangent). Even so, x is determined only up to the sign, but that does not affect the result of this exercise. The distance from the equilibrium position is $A \sin (2\pi (t/T)) = (0.600 \text{ m}) \sin(4\pi/5) = 0.353$ m.

13-14: See Exercise 13-13;

$$t = (\arccos(-1.5/6))(.3/(2\pi)) = 0.0871 \text{ s}.$$

13-15: a) Dividing Eq. (13-17) by ω,

$$x_0 = A \cos \phi, \qquad \frac{v_0}{\omega} = -A \sin \phi.$$

Squaring and adding,

$$x_0^2 + \frac{v_0^2}{\omega^2} = A^2,$$

which is the same as Eq. (13-19). b) At time $t = 0$, Eq. (13-21) becomes

$$\frac{1}{2}kA^2 = \frac{1}{2}mv_0^2 + \frac{1}{2}kx_0^2 = \frac{1}{2}\frac{k}{\omega^2}v_0^2 + \frac{1}{2}kx_0^2,$$

where $m = k\omega^2$ (Eq. (13-10)) has been used. Dividing by $k/2$ gives Eq. (13-19).

13-16: a) $v_{max} = (2\pi f)A = (2\pi(392 \text{ Hz}))(0.60 \times 10^{-3} \text{ m}) = 1.48$ m/s.

b) $K_{max} = \frac{1}{2}m(V_{max})^2 = \frac{1}{2}(2.7 \times 10^{-5} \text{ kg})(1.48 \text{ m/s})^2 = 2.96 \times 10^{-5}$ J.

13-17: a) Setting $\frac{1}{2}mv^2 = \frac{1}{2}kx^2$ in Eq. (13-21) and solving for x gives $x = \pm\frac{A}{\sqrt{2}}$. Eliminating x in favor of v with the same relation gives $v = \pm\sqrt{kA^2/2m} = \pm\frac{\omega A}{\sqrt{2}}$. b) This happens four times each cycle, corresponding the four possible combinations of $+$ and $-$ in the results of part (a). The time between the occurrences is one-fourth of a period or $T/4 = \frac{2\pi}{4\omega} = \frac{\pi}{2\omega}$. c) $U = \frac{1}{4}E, K = \frac{3}{4}E$ ($U = \frac{kA^2}{8}, K = \frac{3kA^2}{8}$)

13-18: a) From Eq. (13-23),

$$v_{max} = \sqrt{\frac{k}{m}} A = \sqrt{\frac{450 \text{ N/m}}{0.500 \text{ kg}}}(0.040 \text{ m}) = 1.20 \text{ m/s}.$$

b) From Eq. (13-22),

$$v = \sqrt{\frac{450 \text{ N}}{0.500 \text{ kg}}} \sqrt{(0.040 \text{ m})^2 - (-0.015 \text{ m})^2} = 1.11 \text{ m/s}.$$

c) The extremes of acceleration occur at the extremes of motion, when $x = \pm A$, and

$$a_{max} = \frac{kA}{m} = \frac{(450 \text{ N/m})(0.040 \text{ m})}{(0.500 \text{ kg})} = 36 \text{ m/s}^2$$

d) From Eq. (13-4), $a = -\frac{(450 \text{ N/m})(-0.015 \text{ m})}{(0.500 \text{ kg})} = 13.5 \text{ m/s}^2$.

e) From Eq. (13-31), $E = \frac{1}{2}(450 \text{ N/m})(0.040 \text{ m})^2 = 0.36 \text{ J}$.

13-19: a) $a_{max} = \omega^2 A = (2\pi f)^2 A = (2\pi(0.85 \text{ Hz}))^2(18.0 \times 10^{-2} \text{ m}) = 5.13 \text{ m/s}^2$. $v_{max} = \omega A = 2\pi f A = 0.961 \text{ m/s}$.   b) $a = -(2\pi f)^2 x = -2.57 \text{ m/s}^2$,

$$v = (2\pi f)\sqrt{A^2 - x^2}$$
$$= (2\pi(0.85 \text{ Hz}))\sqrt{(18.0 \times 10^{-2} \text{ m})^2 - (9.0 \times 10^{-2} \text{ m})^2} = 0.833 \text{ m/s}.$$

c) The fraction of one period is $(1/2\pi)\arcsin(12.0/18.0)$, and so the time is $(T/2\pi) \times \arcsin(12.0/18.0) = 1.37 \times 10^{-1}$ s. Note that this is also $\arcsin(x/A)/\omega$.

d) The conservation of energy can be written $\frac{1}{2}kA^2 = \frac{1}{2}mv^2 + \frac{1}{2}kx^2$. We are given amplitude, frequency in Hz, and various values of x. We could calculate velocity from this information if we use the relationship $k/m = \omega^2 = 4\pi^2 f^2$ and rewrite the conservation equation as $\frac{1}{2}A^2 = \frac{1}{2}\frac{v^2}{4\pi^2 f^2} + \frac{1}{2}x^2$. Using energy principles is generally a good approach when we are dealing with velocities and positions as opposed to accelerations and time when using dynamics is often easier.

13-20: In the example, $A_2 = A_1\sqrt{\frac{M}{M+m}}$ and now we want $A_2 = \frac{1}{2}A_1$. So $\frac{1}{2} = \sqrt{\frac{M}{M+m}}$, or $m = 3M$. For the energy, $E_2 = \frac{1}{2}kA_2^2$, but since $A_2 = \frac{1}{2}A_1$, $E_2 = \frac{1}{4}E_1$, or $\frac{3}{4}E_1$ is lost to heat.

13-21: a) $\frac{1}{2}mv^2 + \frac{1}{2}kx^2 = 0.0284$ J.

b) $\sqrt{x_0^2 + \frac{v_0^2}{\omega^2}} = \sqrt{(0.012 \text{ m})^2 + \frac{(0.300 \text{ m/s})^2}{(300 \text{ N/m})/(0.150 \text{ kg})}} = 0.014$ m.

c) $\omega A = \sqrt{k/m}\, A = 0.615$ m/s.

13-22: Using $k = \frac{F_0}{L_0}$ from the calibration data,

$$m = \frac{(F_0/L_0)}{(2\pi f)^2} = \frac{(200 \text{ N})/(1.25 \times 10^{-1} \text{ m})}{(2\pi(2.60 \text{ Hz}))^2} = 6.00 \text{ kg}.$$

13-23: a) $k = \frac{mg}{\Delta l} = \frac{(65.0 \text{ kg})(9.80 \text{ m/s}^2)}{(0.120 \text{ m})} = 5.31 \times 10^3$ N/m.

b) $T = 2\pi\sqrt{\frac{m}{k}} = 2\pi\sqrt{\frac{\Delta l}{g}} = 2\pi\sqrt{\frac{0.120 \text{ m}}{9.80 \text{ m/s}^2}} = 0.695$ s.

13-24: a) The easy thing to do is to start the end; at the top of the motion, the spring is unstretched and so has no potential energy, the cat is not moving and so has no kinetic energy, and the potential energy relative to the bottom is $2mgA = 2(4.00 \text{ kg})(9.80 \text{ m/s}^2) \times (0.050 \text{ m}) = 3.92$ J. This is the total energy, and is the same total for each part.

b) $U_{\text{grav}} = 0$, $K = 0$, so $U_{\text{spring}} = 3.92$ J.

c) At equilibirum the spring is stretched half as much as it was for part (a), and so $U_{\text{spring}} = \frac{1}{4}(3.92 \text{ J}) = 0.98$ J, $U_{\text{grav}} = \frac{1}{2}(3.92 \text{ J}) = 1.96$ J, and so $K = 0.98$ J.

13-25: The elongation is the weight divided by the spring constant,

$$\Delta l = \frac{w}{k} = \frac{mg}{\omega^2 m} = \frac{gT^2}{4\pi^2} = 3.97 \text{ cm}.$$

13-26: See Exercise 9-34. a) The mass would decrease by a factor of $(1/3)^3 = 1/27$ and so the moment of inertia would decrease by a factor of $(1/27)(1/3)^2 = (1/243)$, and for the same spring constant, the frequency and angular frequency would increase by a factor of $\sqrt{243} = 15.6$. b) The torsion constant would need to be *decreased* by a factor of 243, or changed by a factor of 0.00412 (approximately).

13-27: a) With the approximations given, $I = mR^2 = 2.72 \times 10^{-8}$ kg·m^2, or 2.7×10^{-8} kg·m^2 to two figures. b) $\kappa = (2\pi f)^2 I = (2\pi\, 2 \text{ Hz})^2 (2.72 \times 10^{-8} \text{ kg·m}^2) = 4.3 \times 10^{-6}$ N·m/rad.

13-28: Solving Eq. (13-24) for κ in terms of the period,

$$\kappa = \left(\frac{2\pi}{T}\right)^2 I$$

$$= \left(\frac{2\pi}{1.00 \text{ s}}\right)^2 ((1/2)(2.00 \times 10^{-3} \text{ kg})(2.20 \times 10^{-2} \text{ m})^2)$$

$$= 1.91 \times 10^{-5} \text{ N·m/rad}.$$

13-29:

$$I = \frac{\kappa}{(2\pi f)^2} = \frac{0.450 \text{ N·m/rad}}{(2\pi(125)/(265 \text{ s}))^2} = 0.0512 \text{ kg·m}^2.$$

13-30: The equation $\theta = \Theta \cos(\omega t + \phi)$ describes angular SHM. In this problem, $\phi = 0$.

a) $\frac{d\theta}{dt} = -\omega\Theta \sin(\omega t)$ and $\frac{d^2\theta}{dt^2} = -\omega^2\Theta \cos(\omega t)$.

b) When the angular displacement is Θ, $\Theta = \Theta \cos(\omega t)$, and this occurs at $t = 0$, so

$$\frac{d\theta}{dt} = 0 \text{ since } \sin(0) = 0, \text{ and } \frac{d^2\theta}{dt^2} = -\omega^2\Theta, \text{ since } \cos(0) = 1.$$

c) When the angular displacement is $\Theta/2$, $\frac{\Theta}{2} = \Theta \cos(\omega t)$, or $\frac{1}{2} = \cos(\omega t)$.

$$\frac{d\theta}{dt} = \frac{-\omega\Theta\sqrt{3}}{2} \text{ since } \sin(\omega t) = \frac{\sqrt{3}}{2}, \text{ and } \frac{d^2\theta}{dt^2} = \frac{-\omega^2\Theta}{2}, \text{ since } \cos(\omega t) = 1/2.$$

This corresponds to a displacement of 60°.

13-31: Using the same procedure used to obtain Eq. (13-29), the potential may be expressed as

$$U = U_0 \left[(1 + x/R_0)^{-12} - 2(1 + x/R_0)^{-6} \right].$$

Note that at $r = R_0$, $U = -U_0$. Using the appropriate forms of the binomial theorem for $|x/R_0| \ll 1$,

$$U \approx U_0 \left[\begin{array}{c} \left(1 - 12(x/R_0) + \dfrac{(-12)(-13)}{2}(x/R_0)^2 \right) \\[2mm] -2 \left(1 - 6(x/R_0) + \dfrac{(-6)(-7)}{2}(x/R_0)^2 \right) \end{array} \right]$$

$$= U_0 \left[-1 + \frac{36}{R_0^2} x^2 \right]$$

$$= \frac{1}{2}kx^2 - U_0.$$

where $k = 72U_0/R^2$ has been used. Note that terms in u^2 from Eq. (13-28) must be kept; the fact that the first-order terms vanish is another indication that R_0 is an extreme (in this case a minimum) of U.

13-32: $f = \dfrac{1}{2\pi}\sqrt{\dfrac{k}{(m/2)}} = \dfrac{1}{2\pi}\sqrt{\dfrac{2(580 \text{ N/m})}{(1.008)(1.66 \times 10^{-27} \text{ kg})}} = 1.33 \times 10^{14} \text{ Hz.}$

13-33: $T = 2\pi\sqrt{L/g}$, so for a different acceleration due to gravity g',

$$T' = T\sqrt{g/g'} = (1.60 \text{ s})\sqrt{9.80 \text{ m/s}^2/3.71 \text{ m/s}^2} = 2.60 \text{ s.}$$

13-34: a) To the given precision, the small-angle approximation is valid. The highest speed is at the bottom of the arc, which occurs after a quarter period, $\frac{T}{4} = \frac{\pi}{2}\sqrt{\frac{L}{g}} = 0.25$ s.

b) The same as calculated in (a), 0.25 s. The period is independent of amplitude.

13-35: Besides approximating the pendulum motion as SHM, assume that the angle is sufficiently small that the length of the spring does not change while swinging in the arc. Denote the angular frequency of the vertical motion as $\omega_0 = \sqrt{\frac{k}{m}} = \sqrt{\frac{kg}{w}}$ and $\omega' = \sqrt{\frac{g}{L}} = \frac{1}{2}\omega_0 = \sqrt{\frac{kg}{4w}}$, which is solved for $L = 4w/k$. But L is the length of the stretched spring; the unstretched length is $L_0 = L - w/k = 3w/k = 3(1.00 \text{ N})/(1.50 \text{ N/m}) = 2.00$ m.

13-36:

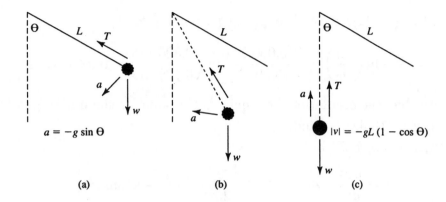

(a)   (b)   (c)

13-37: The period of the pendulum is $T = (136 \text{ s})/100 = 1.36$ s. Then,

$$g = \frac{4\pi^2 \, L}{T^2} = \frac{4\pi^2(.5 \text{ m})}{(1.36 \text{ s})^2} = 10.67 \text{ m/s}^2.$$

13-38: From the parallel axis theorem, the moment of inertia of the hoop about the nail is $I = MR^2 + MR^2 = 2MR^2$, so $T = 2\pi\sqrt{2R/g}$, with $d = R$ in Eq. (13-39). Solving for R, $R = gT^2/8\pi^2 = 0.496$ m.

13-39: For the situation described, $I = mL^2$ and $d = L$ in Eq. (13-39); canceling the factor of m and one factor of L in the square root gives Eq. (13-34).

13-40: a) Solving Eq. (13-39) for I,

$$I = \left(\frac{T}{2\pi}\right)^2 mgd = \left(\frac{0.940 \text{ s}}{2\pi}\right)^2 (1.80 \text{ kg})(9.80 \text{ m/s}^2)(0.250 \text{ m}) = 0.0987 \text{ kg·m}^2.$$

b) The small-angle approximation will not give three-figure accuracy for $\Theta = 0.400$ rad. From energy considerations,

$$mgd(1 - \cos\Theta) = \frac{1}{2}I\Omega_{\max}^2.$$

Expressing $\Omega_{\max}$ in terms of the period of small-angle oscillations, this becomes

$$\Omega_{\max} = \sqrt{2\left(\frac{2\pi}{T}\right)^2 (1 - \cos\Theta)} = \sqrt{2\left(\frac{2\pi}{0.940 \text{ s}}\right)^2 (1 - \cos(0.40 \text{ rad}))} = 2.66 \text{ rad/s}.$$

13-41: Using the given expression for I in Eq. (13-39), with $d = R$ (and of course $m = M$), $T = 2\pi\sqrt{5R/3g} = 0.58$ s.

13-42: From Eq. (13-39),

$$I = mgd\left(\frac{T}{2\pi}\right)^2 = (1.80 \text{ kg})(9.80 \text{ m/s}^2)(0.200 \text{ m})\left(\frac{120 \text{ s}/100}{2\pi}\right)^2 = 0.129 \text{ kg·m}^2.$$

13-43: a) From Eq. (13-43),

$$\omega' = \sqrt{\frac{(2.50 \text{ N/m})}{(0.300 \text{ kg})} - \frac{(0.90 \text{ kg/s})^2}{4(0.300 \text{ kg})^2}} = 2.47 \text{ rad/s, so } f' = \frac{\omega'}{2\pi} = 0.393 \text{ Hz}.$$

b) $b = 2\sqrt{km} = 2\sqrt{(2.50 \text{ N/m})(0.300 \text{ kg})} = 1.73$ kg/s.

13-44:   From Eq. (13-42), $A_2 = A_1 \exp(-\frac{b}{2m}t)$. Solving for b,

$$b = \frac{2m}{t}\ln\left(\frac{A_1}{A_2}\right) = \frac{2(0.050\text{ kg})}{(5.00\text{ s})}\ln\left(\frac{0.300\text{ m}}{0.100\text{ m}}\right) = 0.0220\text{ kg/s}.$$

As a check, note that the oscillation frequency is the same as the undamped frequency to $4.8 \times 10^{-3}\%$, so Eq. (13-42) is valid.

13-45:   a) With $\phi = 0$, $x(0) = A$.

b)
$$v = \frac{dx}{dt} = Ae^{-(b/2m)t}\left[-\frac{b}{2m}\cos\omega't - \omega'\sin\omega't\right],$$

and at $t = 0$, $v = -Ab/2m$; the graph of x versus t near $t = 0$ slopes down.

c)
$$a = \frac{dv}{dt} = Ae^{-(b/2m)t}\left[\left(\frac{b^2}{4m^2} - \omega'^2\right)\cos\omega't + \frac{\omega'b}{2m}\sin\omega't\right],$$

and at $t = 0$,

$$a = A\left(\frac{b^2}{4m^2} - \omega'^2\right) = A\left(\frac{b^2}{2m^2} - \frac{k}{m}\right).$$

(Note that this is $(-bv_0 - kx_0)/m$.) This will be negative if $b < \sqrt{2km}$, zero if $b = \sqrt{2km}$ and positive if $b > \sqrt{2km}$. The graph in the three cases will be curved down, not curved, or curved up, respectively.

13-46:   At resonance, Eq. (13-46) reduces to $A = F_{max}/b\omega_d$.   a) $\frac{A_1}{3}$.   b) $2A_1$. Note that the resonance frequency is independent of the value of b (see Fig. (13-24)).

13-47:   a) The damping constant has the same units as force divided by speed, or $[\text{kg·m/s}^2]/[\text{m/s}] = [\text{kg/s}]$.   b) The units of $\sqrt{km}$ are the same as $[[\text{kg/s}^2][\text{kg}]]^{1/2} = [\text{kg/s}]$, the same as those for b.   c) $\omega_d^2 = k/m$.   (i) $b\omega_d = 0.2k$, so $A = F_{max}/(0.2k) = 5F_{max}/k$.   (ii) $b\omega_d = 0.4k$, so $A = F_{max}/(0.4k) = 2.5F_{max}/k$, as shown in Fig. (13-24).

13-48:   The resonant frequency is

$$\sqrt{k/m} = \sqrt{(2.1 \times 10^6\text{ N/m})/(108\text{ kg})} = 139\text{ rad/s} = 22.2\text{ Hz},$$

and this package does not meet the criterion.

13-49:   a)

$$a = A\omega^2 = \left(\frac{0.100\text{ m}}{2}\right)\left((3500\text{ rev/min})\left(\frac{\pi}{30}\frac{\text{rad/s}}{\text{rev/min}}\right)\right)^2 = 6.72 \times 10^3\text{ m/s}^2.$$

b) $ma = 3.02 \times 10^3$ N.   c) $\omega A = (3500\text{ rev/min})(.05\text{ m})\left(\frac{\pi}{30}\frac{\text{rad/s}}{\text{rev/min}}\right) = 18.3$ m/s. $K = \frac{1}{2}mv^2 = (\frac{1}{2})(.45\text{ kg})(18.3\text{ m/s})^2 = 75.6$ J.   d) At the midpoint of the stroke, $\cos(\omega t) = 0$ and so $\omega t = \pi/2$, thus $t = \pi/2\omega$. $\omega = (3500\text{ rev/min})(\frac{\pi}{30}\frac{\text{rad/s}}{\text{rev/min}}) = \frac{350\pi}{3}$ rad/s, so $t = \frac{3}{2(350)}$ s. Then $P = \Delta K/\Delta t$, or $P = 75.6$ J$/(\frac{3}{2(350)}$ s$) = 1.76 \times 10^4$ W.   e) If the frequency doubles, the acceleration and hence the needed force will quadruple $(12.1 \times 10^3$ N$)$. The maximum speed increases by a factor of 2 since $v \propto \omega$, so the speed will be 36.7 m/s. Because the kinetic energy depends on the square of the velocity, the kinetic energy will increase by a factor of four (302 J). But, because the time to reach the midpoint is halved, due to the doubled velocity, the power increases by a factor of eight (141 kW).

13-50: Denote the mass of the passengers by m and the (unknown) mass of the car by M. The spring constant is then $k = mg/\Delta l$. The period of oscillation of the empty car is $T_E = 2\pi\sqrt{M/k}$ and the period of the loaded car is

$$T_L = 2\pi\sqrt{\frac{M+m}{k}} = \sqrt{T_E^2 + (2\pi)^2\frac{\Delta l}{g}}, \quad \text{so}$$

$$T_E = \sqrt{T_L^2 - (2\pi)^2\frac{\Delta l}{g}} = 1.003 \text{ s.}$$

13-51: a) For SHM, the period, frequency and angular frequency are indpendent of amplitude, and are not changed. b) From Eq. (13-31), the energy is decreased by a factor of $\frac{1}{4}$. c) From Eq. (13-23), the maximum speed is decreased by a factor of $\frac{1}{2}$. d) Initially, the speed at $A_1/4$ was $\frac{\sqrt{15}}{4}\omega A_1$; after the amplitude is reduced, the speed is $\omega\sqrt{(A_1/2)^2 - (A_1/4)^2} = \frac{\sqrt{3}}{4}\omega A_1$, so the speed is decreased by a factor of $\frac{1}{\sqrt{5}}$ (this result is valid at $x = -A_1/4$ as well). e) The potential energy depends on position and is unchanged. From the result of part (d), the kinetic energy is decreased by a factor of $\frac{1}{5}$.

13-52: The distance L is $L = mg/k$; the period of the oscillatory motion is

$$T = 2\pi\sqrt{\frac{m}{k}} = 2\pi\sqrt{\frac{L}{g}},$$

which is the period of oscillation of a simple pendulum of length L.

13-53: a) Rewriting Eq. (13-22) in terms of the period and solving,

$$T = \frac{2\pi\sqrt{A^2 - x^2}}{v} = 1.68 \text{ s.}$$

b) Using the result of part (a),

$$x = \sqrt{A^2 - \left(\frac{vT}{2\pi}\right)^2} = 0.0904 \text{ m.}$$

c) If the block is just on the verge of slipping, the friction force is its maximum, $f = \mu_s \mathcal{N} = \mu_s mg$. Setting this equal to $ma = mA(2\pi/T)^2$ gives $\mu_s = A(2\pi/T)^2/g = 0.143$.

13-54: a) The normal force on the cowboy must always be upward if he is not holding on. He leaves the saddle when the normal force goes to zero (that is, when he is no longer in contact with the saddle, and the contact force vanishes). At this point the cowboy is in free fall, and so his acceleration is $-g$; this must have been the acceleration just before he left contact with the saddle, and so this is also the saddle's acceleration. b) $x = +a/(2\pi f)^2 = +(9.80 \text{ m/s}^2)/(2\pi(1.50 \text{ Hz}))^2 = 0.110 \text{ m.}$ c) The cowboy's speed will be the saddle's speed, $v = (2\pi f)\sqrt{A^2 - x^2} = 2.11$ m/s.   d) Taking $t = 0$ at the time when the cowboy leaves, the position of the saddle as a function of time is given by Eq. (13-13), with $\cos\phi = -\frac{g}{\omega^2 A}$; this is checked by setting $t = 0$ and finding that $x = \frac{g}{\omega^2} = -\frac{a}{\omega^2}$. The cowboy's position is $x_c = x_0 + v_0 t - (g/2)t^2$. Finding the time at which the cowboy and the saddle are again in contact involves a transcendental equation which must be solved numerically; specifically,

$$(0.110 \text{ m}) + (2.11 \text{ m/s})t - (4.90 \text{ m/s}^2)t^2 = (0.25 \text{ m})\cos((9.42 \text{ rad/s})t - 1.11 \text{ rad}),$$

which has as its least non-zero solution $t = 0.538$ s.   e) The speed of the saddle is $(-2.36 \text{ m/s})\cos(\omega t + \phi) = 1.62$ m/s and the cowboy's speed is $(2.11 \text{ m/s}) - (9.80 \text{ m/s}^2) \times (0.538 \text{ s}) = -3.16$ m/s, giving a relative speed of 4.77 m/s (extra figures were kept in the intermediate calculations).

13-55:   The maximum acceleration of both blocks, assuming that the top block does not slip, is $a_{\max} = kA/(m + M)$, and so the maximum force on the top block is $\left(\frac{m}{m+M}\right)kA = \mu_s mg$, and so the maximum amplitude is $A_{\max} = \mu_s(m + M)g/k$.

13-56:   a)

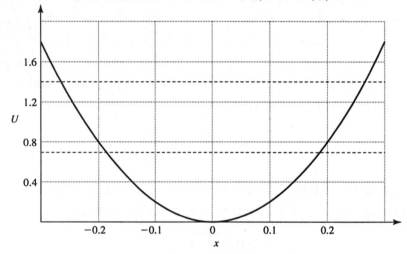

For space considerations, this figure is not precisely to the scale suggested in the problem. The following answers are found algebraically, to be used as a check on the graphical method.

   b)
$$A = \sqrt{\frac{2E}{k}} = \sqrt{\frac{2(0.200 \text{ J})}{(10.0 \text{ N/m})}} = 0.200 \text{ m}.$$

   c) $\frac{E}{4} = 0.050$ J. d) If $U = \frac{1}{2}E$, $x = \frac{A}{\sqrt{2}} = 0.141$ m. e) From Eq. (13-18), using $v_0 = -\sqrt{\frac{2K_0}{m}}$ and $x_0 = \sqrt{\frac{2U_0}{k}}$,

$$-\frac{v_0}{\omega x_0} = \frac{\sqrt{\frac{2K_0}{m}}}{\sqrt{\frac{k}{m}}\sqrt{\frac{2U_0}{k}}} = \sqrt{\frac{K_0}{U_0}} = \sqrt{0.429}$$

and $\phi = \arctan\left(\sqrt{0.429}\right) = 0.580$ rad.

13-57: a) The quantity Δl is the amount that the origin of coordinates has been moved from the unstretched length of the spring, so the spring is stretched a distance $\Delta l - x$ (see Fig. (13-14(c))) and the elastic potential energy is $U_{el} = (1/2)k(\Delta l - x)^2$.

b) $$U = U_{el} + mg(x - x_0) = \frac{1}{2}kx^2 + \frac{1}{2}(\Delta l)^2 - k\Delta l\,x + mgx - mgx_0.$$

Since $\Delta l = mg/k$, the two terms proportional to x cancel, and

$$U = \frac{1}{2}kx^2 + \frac{1}{2}k(\Delta l)^2 - mgx_0.$$

c) An additive constant to the mechanical energy does not change the dependence of the force on x, $F_x = -\frac{dU}{dx}$, and so the relations expressing Newton's laws and the resulting equations of motion are unchanged.

13-58: The "spring constant" for this wire is $k = \frac{mg}{\Delta l}$, so

$$f = \frac{1}{2\pi}\sqrt{\frac{k}{m}} = \frac{1}{2\pi}\sqrt{\frac{g}{\Delta l}} = \frac{1}{2\pi}\sqrt{\frac{9.80 \text{ m/s}^2}{2.00 \times 10^{-3} \text{ m}}} = 11.1 \text{ Hz}.$$

13-59: a) $\frac{2\pi A}{T} = 0.150$ m/s. b) $a = -(2\pi/T)^2 x = -0.112$ m/s^2. c) The time to go from equilibrium to half the amplitude is $\sin\omega t = (1/2)$, or $\omega t = \pi/6$ rad, or one-twelfth of a period. The needed time is twice this, or one-sixth of a period, 0.70 s. d) $\Delta l = \frac{mg}{k} = \frac{g}{\omega^2} = \frac{g}{(2\pi/T)^2} = 4.38$ m.

13-60: Expressing Eq. (13-13) in terms of the frequency, and with $\phi = 0$, and taking two derivatives,

$$x = (0.240 \text{ m})\cos\left(\frac{2\pi t}{1.50 \text{ s}}\right)$$

$$v = -\left(\frac{2\pi(0.240 \text{ m})}{(1.50 \text{ s})}\right)\sin\left(\frac{2\pi t}{1.50 \text{ s}}\right) = -(1.00530 \text{ m/s})\sin\left(\frac{2\pi t}{1.50 \text{ s}}\right)$$

$$a = -\left(\frac{2\pi}{1.50 \text{ s}}\right)^2(0.240 \text{ m})\cos\left(\frac{2\pi t}{1.50 \text{ s}}\right) = -(4.2110 \text{ m/s}^2)\cos\left(\frac{2\pi t}{1.50 \text{ s}}\right).$$

a) Substitution gives $x = -0.120$ m, or using $t = \frac{T}{3}$ gives $x = A\cos 120° = \frac{-A}{2}$.

b) Substitution gives $ma = +(0.0200 \text{ kg})(2.106 \text{ m/s}^2) = 4.21 \times 10^{-2}$ N, in the $+x$-direction.

c) $t = \frac{T}{2\pi}\arccos\left(\frac{-3A/4}{A}\right) = 0.577$ s.

d) Using the time found in part (c), $v = 0.665$ m/s (Eq. (13-22) of course gives the same result).

13-61: a) For the totally inelastic collision, the final speed v in terms of the initial speed $V = \sqrt{2gh}$ is $v = V\frac{M}{m+M} = \sqrt{2(9.80 \text{ m/s}^2)(0.40 \text{ m})}\left(\frac{2.2}{2.4}\right) = 2.57$ m/s, or 2.6 m/s to two figures.   b) When the steak hits, the pan is $\frac{Mg}{k}$ above the new equilibrium position. The ratio $\frac{v_0^2}{\omega^2}$ is $v^2/(k/(m+M)) = 2ghM^2/(k(m+M))$, and so the amplitude of oscillation is

$$A = \sqrt{\left(\frac{Mg}{k}\right)^2 + \frac{2ghM^2}{k(m+M)}}$$

$$= \sqrt{\left(\frac{(2.2\ \text{kg})(9.80\ \text{m/s}^2)}{(400\ \text{N/m})}\right)^2 + \frac{2(9.80\ \text{m/s}^2)(0.40\ \text{m})(2.2\ \text{kg})^2}{(400\ \text{N/m})(2.4\ \text{kg})}}$$

$$= 0.206\ \text{m}.$$

(This avoids the intermediate calculation of the speed.)   c) Using the total mass, $T = 2\pi\sqrt{(m+M)/k} = 0.487$ s.

13-62:   a) Solving Eq. (13-12) for m, and using $k = \frac{F}{\Delta l}$

$$m = \left(\frac{T}{2\pi}\right)^2 \frac{F}{\Delta l} = \left(\frac{1}{2\pi}\right)^2 \frac{40.0\ \text{N}}{0.250\ \text{m}} = 4.05\ \text{kg}.$$

b) $t = (0.35)T$, and so $x = -A\sin 2\pi(0.35) = -0.0405$ m. Since $t > \frac{T}{4}$, the mass has already passed the lowest point of its motion, and is on the way up.   c) Taking upward forces to be positive, $F_{\text{spring}} - mg = -kx$, where x is the displacement from equilibrium, so

$$F_{\text{spring}} = -(160\ \text{N/m})(-0.030\ \text{m}) + (4.05\ \text{kg})(9.80\ \text{m/s}^2) = 44.5\ \text{N}.$$

13-63:   Of the many ways to find the time interval, a convenient method is to take $\phi = 0$ in Eq. (13-13) and find that for $x = A/2$, $\cos\omega t = \cos(2\pi t/T) = \frac{1}{2}$ and so $t = T/6$. The time interval available is from $-t$ to t, and $T/3 = 0.933$ s.

13-64:   See Problem 12-72; using x as the variable instead of r,

$$F(x) = -\frac{dU}{dx} = -\frac{GM_{\text{E}}m}{R_{\text{E}}^3}x, \quad \text{so} \quad \omega^2 = \frac{GM_{\text{E}}}{R_{\text{E}}^3} = \frac{g}{R_{\text{E}}}.$$

The period is then

$$T = \frac{2\pi}{\omega} = 2\pi\sqrt{\frac{R_{\text{E}}}{g}} = 2\pi\sqrt{\frac{6.38 \times 10^6\ \text{m}}{9.80\ \text{m/s}^2}} = 5070\ \text{s},$$

or 84.5 min.

13-65: Take only the positive root (to get the least time), so that

$$\frac{dx}{dt} = \sqrt{\frac{k}{m}}\sqrt{A^2 - x^2}, \quad \text{or}$$

$$\frac{dx}{\sqrt{A^2 - x^2}} = \sqrt{\frac{k}{m}}\, dt$$

$$\int_0^A \frac{dx}{\sqrt{A^2 - x^2}} = \sqrt{\frac{k}{m}} \int_0^{t_1} dt = \sqrt{\frac{k}{m}}(t_1)$$

$$\arcsin(1) = \sqrt{\frac{k}{m}}\, t_1$$

$$\frac{\pi}{2} = \sqrt{\frac{k}{m}}\, t_1,$$

where the integral was taken from Appendix C. The above may be rearranged to show that $t_1 = \frac{\pi}{2\sqrt{\frac{k}{m}}} = \frac{T}{4}$, which is expected.

13-66: a)
$$U = -\int_0^x F\, dx = c\int_0^x x^3\, dx = \frac{c}{4}x^4.$$

b) From conservation of energy, $\frac{1}{2}mv^2 = \frac{c}{4}(A^4 - x^4)$, and using the technique of Problem 13-65, the separated equation is

$$\frac{dx}{\sqrt{A^4 - x^4}} = \sqrt{\frac{c}{2m}}\, dt.$$

Integrating from 0 to A with respect to x and from 0 to $T/4$ with respect to t,

$$\int_0^A \frac{dx}{\sqrt{A^4 - x^4}} = \sqrt{\frac{c}{2m}}\frac{T}{4}.$$

To use the hint, let $u = \frac{x}{A}$, so that $dx = a\, du$ and the upper limit of the u-integral is $u = 1$. Factoring A^2 out of the square root,

$$\frac{1}{A}\int_0^1 \frac{du}{\sqrt{1 - u^4}} = \frac{1.31}{A} = \sqrt{\frac{c}{32m}}\, T,$$

which may be expressed as $T = \frac{7.41}{A}\sqrt{\frac{m}{c}}$.   c) The period does depend on amplitude, and the motion is not simple harmonic.

13-67: As shown in Fig. (13-4(b)), $v = -v_{\text{tan}}\sin\theta$. With $v_{\text{tan}} = A\omega$ and $\theta = \omega t + \phi$, this is Eq. (13-15).

13-68: a) Taking positive displacements and forces to be upward, $N - mg = ma$, $a = -(2\pi f)^2 x$, so

$$N = m\left(g - (2\pi f)^2 A \cos((2\pi f)t + \phi)\right).$$

b) The fact that the ball bounces means that the ball is no longer in contact with the lens, and that the normal force goes to zero periodically. This occurs when the amplitude of the acceleration is equal to g, or when

$$g = (2\pi f_b)^2 A.$$

13-69:   a) For the center of mass to be at rest, the total momentum must be zero, so the momentum vectors must be of equal magnitude but opposite directions, and the momenta can be represented as $\vec{p}$ and $-\vec{p}$.

b) $$K_{\text{tot}} = 2 \times \frac{p^2}{2m} = \frac{p^2}{2(m/2)}.$$

c) The argument of part (a) is valid for any masses. The kinetic energy is

$$K_{\text{tot}} = \frac{p^2}{2m_1} + \frac{p^2}{2m_2} = \frac{p^2}{2}\left(\frac{m_1 + m_2}{m_1 m_2}\right) = \frac{p^2}{2(m_1 m_2/(m_1 + m_2))}.$$

13-70:   a) $$F_r = -\frac{dU}{dr} = A\left[\left(\frac{R_0^7}{r^9}\right) - \frac{1}{r^2}\right].$$

b) Setting the above expression for F_r equal to zero, the term in square brackets vanishes, so that $\frac{R_0^7}{r^9} = \frac{1}{r^2}$, or $R_0^7 = r^7$, and $r = R_0$.

c) $$U(R_0) = -\frac{7A}{8R_0} = -7.57 \times 10^{-19} \text{ J}.$$

d) The above expression for F_r can be expressed as

$$F_r = \frac{A}{R_0^2}\left[\left(\frac{r}{R_0}\right)^{-9} - \left(\frac{r}{R_0}\right)^{-2}\right]$$

$$= \frac{A}{R_0^2}\left[(1 + (x/R_0))^{-9} - (1 + (x/R_0))^{-2}\right]$$

$$\approx \frac{A}{R_0^2}\left[(1 - 9(x/R_0)) - (1 - 2(x/R_0))\right]$$

$$= \frac{A}{R_0^2}(-7x/R_0)$$

$$= -\left(\frac{7A}{R_0^3}\right)x.$$

e) $$f = \frac{1}{2\pi}\sqrt{k/m} = \frac{1}{2\pi}\sqrt{\frac{7A}{R_0^3 m}} = 8.39 \times 10^{12} \text{ Hz}.$$

13-71:   a) $$F_r = -\frac{dU}{dx} = A\left[\frac{1}{r^2} - \frac{1}{(r - 2R_0)^2}\right]$$

b) Setting the term in square brackets equal to zero, and ignoring solutions with $r < 0$ or $r > 2R_0$, $r = 2R_0 - r$, or $r = R_0$. c) The above expression for F_r may be written as

$$F_r = \frac{A}{R_0^2}\left[\left(\frac{r}{R_0}\right)^{-2} - \left(\frac{r}{R_0} - 2\right)^{-2}\right]$$

$$= \frac{A}{R_0^2}\left[(1 + (x/R_0))^{-2} - (1 - (x/R_0))^{-2}\right]$$

$$\approx \frac{A}{R_0^2}\left[(1 - 2(x/R_0)) - (1 - (-2)(x/R_0))\right]$$

$$= -\left(\frac{4A}{R_0^3}\right)x,$$

corresponding to a force constant of $k = 4A/R_0^3$. d) The frequency of small oscillations would be $f = (1/2\pi)\sqrt{k/m} = (1/\pi)\sqrt{A/mR_0^3}$.

13-72: a) As the mass approaches the origin, the motion is that of a mass attached to a spring of spring constant k, and the time to reach the origin is $\frac{\pi}{2}\sqrt{m/k}$. After passing through the origin, the motion is that of a mass attached to a spring of spring constant $2k$ and the time it takes to reach the other extreme of the motion is $\frac{\pi}{2}\sqrt{m/2k}$. The period is twice the sum of these times, or $T = \pi\sqrt{\frac{m}{k}}\left(1 + \frac{1}{\sqrt{2}}\right)$. The period does not depend on the amplitude, but the motion is not simple harmonic. b) From conservation of energy, if the negative extreme is A', $\frac{1}{2}kA^2 = \frac{1}{2}(2k)A'^2$, so $A' = -\frac{A}{\sqrt{2}}$; the motion is not symmetric about the origin.

13-73: There are many equivalent ways to find the period of this oscillation. Energy considerations give an elegant result. Using the force and torque equations, taking torques about the contact point, saves a few intermediate steps. Following the hint, take torques about the cylinder axis, with positive torques counterclockwise; the direction of positive rotation is then such that $\alpha = Ra$, and the friction force f that causes this torque acts in the $-x$-direction. The equations to solve are then

$$Ma = -f - kx, \quad fR = I_{cm}\alpha, \quad a = R\alpha,$$

which are solved for

$$a = -\frac{kx}{M + I/R^2} = -\frac{k}{(3/2)M}x,$$

where $I = I_{cm} = (1/2)MR^2$ has been used for the combination of cylinders. Comparison with Eq. (13-8) gives $T = \frac{2\pi}{\omega} = 2\pi\sqrt{3M/2k}$.

13-74: The torque on the rod about the pivot (with angles positive in the direction indicated in the figure) is $\tau = -\left(k\frac{L}{2}\theta\right)\frac{L}{2}$. Setting this equal to the rate of change of angular momentum, $I\alpha = I\frac{d^2\theta}{dt^2}$,

$$\frac{d^2\theta}{dt^2} = -k\frac{L^2/4}{I}\theta = -\frac{3k}{M}\theta,$$

where the moment of inertia for a slender rod about its center, $I = \frac{1}{12}ML^2$ has been used. It follows that $\omega^2 = \frac{3k}{M}$, and $T = \frac{2\pi}{\omega} = 2\pi\sqrt{\frac{M}{3k}}$.

13-75:   The period of the simple pendulum (the clapper) must be the same as that of the bell; equating the expression in Eq. (13-34) to that in Eq. (13-39) and solving for L gives $L = I/md = (18.0 \text{ kg·m}^2)/((34.0 \text{ kg})(0.60 \text{ m})) = 0.882 \text{ m}$. Note that the mass of the bell, not the clapper, is used. As with any simple pendulum, the period of small oscillations of the clapper is independent of its mass.

13-76:   The moment of inertia about the pivot is $2(1/3)ML^2 = (2/3)ML^2$, and the center of gravity when balanced is a distance $d = L/(2\sqrt{2})$ below the pivot (see Problem 8-83). From Eq. (13-39), the frequency is

$$f = \frac{1}{T} = \frac{1}{2\pi}\sqrt{\frac{3g}{4\sqrt{2}L}} = \frac{1}{4\pi}\sqrt{\frac{3g}{\sqrt{2}L}}.$$

13-77:   a) $L = g(T/2\pi)^2 = 3.97 \text{ m}$.   b) There are many possibilities. One is to have a uniform thin rod pivoted about an axis perpendicular to the rod a distance d from its center. Using the desired period in Eq. (13-39) gives a quadratic in d, and using the maximum size for the length of the rod gives a pivot point a distance of 5.25 mm, which is on the edge of practicality. Using a "dumbbell," two spheres separated by a light rod of length L gives a slight improvement to $d = 1.6$ cm (neglecting the radii of the spheres in comparison to the length of the rod; see Problem 13-80).

13-78:   Using the notation $\frac{b}{2m} = \gamma$, $\frac{k}{m} = \omega^2$ and taking derivatives of Eq. (13-42) (setting the phase angle $\phi = 0$ does not affect the result),

$$x = Ae^{-\gamma t}\cos\omega't$$
$$v = -Ae^{-\gamma t}\left(\omega'\sin\omega't + \gamma\cos\omega't\right)$$
$$a = -Ae^{-\gamma t}\left(\left(\omega'^2 - \gamma^2\right)\cos\omega't - 2\omega'\gamma\sin\omega't\right).$$

Using these expressions in the left side of Eq. (13-41),

$$-kx - bv = Ae^{-\gamma t}\left(-k\cos\omega't + (2\gamma m)\omega'\sin\omega't + 2m\gamma^2\cos\omega't\right)$$
$$= mAe^{-\gamma t}\left(\left(2\gamma^2 - \omega^2\right)\cos\omega't + 2\gamma\omega'\sin\omega't\right).$$

The factor $(2\gamma^2 - \omega^2)$ is $\gamma^2 - \omega'^2$ (this is Eq. (13-43)), and so

$$-kx - bv = mAe^{-\gamma t}\left(\left(\gamma^2 - \omega'^2\right)\cos\omega't + 2\gamma\omega'\sin\omega't\right) = ma.$$

13-79:   a) In Eq. (13-38), $d = x$ and from the parallel axis theorem, $I = m(L^2/12 + x^2)$, so $\omega^2 = \frac{gx}{(L^2/12) + x^2}$.   c) Differentiating the ratio $\omega^2/g = \frac{x}{(L^2/12) + x^2}$ with respect to x and setting the result equal to zero gives

$$\frac{1}{(L^2/12) + x^2} = \frac{2x^2}{((L^2/12) + x^2)^2}, \quad \text{or} \quad 2x^2 = x^2 + L^2/12,$$

which is solved for $x = L/\sqrt{12}$. c) When x is the value that maximizes ω, the ratio $\frac{\omega^2}{g} = \frac{L/\sqrt{12}}{2(L^2/12)} = \frac{6}{L\sqrt{12}} = \frac{\sqrt{3}}{L}$, so the length is $L = \frac{\sqrt{3}g}{\omega^2} = 0.430$ m.

13-80: a) From the parallel axis theorem, the moment of inertia about the pivot point is $M(L^2 + (2/5)R^2)$. Using this in Eq. (13-39), with $d = L$ gives

$$T = 2\pi\sqrt{\frac{L^2 + (2/5)R^2}{gL}} = 2\pi\sqrt{\frac{L}{g}}\sqrt{1 + 2R^2/5L^2} = T_{\text{sp}}\sqrt{1 + 2R^2/5L^2}.$$

b) Letting $\sqrt{1 + 2R^2/5L^2} = 1.001$ and solving for the ratio L/R (or approximating the square root as $1 + R^2/5L^2$) gives $\frac{L}{R} = 14.1$. c) $(14.1)(1.270 \text{ cm}) = 18.0$ cm.

13-81: a) The net force on the block at equilibrium is zero, and so one spring (the one with $k_1 = 2.00$ N/m) must be stretched three times as much as the one with $k_2 = 6.00$ N/m. The sum of the elongations is 0.200 m, and so one spring stretches 0.150 m and the other stretches 0.050 m, and so the equilibrium lengths are 0.350 m and 0.250 m. b) There are many ways to approach this problem, all of which of course lead to the result of Problem 13-82(b). The most direct way is to let $\Delta x_1 = 0.150$ m and $\Delta x_2 = 0.050$ m, the results of part (a). When the block in Fig. (13-35) is displaced a distance x to the right, the net force on the block is

$$-k_1(\Delta x_1 + x) + k_2(\Delta x_2 - x) = [k_1\Delta x_1 - k_2\Delta x_2] - (k_1 + k_2)x.$$

From the result of part (a), the term in square brackets is zero, and so the net force is $-(k_1 + k_2)x$, the effective spring constant is $k_{\text{eff}} = k_1 + k_2$ and the period of vibration is $T = 2\pi\sqrt{\frac{0.100 \text{ kg}}{8.00 \text{ N/m}}} = 0.702$ s.

13-82: In each situation, imagine the mass moves a distance Δx, the springs move distances Δx_1 and Δx_2, with forces $F_1 = -k_1\Delta x_1$, $F_2 = -k_2\Delta x_2$. a) $\Delta x = \Delta x_1 = \Delta x_2$, $F = F_1 + F_2 = -(k_1 + k_2)\Delta x$, so $k_{\text{eff}} = k_1 + k_2$. b) Despite the orientation of the springs, and the fact that one will be compressed when the other is extended, $\Delta x = \Delta x_1 + \Delta x_2$, and the above result is still valid; $k_{\text{eff}} = k_1 + k_2$. c) For massless springs, the force on the block must be equal to the tension in any point of the spring combination, and $F = F_1 = F_2$, and so $\Delta x_1 = -\frac{F}{k_1}$, $\Delta x_2 = -\frac{F}{k_2}$, and

$$\Delta x = -\left(\frac{1}{k_1} + \frac{1}{k_2}\right)F = -\frac{k_1 + k_2}{k_1 k_2}F$$

and $k_{\text{eff}} = \frac{k_1 k_2}{k_1 + k_2}$. d) The result of part (c) shows that when a spring is cut in half, the effective spring constant doubles, and so the frequency increases by a factor of $\sqrt{2}$.

13-83: a) Using the hint,

$$T + \Delta T \approx 2\pi\sqrt{L}\left(g^{-1/2} - \frac{1}{2}g^{-3/2}\Delta g\right) = T - T\frac{\Delta g}{2g},$$

so $\Delta T = -(1/2)(T/g)\Delta g$. This result can also be obtained from $T^2 g = 4\pi^2 L$, from which $(2T\Delta T)g + T^2\Delta g = 0$. Therefore, $\frac{\Delta T}{T} = -\frac{1}{2}\frac{\Delta g}{g}$. b) The clock runs slow; $\Delta T > 0$, $\Delta g < 0$

and

$$g + \Delta g = g\left(1 - \frac{2\Delta T}{T}\right) = (9.80 \text{ m/s}^2)\left(1 - \frac{2(4.00 \text{ s})}{(86{,}400 \text{ s})}\right) = 9.7991 \text{ m/s}^2.$$

13-84: Denote the position of a piece of the spring by l; $l = 0$ is the fixed point and $l = L$ is the moving end of the spring. Then the velocity of the point corresponding to l, denoted u, is $u(l) = v\frac{l}{L}$ (when the spring is moving, l will be a function of time, and so u is an implicit function of time). a) $dm = \frac{M}{L}dl$, and so

$$dK = \frac{1}{2}dm\,u^2 = \frac{1}{2}\frac{Mv^2}{L^3}l^2\,dl,$$

and

$$K = \int dK = \frac{Mv^2}{2L^3}\int_0^L l^2\,dl = \frac{Mv^2}{6}.$$

b) $mv\frac{dv}{dt} + kx\frac{dx}{dt} = 0$, or $ma + kx = 0$, which is Eq. (13-4). c) m is replaced by $\frac{M}{3}$, so $\omega = \sqrt{\frac{3k}{M}}$ and $M' = \frac{M}{3}$.

13-85: a) With $I = (1/3)ML^2$ and $d = L/2$ in Eq. (13-39), $T_0 = 2\pi\sqrt{2L/3g}$. With the added mass, $I = M((L^2/3) + y^2)$, $m = 2M$ and $d = (L/4) + y/2$, $T = 2\pi \times \sqrt{(L^2/3 + y^2)/(g(L/2 + y))}$ and

$$r = \frac{T}{T_0} = \sqrt{\frac{L^2 + 3y^2}{L^2 + 2yL}}.$$

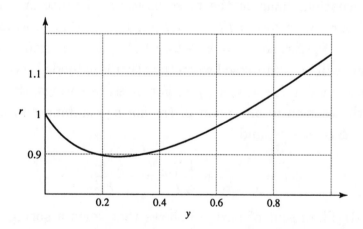

b) From the expression found in part (a), $T = T_0$ when $y = \frac{2}{3}L$. At this point, a simple pendulum with length y would have the same period as the meter stick without the added mass; the two bodies oscillate with the same period and do not affect the other's motion.

13-86: Let the two distances from the center of mass be d_1 and d_2. There are then two relations of the form of Eq. (13-39); with $I_1 = I_{\text{cm}} + md_1^2$ and $I_2 = I_{\text{cm}} + md_2^2$, these relations may be rewritten as

$$mgd_1 T^2 = 4\pi^2 \left(I_{\text{cm}} + md_1^2 \right)$$
$$mgd_2 T^2 = 4\pi^2 \left(I_{\text{cm}} + md_2^2 \right).$$

Subtracting the expressions gives

$$mg(d_1 - d_2)T^2 = 4\pi^2 m \left(d_1^2 - d_2^2 \right) = 4\pi^2 m (d_1 - d_2)(d_1 + d_2),$$

and dividing by the common factor of $m(d_1 - d_2)$ and letting $d_1 + d_2 = L$ gives the desired result.

13-87: a) The spring, when stretched, provides an inward force; using $\omega'^2 l$ for the magnitude of the inward radial acceleration,

$$m\omega'^2 l = k(l - l_0), \quad \text{or} \quad l = \frac{kl_0}{k - m\omega'^2}.$$

b) The spring will tend to become unboundedly long.

13-88: Let $r = R_0 + x$, so that $r - R_0 = x$ and

$$F = A \left[e^{-2bx} - e^{-bx} \right].$$

When x is small compared to b^{-1}, expanding the exponential function gives

$$F \approx A \left[(1 - 2bx) - (1 - bx) \right] = -Abx,$$

corresponding to a force constant of $Ab = 579.2$ N/m or 579 N/m to three figures. This is close to the value given in Exercise 13-32.

Chapter 14   Fluid Mechanics

14-1: $w = mg = \rho V g$

$$= (7.8 \times 10^3 \text{ kg/m}^3)(0.858 \text{ m})\pi(1.43 \times 10^{-2} \text{ m})^2(9.80 \text{ m/s}^2) = 41.8 \text{ N},$$

or 42 N to two places. A cart is not necessary.

14-2: $\rho = \dfrac{m}{V} = \dfrac{m}{\frac{4}{3}\pi r^3} = \dfrac{(7.35 \times 10^{22} \text{ kg})}{\frac{4}{3}\pi(1.74 \times 10^6 \text{ m})^3} = 3.33 \times 10^3 \text{ kg/m}^3.$

14-3: $\rho = \frac{m}{V} = \dfrac{(0.0158 \text{ kg})}{(5.0 \times 15.0 \times 30.0) \text{ mm}^3} = 7.02 \times 10^3 \text{ kg/m}^3.$ You were cheated.

14-4:   The length L of a side of the cube is

$$L = V^{\frac{1}{3}} = \left(\frac{m}{\rho}\right)^{\frac{1}{3}} = \left(\frac{40.0 \text{ kg}}{21.4 \times 10^3 \text{ kg/m}^3}\right)^{\frac{1}{3}} = 12.3 \text{ cm}.$$

14-5:   a) $\rho g h = (600 \text{ kg/m}^3)(9.80 \text{ m/s}^2)(0.12 \text{ m}) = 706 \text{ Pa}.$
     b) $706 \text{ Pa} + (1000 \text{ kg/m}^3)(9.80 \text{ m/s}^2)(0.250 \text{ m}) = 3.16 \times 10^3 \text{ Pa}.$

14-6:   a) The pressure used to find the area is the absolute pressure, and so the total area is

$$\frac{(16.5 \times 10^3 \text{ N})}{(205 \times 10^3 \text{ Pa} + 1.013 \times 10^5 \text{ Pa})} = 539 \text{ cm}^2.$$

b) With the extra weight, repeating the above calculation gives 836 cm^2.

14-7:   a) $\rho g h = (1.03 \times 10^3 \text{ kg/m}^3)(9.80 \text{ m/s}^2)(250 \text{ m}) = 2.52 \times 10^6 \text{ Pa}.$ b) The pressure difference is the gauge pressure, and the net force due to the water and the air is $(2.52 \times 10^6 \text{ Pa})(\pi(0.15 \text{ m})^2) = 1.78 \times 10^5 \text{ N}.$

14-8: $p = \rho g h = (1.00 \times 10^3 \text{ kg/m}^3)(9.80 \text{ m/s}^2)(640 \text{ m}) = 6.27 \times 10^6 \text{ Pa} = 61.9 \text{ atm}.$

14-9:   a) $p_a + \rho g y_2 = 980 \times 10^2 \text{ Pa} + (13.6 \times 10^3 \text{ kg/m}^3)(9.80 \text{ m/s}^2)(7.00 \times 10^{-2} \text{ m}) = 1.07 \times 10^5 \text{ Pa}.$ b) Repeating the calcultion with $y = y_2 - y_1 = 4.00$ cm instead of y_2 gives 1.03×10^5 Pa.   c) The absolute pressure is that found in part (b), 1.03×10^5 Pa.
d) $(y_2 - y_1)\rho g = 5.33 \times 10^3$ Pa (this is not the same as the difference between the results of parts (a) and (b) due to roundoff error).

14-10: $\rho g h = (1.00 \times 10^3 \text{ kg/m}^3)(9.80 \text{ m/s}^2)(6.1 \text{ m}) = 6.0 \times 10^4 \text{ Pa}.$

14-11:   The force is the difference between the upward force of the water and the downward forces of the air and the weight. The difference between the pressure inside and out is the gauge pressure, so

$$F = (\rho g h)A - w = (1.03 \times 10^3)(9.80 \text{ m/s}^2)(30 \text{ m})(0.75 \text{ m}^2) - 300 \text{ N} = 2.27 \times 10^5 \text{ N}.$$

14-12: $\left[130 \times 10^3 \text{ Pa} + (1.00 \times 10^3 \text{ kg/m}^3)(3.71 \text{ m/s}^2)(14.2 \text{ m}) - 93 \times 10^3 \text{ Pa}\right](2.00 \text{ m}^2)$
$= 1.79 \times 10^5 \text{ N}.$

14-13: The depth of the kerosene is the difference in pressures, divided by the product $\rho g = \frac{mg}{V}$,

$$h = \frac{(16.4 \times 10^3 \text{ N})/(0.0700 \text{ m}^2) - 2.01 \times 10^5 \text{ Pa}}{(205 \text{ kg})(9.80 \text{ m/s}^2)/(0.250 \text{ m}^3)} = 4.14 \text{ m}.$$

14-14: $p = \dfrac{F}{A} = \dfrac{mg}{\pi(d/2)^2} = \dfrac{(1200 \text{ kg})(9.80 \text{ m/s}^2)}{\pi(0.15 \text{ m})^2} = 1.66 \times 10^5 \text{ Pa} = 1.64 \text{ atm}.$

14-15: The buoyant force must be equal to the total weight; $\rho_{\text{water}} V g = \rho_{\text{ice}} V g + mg$, so

$$V = \frac{m}{\rho_{\text{water}} - \rho_{\text{ice}}} = \frac{45.0 \text{ kg}}{1000 \text{ kg/m}^3 - 920 \text{ kg/m}^3} = 0.563 \text{ m}^3,$$

or 0.56 m^3 to two figures.

14-16: The bouyant force is $B = 17.50 \text{ N} - 11.20 \text{ N} = 6.30 \text{ N}$, and

$$V = \frac{B}{\rho_{\text{water}} g} = \frac{(6.30 \text{ N})}{(1.00 \times 10^3 \text{ kg/m}^3)(9.80 \text{ m/s}^2)} = 6.43 \times 10^{-4} \text{ m}^3.$$

The density is

$$\rho = \frac{m}{V} = \frac{w/g}{B/\rho_{\text{water}} g} = \rho_{\text{water}} \frac{w}{B} = (1.00 \times 10^3 \text{ kg/m}^3)\left(\frac{17.50}{6.30}\right) = 2.78 \times 10^3 \text{ kg/m}^3.$$

14-17: a) The displaced fluid must weigh more than the object, so $\rho < \rho_{\text{fluid}}$. b) If the ship does not leak, much of the water will be displaced by air or cargo, and the average density of the floating ship is less than that of water. c) Let the portion submerged have volume V, and the total volume be V_0. Then, $\rho V_0 = \rho_{\text{fluid}} V$, so $\frac{V}{V_0} = \frac{\rho}{\rho_{\text{fluid}}}$. The fraction above the fluid is then $1 - \frac{\rho}{\rho_{\text{fluid}}}$. If $\rho \to 0$, the entire object floats, and if $\rho \to \rho_{\text{fluid}}$, none of the object is above the surface. d) Using the result of part (c),

$$1 - \frac{\rho}{\rho_{\text{fluid}}} = 1 - \frac{(0.042 \text{ kg})/(5.0 \times 4.0 \times 3.0 \times 10^{-6} \text{ m}^3)}{1030 \text{ kg/m}^3} = 0.32 = 32\%.$$

14-18: a) $B = \rho_{\text{water}} g V = (1.00 \times 10^3 \text{ kg/m}^3)(9.80 \text{ m/s}^2)(0.650 \text{ m}^3) = 6370 \text{ N}.$

b) $m = \frac{w}{g} = \frac{B - T}{g} = \frac{6370 \text{ N} - 900 \text{ N}}{9.80 \text{ m/s}^2} = 558 \text{ kg}.$

c) (See Exercise 14-17.) If the submerged volume is V',

$$V' = \frac{w}{\rho_{\text{water}} g} \quad \text{and} \quad \frac{V'}{V} = \frac{w}{\rho_{\text{water}} g V} = \frac{5470 \text{ N}}{6370 \text{ N}} = 0.859 = 85.9\%.$$

14-19: a) $\rho_{\text{oil}} g h_{\text{oil}} = 116 \text{ Pa}.$

b) $((790 \text{ kg/m}^3)(0.100 \text{ m}) + (1000 \text{ kg/m}^3)(0.0150 \text{ m}))(9.80 \text{ m/s}^2) = 921 \text{ Pa}.$

c) $\qquad m = \dfrac{w}{g} = \dfrac{(p_{\text{bottom}} - p_{\text{top}})A}{g} = \dfrac{(805 \text{ Pa})(0.100 \text{ m})^2}{(9.80 \text{ m/s}^2)} = 0.822 \text{ kg}.$

The density of the block is $p = \frac{0.822 \text{ kg}}{(0.10 \text{ m})^3} = 822 \frac{\text{kg}}{\text{m}^3}$. Note that is is the same as the average density of the fluid displaced, $(0.85)(790 \text{ kg/m}^3) + (0.15)(1000 \text{ kg/m}^3)$.

14-20: a) Neglecting the density of the air,

$$V = \frac{m}{\rho} = \frac{w/g}{\rho} = \frac{w}{g\rho} = \frac{(89 \text{ N})}{(9.80 \text{ m/s}^2)(2.7 \times 10^3 \text{ kg/m}^3)} = 3.36 \times 10^{-3} \text{ m}^3,$$

or 3.4×10^{-3} m^3 to two figures.

b) $T = w - B = w - g\rho_{\text{water}}V = w \left(1 - \frac{\rho_{\text{water}}}{\rho_{\text{aluminum}}}\right) = (89 \text{ N})\left(1 - \frac{1.00}{2.7}\right) = 56.0$ N.

14-21: $\frac{4\gamma}{R} = \frac{8\gamma}{D} = 6.67$ Pa.

14-22: Using Eq. (14-13), $p_g = \frac{2\gamma}{R}$, and $\gamma = 72.8 \times 10^{-3}$ N/m gives

a) 146 Pa, b) 1.46×10^4 Pa (note that this is 100 times the answer to part (a)).

14-23: Individual foot size will vary, but the product of the circumference of a typical pair of feet and the surface tension of water is about 0.1 N, the weight of a body mass of about 10 g.

14-24: The analysis leading to Eq. (14-13) is valid for the pores;

$$\frac{2\gamma}{R} = \frac{4\gamma}{D} = 2.9 \times 10^7 \text{ Pa}.$$

14-25: $T = F - w = 2l\gamma - w = 2(0.220 \text{ m})(25 \times 10^{-3} \text{ N/m}) - (0.70 \times 10^{-3} \text{ kg}) \times (9.80 \text{ m/s}^2) = 4.14 \times 10^{-3}$ N. (Some editions may have an incorrect mass given in the problem.)

14-26: $v_2 = v_1\frac{A_1}{A_2} = \frac{(3.50 \text{ m/s})(0.0700 \text{ m}^2)}{A_2} = \frac{0.245 \text{ m}^3/\text{s}}{A_2}.$

a) (i) $A_2 = 0.1050$ m^2, $v_2 = 2.33$ m/s. (ii) $A_2 = 0.047$ m^2, $v_2 = 5.21$ m/s.
b) $v_1A_1t = v_2A_2t = (0.245 \text{ m}^3/\text{s})(3600 \text{ s}) = 882$ m^3.

14-27: a) $v = \frac{dV/dt}{A} = \frac{(1.20 \text{ m}^3/\text{s})}{\pi(0.150 \text{ m})^2} = 16.98$ m/s.

b) $r_2 = r_1\sqrt{v_1/v_2} = \sqrt{(dV/dt)/\pi v_2} = 0.317$ m.

14-28: a) From the equation preceding Eq. (14-14), dividing by the time interval dt gives Eq. (14-16). b) The volume flow rate decreases by 1.50% (to two figures).

14-29: The hole is given as being "small," and this may be taken to mean that the velocity of the seawater at the top of the tank is zero, and Eq. (14-22) gives

$$v = \sqrt{2(gy + (p/\rho))}$$

$$= \sqrt{2((9.80 \text{ m/s}^2)(11.0 \text{ m}) + (3.00)(1.013 \times 10^5 \text{ Pa})/(1.03 \times 10^3 \text{ kg/m}^3))}$$

$$= 28.4 \text{ m/s}.$$

Note that $y = 0$ and $p = p_a$ were used at the bottom of the tank, so that p was the given gauge pressure at the top of the tank.

14-30: a) From Eq. (14-22), $v = \sqrt{2gh} = \sqrt{2(9.80 \text{ m/s}^2)(14.0 \text{ m})} = 16.6$ m/s.

b) $vA = (16.57 \text{ m/s})\,(\pi(0.30 \times 10^{-2} \text{ m})^2) = 4.69 \times 10^{-4} \text{ m}^3/\text{s}$. Note that an extra figure was kept in the intermediate calculation.

14-31: The assumption may be taken to mean that $v_1 = 0$ in Eq. (14-21). At the maximum height, $v_2 = 0$, and using gauge pressures for p_1 and p_2, $p_2 = 0$ (the water is open to the atmosphere), $p_1 = \rho g y_2 = 1.47 \times 10^5$ Pa.

14-32: Using $v_2 = \frac{1}{4}v_1$ in Eq. (14-21),

$$p_2 = p_1 + \frac{1}{2}\rho\left(v_1^2 - v_2^2\right) + \rho g(y_1 - y_2) = p_1 + \rho\left[\left(\frac{15}{32}\right)v_1^2 + g(y_1 - y_2)\right]$$

$$= 5.00 \times 10^4 \text{ Pa} + (1.00 \times 10^3 \text{ kg/m}^3)\left(\frac{15}{32}(3.00 \text{ m/s})^2 + (9.80 \text{ m/s}^2)(11.0 \text{ m})\right)$$

$$= 1.62 \text{ Pa.}$$

14-33: Neglecting the thickness of the wing (so that $y_1 = y_2$ in Eq. (14-21)), the pressure difference is $\Delta p = (1/2)\rho(v_2^2 - v_1^2) = 780$ Pa. The net upward force is then $(780 \text{ Pa}) \times (16.2 \text{ m}^2) - (1340 \text{ kg})(9.80 \text{ m/s}^2) = -496$ N.

14-34: a) $\frac{(220)(0.355 \text{ kg})}{60.0 \text{ s}} = 1.30$ kg/s. b) The density of the liquid is $\frac{0.355 \text{ kg}}{0.355 \times 10^{-3} \text{ m}^3} = 1000 \text{ kg/m}^3$, and so the volume flow rate is $\frac{1.30 \text{ kg/s}}{1000 \text{ kg/m}^3} = 1.30 \times 10^{-3} \text{ m}^3/\text{s} = 1.30$ L/s. This result may also be obtained from $\frac{(220)(0.355 \text{ L})}{60.0 \text{ s}} = 1.30$ L/s. c) $v_1 = \frac{1.30 \times 10^{-3} \text{ m}^3/\text{s}}{2.00 \times 10^{-4} \text{ m}^2} = 6.50$ m/s, $v_2 = v_1/4 = 1.63$ m/s.

d)
$$p_1 = p_2 + \frac{1}{2}\rho\left(v_2^2 - v_1^2\right) + \rho g(y_2 - y_1)$$

$$= 152 \text{ kPa} + (1/2)(1000 \text{ kg/m}^3)\left((1.63 \text{ m/s})^2 - (6.50 \text{ m/s})^2\right)$$

$$+ (1000 \text{ kg/m}^3)(9.80 \text{ m/s}^2)(-1.35 \text{ m})$$

$$= 119 \text{ kPa.}$$

14-35: The water is discharged at a rate of $v_1 = \frac{4.65 \times 10^{-4} \text{ m}^3/\text{s}}{1.32 \times 10^{-3} \text{ m}^2} = 0.352$ m/s. The pipe is given as horizontal, so the speed at the constriction is $v_2 = \sqrt{v_1^2 + 2\Delta p/\rho} = 8.95$ m/s, keeping an extra figure, so the cross-section area at the constriction is $\frac{4.65 \times 10^{-4} \text{ m}^3/\text{s}}{8.95 \text{ m/s}} = 5.19 \times 10^{-5} \text{ m}^2$, and the radius is $r = \sqrt{A/\pi} = 0.41$ cm.

14-36: From Eq. (14-21), with $y_1 = y_2$,

$$p_2 = p_1 + \frac{1}{2}\rho\left(v_1^2 - v_2^2\right) = p_1 + \frac{1}{2}\rho\left(v_1^2 - \frac{v_1^2}{4}\right) = p_1 + \frac{3}{8}\rho v_1^2$$

$$= 1.80 \times 10^4 \text{ Pa} + \frac{3}{8}(1.00 \times 10^3 \text{ kg/m}^3)(2.50 \text{ m/s})^2 = 2.03 \times 10^4 \text{ Pa,}$$

where the continuity relation $v_2 = \frac{v_1}{2}$ has been used.

14-37: a) From Eq. (14-25), the speed at $r = R/2$ is $\frac{3}{4}$ of the maximum at $r = 0$, or 1.88 m/s. b) From Eq. (14-25), the speed of the fluid at the boundary (the wall of the pipe) is zero.

14-38: At the center, $r = 0$ in Eq. (14-25), and solving for $p_1 - p_2 = \Delta p$,

$$\Delta p = \frac{4\eta L v_{\max}}{R^2} = \frac{4(1.005 \times 10^{-3} \text{ N·s/m}^2)(3.00 \text{ m})(0.200 \text{ m/s})}{(0.85 \times 10^{-2} \text{ m})^2} = 33.4 \text{ Pa.}$$

14-39: a) From Eq. (14-26),

$$\frac{dV}{dt} = \frac{\pi}{8}\left(\frac{(4.50 \times 10^{-2} \text{ m})^4}{(1.005 \times 10^{-3} \text{ N·s/m}^2)}\right)\left(\frac{1200 \text{ Pa}}{15.0 \text{ m}}\right) = 1.28 \times 10^{-1} \text{ m}^3/\text{s.}$$

b) Reducing the diameter to one-third its original value would mean that the pressure difference would need to be increased by a factor of 81, to 9.72×10^4 Pa. c) All other variables being the same, the flow rate is inversely proportional to the viscosity; $(1.28 \times 10^{-1} \text{ m}^3/\text{s})\left(\frac{1.005}{0.469}\right) = 0.275 \text{ m}^3/\text{s.}$

14-40: a) Solving Eq. (14-26) for the gauge pressure $\Delta p = p_1 - p_2$,

$$\Delta p = \frac{8\eta L(dV/dt)}{\pi R^4}$$

$$\frac{8(1.0 \times 10^{-3} \text{ N·s/m}^2)(0.20 \times 10^{-3} \text{ m})(0.25 \times 10^{-6} \text{ m}^3)/(15 \times 60 \text{ s})}{\pi(5 \times 10^{-6} \text{ m})^4}$$

$$= 2.3 \times 10^5 \text{ Pa} = 2.2 \text{ atm.}$$

This is the amount by which the pressure in the bug's mouth is *lower* than the atmosphere; the gauge pressure is negative. b) The pressure difference is proportional to the negative fourth power of the diameter, and so the largest contribution to the pressure difference is due to the narrowest part of the insect's mouth.

14-41: Using $F = (1/4)mg = (1/4)\rho V g = (1/4)\rho(4/3)\pi r^3 g$ in Eq. (14-27) and solving for v,

$$v = \frac{r^2 \rho g}{18\eta} = \frac{(2.00 \times 10^{-3} \text{ m})^2(2.70 \times 10^3 \text{ kg/m}^3)(9.80 \text{ m/s}^2)}{18(0.986 \text{ N·s/m}^2)} = 0.596 \text{ cm/s.}$$

Note that the density used is that of aluminum; any buoyant force is not considered in this problem.

14-42: At terminal velocity, Eq. (14-27) gives the difference between the weight and the buoyant force;

$$6\pi\eta r v_t = mg - B = mg\left(1 - \frac{\rho_{\text{liquid}}}{\rho_{\text{brass}}}\right),$$

and solving for the viscosity gives

$$\eta = \frac{mg\left(1 - \frac{\rho_1}{\rho_c}\right)}{6\pi r v_t}.$$

The radius is found from $V = \frac{m}{\rho_c} = \frac{4}{3}\pi r^3$, which can be solved for $r = 2.134 \times 10^{-3}$ m. Substituting this value and the data given in the problem yields $\eta = 1.13$ N·s/m^2, which is 11 poise to two figures.

14-43: From Eq. (14-26) for all parts, a) $(2)^4 = 16$; the flow rate would increase by a factor of 16. b) The flow rate would decrease by a factor of 2. c) Increase by a factor of 2. d) Increase by a factor of 2. e) Decrease by a factor of 2.

14-44: From Eq. (14-27), the Stokes's law force is

$$6\pi(181 \times 10^{-7}\ \text{N·s/m}^2)(0.124\ \text{m})(5.00\ \text{m/s}) = 2.12 \times 10^{-4}\ \text{N}$$

and the weight is 5.88 N; the ratio is 3.60×10^{-5}.

14-45: a) Using $\eta = 181 \times 10^{-7}\ \text{N·s/m}^2$ for air, Eq. (14-25) gives for $r = 0$

$$v = \frac{\Delta p}{4\eta L}R^2 = \frac{(1.013 \times 10^5\ \text{Pa})}{4(181 \times 10^{-7}\ \text{N·s/m}^2)(0.180\ \text{m})}(50 \times 10^{-6}\ \text{m})^2 = 19.4\ \text{m/s}.$$

The speed at the edge is zero, and the speed halfway between is $(3/4)$ of the speed at the center, 14.6 m/s. (See Exercise 14-37.) b) Using $\frac{dV}{dt} = \frac{\Delta V}{\Delta t}$ in Eq. (14-26) and solving for Δt,

$$\Delta t = \frac{8\eta L\,\Delta V}{\pi R^4 \Delta p} = \frac{8(181 \times 10^{-7}\ \text{N·s/m}^2)(0.180\ \text{m})(1.00\ \text{m}^3)}{\pi(50 \times 10^{-6}\ \text{m})^4(1.013 \times 10^5\ \text{Pa})} = 1.31 \times 10^7\ \text{s} = 152\ \text{d}.$$

c) The speeds are increased by a factor of $(2)^2 = 4$ and the time is decreased by a factor of $(2)^4 = 16$.

14-46: a) The cross-sectional area presented by a sphere is $\pi\frac{D^2}{4}$, therefore $F = (p_0 - p)\pi\frac{D^2}{4}$. b) The force on each hemisphere due to the atmosphere is $\pi(5.00 \times 10^{-2}\ \text{m})^2$ $(1.013 \times 10^5\ \text{Pa})(0.975) = 776\ \text{N}$.

14-47: a) $\rho g h = (1.03 \times 10^3\ \text{kg/m}^3)(9.80\ \text{m/s}^2)(10.92 \times 10^3\ \text{m}) = 1.10 \times 10^8\ \text{Pa}$. b) The fractional change in volume is the negative of the fractional change in density. The density at that depth is then

$$\rho = \rho_0\,(1 + k\Delta p) = (1.03 \times 10^3\ \text{kg/m}^3)(1 + (1.16 \times 10^8\ \text{Pa})(45.8 \times 10^{-11}\ \text{Pa}^{-1}))$$
$$= 1.08 \times 10^3\ \text{kg/m}^3,$$

a fractional increase of 5.0%. Note that to three figures, the gauge pressure and absolute pressure are the same.

14-48: a) The weight of the water is

$$\rho g V = (1.00 \times 10^3\ \text{kg/m}^3)(9.80\ \text{m/s}^2)((5.00\ \text{m})(4.0\ \text{m})(3.0\ \text{m})) = 5.88 \times 10^5\ \text{N},$$

or 5.9×10^5 N to two figures. b) Integration gives the expected result that the force is what it would be if the pressure were uniform and equal to the pressure at the midpoint;

$$F = \rho g A \frac{d}{2}$$
$$= (1.00 \times 10^3\ \text{kg/m}^3)(9.80\ \text{m/s}^2)((4.0\ \text{m})(3.0\ \text{m}))(1.50\ \text{m}) = 1.76 \times 10^5\ \text{N},$$

or 1.8×10^5 N to two figures.

14-49: Let the width be w and the depth at the bottom of the gate be H. The force on a strip of vertical thickness dh at a depth h is then $dF = \rho gh(w\,dh)$ and the torque about the hinge is $d\tau = \rho gwh(h - H/2)dh$; integrating from $h = 0$ to $h = H$ gives $\tau = \rho gwH^3/12 = 2.61 \times 10^4$ N·m.

14-50: a) See Problem 14-49; the net force is $\int dF$ from $h = 0$ to $h = H$, $F = \rho gwH^2/2 = \rho gAH/2$, where $A = wH$. b) The torque on a strip of vertical thickness dh about the bottom is $d\tau = dF(H - h) = \rho gwh(H - h)\,dh$, and integrating from $h = 0$ to $h = H$ gives $\tau = \rho gwH^3/6 = \rho gAH^2/6$. c) The force depends on the width and the square of the depth, and the torque about the bottom depends on the width and the cube of the depth; the surface area of the lake does not affect either result (for a given width).

14-51: The acceleration due to gravity on the planet is

$$g = \frac{\Delta p}{\rho d} = \frac{\Delta p}{\frac{m}{V}d}$$

and so the planet's mass is

$$M = \frac{gR^2}{G} = \frac{\Delta pVR^2}{mGd}$$

14-52: The cylindrical rod has mass M, radius R, and length L with a density that is proportional to the square of the distance from one end, $\rho = Cx^2$.

a) $M = \int \rho dV = \int Cx^2 dV$. The volume element $dV = \pi R^2 dx$. Then the integral becomes $M = \int_0^L Cx^2 \pi R^2 dx$. Integrating gives $M = C\pi R^2 \int_0^L x^2 dx = C\pi R^2 \frac{L^3}{3}$. Solving for C, $C = 3M/\pi R^2 L^3$.

b) The density at the $x = L$ end is $\rho = Cx^2 = \left(\frac{3M}{\pi R^2 L^3}\right)(L^2) = \left(\frac{3M}{\pi R^2 L}\right)$. The denominator is just the total volume V, so $\rho = 3M/V$, or three times the average density, M/V. So the average density is one-third the density at the $x = L$ end of the rod.

14-53: a) At $r = 0$, the model predicts $\rho = A = 12{,}700$ kg/m^3 and at $r = R$, the model predicts $\rho = A - BR = 12{,}700$ kg/m$^3 - (1.50 \times 10^{-3}$ kg/m$^4)(6.37 \times 10^6$ m$) = 3.15 \times 10^3$ kg/m^3. b), c)

$$M = \int dm = 4\pi \int_0^R [A - Br]r^2\, dr = 4\pi\left[\frac{AR^3}{3} - \frac{BR^4}{4}\right] = \left(\frac{4\pi R^3}{3}\right)\left[A - \frac{3BR}{4}\right]$$

$$= \left(\frac{4\pi(6.37 \times 10^6 \text{ m})^3}{3}\right)\left[12{,}700 \text{ kg/m}^3 - \frac{3(1.50 \times 10^{-3} \text{ kg/m}^4)(6.37 \times 10^6 \text{ m})}{4}\right]$$

$$= 5.99 \times 10^{24} \text{ kg,}$$

which is within 0.36% of the earth's mass. d) If $m(r)$ is used to denote the mass contained in a sphere of radius r, then $g = Gm(r)/r^2$. Using the same integration as that in part (b), with an upper limit of r instead of R gives the result. e) $g = 0$ at $r = 0$, and g at $r = R$,

$g = Gm(R)/R^2 = (6.673 \times 10^{-11} \text{ N·m}^2/\text{kg}^2)(5.99 \times 10^{24} \text{ kg})/(6.37 \times 10^6 \text{ m})^2 = 9.85 \text{ m/s}^2.$

f)
$$\frac{dg}{dr} = \left(\frac{4\pi G}{3}\right) \frac{d}{dr}\left[Ar - \frac{3Br^2}{4}\right] = \left(\frac{4\pi G}{3}\right)\left[A - \frac{3Br}{2}\right];$$

setting this equal to zero gives $r = 2A/3B = 5.64 \times 10^6$ m, and at this radius

$$g = \left(\frac{4\pi G}{3}\right)\left(\frac{2A}{3B}\right)\left[A - \left(\frac{3}{4}\right)B\left(\frac{2A}{3B}\right)\right]$$

$$= \frac{4\pi G A^2}{9B}$$

$$= \frac{4\pi(6.673 \times 10^{-11} \text{ N·m}^2/\text{kg}^2)(12{,}700 \text{ kg/m}^3)^2}{9(1.50 \times 10^{-3} \text{ kg/m}^4)} = 10.02 \text{ m/s}^2.$$

(Using a less precise value for G gives $g = 10.01$ m/s^2.)

14-54: a) Equation (14-4), with the radius r instead of height y, becomes $dp = -\rho g\, dr = -\rho g_s(r/R)\, dr$. This form shows that the pressure decreases with increasing radius. Integrating, with $p = 0$ at $r = R$,

$$p = -\frac{\rho g_s}{R}\int_R^r r\, dr = \frac{\rho g_s}{R}\int_r^R r\, dr = \frac{\rho g_s}{2R}(R^2 - r^2).$$

b) Using the above expression with $r = 0$ and $\rho = \frac{M}{V} = \frac{3M}{4\pi R^3}$,

$$p(0) = \frac{3(5.97 \times 10^{24} \text{ kg})(9.80 \text{ m/s}^2)}{8\pi(6.38 \times 10^6 \text{ m})^2} = 1.71 \times 10^{11} \text{ Pa}.$$

c) While the same order of magnitude, this is not in very good agreement with the esitmated value. In more realistic density models (see Problem 14-53 or Problem 9-85), the concentration of mass at lower radii leads to a higher pressure.

14-55: a) $\rho_{\text{water}} g h_{\text{water}} = (1.00 \times 10^3 \text{ kg/m}^3)(9.80 \text{ m/s}^2)(15.0 \times 10^{-2} \text{ m}) = 1.47 \times 10^3$ Pa.
b) The gauge pressure at a depth of 15.0 cm $- h$ below the top of the mercury column must be that found in part (a); $\rho_{\text{Hg}} g(15.0 \text{ cm} - h) = \rho_{\text{water}} g(15.0 \text{ cm})$, which is solved for $h = 13.9$ cm.

14-56: Following the hint,

$$F = \int_0^h (\rho g y)(2\pi R)\, dy = \rho g \pi R h^2,$$

where R and h are the radius and height of the tank (the fact that $2R = h$ is more or less coincidental). Using the given numerical values gives $F = 5.07 \times 10^8$ N.

14-57: For the barge to be completely submerged, the mass of water displaced would need to be $\rho_{\text{water}} V = (1.00 \times 10^3 \text{ kg/m}^3)(22 \times 40 \times 12 \text{ m}^3) = 1.056 \times 10^7$ kg. The mass of the barge itself is

$$(7.8 \times 10^3 \text{ kg/m}^3)((2(22 + 40) \times 12 + 22 \times 40) \times 4.0 \times 10^{-2} \text{ m}^3) = 7.39 \times 10^5 \text{ kg},$$

so the barge can hold 9.82×10^6 kg of coal. This mass of coal occupies a solid volume of 6.55×10^3 m^3, which is less than the volume of the interior of the barge (1.06×10^4 m^3), but the coal must not be too loosely packed.

14-58: The difference between the densities must provide the "lift" of 5800 N (see Problem 14-63). The average density of the gases in the balloon is then

$$\rho_{ave} = 1.23 \text{ kg/m}^3 - \frac{(5800 \text{ N})}{(9.80 \text{ m/s}^2)(2200 \text{ m}^3)} = 0.96 \text{ kg/m}^3.$$

14-59: a) The submerged volume V' is $\frac{w}{\rho_{water}g}$, so

$$\frac{V'}{V} = \frac{w/\rho_{water}g}{V} = \frac{m}{\rho_{water}V} = \frac{(900 \text{ kg})}{(1.00 \times 10^3 \text{ kg/m}^3)(3.0 \text{ m}^3)} = 0.30 = 30\%.$$

b) As the car is about to sink, the weight of the water displaced is equal to the weight of the car plus the weight of the water inside the car. If the volume of water inside the car is V'',

$$V\rho_{water}g = w + V''\rho_{water}g, \quad \text{or} \quad \frac{V''}{V} = 1 - \frac{w}{V\rho_{water}g} = 1 - 0.30 = 0.70 = 70\%.$$

14-60: a) The volume displaced must be that which has the same weight and mass as the ice, $\frac{9.70 \text{ gm}}{1.00 \text{ gm/cm}^3} = 9.70$ cm^3 (note that the choice of the form for the density of water avoids conversion of units).   b) No; when melted, it is as if the volume displaced by the 9.70 gm of melted ice displaces the same volume, and the water level does not change. c) $\frac{9.70 \text{ gm}}{1.05 \text{ gm/cm}^3} = 9.24$ cm^3.   d) The melted water takes up more volume than the salt water displaced, and so 0.46 cm^3 flows over. A way of considering this situation (as a thought experiment only) is that the less dense water "floats" on the salt water, and as there is insufficient volume to contain the melted ice, some spills over.

14-61: The total mass of the lead and wood must be the mass of the water displaced, or

$$V_{Pb}\rho_{Pb} + V_{wood}\rho_{wood} = (V_{Pb} + V_{wood})\rho_{water};$$

solving for the volume V_{Pb},

$$V_{Pb} = V_{wood}\frac{\rho_{water} - \rho_{wood}}{\rho_{Pb} - \rho_{water}}$$

$$= (1.2 \times 10^{-2} \text{ m}^3)\frac{1.00 \times 10^3 \text{ kg/m}^3 - 600 \text{ kg/m}^3}{11.3 \times 10^3 \text{ kg/m}^3 - 1.00 \times 10^3 \text{ kg/m}^3}$$

$$= 4.66 \times 10^{-4} \text{ m}^3,$$

which has a mass of 5.27 kg.

14-62: The fraction f of the volume that floats above the fluid is $f = 1 - \frac{\rho}{\rho_{fluid}}$, where ρ is the average density of the hydrometer (see Problem 14-17 or Problem 14-59), which can be expressed as $\rho_{fluid} = \rho\frac{1}{1-f}$. Thus, if two fluids are observed to have floating fraction f_1 and f_2, $\rho_2 = \rho_1\frac{1-f_1}{1-f_2}$. In this form, it's clear that a larger f_2 corresponds to a larger density; more of the stem is above the fluid. Using $f_1 = \frac{(8.00 \text{ cm})(0.400 \text{ cm}^2)}{(13.2 \text{ cm}^3)} = 0.242$, $f_2 = \frac{(3.20 \text{ cm})(0.400 \text{ cm}^2)}{(13.2 \text{ cm}^3)} = 0.097$ gives $\rho_{alcohol} = (0.839)\rho_{water} = 839 \text{ kg/m}^3$.

14-63: a) The "lift" is $V(\rho_{air} - \rho_{H_2})g$, from which

$$V = \frac{120,000 \text{ N}}{(1.20 \text{ kg/m}^3 - 0.0899 \text{ kg/m}^3)(9.80 \text{ m/s}^2)} = 11.0 \times 10^3 \text{ m}^3.$$

b) For the same volume, the "lift" would be different by the ratio of the density differences,

$$(120,000 \text{ N})\left(\frac{\rho_{air} - \rho_{He}}{\rho_{air} - \rho_{H_2}}\right) = 11.2 \times 10^4 \text{ N}.$$

This increase in lift is not worth the hazards associated with use of hydrogen.

14-64: a) Archimedes' principle states $\rho g L A = Mg$, so $L = \frac{M}{\rho A}$.

b) The buoyant force is $\rho g A(L + x) = Mg + F$, and using the result of part (a) and solving for x gives $x = \frac{F}{\rho g A}$.

c) The "spring constant," that is, the proportionality between the displacement x and the applied force F, is $k = \rho g A$, and the period of oscillation is

$$T = 2\pi\sqrt{\frac{M}{k}} = 2\pi\sqrt{\frac{M}{\rho g A}}.$$

14-65: a) $\quad x = \dfrac{w}{\rho g A} = \dfrac{mg}{\rho g A} = \dfrac{m}{\rho A} = \dfrac{(70.0 \text{ kg})}{(1.03 \times 10^3 \text{ kg/m}^3)\pi(0.450 \text{ m})^2} = 0.107 \text{ m}.$

b) Note that in part (c) of Problem 14-64, M is the mass of the buoy, not the mass of the man, and A is the cross-section area of the buoy, not the amplitude. The period is then

$$T = 2\pi\sqrt{\frac{(950 \text{ kg})}{(1.03 \times 10^3 \text{ kg/m}^3)(9.80 \text{ m/s}^2)\pi(0.450 \text{ m})^2}} = 2.42 \text{ s}.$$

14-66: To save some intermediate calculation, let the density, mass and volume of the life preserver be ρ_0, m and v, and the same quantities for the person be ρ_1, M and V. Then, equating the buoyant force and the weight, and dividing out the common factor of g,

$$\rho_{water}((0.80)V + v) = \rho_0 v + \rho_1 V,$$

Eliminating V in favor of ρ_1 and M, and eliminating m in favor of ρ_0 and v,

$$\rho_0 v + M = \rho_{water}\left((0.80)\frac{M}{\rho_1} + v\right).$$

Solving for ρ_0,

$$\rho_0 = \frac{1}{v}\left(\rho_{\text{water}}\left((0.80)\frac{M}{\rho_1} + v\right) - M\right)$$

$$= \rho_{\text{water}} - \frac{M}{v}\left(1 - (0.80)\frac{\rho_{\text{water}}}{\rho_1}\right)$$

$$= 1.03 \times 10^3 \text{ kg/m}^3 - \frac{75.0 \text{ kg}}{0.0400 \text{ m}^3}\left(1 - (0.80)\frac{1.03 \times 10^3 \text{ kg/m}^3}{980 \text{ kg/m}^3}\right)$$

$$= 732 \text{ kg/m}^3.$$

14-67: To the given precision, the density of air is negligible compared to that of brass, but not compared to that of the wood. The fact that the density of brass may not be known the three-figure precision does not matter; the mass of the brass is given to three figures. The weight of the brass is the difference between the weight of the wood and the buoyant force of the air on the wood, and canceling a common factor of g, $V_{\text{wood}}(\rho_{\text{wood}} - \rho_{\text{air}}) = M_{\text{brass}}$, and

$$M_{\text{wood}} = \rho_{\text{wood}}V_{\text{wood}} = M_{\text{brass}}\frac{\rho_{\text{wood}}}{\rho_{\text{wood}} - \rho_{\text{air}}} = M_{\text{brass}}\left(1 - \frac{\rho_{\text{air}}}{\rho_{\text{wood}}}\right)^{-1}$$

$$= (0.0950 \text{ kg})\left(1 - \frac{1.20 \text{ kg/m}^3}{150 \text{ kg/m}^3}\right)^{-1} = 0.0958 \text{ kg}.$$

14-68: The buoyant force on the mass A, divided by g, must be $7.50 \text{ kg} - 1.00 \text{ kg} - 1.80 \text{ kg} = 4.70 \text{ kg}$ (see Example 14-6), so the mass of the block is $4.70 \text{ kg} + 3.50 \text{ kg} = 8.20 \text{ kg}$. a) The mass of the liquid displaced by the block is 4.70 kg, so the density of the liquid is $\frac{4.70 \text{ kg}}{3.80 \times 10^{-3} \text{ m}^3} = 1.24 \times 10^3 \text{ kg/m}^3$. b) Scale D will read the mass of the block, 8.20 kg, as found above. Scale E will read the sum of the masses of the beaker and liquid, 2.80 kg.

14-69: Neglecting the buoyancy of the air, the weight in air is

$$g(\rho_{\text{Au}}V_{\text{Au}} + \rho_{\text{Al}}V_{\text{Al}}) = 45.0 \text{ N}$$

and the buoyant force when suspended in water is

$$\rho_{\text{water}}(V_{\text{Au}} + V_{\text{Al}})g = 45.0 \text{ N} - 39.0 \text{ N} = 6.0 \text{ N}.$$

These are two equations in the two unknowns V_{Au} and V_{Al}. Multiplying the second by ρ_{Al} and the first by ρ_{water} and subtracting to eliminate the V_{Al} term gives

$$\rho_{\text{water}}V_{\text{Au}}g(\rho_{\text{Au}} - \rho_{\text{Al}}) = \rho_{\text{water}}(45.0 \text{ N}) - \rho_{\text{Al}}(6.0 \text{ N})$$

$$w_{\text{Au}} = \rho_{\text{Au}}gV_{\text{Au}} = \frac{\rho_{\text{Au}}}{\rho_{\text{water}}(\rho_{\text{Au}} - \rho_{\text{Al}})}(\rho_{\text{water}}(45.0 \text{ N}) - \rho_{\text{Al}}(6.0))$$

$$= \frac{(19.3)}{(1.00)(19.3 - 2.7)}((1.00)(45.0 \text{ N}) - (2.7)(6.0 \text{ N}))$$

$$= 33.5 \text{ N}.$$

Note that in the numerical determination of w_{Au}, specific gravities were used instead of densities.

14-70: (Note that increasing x corresponds to moving toward the back of the car.) a) The mass of air in the volume element is $\rho \, dV = \rho A \, dx$, and the net force on the element in the forward direction is $(p + dp)A - pA = A \, dp$. From Newton's second law, $A \, dp = (\rho A \, dx)a$, from which $dp = \rho a \, dx$. b) With ρ given to be constant, and with $p = p_0$ at $x = 0$, $p = p_0 + \rho ax$. c) Using $\rho = 1.2 \text{ kg/m}^3$ in the result of part (b) gives $(1.2 \text{ kg/m}^3)(5.0 \text{ m/s}^2)(2.5 \text{ m}) = 15.0 \text{ Pa} \sim 15 \times 10^{-5} p_{atm}$, so the fractional pressure difference is negligible. d) Following the argument in Section 14-4, the force on the balloon must be the same as the force on the same volume of air; this force is the product of the mass ρV and the acceleration, or $\rho V a$. e) The acceleration of the balloon is the force found in part (d) divided by the mass $\rho_{bal} V$, or $(\rho/\rho_{bal})a$. The acceleration relative to the car is the difference between this acceleration and the car's acceleration, $a_{rel} = [(\rho/\rho_{bal}) - 1] a$. f) For a balloon filled with air, $(\rho/\rho_{bal}) < 1$ (air balloons tend to sink in still air), and so the quantity in square brackets in the result of part (e) is negative; the balloon moves to the back of the car. For a helium balloon, the quantity in square brackets is positive, and the balloon moves to the front of the car.

14-71: a) The weight of the crown in terms of its volume V is $w = \rho_{crown} g V$, and when suspended the apparent weight is the difference between the weight and the buoyant force,

$$fw = f\rho_{crown} g V = (\rho_{crown} - \rho_{water}) \, gV.$$

Dividing by the common factors leads to

$$-\rho_{water} + \rho_{crown} = f\rho_{crown} \quad \text{or} \quad \frac{\rho_{crown}}{\rho_{water}} = \frac{1}{1 - f}.$$

As $f \to 0$, the apparent weight approaches zero, which means the crown tends to float; from the above result, the specific gravity of the crown tends to 1. As $f \to 1$, the apparent weight is the same as the weight, which means that the buoyant force is negligble compared to the weight, and the specific gravity of the crown is very large, as reflected in the above expression. b) Solving the above equations for f in terms of the specific gravity, $f = 1 - \frac{\rho_{water}}{\rho_{crown}}$, and so the weight of the crown would be $(1 - (1/19.3))(12.9 \text{ N}) = 12.2 \text{ N}$. c) Approximating the average density by that of lead for a "thin" gold plate, the apparent weight would be $(1 - (1/11.3))(12.9 \text{ N}) = 11.8 \text{ N}$.

14-72: a) See Problem 14-71. Replacing f with, respectively, w_{water}/w and w_{fluid}/w gives

$$\frac{\rho_{steel}}{\rho_{fluid}} = \frac{w}{w - w_{fluid}}, \quad \frac{\rho_{steel}}{\rho_{water}} = \frac{w}{w - w_{water}},$$

and dividing the second of these by the first gives

$$\frac{\rho_{fluid}}{\rho_{water}} = \frac{w - w_{fluid}}{w - w_{water}}.$$

b) When w_{fluid} is greater than w_{water}, the term on the right in the above expression is less than one, indicating that the fluid is less dense than water, and this is consistent with the buoyant force when suspended in liquid being less than that when suspended in water. If the density of the fluid is the same as that of water, $w_{\text{fluid}} = w_{\text{water}}$, as expected. Similarly, if w_{fluid} is less than w_{water}, the term on the right in the above expression is greater than one, indicating the the fluid is denser than water.   c) Writing the result of part (a) as

$$\frac{\rho_{\text{fluid}}}{\rho_{\text{water}}} = \frac{1 - f_{\text{fluid}}}{1 - f_{\text{water}}}$$

and solving for f_{fluid},

$$f_{\text{fluid}} = 1 - \frac{\rho_{\text{fluid}}}{\rho_{\text{water}}}(1 - f_{\text{water}}) = 1 - (1.220)(0.128) = 0.844 = 84.4\%.$$

14-73:   a) Let the total volume be V; neglecting the density of the air, the buoyant force in terms of the weight is

$$B = \rho_{\text{water}}gV = \rho_{\text{water}}g\left(\frac{(w/g)}{\rho_{\text{m}}} + V_0\right),$$

or

$$V_0 = \frac{B}{\rho_{\text{water}}g} - \frac{w}{\rho_{\text{w}}g}.$$

b) $\frac{B}{\rho_{\text{water}}g} - \frac{w}{\rho_{\text{Cu}}g} = 2.52 \times 10^{-4}$ m^3. Since the total volume of the casting is $\frac{B}{\rho_{\text{water}}g}$, the cavities are 12.4% of the total volume.

14-74:   a) Let d be the depth of the oil layer, h the depth that the cube is submerged in the water, and L be the length of a side of the cube. Then, setting the buoyant force equal to the weight, canceling the common factors of g and the cross-section area and supressing units, $(1000)h + (750)d = (550)L$. d, h and L are related by $d + h + (0.35)L = L$, so $h = (0.65)L - d$. Substitution into the first relation gives $d = L\frac{(0.65)(1000) - (550)}{(1000) - (750)} = \frac{2L}{5.00} = 0.040$ m. b) The gauge pressure at the lower face must be sufficient to support the block (the oil exerts only sideways forces directly on the block), and $p = \rho_{\text{wood}}gL = (550 \text{ kg/m}^3)(9.80 \text{ m/s}^2)(0.100 \text{ m}) = 539$ Pa. As a check, the gauge pressure, found from the depths and densities of the fluids, is $((0.040 \text{ m})(750 \text{ kg/m}^3) + (0.025 \text{ m})(1000 \text{ kg/m}^3))(9.80 \text{ m/s}^2) = 539$ Pa.

14-75:   The ship will rise; the total mass of water displaced by the barge-anchor combination must be the same, and when the anchor is dropped overboard, it displaces some water and so the barge itself displaces less water, and so rises.

To find the amount the barge rises, let the original depth of the barge in the water be $h_0 = (m_{\text{b}} + m_{\text{a}})/(\rho_{\text{water}}A)$, where m_{b} and m_{a} are the masses of the barge and the anchor, and A is the area of the bottom of the barge. When the anchor is dropped, the buoyant force on the barge is less than what it was by an amount equal to the buoyant force on the anchor; symbolically,

$$h'\rho_{\text{water}}Ag = h_0\rho_{\text{water}}Ag - (m_{\text{a}}/\rho_{\text{steel}})\rho_{\text{water}}g,$$

which is solved for

$$\Delta h = h_0 - h' = \frac{m_a}{\rho_{steel} A} = \frac{(35.0 \text{ kg})}{(7860 \text{ kg/m}^3)(8.00 \text{ m}^2)} = 5.57 \times 10^{-4} \text{ m},$$

or about 0.56 mm.

14-76: a) The average density of a filled barrel is $\rho_{oil} + \frac{m}{V} = 750 \text{ kg/m}^3 + \frac{15.0 \text{ kg}}{0.120 \text{ m}^3} = 875 \text{ kg/m}^3$, which is less than the density of seawater, so the barrel floats.

b) The fraction that floats (see Problem 14-17) is

$$1 - \frac{\rho_{ave}}{\rho_{water}} = 1 - \frac{875 \text{ kg/m}^3}{1030 \text{ kg/m}^3} = 0.150 = 15.0\%.$$

c) The average density is $910 \frac{\text{kg}}{\text{m}^3} + \frac{32.0 \text{ kg}}{0.120 \text{ m}^3} = 1172 \frac{\text{kg}}{\text{m}^3}$ which means the barrel sinks. In order to lift it, a tension $T = (1177 \frac{\text{kg}}{\text{m}^3})(0.120 \text{ m}^3)(9.80 \frac{\text{m}}{\text{s}^2}) - (1030 \frac{\text{kg}}{\text{m}^3})(0.120 \text{ m}^3)(9.80 \frac{\text{m}}{\text{s}^2}) = 173 \text{ N}$ is required.

14-77: a) See Exercise 14-17; the fraction of the volume that remains unsubmerged is $1 - \frac{\rho_B}{\rho_L}$. b) Let the depth of the liquid be x and the depth of the water be y. Then $\rho_L g x + \rho_w g y = \rho_B g L$ and $x + y = L$. Therefore $x = L - y$ and $y = \frac{(\rho_L - \rho_B)L}{\rho_L - \rho_w}$. c) $y = \frac{13.6 - 7.8}{13.6 - 1.0}(0.10 \text{ m}) = 0.046 \text{ m}$.

14-78: a) The change in height Δy is related to the displaced volume ΔV by $\Delta y = \frac{\Delta V}{A}$, where A is the surface area of the water in the lock. ΔV is the volume of water that has the same weight as the metal, so

$$\Delta y = \frac{\Delta V}{A} = \frac{w/\rho_{water} g}{A} = \frac{w}{\rho_{water} g A}$$

$$= \frac{(2.50 \times 10^6 \text{ N})}{(1.00 \times 10^3 \text{ kg/m}^3)(9.80 \text{ m/s}^2)((60.0 \text{ m})(20.0 \text{ m}))} = 0.213 \text{ m}.$$

b) In this case, ΔV is the volume of the metal; in the above expression, ρ_{water} is replaced by $\rho_{metal} = 9.00 \rho_{water}$, which gives $\Delta y' = \frac{\Delta y}{9}$, and $\Delta y - \Delta y' = \frac{8}{9}\Delta y = 0.189 \text{ m}$; the water sinks by this amount.

14-79: a) Consider the fluid in the horizontal part of the tube. This fluid, with mass $\rho A l$, is subject to a net force due to the pressure difference between the ends of the tube, which is the difference between the gauge pressures at the bottoms of the ends of the tubes. This difference is $\rho g(y_L - y_R)$, and the net force on the horizontal part of the fluid is

$$\rho g(y_L - y_R)A = \rho A l a,$$

or

$$(y_L - y_R) = \frac{a}{g} l.$$

b) Again consider the fluid in the horizontal part of the tube. As in part (a), the fluid is accelerating; the center of mass has a radial acceleration of magnitude $a_{rad} = \omega^2 l/2$, and so the difference in heights between the columns is $(\omega^2 l/2)(l/g) = \omega^2 l^2/2g$.

Anticipating Problem 14-81, an equivalent way to do part (b) is to break the fluid in the horizontal part of the tube into elements of thickness dr; the pressure difference between the sides of this piece is $dp = \rho(\omega^2 r)\, dr$ (see Problem 14-70), and integrating from $r = 0$ to $r = l$ gives $\Delta p = \rho\omega^2 l^2/2$, giving the same result.

c) At any point, Newton's second law gives $dpA = pAdla$ from which the area A cancels out. Therefore the cross-sectional area does not affect the result, even if it varies. Integrating the above result from 0 to l gives $\Delta p = pal$ between the ends. This is related to the height of the columns through $\Delta p = pg\Delta y$ from which p cancels out.

14-80: a) The change in pressure with respect to the vertical distance supplies the force necessary to keep a fluid element in vertical equilibrium (opposing the weight). For the rotating fluid, the change in pressure with respect to radius supplies the force necessary to keep a fluid element accelerating toward the axis; specifically, $dp = \frac{\partial p}{\partial r}\, dr = \rho a\, dr$, and using $a = \omega^2 r$ gives $\frac{\partial p}{\partial r} = \rho\omega^2 r$.   b) Let the pressure at $y = 0$, $r = 0$ be p_a (atmospheric pressure); integrating the expression for $\frac{\partial p}{\partial r}$ from part (a) gives

$$p(r,\, y = 0) = p_a + \frac{\rho\omega^2}{2} r^2.$$

c) In Eq. (14-5), $p_2 = p_a$, $p_1 = p(r, y = 0)$ as found in part (b), $y_1 = 0$ and $y_2 = h(r)$, the height of the liquid above the $y = 0$ plane. Using the result of part (b) gives $h(r) = \omega^2 r^2/2g$.

14-81: a) The net inward force is $(p + dp)A - pA = A\, dp$, and the mass of the fluid element is $\rho A\, dr'$. Using Newton's second law, with the inward radial acceleration of $\omega^2 r'$, gives $dp = \rho\omega^2 r'\, dr'$.   b) Integrating the above expression,

$$\int_{p_0}^{p} dp = \int_{r_0}^{r} \rho\omega^2 r'\, dr'$$

$$p - p_0 = \left(\frac{\rho\omega^2}{2}\right)\left(r^2 - r_0^2\right),$$

which is the desired result.   c) Using the same reasoning as in Section 14-4 (and Problem 14-70), the net force on the object must be the same as that on a fluid element of the same shape. Such a fluid element is accelerating inward with an acceleration of magnitude $\omega^2 R_{cm}$, and so the force on the object is $\rho V \omega^2 R_{cm}$.   d) If $\rho R_{cm} > \rho_{ob} R_{cmob}$, the inward force is greater than that needed to keep the object moving in a circle with radius R_{cmob} at angular frequency ω, and the object moves inward. If $\rho R_{cm} < \rho_{ob} R_{cmob}$, the net force is insufficient to keep the object in the circular motion at that radius, and the object moves outward.   e) Objects with lower densities will tend to move toward the center, and objects with higher densities will tend to move away from the center.

14-82: Solving Eq. (14-13) for R gives

$$R = \frac{2\gamma}{\Delta p} = \frac{2(72.8 \times 10^{-3} \text{ N·s/m}^2)}{(0.0250 \text{ atm})(1.013 \times 10^5 \text{ Pa})} = 5.75 \times 10^{-5} \text{ m}.$$

14-83: If the block were uniform, the buoyant force would be along a line directed through its geometric center, and the fact that the center of gravity is not at the geometric center does not affect the buoyant force. This means that the torque about the geometric center is due to the offset of the center of gravity, and is equal to the product of the block's weight and the horizontal displacement of the center of gravity from the geometric center, $(0.075 \text{ m})/\sqrt{2}$. The block's mass is half of its volume times the density of water, so the net torque is

$$\frac{(0.30 \text{ m})^3(1000 \text{ kg/m}^3)}{2}(9.80 \text{ m/s}^2)\frac{0.075 \text{ m}}{\sqrt{2}} = 7.02 \text{ N·m},$$

or 7.0 N·m to two figures. Note that the buoyant force and the block's weight form a couple, and the torque is the same about any axis.

14-84: a) As in Example 14-9, the speed of efflux is $\sqrt{2gh}$. After leaving the tank, the water is in free fall, and the time it takes any portion of the water to reach the ground is $t = \sqrt{\frac{2(H-h)}{g}}$, in which time the water travels a horizontal distance $R = vt = 2\sqrt{h(H-h)}$.

b) Note that if $h' = H - h$, $h'(H - h') = (H - h)h$, and so $h' = H - h$ gives the same range.

14-85: The water will rise until the rate at which the water flows out of the hole is the rate at which water is added;

$$A\sqrt{2gh} = \frac{dV}{dt},$$

which is solved for

$$h = \left(\frac{dV/dt}{A}\right)^2 \frac{1}{2g} = \left(\frac{2.40 \times 10^{-4} \text{ m}^3/\text{s}}{1.50 \times 10^{-4} \text{ m}^2}\right)^2 \frac{1}{2(9.80 \text{ m/s}^2)} = 13.1 \text{ cm}.$$

Note that the result is independent of the diameter of the bucket.

14-86: a)

$$v_3 A_3 = \sqrt{2g(y_1 - y_3)}A_3 = \sqrt{2(9.80 \text{ m/s}^2)(8.00 \text{ m})}\left(0.0160 \text{ m}^2\right) = 0.200 \text{ m}^3/\text{s}.$$

b) Since p_3 is atmospheric, the gauge pressure at point 2 is

$$p_2 = \frac{1}{2}\rho\left(v_3^2 - v_2^2\right) = \frac{1}{2}\rho v_3^2\left(1 - \left(\frac{A_3}{A_2}\right)^2\right) = \frac{8}{9}\rho g(y_1 - y_3),$$

using the expression for v_3 found above. Subsitition of numerical values gives $p_2 = 6.97 \times 10^4$ Pa.

14-87: The pressure difference, neglecting the thickness of the wing, is $\Delta p = (1/2)\rho(v_{\text{top}}^2 - v_{\text{bottom}}^2)$, and solving for the speed on the top of the wing gives

$$v_{\text{top}} = \sqrt{(120 \text{ m/s})^2 + 2(2000 \text{ Pa})/(1.20 \text{ kg/m}^3)} = 133 \text{ m/s}.$$

The pressure difference is comparable to that due to an altitude change of about 200 m, so ignoring the thickness of the wing is valid.

14-88: a) Using the constancy of angular momentum, the product of the radius and speed is constant, so the speed at the rim is about $(200 \text{ km/h})\left(\frac{30}{350}\right) = 17 \text{ km/h}$. b) The pressure is lower at the eye, by an amount

$$\Delta p = \frac{1}{2}(1.2 \text{ kg/m}^3)\left((200 \text{ km/h})^2 - (17 \text{ km/h})^2\right)\left(\frac{1 \text{ m/s}}{3.6 \text{ km/h}}\right)^2 = 1.8 \times 10^3 \text{ Pa}.$$

c) $\frac{v^2}{2g} = 160$ m to two figures. d) The pressure at higher altitudes is even lower.

14-89: The speed of efflux at point D is $\sqrt{2gh_1}$, and so is $\sqrt{8gh_1}$ at C. The gauge pressure at C is then $\rho g h_1 - 4\rho g h_1 = -3\rho g h_1$, and this is the gauge pressure at E. The height of the fluid in the column is $3h_1$.

14-90: a) $v = \frac{dV/dt}{A}$, so the speeds are

$$\frac{6.00 \times 10^{-3} \text{ m}^3/\text{s}}{10.0 \times 10^{-4} \text{ m}^2} = 6.00 \text{ m/s} \quad \text{and} \quad \frac{6.00 \times 10^{-3} \text{ m}^3/\text{s}}{40.0 \times 10^{-4} \text{ m}^2} = 1.50 \text{ m/s}.$$

b) $\Delta p = \frac{1}{2}\rho\left(v_1^2 - v_2^2\right) = 1.688 \times 10^4$ Pa, or 1.69×10^4 Pa to three figures.

c) $\Delta h = \frac{\Delta p}{\rho_{\text{Hg}} g} = \frac{(1.688 \times 10^4 \text{ Pa})}{(13.6 \times 10^3 \text{ kg/m}^3)(9.80 \text{ m/s}^2)} = 12.7$ cm.

14-91: a) The speed of the liquid as a function of the distance y that it has fallen is $v = \sqrt{v_0^2 + 2gy}$, and the cross-section area of the flow is inversely proportional to this speed. The radius is then inversely proportional to the square root of the speed, and if the radius of the pipe is r_0, the radius r of the stream a distance y below the pipe is

$$r = \frac{r_0\sqrt{v_0}}{(v_0^2 + 2gy)^{1/4}} = r_0\left(1 + \frac{2gy}{v_0^2}\right)^{-1/4}.$$

b) From the result of part (a), the height is found from $(1 + 2gy/v_0^2)^{1/4} = 2$, or

$$y = \frac{15v_0^2}{2g} = \frac{15(1.2 \text{ m/s})^2}{2(9.80 \text{ m/s}^2)} = 1.10 \text{ m}.$$

14-92: a) The net force on the ball is the sum of the gravitational , buoyant and viscous forces; from $F = ma$, $mg - B - F_d = \frac{mg}{2}$, so $F_d = \frac{mg}{2} - B$. Substitution of F_d from Eq. (14-27) and solving for v_t in terms of the densities gives the expression for v_t as found in Example 14-13, but with ρ replaced by $\frac{\rho}{2}$; specifically,

$$v_t = \frac{2}{9}\frac{r^2 g}{\eta}\left(\frac{\rho}{2} - \rho'\right)$$

$$= \frac{2}{9}\frac{(2.50 \times 10^{-3} \text{ m})^2(9.80 \text{ m/s}^2)}{(0.830 \text{ N·s/m}^2)}\left(4.3 \times 10^3 \text{ kg/m}^3 - 1.26 \times 10^3 \text{ kg/m}^3\right)$$

$$= 4.99 \times 10^{-2} \text{ m/s}.$$

b) Repeating the calculation without the factor of $\frac{1}{2}$ multiplying ρ gives $v_t = 0.120$ m/s.

14-93: a) See Example 14-13; in this case $\rho = \rho_{air}$ and $\rho' = \rho_{liq}$, and

$$v_t = \frac{2(1.00 \times 10^{-3}\text{ m})^2(9.80\text{ m/s}^2)}{9(0.150\text{ N·s/m}^2)}\left(1.2\text{ kg/m}^3 - 900\text{ kg/m}^3\right) = -1.30\text{ cm/s},$$

with the minus sign indicating an upward velocity (v_t was considered to be positive for a downward velocity in the example). b) Repeating the calculation of part (a) with $\rho' = 1000$ kg/m^3 and $\eta = 1.005 \times 10^{-3}$ N·s/m^2 gives $v_t = -2.16$ m/s.

14-94: a) Solving Eq. (14-29) for $p_1 - p_2 = \Delta p$ and setting the change in height equal to 0,

$$\Delta p = \rho g h + \frac{dV}{dt}\frac{8\eta L}{\pi R^4}$$

$$= (0.0600\text{ m}^3/\text{s})\left(\frac{8(0.300\text{ N·s/m}^2)(1.50 \times 10^3\text{ m})}{\pi(0.055\text{ m})^4}\right)$$

$$= 7.51 \times 10^6\text{ Pa} = 74.2\text{ atm}.$$

b) $P = \Delta p\frac{dV}{dt} = (7.51 \times 10^6\text{ Pa})(0.0600\text{ m}^3/\text{s}) = 4.51 \times 10^5$ W. The work done is $\Delta p dV$.

14-95: a)
$$\frac{dV}{dt} = v_h A_h = \sqrt{2gy}A_h$$
$$= \sqrt{2(9.80\text{ m/s}^2)(0.600\text{ m})}\left(0.20 \times 10^{-4}\text{ m}^2\right)$$
$$= 6.86 \times 10^{-5}\text{ m}^3/\text{s}.$$

b) The speed is inversely proportional to the area; $v_h = \sqrt{2gy} = 3.43$ m/s $= v_g$, and the speed at c and d is one-fifth of this, or 0.69 m/s, and the speed at e and f is $\frac{1}{2}v_h = 1.72$ m/s.

c) The gauge pressure at the bottom of each pipe is $\rho g x$, where x is the height of the column. Using this in Eq. (14-22), with $y_1 = 0$, $y_2 = y$, $v_2 = 0$ and v_1 the speed found in part (b), $y - x = \frac{v^2}{2g}$. Rather than substituting, note that v is related to y by

$$v^2 = v_h^2\left(\frac{A_h}{A}\right)^2 = 2gy\left(\frac{A_h}{A}\right)^2,$$

and so $y - x = y\left(\frac{A_h}{A}\right)^2$, or $x = y\left(1 - \left(\frac{A_h}{A}\right)^2\right)$. The heights of the columns are then .96y, .75y and zero, or 0.576 m, 0.450 m and zero.

d) & e) Note that for a circular cross-section, Eq. (14-26) may be written

$$\Delta p = \frac{dV}{dt}\frac{8\pi\eta L}{A^2}.$$

The difference between levels Δx is related to the difference in pressure by $\Delta x = \frac{\Delta p}{\rho g}$, so

$$\Delta x = \frac{dV}{dt}\frac{8\eta\pi L}{\rho g}\frac{1}{A^2} = (2.638 \times 10^{-10} \text{ m}^3)\frac{1}{A^2}$$

The flow rate $\frac{dV}{dt}$ is that found in part (a), and subsitution of numerical values gives d) 0.0264 m and e) 0.165 m.

f) In Eq. (14-25), the speed on the axis, $v = v_{max}$ at $r = 0$, may be substituted into Eq. (14-26) to obtain

$$\frac{dV}{dt} = \frac{\pi R^2 v_{max}}{2} = \frac{A v_{max}}{2},$$

and so the speeds will be twice those found in part (a), or 6.86 m/s, 3.44 m/s and 1.37 m/s.

14-96: a) The volume V of the rock is

$$V = \frac{B}{\rho_{water}g} = \frac{w - T}{\rho_{water}g} = \frac{((3.00 \text{ kg})(9.80 \text{ m/s}^2) - 21.0 \text{ N})}{(1.00 \times 10^3 \text{ kg/m}^3)(9.80 \text{ m/s}^2)} = 8.57 \times 10^{-4} \text{ m}^3.$$

In the accelerated frames, all of the quantities that depend on g (weights, buoyant forces, gauge pressures and hence tensions) may be replaced by $g' = g + a$, with the positive direction taken upward. Thus, the tension is $T = mg' - B' = (m - \rho V)g' = T_0\frac{g'}{g}$, where $T_0 = 21.0$ N.

b) $g' = g + a$; for $a = 2.50$ m/s^2, $T = (21.0 \text{ N})\frac{9.80 + 2.50}{9.80} = 26.4$ N.

c) For $a = -2.50$ m/s^2, $T = (21.0 \text{ N})\frac{9.80 - 2.50}{9.80} = 15.6$ N.

d) If $a = -g$, $g' = 0$ and $T = 0$.

14-97: a) The tension in the cord plus the weight must be equal to the buoyant force, so

$$T = Vg(\rho_{water} - \rho_{foam})$$
$$= (1/2)(0.20 \text{ m})^2(0.50 \text{ m})(9.80 \text{ m/s}^2)(1000 \text{ kg/m}^3 - 180 \text{ kg/m}^3)$$
$$= 80.4 \text{ N}.$$

b) The depth of the bottom of the styrofoam is not given; let this depth be h_0. Denote the length of the piece of foam by L and the length of the two sides by l. The pressure force on the bottom of the foam is then $(p_0 + \rho g h_0)L(\sqrt{2}l)$ and is directed up. The pressure on each side is not constant; the force can be found by integrating, or using the result of Problem 14-48 or Problem 14-50. Although these problems found forces on vertical surfaces, the result that the force is the product of the average pressure and the area is valid. The average pressure is $p_0 + \rho g(h_0 - (l/(2\sqrt{2})))$, and the force on one side has magnitude

$$(p_0 + \rho g(h_0 - l/(2\sqrt{2})))Ll$$

and is directed perpendicular to the side, at an angle of 45.0° from the vertical. The force on the other side has the same magnitude, but has a horizontal component that is opposite that of the other side. The horizontal component of the net buoyant force is zero, and the vertical component is

$$B = (p_0 + \rho g h_0) L l \sqrt{2} - 2(\cos 45.0°)(p_0 + \rho g(h_0 - l/(2\sqrt{2}))) L l = \rho g \frac{L l^2}{2},$$

the weight of the water displaced.

14-98: When the level of the water is a height y above the opening, the efflux speed is $\sqrt{2gy}$, and $\frac{dV}{dt} = \pi(d/2)^2\sqrt{2gy}$. As the tank drains, the height decreases, and

$$\frac{dy}{dt} = -\frac{dV/dt}{A} = -\frac{\pi(d/2)^2\sqrt{2gy}}{\pi(D/2)^2} = -\left(\frac{d}{D}\right)^2\sqrt{2gy}.$$

This is a separable differential equation, and the time T to drain the tank is found from

$$\frac{dy}{\sqrt{y}} = -\left(\frac{d}{D}\right)^2\sqrt{2g}\,dt,$$

which integrates to

$$[2\sqrt{y}]_H^0 = -\left(\frac{d}{D}\right)^2\sqrt{2g}T,$$

or

$$T = \left(\frac{D}{d}\right)^2\frac{2\sqrt{H}}{\sqrt{2g}} = \left(\frac{D}{d}\right)^2\sqrt{\frac{2H}{g}}.$$

14-99: a) The fact that the water first moves upwards before leaving the siphon does not change the efflux speed, $\sqrt{2gh}$. b) Water will not flow if the absolute (not gauge) pressure would be negative. The hose is open to the atmosphere at the bottom, so the pressure at the top of the siphon is $p_a - \rho g(H+h)$, where the assumption that the cross-section area is constant has been used to equate the speed of the liquid at the top and bottom. Setting $p = 0$ and solving for H gives $H = (p_a/\rho g) - h$.

14-100: Any bubbles will cause inaccuracies. At the bubble, the pressure at the surfaces of the water will be the same, but the levels need not be the same. The use of a hose as a level assumes that pressure is the same at all point that are at the same level, an assumption that is invalidated by the bubble.

Chapter 15   Temperature and Heat

15-1:   From Eq. (15-1), a) $(9/5)(-62.8) + 32 = -81.0°F$. b) $(9/5)(56.7) + 32 = 134.1°F$.
c) $(9/5)(31.1) + 32 = 88.0°F$.

15-2:   From Eq. (15-2), a) $(5/9)(41.0 - 32) = 5.0°C$. b) $(5/9)(107 - 32) = 41.7°C$.
c) $(5/9)(-18 - 32) = -27.8°C$.

15-3:   Setting $T_C = T_F = T$ in either of Eq. (15-1) or Eq. (15-2) gives $9T_C = 5T_F - 160$,
which is solved for $T = -40.0°C = -40.0°F$.

15-4:   a) $(5/9)(45.0 - (-4.0)) = 27.2°C$. b) $(5/9)(-56.0 - 44) = -55.6°C$.

15-5:   a) From Eq. (15-1), $(9/5)(40.2) + 32 = 104.4°F$, which is cause for worry.
  b) $(9/5)(12) + 32 = 53.6°F$, or $54°F$ to two figures.

15-6: $(9/5)(11.8) = 21.2F°$

15-7:   Combining Eq. (15-2) and Eq. (15-3),

$$T_K = \frac{5}{9}(T_F - 32°) + 273.15,$$

and subsitution of the given Fahrenheit temperatures gives a) 216.5 K, b) 325.9 K,
c) 205.4 K.

15-8:   (In these calculations, extra figures were kept in the intermediate calculations to
arrive at the numerical results.) a) $T_C = 400 - 273.15 = 127°C$, $T_F = (9/5)(126.85) + 32 = 260°F$. b) $T_C = 95 - 273.15 = -178°C$, $T_F = (9/5)(-178.15) + 32 = -289°F$.
c) $T_C = 1.55 \times 10^7 - 273.15 = 1.55 \times 10^7 °C$, $T_F = (9/5)(1.55 \times 10^7) + 32 = 2.79 \times 10^7 °F$.

15-9:   From Eq. (15-3), $T_K = (-245.92°C) + 273.15 = 27.23$ K.

15-10:   From Eq. (15-4), $(7.476)(273.16 \text{ K}) = 2042.14$ K $- 273.15 = 1769°C$.

15-11:   From Eq. (15-4), $(325.0 \text{ mm}) \left(\frac{373.15 \text{ K}}{273.16 \text{ K}}\right) = 444$ mm.

15-12:   On the Kelvin scale, the triple point is 273.16 K, so $°R = (9/5)\,273.15$ K $= 491.69°R$. One could also look at Figure 15-5 and note that the Fahrenheit scale extends
from $-460°F$ to $+32°F$ and conclude that the triple point is about $492°R$.

15-13:   From the point-slope formula for a straight line (or linear regression, which, while
perhaps not appropriate, may be convenient for some calculators),

$$(0.01°C) - (100.0°C)\frac{4.80 \times 10^4 \text{ Pa}}{6.50 \times 10^4 \text{ Pa} - 4.80 \times 10^4 \text{ Pa}} = -282.33°C,$$

which is $-282°C$ to three figures.

  b) Equation (15-4) was not obeyed precisely. If it were, the pressure at the triple
point would be $P = (273.16)(\frac{6.50 \times 10^4 \text{ Pa}}{373.15}) = 4.76 \times 10^4$ Pa.

15-14: $\Delta T = (\Delta L)/(\alpha L_0) = (25 \times 10^{-2} \text{ m})/((2.4 \times 10^{-5} \text{ (C°)}^{-1})(62.1 \text{ m})) = 168$ C°, so
the temperature is $183°C$.

15-15: $\alpha L_0 \Delta T = (1.2 \times 10^{-5} \text{ (C°)}^{-1})(1410 \text{ m})(18.0°C - (-5.0)°C) = +0.39$ m.

15-16: $d + \Delta d = d(1 + \alpha \Delta T)$

$$= (0.4500 \text{ cm})\left(1 + (2.4 \times 10^{-5} \text{ (C}^\circ)^{-1})(23.0^\circ\text{C} - (-78.0^\circ\text{C}))\right)$$

$$= 0.4511 \text{ cm} = 4.511 \text{ mm}.$$

15-17: a) $\alpha D_0 \Delta T = (2.6 \times 10^{-5} \text{ (C}^\circ)^{-1})(1.90 \text{ cm})(28.0^\circ\text{C}) = 1.4 \times 10^{-3}$ cm, so the diameter is 1.901 cm (unchanged to two figures). b) $\alpha D_0 \Delta T = -3.6 \times 10^{-3}$ cm, so the diameter is 1.896 cm (unchanged to two figures).

15-18: $\alpha \Delta T = (2.0 \times 10^{-5} \text{ (C}^\circ)^{-1})(5.00^\circ\text{C} - 19.5^\circ\text{C}) = -2.9 \times 10^{-4}$.

15-19: $\alpha = (\Delta L)/(L_0 \Delta T) = (2.3 \times 10^{-4} \text{ m})/((40.125 \times 10^{-2} \text{ m})(25.0 \text{ C}^\circ)) = 2.3 \times 10^{-5} \text{ (C}^\circ)^{-1}$.

15-20: From Eq. (15-8), $\Delta T = \frac{\Delta V/V_0}{\beta} = \frac{1.50 \times 10^{-3}}{5.1 \times 10^{-5} \text{ K}^{-1}} = 29.4^\circ\text{C}$, so $T = 49.4^\circ\text{C}$.

15-21: $\beta V_0 \Delta T = (75 \times 10^{-5} \text{ (C}^\circ)^{-1})(1700 \text{ L})(-9.0^\circ\text{C}) = -11$ L, so there is 11 L of air.

15-22: The coefficient can be found from the slope of the curve at 9°C.

$$\beta = \frac{1}{V_0}\frac{\Delta V}{\Delta T} = \frac{1}{1.00025 \text{ cm}^3}\frac{(1.0003 \text{ cm}^3 - 1.0000 \text{ cm}^3)}{(10^\circ\text{C} - 6^\circ\text{C})}, \beta = .000075/^\circ\text{C} = 7.5 \times 10^{-5}/^\circ\text{C}.$$

15-23: The amount of mercury that overflows is the difference between the volume change of the mercury and that of the glass;

$$\beta_{\text{glass}} = 18.0 \times 10^{-5} \text{ K}^{-1} - \frac{(8.95 \text{ cm}^3)}{(1000 \text{ cm}^3)(55.0^\circ\text{C})} = 1.7 \times 10^{-5} \text{ (C}^\circ)^{-1}.$$

15-24: a) $A = L^2$, $\Delta A = 2L\Delta L = 2\frac{\Delta L}{L}L^2 = 2\frac{\Delta L}{L}A_0$. But $\frac{\Delta L}{L} = \alpha \Delta T$, and so $\Delta A = 2\alpha \Delta T A_0 = (2\alpha)A_0 \Delta T$. b) $\Delta A = (2\alpha)A_o \Delta T = (2)(2.4 \times 10^{-5}(\text{C}^\circ)^{-1})(\pi \times (.275 \text{ m})^2)(12.5^\circ\text{C}) = 1.4 \times 10^{-4} \text{ m}^2$.

15-25: a) $A_0 = \frac{\pi D^2}{4} = \frac{\pi}{4}(1.350 \text{ cm})^2 = 1.431 \text{ cm}^2$.

b) $A = A_0(1 + 2\alpha \Delta T) = (1.431 \text{ cm}^2)(1 + (2)(1.20 \times 10^{-5}{}^\circ\text{C})(150^\circ\text{C})) = 1.437 \text{ cm}^2$.

15-26: From Eq. (15-12),

$$F = -Y\alpha \Delta T A$$

$$= -(0.9 \times 10^{11} \text{ Pa})(2.0 \times 10^{-5} \text{ (C}^\circ)^{-1})(-110^\circ\text{C})(2.01 \times 10^{-4} \text{ m}^2)$$

$$= 4.0 \times 10^4 \text{ N}.$$

15-27: a) $\alpha = (\Delta L)/(L_0 \Delta T) = (1.9 \times 10^{-2} \text{ m})/((1.50 \text{ m})(400 \text{ C}^\circ)) = 3.2 \times 10^{-5} \text{ (C}^\circ)^{-1}$.

b) $Y\alpha \Delta T = Y\Delta L/L_0 = (2.0 \times 10^{11} \text{ Pa})(1.9 \times 10^{-2} \text{ m})/(1.50 \text{ m}) = 2.5 \times 10^9$ Pa.

15-28: a) $\Delta L = \alpha \Delta T L = (1.2 \times 10^{-5} \text{ K}^{-1})(35.0 \text{ K})(12.0 \text{ m}) = 5.0 \times 10^{-3}$ m.

b) Using absolute values in Eq. (15-12),

$$\frac{F}{A} = Y\alpha \Delta T = (2.0 \times 10^{11} \text{ Pa})(1.2 \times 10^{-5} \text{ K}^{-1})(35.0 \text{ K}) = 8.4 \times 10^7 \text{ Pa}.$$

15-29: a) $(37^\circ\text{C} - (-20^\circ\text{C}))(0.50 \text{ L})(1.3 \times 10^{-3} \text{ kg/L})(1020 \text{ J/kg·K}) = 38$ J.

b) There will be 1200 breaths per hour, so the heat lost is $(1200)(38 \text{ J}) = 4.5 \times 10^4$ J.

15-30: $$t = \frac{Q}{P} = \frac{mc\Delta T}{P} = \frac{(70 \text{ kg})(3480 \text{ J/kg·K})(7 \text{ C}°)}{(1200 \text{ W})} = 1.4 \times 10^3 \text{ s},$$

about 24 min.

15-31: Using $Q = mgh$ in Eq. (15-13) and solving for ΔT gives

$$\Delta T = \frac{gh}{c} = \frac{(9.80 \text{ m/s}^2)(225 \text{ m})}{(4190 \text{ J/kg·K})} = 0.53 \text{ C}°.$$

15-32: a) The work done by friction is the loss of mechanical energy,

$$mgh + \frac{1}{2}m\left(v_1^2 - v_2^2\right) = (35.0 \text{ kg})\left((9.80 \text{ m/s}^2)(8.00 \text{ m})\sin 36.9° - \frac{1}{2}(2.50 \text{ m/s})^2\right)$$
$$= 1.54 \times 10^3 \text{ J}.$$

b) Using the result of part (a) for Q in Eq. (15-13) gives

$$\Delta T = (1.54 \times 10^3 \text{ J})/((35.0 \text{ kg})(3650 \text{ J/kg·K})) = 1.21 \times 10^{-2} \text{ C}°.$$

15-33: $(210°\text{C} - 20°\text{C})((1.60 \text{ kg})(910 \text{ J/kg·K}) + (0.30 \text{ kg})(470 \text{ J/kg·K})) = 3.03 \times 10^5 \text{ J}.$

15-34: Assuming $Q = (0.60) \times 10 \times K$,

$$\Delta T = (0.60) \times 10 \times \frac{K}{mc} = 6\frac{\frac{1}{2}MV^2}{mc} = \frac{(6)\frac{1}{2}(1.80 \text{ kg})(7.80 \text{ m/s})^2}{(8.00 \times 10^{-3} \text{ kg})(910 \text{ J/kg·K})} = 45.1 \text{ C}°.$$

15-35: $(85.0°\text{C} - 20.0°\text{C})((1.50 \text{ kg})(910 \text{ J/kg·K}) + (1.80 \text{ kg})(4190 \text{ J/kg·K}))$
$$= 5.79 \times 10^5 \text{ J}.$$

15-36: a) $Q = mc\Delta T = (0.320 \text{ kg})(4190 \text{ J/kg·K})(60.0 \text{ K}) = 8.05 \times 10^4 \text{ J}.$
  b) $t = \frac{Q}{P} = \frac{8.05 \times 10^4 \text{ J}}{200 \text{ W}} = 402 \text{ s}.$

15-37: a) $c = \dfrac{Q}{m\Delta T} = \dfrac{(120 \text{ s})(65.0 \text{ W})}{(0.780 \text{ kg})(22.54°\text{C} - 18.55°\text{C})} = 2.51 \times 10^3 \text{ J/kg·K}.$

b) An overestimate; the heat Q is in reality less than the power times the time interval.

15-38: a) Let the man be designated by the subscript m and the "'water" by w, and T is the final equilibrium temperature.

$$-m_m C_m \Delta T_m = m_w C_w \Delta T_w$$
$$-m_m C_m (T - T_m) = m_w C_w (T - T_w)$$
$$m_m C_m (T_m - T) = m_w C_w (T - T_w)$$

Or solving for T, $T = \frac{m_m C_m T_m + m_w C_w T_w}{m_m C_m + m_w C_w}$. Inserting numbers, and realizing we can change K to °C, and the mass of water is .355 kg, we get

$$T = \frac{(70.0 \text{ kg})(3480 \text{ J/kg · K})(37.0°\text{C}) + (0.355 \text{ kg})(4190 \text{ J/kg ·°C})(12.0°\text{C})}{(70.0 \text{ kg})(3480 \text{ J/kg ·°C}) + (0.355 \text{ kg})(4190 \text{ J/kg ·°C})}.$$

Thus, $T = 36.85°\text{C}$.

b) It is possible a sensitive digital thermometer could measure this change since they can read to .1°C. It is best to refrain from drinking cold fluids prior to orally measuring a body temperature due to cooling of the mouth.

15-39: The rate of heat loss is $\Delta Q/\Delta t$. $\left(\frac{\Delta Q}{\Delta t}\right) = \frac{mC\Delta T}{\Delta t}$, or $\Delta t = \frac{mC\Delta t}{\left(\frac{\Delta Q}{\Delta t}\right)}$. Inserting numbers, $\Delta t = \frac{(70.355 \text{ kg})(3480 \text{ J/kg·°C})(.15°\text{C})}{7 \times 10^6 \text{ J/day}} = .005$ d, or $\Delta t = 7.6$ minutes. This may account for mothers taking the temperature of a sick child several minutes *after* the child has something to drink.

15-40:
$$Q = m\left(c\Delta T + L_f\right)$$
$$= (0.350 \text{ kg})\left((4190 \text{ J/kg·K})(18.0 \text{ K}) + 334 \times 10^3 \text{ J/kg}\right)$$
$$= 1.43 \times 10^5 \text{ J} = 34.2 \text{ kcal} = 136 \text{ Btu}.$$

15-41:
$$Q = m\left(c_{\text{ice}}\Delta T_{\text{ice}} + L_f + c_{\text{water}}\Delta T_{\text{water}} + L_v\right)$$
$$= (12.0 \times 10^{-3} \text{ kg})\left(\begin{array}{c}(2100 \text{ J/kg·K})(10.0 \text{ C}°) + 334 \times 10^3 \text{ J/kg} \\ +(100 \text{ C}°)(4190 \text{ J/kg·K}) + 2256 \times 10^3 \text{ J/kg}\end{array}\right)$$
$$= 3.64 \times 10^4 \text{ J} = 8.69 \text{ kcal} = 34.5 \text{ Btu}.$$

15-42: a)
$$t = \frac{Q}{P} = \frac{mc\Delta T}{P} = \frac{(0.550 \text{ kg})(2100 \text{ J/kg·K})(15.0 \text{ K})}{(800 \text{ J/min})} = 21.7 \text{ min}.$$

b) $\frac{mL_f}{P} = \frac{(0.550 \text{ kg})(334 \times 10^3 \text{ J/kg})}{(800 \text{ J/min})} = 230$ min, so the time until the ice has melted is 21.7 min + 230 min = 252 min.

c)

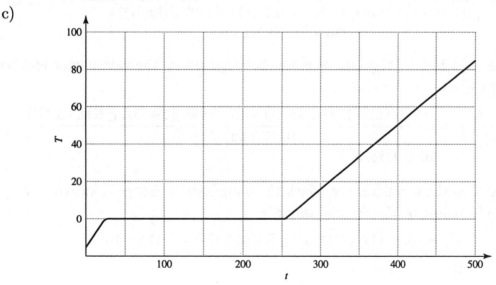

15-43:
$$\frac{((4000 \text{ lb})/(2.205 \text{ lb/kg}))(334 \times 10^3 \text{ J/kg})}{(86,400 \text{ s})} = 7.01 \text{ kW} = 2.40 \times 10^4 \text{ Btu/hr}.$$

15-44: a) $m(c\Delta T + L_v) = (25.0 \times 10^{-3} \text{ kg})((4190 \text{ J/kg·K})(66.0 \text{ K}) + 2256 \times 10^3 \text{ J/kg}) = 6.33 \times 10^4$ J. b) $mc\Delta T = (25.0 \times 10^{-3} \text{ kg})(4190 \text{ J/kg·K})(66.0 \text{ K}) = 6.91 \times 10^3$ J. c) Steam burns are far more severe than hot-water burns.

15-45: With $Q = m(c\Delta T + L_f)$ and $K = (1/2)mv^2$, setting $Q = K$ and solving for v gives

$$v = \sqrt{2\left((130 \text{ J/kg·K})(302.3 \text{ C°}) + 24.5 \times 10^3 \text{ J/kg}\right)} = 357 \text{ m/s.}$$

15-46: a) $\quad m_{\text{sweat}} = \dfrac{Mc\Delta T}{L_v} = \dfrac{(70.0 \text{ kg})(3480 \text{ J/kg·K})(1.00 \text{ K})}{(2.42 \times 10^6 \text{ J/kg})} = 101 \text{ g.}$

b) This much water has a volume of 101 cm³, about a third of a can of soda.

15-47: The mass of water that the camel saves is

$$\frac{Mc\Delta T}{L_v} = \frac{(400 \text{ kg})(3480 \text{ J/kg·K})(6.0 \text{ K})}{(2.42 \times 10^6 \text{ J/kg})} = 3.45 \text{ kg,}$$

which is a volume of 3.45 L.

15-48: For this case, the algebra reduces to

$$T = \frac{\left(\begin{array}{c}((200)(3.00 \times 10^{-3} \text{ kg}))(390 \text{ J/kg·K})(100.0 \text{ C°}) \\ +(0.240 \text{ kg})(4190 \text{ J/kg·K})(20.0 \text{ C°})\end{array}\right)}{\left(\begin{array}{c}((200)(3.00 \times 10^{-3} \text{ kg}))(390 \text{ J/kg·K}) \\ +(0.240 \text{ kg})(4190 \text{ J/kg·K})\end{array}\right)} = 35.1°C.$$

15-49: The algebra reduces to

$$T = \frac{\left(\begin{array}{c}((0.500 \text{ kg})(390 \text{ J/kg·K}) + (0.170 \text{ kg})(4190 \text{ J/kg·K}))(20.0°C) \\ +(0.250 \text{ kg})(470 \text{ J/kg·K})(85.0°C)\end{array}\right)}{\left(\begin{array}{c}((0.500 \text{ kg})(390 \text{ J/kg·K}) + (0.170 \text{ kg})(4190 \text{ J/kg·K})) \\ +(0.250 \text{ kg})(470 \text{ J/kg·K})\end{array}\right)} = 27.5°C.$$

15-50: The heat lost by the sample is the heat gained by the calorimeter and water, and the heat capacity of the sample is

$$c = \frac{Q}{m\Delta T} = \frac{((0.200 \text{ kg})(4190 \text{ J/kg·K}) + (0.150 \text{ kg})(390 \text{ J/kg·K}))(7.1 \text{ C°})}{(0.0850 \text{ kg})(73.9 \text{ C°})}$$

$$= 1010 \text{ J/kg·K,}$$

or 1000 J/kg·K to the two figures to which the temperature change is known.

15-51: The heat lost by the original water is

$$-Q = (0.250 \text{ kg})(4190 \text{ J/kg·K})(45.0 \text{ C°}) = 4.174 \times 10^4 \text{ J,}$$

and the mass of the ice needed is

$$m_{\text{ice}} = \frac{-Q}{c_{\text{ice}}\Delta T_{\text{ice}} + L_f + c_{\text{water}}\Delta T_{\text{water}}}$$

$$= \frac{(4.174 \times 10^4) \text{ J}}{(2100 \text{ J/kg·K})(20.0 \text{ C°}) + (334 \times 10^3 \text{ J/kg}) + (4190 \text{ J/kg·K})(30.0 \text{ C°})}$$

$$= 9.40 \times 10^{-2} \text{ kg} = 94.0 \text{ g.}$$

15-52: The heat lost by the sample (and vial) melts a mass m, where

$$m = \frac{Q}{L_f} = \frac{((16.0 \text{ g})(2250 \text{ J/kg·K}) + (6.0 \text{ g})(2800 \text{ J/kg·K}))(19.5 \text{ K})}{(334 \times 10^3 \text{ J/kg})} = 3.08 \text{ g}.$$

Since this is less than the mass of ice, not all of the ice melts, and the sample is indeed cooled to 0°C. Note that conversion from grams to kilograms was not necessary.

15-53:
$$\frac{(4.00 \text{ kg})(234 \text{ J/kg·K})(750 \text{ C°})}{(334 \times 10^3 \text{ J/kg})} = 2.10 \text{ kg}.$$

15-54: Equating the heat lost by the lead to the heat gained by the calorimeter (including the water-ice mixture),

$$m_{Pb}c_{Pb}(200°C - T) = (m_w + m_{ice})\,c_w T + m_{Cu}c_{Cu}T + m_{ice}L_f.$$

Solving for the final temperature T and using numerical values,

$$T = \frac{\left(\begin{array}{c}(0.750 \text{ kg})(130 \text{ J/kg·K})(255 \text{ C°}) \\ -(0.018 \text{ kg})(334 \times 10^3 \text{ J/kg})\end{array}\right)}{\left(\begin{array}{c}(0.750 \text{ kg})(130 \text{ J/kg·K}) \\ +(0.178 \text{ kg})(4190 \text{ J/kg·K}) \\ +(0.100 \text{ kg})(390 \text{ J/kg·K})\end{array}\right)} = 21.4°C.$$

(The fact that a positive Celsius temperature was obtained indicates that all of the ice does indeed melt.)

15-55: The steam both condenses and cools, and the ice melts and heats up along with the original water; the mass of steam needed is

$$m_{steam} = \frac{(0.450 \text{ kg})(334 \times 10^3 \text{ J/kg}) + (2.85 \text{ kg})(4190 \text{ J/kg·K})(28.0 \text{ C°})}{2256 \times 10^3 \text{ J/kg} + (4190 \text{ J/kg·K})(72.0 \text{ C°})}$$
$$= 0.190 \text{ kg}.$$

15-56: The SI units of H and $\frac{dQ}{dt}$ are both watts, the units of area are m², temperature difference is in K, length in meters, so the SI units for thermal conductivity are

$$\frac{[\text{W}][\text{m}]}{[\text{m}^2][\text{K}]} = \frac{\text{W}}{\text{m·K}}.$$

15-57: a) $\frac{100 \text{ K}}{0.450 \text{ m}} = 222 \text{ K/m}.$ b) $(385 \text{ W/m·K})(1.25 \times 10^{-4} \text{ m}^2)(400 \text{ K/m}) = 10.7 \text{ W}.$
c) $100.0°C - (222 \text{ K/m})(12.00 \times 10^{-2} \text{ m}) = 73.3°C.$

15-58: Using the chain rule, $H = \frac{dQ}{dt} = L_f \frac{dm}{dt}$ and solving Eq. (15-21) for k,

$$k = L_f \frac{dm}{dt} \frac{L}{A\Delta T}$$
$$= (334 \times 10^3 \text{ J/kg}) \frac{(8.50 \times 10^{-3} \text{ kg})}{(600 \text{ s})} \frac{(60.0 \times 10^{-2} \text{ m})}{(1.250 \times 10^{-4} \text{ m}^2)(100 \text{ K})}$$
$$= 227 \text{ W/m·K}.$$

15-59: (Although it may be easier for some to solve for the heat flow per unit area, part (b), first, the method presented here follows the order in the text.)   a) See Example 15-14; as in that example, the area may be divided out, and solving for temperature T at the boundary,

$$T = \frac{(k_{\text{foam}}/L_{\text{foam}})T_{\text{in}} + (k_{\text{wood}}/L_{\text{wood}})T_{\text{out}}}{N(k_{\text{foam}}/L_{\text{foam}}) + (k_{\text{wood}}/L_{\text{wood}})}$$

$$\frac{((0.010 \text{ W/m·K})/(2.2 \text{ cm}))(19.0°C) + ((0.080 \text{ W/m·K})/(3.0 \text{ cm}))(-10.0°C)}{((0.010 \text{ W/m·K})/(2.2 \text{ cm})) + ((0.080 \text{ W/m·K})/(3.0 \text{ cm}))}$$

$$= -5.8°C.$$

Note that the conversion of the thicknesses to meters was not necessary.   b) Keeping extra figures for the result of part (a), and using that result in the temperature difference across either the wood or the foam gives

$$\frac{H_{\text{foam}}}{A} = \frac{H_{\text{wood}}}{A} = (0.010 \text{ W/m·K})\frac{(19.0°C - (-5.767°C))}{2.2 \times 10^{-2} \text{ m}}$$

$$= (0.080 \text{ W/m·K})\frac{(-5.767°C - (-10.0°C))}{3.0 \times 10^{-2} \text{ m}}$$

$$= 11 \text{ W/m}^2.$$

15-60:   a) From Eq. (15-21),

$$H = (0.040 \text{ W/m·K})(1.40 \text{ m}^2)\frac{(140 \text{ K})}{(4.0 \times 10^{-2} \text{ m})} = 196 \text{ W},$$

or 200 W to two figures.   b) The result of part (a) is the needed power input.

15-61:   From Eq. (15-23), the energy that flows in time Δt is

$$H\Delta t = \frac{A\Delta T}{R}\Delta t = \frac{(125 \text{ ft}^2)(34\text{F}°)}{(30 \text{ ft}^2 \cdot \text{F}° \cdot \text{h/Btu})}(5.0 \text{ h}) = 708 \text{ Btu} = 7.5 \times 10^5 \text{ J}.$$

15-62:   a) The heat current will be the same in both metals; since the length of the copper rod is known,

$$H = (385.0 \text{ W/m·K})(4.00 \times 10^{-4} \text{ m}^2)\frac{(35.0 \text{ K})}{(1.00 \text{ m})} = 5.39 \text{ W}.$$

b) The length of the steel rod may be found by using the above value of H in Eq. (15-21) and solving for L_2, or, since H and A are the same for the rods,

$$L_2 = L\frac{k_2}{k}\frac{\Delta T_2}{\Delta T} = (1.00 \text{ m})\frac{(50.2 \text{ W/m·K})}{(385.0 \text{ W/m·K})}\frac{(65.0 \text{ K})}{(35.0 \text{ K})} = 0.242 \text{ m}.$$

15-63: Using $H = L_v \frac{dm}{dt}$ (see Problem 15-58) in Eq. (15-21),

$$\Delta T = L_v \frac{dm}{dt} \frac{L}{kA}$$

$$= (2256 \times 10^3 \text{ J/kg}) \frac{(0.390 \text{ kg})}{(180 \text{ s})} \frac{(0.85 \times 10^{-2} \text{ m})}{(50.2 \text{ W/m·K})(0.150 \text{ m}^2)} = 5.5 \text{ C}°,$$

and the temperature of the bottom of the pot is $100°C + 6 \text{ C}° = 106°C$.

15-64: From Eq. (15-25), with $e = 1$,

a) $(5.67 \times 10^{-8} \text{ W/m}^2\text{·K}^4)(273 \text{ K})^4 = 315 \text{ W/m}^2$.

b) A factor of ten increase in temperature results in a factor of 10^4 increase in the output; $3.15 \times 10^6 \text{ W/m}^2$.

15-65: Repeating the calculation with $T_s = 273 \text{ K} + 5.0 \text{ C}° = 278 \text{ K}$ gives $H = 167 \text{ W}$.

15-66: The power input will be equal to H_{net} as given in Eq. (15-26);

$$P = Ae\sigma \left(T^4 - T_s^4 \right)$$

$$= (4\pi(1.50 \times 10^{-2} \text{ m})^2)(0.35)(5.67 \times 10^{-8} \text{ W/m}^2\text{·K}^4)((3000 \text{ K})^4 - (290 \text{ K})^4)$$

$$= 4.54 \times 10^3 \text{ W}.$$

15-67:
$$A = \frac{H}{e\sigma T^4} = \frac{150 \text{ W}}{(0.35)(5.67 \times 10^{-8} \text{ W/m}^2 \cdot \text{K}^4)(2450 \text{ K})^4} = 2.10 \text{ cm}^2$$

15-68: The radius is found from

$$R = \sqrt{\frac{A}{4\pi}} = \sqrt{\frac{H/(\sigma T^2)}{4\pi}} = \sqrt{\frac{H}{4\pi\sigma}} \frac{1}{T^2}.$$

Using the numerical values, the radius for parts (a) and (b) are

$$R_a = \sqrt{\frac{(2.7 \times 10^{32} \text{ W})}{4\pi(5.67 \times 10^{-8} \text{ W/m}^2 \cdot \text{K}^4)} \frac{1}{(11{,}000 \text{ K})^2}} = 1.61 \times 10^{11} \text{ m}$$

$$R_b = \sqrt{\frac{(2.10 \times 10^{23} \text{ W})}{4\pi(5.67 \times 10^{-8} \text{ W/m}^2 \cdot \text{K}^4)} \frac{1}{(10{,}000 \text{ K})^2}} = 5.43 \times 10^6 \text{ m}$$

c) The radius of Procyon B is comparable to that of the earth, and the radius of Rigel is comparable to the earth-sun distance.

15-69: $12 \text{ W}/20 \times 10^{-4} \text{ cm}^2 = 6.0 \times 10^3 \text{ W/m}^2$.

15-70: Insertion of the given values into Eq. (15-28) yields

$$18°C + (5.0 \text{ K/W})(12 \text{ W}) = 78°C.$$

15-71: Solving Eq. (15-28) for the ambient temperature,

$$T_{amb} = (120°C) - (4.5 \text{ K/W})(28 \text{ W}) = -6°C.$$

15-72: Solving Eq. (15-28) for $\Delta T = T_{\text{ic}} - T_{\text{amb}}$, $r_{\text{th}} = \frac{\Delta T}{P}$.

   a) $\frac{88 \text{ C}°}{22 \text{ W}} = 4.0$ K/W. b) $\frac{58 \text{ C}°}{22 \text{ W}} = 2.6$ K/W.

15-73: See Exercise 15-72: $\frac{80.0 \text{ C}°}{14 \text{ W}} = 5.7$ K/W.

15-74: All linear dimensions of the hoop are increased by the same factor of $\alpha \Delta T$, so the increase in the radius of the hoop would be

$$R\alpha\Delta T = (6.38 \times 10^6 \text{ m})(1.2 \times 10^{-5} \text{ K}^{-1})(0.5 \text{ K}) = 38 \text{ m}.$$

15-75: The tube is initially at temperature T_0, has sides of length L_0, volume V_0, density ρ_0, and coefficient of volume expansion β.

   a) When the temperature increases to $T_0 + \Delta T$, the volume changes by an amount ΔV, where $\Delta V = \beta V_0 \Delta T$. Then, $\rho = \frac{m}{V_0 + \Delta V}$, or eliminating ΔV, $\rho = \frac{m}{V_0 + \beta V_0 \Delta T}$. Divide the top and bottom by V_0 and substitute $\rho_0 = m/V_0$. Then $\rho = \frac{m/V_0}{V_0/V_0 + \beta V_0 \Delta T/V_0}$ or $\rho = \frac{\rho_0}{1 + \beta \Delta T}$. This can be rewritten as $\rho = \rho_0 (1 + \beta \Delta T)^{-1}$. Then using the expression $(1+x)^n \approx 1 + nx$, where $n = -1$, $\rho = \rho_0(1 - \beta\Delta T)$.

   b) The copper cube has sides of length 1.25 cm = .0125 m, and $\Delta T = 70.0°\text{C} - 20.0°\text{C} = 50.0°\text{C}$. $\Delta V = \beta V_0 \Delta T = (5.1 \times 10^{-5}/°\text{C})(.0125 \text{ m})^3(50.0°\text{C}) = 5 \times 10^{-9} \text{ m}^3$. Similarly, $\rho = 8.9 \times 10^3 \text{ kg/m}^3(1 - (5.1 \times 10^{-5}/°\text{C})(50.0°\text{C}))$, or $\rho = 8.877 \times 10^3 \text{ kg/m}^3$; extra significant figures have been kept. So $\Delta\rho = 23 \frac{\text{kg}}{\text{m}^3}$.

15-76: a) The change in height will be the difference between the changes in volume of the liquid and the glass, divided by the area. The liquid is free to expand along the column, but not across the diameter of the tube, so the increase in volume is reflected in the change in the length of the columns of liquid in the stem.

   b)
$$\Delta h = \frac{\Delta V_{\text{liquid}} - \Delta V_{\text{glass}}}{A} = \frac{V}{A}\left(\beta_{\text{liquid}} - \beta_{\text{glass}}\right)\Delta T$$

$$= \frac{(100 \times 10^{-6} \text{ m}^3)}{(50.0 \times 10^{-6} \text{ m}^2)}(8.00 \times 10^{-4} \text{ K}^{-1} - 2.00 \times 10^{-5} \text{ K}^{-1})(30.0 \text{ K})$$

$$= 4.68 \times 10^{-2} \text{ m}.$$

15-77: To save some intermediate calculation, let the third rod be made of fractions f_1 and f_2 of the original rods; then $f_1 + f_2 = 1$ and $f_1(0.0650) + f_2(0.0350) = 0.0580$. These two equations in f_1 and f_2 are solved for

$$f_1 = \frac{0.0580 - 0.0350}{0.0650 - 0.0350}, \quad f_2 = 1 - f_1,$$

and the lengths are $f_1(30.0 \text{ cm}) = 23.0$ cm and $f_2(30.0 \text{ cm}) = 7.00$ cm.

15-78: a) The lost volume, 2.6 L, is the difference between the expanded volume of the fuel and the tanks, and the maximum temperature difference is

$$\Delta T = \frac{\Delta V}{(\beta_{\text{fuel}} - \beta_{\text{Al}})V_0}$$

$$= \frac{(2.6 \times 10^{-3} \text{ m}^3)}{(9.5 \times 10^{-4} \text{ (C}°)^{-1} - 7.2 \times 10^{-5} \text{ (C}°)^{-1})(106.0 \times 10^{-3} \text{ m}^3)}$$

$$= 27.8 \text{ C}°,$$

or 28 C° to two figures; the maximum temperature was 32°C.   b) No fuel can spill if the tanks are filled just before takeoff.

15-79: a) The change in length is due to the tension and heating: $\frac{\Delta L}{L_0} = \frac{F}{AY} + \alpha\Delta T$. Solving for F/A, $\frac{F}{A} = Y\left(\frac{\Delta L}{L_0} - \alpha\Delta T\right)$.

b) The brass bar is given as "heavy" and the wires are given as "fine," so it may be assumed that the stress in the bar due to the fine wires does not affect the amount by which the bar expands due to the temperature increase. This means that in the equation preceding Eq. (15-12), ΔL is not zero, but is the amount $\alpha_{\text{brass}}L_0\Delta T$ that the brass expands, and so

$$\frac{F}{A} = Y_{\text{steel}}(\alpha_{\text{brass}} - \alpha_{\text{steel}})\Delta T$$

$$= (20 \times 10^{10} \text{ Pa})(2.0 \times 10^{-5} \text{ (C}°)^{-1} - 1.2 \times 10^{-5} \text{ (C}°)^{-1})(120°C)$$

$$= 1.92 \times 10^8 \text{ Pa}.$$

15-80: In deriving Eq. (15-12), it was assumed that $\Delta L = 0$; if this is not the case when there are both thermal and tensile stresses, Eq. (15-12) becomes

$$\Delta L = L_0\left(\alpha\Delta T + \frac{F}{AY}\right).$$

For the situation in this problem, there are two length changes which must sum to zero, and so Eq. (15-12) may be extended to two materials a and b in the form

$$L_{0a}\left(\alpha_a\Delta T + \frac{F}{AY_a}\right) + L_{0b}\left(\alpha_b\Delta T + \frac{F}{AY_b}\right) = 0.$$

Note that in the above, ΔT, F and A are the same for the two rods. Solving for the stress F/A,

$$\frac{F}{A} = -\frac{\alpha_a L_{0a} + \alpha_b L_{0b}}{((L_{0a}/Y_a) + (L_{0b}/Y_b))}\Delta T$$

$$= -\frac{(1.2 \times 10^{-5} \text{ (C}°)^{-1})(0.350 \text{ m}) + (2.4 \times 10^{-5} \text{ (C}°)^{-1})(0.250 \text{ m})}{((0.350 \text{ m})/(20 \times 10^{10} \text{ Pa}) + (0.250 \text{ m}/7 \times 10^{10} \text{ Pa}))}(60.0 \text{ C}°)$$

$$= -1.2 \times 10^8 \text{ Pa}$$

to two figures.

15-81: a) $\Delta T = \frac{\Delta R}{\alpha R_0} = \frac{(0.0020 \text{ in.})}{(1.2 \times 10^{-5} \text{ (C°)}^{-1})(2.5000 \text{ in.})} = 67 \text{ C°}$ to two figures, so the ring should be warmed to 87°C.   b) The difference in the radii was initially 0.0020 in., and this must be the difference between the amounts the radii have shrunk. Taking R_0 to be the same for both rings, the temperature must be lowered by an amount

$$\Delta T = \frac{\Delta R}{(\alpha_{\text{brass}} - \alpha_{\text{steel}})R_0}$$
$$= \frac{(0.0020 \text{ in.})}{(2.0 \times 10^{-5} \text{ (C°)}^{-1} - 1.2 \times 10^{-5} \text{ (C°)}^{-1})(2.50 \text{ in.})} = 100 \text{ C°}$$

to two figures, so the final temperature would be $-80°$C.

15-82: a) The change in volume due to the temperature increase is $\beta V \Delta T$, and the change in volume due to the pressure increase is $-\frac{V}{B}\Delta p$ (Eq. (11-13)). Setting the net change equal to zero, $\beta V \Delta T = V\frac{\Delta p}{B}$, or $\Delta p = B\beta\Delta V$. b) From the above, $\Delta p = (1.6 \times 10^{11} \text{ Pa})(3.0 \times 10^{-5} \text{ K}^{-1})(15.0 \text{ K}) = 8.64 \times 10^7 \text{ Pa}$.

15-83: As the liquid is compressed, its volume changes by an amount $\Delta V = -\Delta pkV_0$. When cooled, the difference between the decrease in volume of the liquid and the decrease in volume of the metal must be this change in volume, or $(\alpha_l - \alpha_m)V_0\Delta T = \Delta V$. Setting the expressions for ΔV equal and solving for ΔT gives

$$\Delta T = \frac{\Delta pk}{\alpha_m - \alpha_l} = \frac{(5.065 \times 10^6 \text{ Pa})(8.50 \times 10^{-10} \text{ Pa}^{-1})}{(3.90 \times 10^{-5} \text{ K}^{-1} - 4.8 \times 10^{-4} \text{ K}^{-1})} = -9.76 \text{ C°},$$

so the temperature is 20.2°C.

15-84: Equating the heat lost by the soda and mug to the heat gained by the ice and solving for the final temperature $T =$

$$\frac{\left(\begin{array}{c}((2.00 \text{ kg})(4190 \text{ J/kg·K}) + (0.257 \text{ kg})(910 \text{ J/kg·K}))(20.0 \text{ C°}) \\ -(0.120 \text{ kg})((2100 \text{ J/kg·K})(15.0 \text{ C°}) + 334 \times 10^3 \text{ J/kg})\end{array}\right)}{(2.00 \text{ kg})(4190 \text{ J/kg·K}) + (0.257 \text{ kg})(910 \text{ J/kg·K}) + (0.120 \text{ kg})(4190 \text{ J/kg·K})}$$

$= 14.1°$C. Note that the mass of the ice (0.120 kg) appears in the denominator of this expression multiplied by the heat capacity of water; after the ice melts, the mass of the melted ice must be raised further to T.

15-85: a) $\frac{K}{Q} = \frac{(1/2)mv^2}{cm\Delta T} = \frac{v^2}{2c\Delta T} = \frac{(7700 \text{ m/s})^2}{2(910 \text{ J/kg·K})(600 \text{ C°})} = 54.3.$

b) Unless the kinetic energy can be converted into forms other than the increased heat of the satellite, the satellite cannot return intact.

15-86: a) The capstan is doing work on the rope at a rate

$$P = \tau\omega = \Delta Fr\frac{2\pi}{T} = (520 \text{ N})(5.0 \times 10^{-2} \text{ m})\frac{2\pi}{(0.90 \text{ s})} = 182 \text{ W},$$

or 180 W to two figures. The net torque that the rope exerts on the capstan, and hence the net torque that the capstan exerts on the rope, is the difference between the forces of the ends times the radius. A larger number of turns might increase the force, but for given forces, the torque is independent of the number of turns.

b) $$\frac{dT}{dt} = \frac{dQ/dt}{mc} = \frac{P}{mc} = \frac{(182 \text{ W})}{(6.00 \text{ kg})(470 \text{ J/mol·K})} = 0.064 \text{ C}°/\text{s}.$$

15-87: a) Replacing m with nM and nMc with nC,

$$Q = \int dQ = \frac{nk}{\Theta^3} \int_{T_1}^{T_2} T^3 \, dT = \frac{nk}{4\Theta^3} \left(T_2^4 - T_1^4 \right).$$

For the given temperatures,

$$Q = \frac{(1.50 \text{ mol})(1940 \text{ J/mol·K})}{4(281 \text{ K})^3} ((40.0 \text{ K})^4 - (10.0 \text{ K})^4) = 83.6 \text{ J}.$$

b) $\frac{Q}{n\Delta T} = \frac{(83.6 \text{ J})}{(1.50 \text{ mol})(30.0 \text{ K})} = 1.86 \text{ J/mol·K}.$

c) $C = (1940 \text{ J/mol·K})(40.0 \text{ K}/281 \text{ K})^3 = 5.60 \text{ J/mol·K}.$

15-88: Setting the decrease in internal energy of the water equal to the final gravitational potential energy, $L_f \rho_w V_w + C_w \rho_w V_w \Delta T = mgh$. Solving for h, and inserting numbers:

$$h = \frac{\rho_w V_w (L_f + C_w \Delta T)}{mg}$$
$$= \frac{(1000 \text{ kg/m}^3)(1.9 \times .8 \times .1 \text{ m}^3)[334 \times 10^3 \text{ J/kg} + (4190 \text{ J/kg} \cdot° \text{C})(37°\text{C})]}{(70 \text{ kg})(9.8 \text{ m/s}^2)}$$
$$= 1.08 \times 10^5 \text{ m} = 108 \text{ km}.$$

15-89: a) $(90)(100 \text{ W})(3000 \text{ s}) = 2.7 \times 10^7 \text{ J}.$

b) $$\Delta T = \frac{Q}{cm} = \frac{Q}{c\rho V} = \frac{2.7 \times 10^7 \text{ J}}{(1020 \text{ J/kg·K})(1.20 \text{ kg/m}^3)(3200 \text{ m}^3)} = 6.89 \text{ C}°,$$

or 6.9 C° to the more appropriate two figures. c) The answers to both parts (a) and (b) are multiplied by 2.8, and the temperature rises by 19.3 C°.

15-90: See Problem 15-87. Denoting C by $C = a + bT$, a and b independent of temperature, integration gives

$$Q = n \left(a \left(T_2 - T_1 \right) + \frac{b}{2} \left(T_2^2 - T_1^2 \right) \right).$$

In this form, the temperatures for the linear part may be expressed in terms of Celsius temperatures, but the quadratic terms *must* be converted to Kelvin temperatures,

$T_1 = 300$ K and $T_2 = 500$ K. Insertion of the given values yields

$$Q = (3.00 \text{ mol})((29.5 \text{ J/mol·K})(500 \text{ K} - 300 \text{ K})$$
$$+ (4.10 \times 10^{-3} \text{ J/mol·K}^2)((500 \text{ K})^2 - (300 \text{ K})^2))$$
$$= 1.97 \times 10^4 \text{ J}.$$

15-91: a) To heat the ice cube to 0.0°C, heat must be lost by the water, which means that some of the water will freeze. The mass of this water is

$$m_{\text{water}} = \frac{m_{\text{ice}} c_{\text{ice}} \Delta T_{\text{ice}}}{L_f} = \frac{(0.075 \text{ kg})(2100 \text{ J/kg·K})(10.0 \text{ C}°)}{(334 \times 10^3 \text{ J/kg})} = 4.72 \times 10^{-3} \text{ kg} = 4.72 \text{ g}.$$

b) In theory, yes, but it takes 16.7 kg of ice to freeze 1 kg of water, so this is impractical.

15-92: The ratio of the masses is

$$\frac{m_s}{m_w} = \frac{c_w \Delta T_w}{c_w \Delta T_s + L_v} = \frac{(4190 \text{ J/kg·K})(42.0 \text{ K})}{(4190 \text{ J/kg·K})(65.0 \text{ K}) + 2256 \times 10^3 \text{ J/kg}} = 0.0730,$$

so 0.0696 kg of steam supplies the same heat as 1.00 kg of water. Note the heat capacity of water is used to find the heat lost by the condensed steam.

15-93: a) The possible final states are steam, water and copper at 100°C, water, ice and copper at 0.0°C or water and copper at an intermediate temperature. Assume the last possibility; the final temperature would be

$$T = \frac{\left(\begin{array}{c}(0.0350 \text{ kg})((4190 \text{ J/kg·K})(100 \text{ C}°) + 2256 \times 10^3 \text{ J/kg}) \\ -(0.0950 \text{ kg})(334 \times 10^3 \text{ J/kg})\end{array}\right)}{\left(\begin{array}{c}(0.0350 \text{ kg})(4190 \text{ J/kg·K}) + (0.446 \text{ kg})(390 \text{ J/kg·K}) \\ +(0.0950 \text{ kg})(4190 \text{ J/kg·K})\end{array}\right)} = 86.1°\text{C}.$$

This is indeed a temperature intermediate between the freezing and boiling points, so the reasonable assumption was a valid one. b) There are 0.13 kg of water.

15-94: a) The three possible final states are ice at a temperature below 0.0°C, an ice-water mixture at 0.0°C or water at a temperature above 0.0°C. To make an educated guess at the final possibility, note that $(0.140 \text{ kg})(2100 \text{ J/kg·K})(15.0 \text{ C}°) = 4.41$ kJ are needed to heat the ice to 0.0°C, and $(0.190 \text{ kg})(4190 \text{ J/kg·K})(35.0 \text{ C}°) = 27.9$ kJ must removed to cool the water to 0.0°C, so the water will not freeze. Melting all of the ice would require an additional $(0.140 \text{ kg})(334 \times 10^3 \text{ J/kg}) = 46.8$ kJ, so some of the ice melts but not all; the final temperature of the system is 0.0°C.

Considering the other possibilities would lead to contradictions, as either water at a temperature below freezing or ice at a temperature above freezing.

b) The ice will give up 27.9 kJ of heat energy to cool the water to 0°C. Then, $m = \frac{(27.9 \text{ kJ} - 4.41 \text{ kJ})}{334 \times 10^3 \text{ J/kg}} = .070$ kg will be converted to water. There will be .070 kg of ice and .260 kg of water.

15-95: a) If all of the steam were to condense, the energy available to heat the water would be $(0.0400 \text{ kg})(2256 \times 10^3 \text{ J/kg}) = 9.02 \times 10^4$ J. If all of the water were to be heated to 100.0°C, the needed heat would be $(0.200 \text{ kg})(4190 \text{ J/kg·K})(50.0 \text{ C°}) = 4.19 \times 10^4$ J. Thus, the water heats to 100.0°C and some of the steam condenses; the temperature of the final state is 100°C.

b) Because the steam has more energy to give up than it takes to raise the water temperature, we can assume that some of the steam is converted to water:

$$m = \frac{4.19 \times 10^4 \text{ J}}{2256 \times 10^3 \text{ J/kg}} = .019 \text{ kg}.$$

Thus in the final state, there are .219 kg of water and .021 kg of steam.

15-96: The mass of the steam condensed is $0.525 \text{ kg} - 0.490 \text{ kg} = 0.035$ kg. The heat lost by the steam as it condenses and cools is

$$(0.035 \text{ kg})L_v + (0.035 \text{ kg})(4190 \text{ J/kg·K})(29.0 \text{ K}),$$

and the heat gained by the original water and calorimeter is

$$((0.150 \text{ kg})(420 \text{ J/kg·K}) + (0.340 \text{ kg})(4190 \text{ J/kg·K}))(56.0 \text{ K}) = 8.33 \times 10^4 \text{ J}.$$

Setting the heat lost equal to the heat gained and solving for L_v gives 2.26×10^6 J/kg, or 2.3×10^6 J/kg to two figures (the mass of steam condensed is known to only two figures).

15-97: a) The possible final states are an ice-water mix at 0.0°C, a water-steam mix at 100.0°C or water at an intermediate temperature. Due to the large latent heat of vaporization, it is reasonable to make an initial guess that the final state is at 100.0°C. To check this, the energy lost by the steam if all of it were to condense would be $(0.0950 \text{ kg})(2256 \times 10^3 \text{ J/kg}) = 2.14 \times 10^5$ J. The energy required to melt the ice and heat it to 100°C is $(0.150 \text{ kg})(334 \times 10^3 \text{ J/kg} + (4190 \text{ J/kg·K})(100 \text{ C°})) = 1.13 \times 10^5$ J, and the energy required to heat the original water to 100°C is $(0.200 \text{ kg})(4190 \text{ J/kg·K})(50.0 \text{ C°}) = 4.19 \times 10^4$ J. Thus, some of the steam will condense, and the final state of the system will be a water-steam mixture at 100.0°C.

b) All of the ice is converted to water, so it adds .150 kg to the mass of water. Some of the steam condenses giving up 1.55×10^5 J of energy to melt the ice and raise the temperature. Thus, $m = \frac{1.55 \times 10^5 \text{ J}}{2256 \times 10^3 \text{ J/kg}} = .069$ kg and the final mass of steam is .026 kg, and of the water, $.150 \text{ kg} + .069 \text{ kg} + .20 \text{ kg} = .419$ kg.

c) Due to the much larger quantity of ice, a reasonable initial guess is an ice-water mix at 0.0°C. The energy required to melt all of the ice would be $(0.350 \text{ kg})(334 \times 10^3 \text{ J/kg}) = 1.17 \times 10^5$ J. The maximum energy that could be transfered to the ice would be if all of the steam would condense and cool to 0.0°C and if all of the water would cool to 0.0°C,

$$(0.0120) \text{ kg } (2256 \times 10^3 \text{ J/kg} + (4190 \text{ J/kg·K})(100.0 \text{ C°}))$$
$$+(0.200 \text{ kg})(4190 \text{ J/kg·K})(40.0 \text{ C°}) = 6.56 \times 10^4 \text{ J}.$$

This is insufficient to melt all of the ice, so the final state of the system is an ice-water mixture at 0.0°C. 6.56×10^4 J of energy goes into melting the ice. So, $m = \frac{6.56 \times 10^4 \text{ J}}{334 \times 10^3 \text{ J/kg}} =$ .196 kg. So there is .154 kg of ice, and .012 kg + .196 kg + .20 kg = .408 kg of water.

15-98: Solving Eq. (15-21) for k,

$$k = H\frac{L}{A\Delta T} = (180 \text{ W})\frac{(3.9 \times 10^{-2} \text{ m})}{(2.18 \text{ m}^2)(65.0 \text{ K})} = 5.0 \times 10^{-2} \text{ W/m·K}.$$

15-99: a) $H = kA\frac{\Delta T}{L} = (0.120 \text{ J/mol·K})(2.00 \times 0.95 \text{ m}^2)\left(\frac{28.0 \text{ C}°}{5.0 \times 10^{-2} \text{ m} + 1.8 \times 10^{-2} \text{ m}}\right)$
$$= 93.9 \text{ W}.$$

b) The flow through the wood part of the door is reduced by a factor of $1 - \frac{(0.50)^2}{(2.00 \times 0.95)} =$ 0.868 to 81.5 W. The heat flow through the glass is

$$H_{\text{glass}} = (0.80 \text{ J/mol·K})(0.50 \text{ m})^2\left(\frac{28.0 \text{ C}°}{12.45 \times 10^{-2} \text{ m}}\right) = 45.0 \text{ W},$$

and so the ratio is $\frac{81.5 + 45.0}{93.9} = 1.35.$

15-100: $R_1 = \frac{L}{k_1}$, $R_2 = \frac{L}{k_2}$, $H_1 = H_2$, and so $\Delta T_1 = \frac{H}{A}R_1$, $\Delta T_2 = \frac{H}{A}R_2$. The temperature difference across the combination is

$$\Delta T = \Delta T_1 + \Delta T_2 = \frac{H}{A}(R_1 + R_2) = \frac{H}{A}R,$$

so $R = R_1 + R_2$.

15-101: The ratio will be the inverse of the ratio of the total thermal resistance, as given by Eq. (15-24). With two panes of glass with the air trapped in between, compared to the single pane, the ratio of the heat flows is

$$\frac{(2(L_{\text{glass}}/k_{\text{glass}}) + R_0 + (L_{\text{air}}/k_{\text{air}})}{(L_{\text{glass}}/k_{\text{glass}}) + R_0},$$

where R_0 is the thermal resistance of the air films. Numerically, the ratio is

$$\frac{\left(\begin{array}{c}2((4.2 \times 10^{-3} \text{ m})/(0.80 \text{ W/m·K})) + 0.15 \text{ m}^2\text{·K/W} \\ +((7.0 \times 10^{-3} \text{ m})/(0.024 \text{ W/m·K}))\end{array}\right)}{(4.2 \times 10^{-3} \text{ m})/(0.80 \text{ W/m·K}) + 0.15 \text{ m}^2\text{·K/W}} = 2.9.$$

15-102: Denote the quantities for copper, brass and steel by 1, 2 and 3, respectively, and denote the temperature at the junction by T_0.

a) $H_1 = H_2 + H_3$, and using Eq. (15-21) and dividing by the common area,

$$\frac{k_1}{L_1}(100°\text{C} - T_0) = \frac{k_2}{L_2}T_0 + \frac{k_3}{L_3}T_0.$$

Solving for T_0 gives

$$T_0 = \frac{(k_1/L_1)}{(k_1/L_1) + (k_2/L_2) + (k_3/L_3)}(100°\text{C}).$$

Substitution of numerical values gives $T_0 = 78.4°\text{C}$.

b) Using $H = \frac{kA}{L}\Delta T$ for each rod, with $\Delta T_1 = 21.6\ \text{C}°$, $\Delta T_2 = \Delta T_3 = 78.4°\text{C}$ gives $H_1 = 12.8$ W, $H_2 = 9.50$ W and $H_3 = 3.30$ W. If higher precision is kept, H_1 is seen to be the sum of H_2 and H_3.

15-103: a) See Figure 15-7. As the temperature approaches $0.0°\text{C}$, the coldest water rises to the top and begins to freeze while the slightly warmer water, which is more dense, will be beneath the surface. b) (As in part (c), a constant temperature difference is assumed.) Let the thickness of the sheet be x, and the amount the ice thickens in time dt be dx. The mass of ice added per unit area is then $\rho_{\text{ice}}\,dx$, meaning a heat transfer of $\rho_{\text{ice}}L_f\,dx$. This must be the product of the heat flow per unit area times the time, $(H/A)\,dt = (k\Delta T/x)\,dt$. Equating these expressions,

$$\rho_{\text{ice}}L_f\,dx = \frac{k\Delta T}{x}\,dt \quad \text{or} \quad x\,dx = \frac{k\Delta T}{\rho_{\text{ice}}L_f}\,dt.$$

This is a separable differential equation; integrating both sides, setting $x = 0$ at $t = 0$, gives

$$x^2 = \frac{2k\Delta T}{\rho_{\text{ice}}L_f}t.$$

The square of the thickness is proportional to the time, so the thickness is proportional to the square root of the time. c) Solving for the time in the above expression,

$$t = \frac{(920\ \text{kg/m}^3)(334 \times 10^3\ \text{J/kg})}{2(1.6\ \text{J/mol·K})(10°\ \text{C})}(0.25\ \text{m})^2 = 6.0 \times 10^5\ \text{s}.$$

d) Using $x = 40$ m in the above calculation gives $t = 1.5 \times 10^{10}$ s, about 500 y, a very long cold spell.

15-104: Equation (15-21) becomes $H = kA\frac{\partial T}{\partial x}$.

a) $H = (380\ \text{J/kg·K})(2.50 \times 10^{-4}\ \text{m}^2)(140\ \text{C}°/\text{m}) = 13.3$ W.

b) Denoting the points as 1 and 2, $H_2 - H_1 = \frac{dQ}{dt} = mc\frac{\partial T}{\partial t}$. Solving for $\frac{\partial T}{\partial x}$ at 2,

$$\left.\frac{\partial T}{\partial x}\right|_2 = \left.\frac{\partial T}{\partial x}\right|_1 + \frac{mc}{kA}\frac{\partial T}{\partial t}.$$

The mass m is $\rho A\Delta x$, so the factor multiplying $\frac{\partial T}{\partial t}$ in the above expression is $\frac{c\rho}{k}\Delta x = 137$ s/m. Then,

$$\left.\frac{\partial T}{\partial x}\right|_2 = 140\ \text{C}°/\text{m} + (137\ \text{s/m})(0.250\ \text{C}°/\text{s}) = 174\ \text{C}°/\text{m}.$$

15-105: The mass of ice per unit area will be the product of the density and the thickness x, and the energy needed per unit area to melt the ice is the product of the mass per unit area and the heat of fusion. The time is then

$$t = \frac{\rho x L_f}{P/A} = \frac{(920 \text{ kg/m}^3)(2.50 \times 10^{-2} \text{ m})(334 \times 10^3 \text{ J/kg})}{(0.70)(600 \text{ W/m}^2)}$$
$$= 18.3 \times 10^3 \text{ s} = 305 \text{ min}.$$

15-106: a) Assuming no substantial energy loss in the region between the earth and the sun, the power per unit area will be inversely proportional to the square of the distance from the center of the sun, and so the energy flux at the surface of the sun is $(1.50 \times 10^3 \text{ W/m}^2) \left(\frac{1.50 \times 10^{11} \text{ m}}{6.96 \times 10^8 \text{ m}}\right)^2 = 6.97 \times 10^7 \text{ W/m}^2$. b) Solving Eq. (15-25) with $e = 1$,

$$T = \left[\frac{H}{A}\frac{1}{\sigma}\right]^{\frac{1}{4}} = \left[\frac{6.97 \times 10^7 \text{ W/m}^2}{5.67 \times 10^{-8} \text{ W/m}^2\cdot\text{K}^4}\right]^{\frac{1}{4}} = 5920 \text{ K}.$$

15-107: The rate at which the helium evaporates is the heat gained from the surroundings by radiation divided by the heat of vaporization. The heat gained from the surroundings comes from both the side and the ends of the cylinder, and so the rate at which the mass is lost is

$$\frac{(h\pi d + 2\pi(d/2)^2)\sigma e\, (T_s^4 - T^4)}{L_v}$$
$$= \frac{\left(\begin{array}{c}((0.250 \text{ m})\pi(0.090 \text{ m}) + 2\pi(0.045 \text{ m})^2)(.200) \\ \times (5.67 \times 10^{-8} \text{ W/m}^2 \cdot \text{K}^4)((77.3 \text{ K})^4 - (4.22 \text{ K})^4)\end{array}\right)}{(2.09 \times 10^4 \text{ J/kg})}$$
$$= 1.62 \times 10^{-6} \text{ kg/s},$$

which is 5.82 g/h.

15-108: a) With $\Delta p = 0$,

$$p\Delta V = nR\Delta T = \frac{pV}{T}\Delta T,$$

or

$$\frac{\Delta V}{V} = \frac{\Delta T}{T}, \quad \text{and} \quad \beta = \frac{1}{T}.$$

b)
$$\frac{\beta_{\text{air}}}{\beta_{\text{copper}}} = \frac{1}{(293 \text{ K})(5.1 \times 10^{-5} \text{ K}^{-1})} = 67.$$

15-109: a) At steady state, the input power all goes into heating the water, so $P = H = \frac{dm}{dt}c\Delta T$ and

$$\Delta T = \frac{P}{c(dm/dt)} = \frac{(1800 \text{ W})}{(4190 \text{ J/kg}\cdot\text{K})(0.500 \text{ kg/min})/(60 \text{ s/min})} = 51.6 \text{ K},$$

and the output temperature is $18.0°C + 51.6\ C° = 69.6°C$. b) At steady state, the apparatus will neither remove heat from nor add heat to the water.

15-110: a) The heat generated by the hamster is the heat added to the box;

$$P = mc\frac{dT}{dt} = (1.20\ \text{kg/m}^3)(0.0500\ \text{m}^3)(1020\ \text{J/mol·K})(1.60\ \text{C°/h}) = 97.9\ \text{J/h}.$$

b) Taking the efficiency into account,

$$\frac{M}{t} = \frac{P_0}{L_c} = \frac{P/(10\%)}{L_c} = \frac{979\ \text{J/h}}{24\ \text{J/g}} = 40.8\ \text{g/h}.$$

15-111: For a spherical or cylindrical surface, the area A in Eq. (15-21) is not constant, and the material must be considered to consist of shells with thickness dr and a temperature difference between the inside and outside of the shell dT. The heat current will be a constant, and must be found by integrating a differential equation. a) Equation (15-21) becomes

$$H = k(4\pi r^2)\frac{dT}{dr} \quad \text{or} \quad \frac{H\,dr}{4\pi r^2} = k\,dT.$$

Integrating both sides between the appropriate limits,

$$\frac{H}{4\pi}\left(\frac{1}{a} - \frac{1}{b}\right) = k\,(T_2 - T_1).$$

In this case the "appropriate limits" have been chosen so that if the inner temperature T_2 is at the higher temperature T_1, the heat flows outward; that is, $\frac{dT}{dr} < 0$. Solving for the heat current,

$$H = \frac{k4\pi ab(T_2 - T_1)}{b - a}.$$

b) Of the many ways to find the temperature, the one presented here avoids some intermediate calculations and avoids (or rather sidesteps) the sign ambiguity mentioned above. From the model of heat conduction used, the rate of change of temperature with radius is of the form $\frac{dT}{dr} = \frac{B}{r^2}$, with B a constant. Integrating from $r = a$ to r and from $r = a$ to $r = b$ gives

$$T(r) - T_2 = B\left(\frac{1}{a} - \frac{1}{r}\right) \quad \text{and} \quad T_1 - T_2 = B\left(\frac{1}{a} - \frac{1}{b}\right).$$

Using the second of these to eliminate B and solving for $T(r)$ (and rearranging to eliminate compound fractions) gives

$$T(r) = T_2 - (T_2 - T_1)\left(\frac{r - a}{b - a}\right)\left(\frac{b}{r}\right).$$

There are, of course, many equivalent forms. As a check, note that at $r = a$, $T = T_2$ and at $r = b$, $T = T_1$. c) As in part (a), the expression for the heat current is

$$H = k(2\pi r L)\frac{dT}{dr} \quad \text{or} \quad \frac{H\,dr}{2\pi r} = kL\,dT,$$

which integrates, with the same condition on the limits, to

$$\frac{H}{2\pi}\ln(b/a) = kL(T_2 - T_1) \quad \text{or} \quad H = \frac{2\pi kL(T_2 - T_1)}{\ln(b/a)}.$$

d) A method similar (but slightly simpler) than that used in part (b) gives

$$T(r) = T_2 + (T_1 - T_2)\frac{\ln(r/a)}{\ln(b/a)}.$$

e) For the sphere: Let $b - a = l$, and approximate $b \sim a$, with a the common radius. Then the surface area of the sphere is $A = 4\pi a^2$, and the expression for H is that of Eq. (15-21) (with l instead of L, which has another use in this problem). For the cylinder: with the same notation, consider

$$\ln\left(\frac{b}{a}\right) = \ln\left(1 + \frac{l}{a}\right) \sim \frac{l}{a},$$

where the Taylor series approximation for $\ln(1 + \epsilon)$ for small ϵ has been used. The expression for H then reduces to $k(2\pi La)(\Delta T/l)$, which is Eq. (15-21) with $A = 2\pi La$.

15-112: From the result of Problem 15-111, the heat current through each of the jackets is related to the temperature difference by $H = \frac{2\pi lk}{\ln(b/a)}\Delta T$, where l is the length of the cylinder and b and a are the inner and outer radii of the cylinder. Let the temperature across the cork be ΔT_1 and the temperature across the styrofoam be ΔT_2, with similar notation for the thermal conductivities and heat currents. Then, $\Delta T_1 + \Delta T_2 = \Delta T = 125$ C°. Setting $H_1 = H_2 = H$ and canceling the common factors,

$$\frac{\Delta T_1 k_1}{\ln 2} = \frac{\Delta T_2 k_2}{\ln 1.5}.$$

Eliminating ΔT_2 and solving for ΔT_1 gives

$$\Delta T_1 = \Delta T\left(1 + \frac{k_1 \ln 1.5}{k_2 \ln 2}\right)^{-1}.$$

Substitution of numerical values gives $\Delta T_1 = 37$ C°, and the temperature at the radius where the layers meet is $140°C - 37°C = 103°C$. b) Substitution of this value for ΔT_1 into the above expression for $H_1 = H$ gives

$$H = \frac{2\pi(2.00\text{ m})(0.04\text{ J/mol·K})}{\ln 2}(37\text{ C°}) = 27\text{ W}.$$

15-113: a)

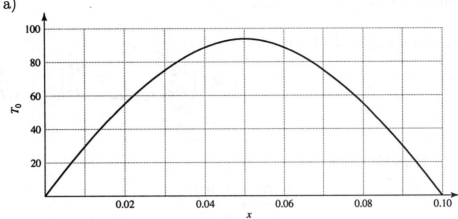

b) After a very long time, no heat will flow, and the entire rod will be at a uniform temperature which must be that of the ends, 0°C.

c)

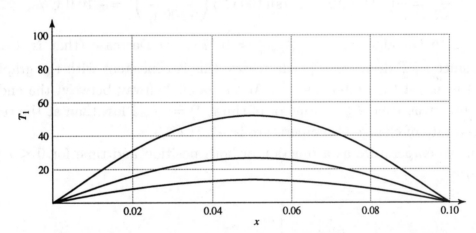

d) $\frac{\partial T}{\partial x} = (100°C)(\pi/L)\cos \pi x/L$. At the ends, $x = 0$ and $x = L$, the cosine is ± 1 and the temperature gradient is $\pm(100°C)(\pi/0.100 \text{ m}) = \pm 3.14 \times 10^3$ C°/m.   e) Taking the phrase "into the rod" to mean an absolute value, the heat current will be $kA\frac{\partial T}{\partial x} = (385.0 \text{ W/m·K})(1.00 \times 10^{-4} \text{ m}^2)(3.14 \times 10^3 \text{ C°/m}) = 121$ W.   f) Either by evaluating $\frac{\partial T}{\partial x}$ at the center of the rod, where $\pi x/L = \pi/2$ and $\cos(\pi/2) = 0$, or by checking the figure in part (a), the temperature gradient is zero, and no heat flows through the center; this is consistent with the symmetry of the situation. There will not be any heat current at the center of the rod at any later time.   g) See Problem 15-104;

$$\frac{k}{\rho c} = \frac{(385 \text{ W/m·K})}{(8.9 \times 10^3 \text{ kg/m}^3)(390 \text{ J/kg·K})} = 1.1 \times 10^{-4} \text{ m}^2/\text{s}.$$

h) Although there is no net heat current, the temperature of the center of the rod is decreasing; by considering the heat current at points just to either side of the center, where there is a non-zero temperature gradient, there must be a net flow of heat out of the region around the center. Specifically,

$$H((L/2) + \Delta x) - H((L/2) - \Delta x) = \rho A \Delta x c \frac{\partial T}{\partial t}$$

$$= kA \left(\frac{\partial T}{\partial x} \bigg|_{(L/2)+\Delta x} - \frac{\partial T}{\partial x} \bigg|_{(L/2)-\Delta x} \right)$$

$$= kA \frac{\partial^2 T}{\partial x^2} \Delta x,$$

from which the Heat Equation,

$$\frac{\partial T}{\partial t} = \frac{k}{\rho c} \frac{\partial^2 T}{\partial x^2}$$

is obtained. At the center of the rod, $\frac{\partial^2 T}{\partial x^2} = -(100 \text{ C}°)(\pi/L)^2$, and so

$$\frac{\partial T}{\partial t} = -(1.11 \times 10^{-4} \text{ m}^2/\text{s})(100 \text{ C}°) \left(\frac{\pi}{0.100 \text{ m}} \right)^2 = -10.9 \text{ C}°/\text{s},$$

or -11 C°/s to two figures.   i) $\frac{100 \text{ C}°}{10.9 \text{ C}°/\text{s}} = 9.17$ s.   j) Decrease (that is, become less negative), since as T decreases, $\frac{\partial^2 T}{\partial x^2}$ decreases; this is consistent with the graphs, which correspond to equal time intervals.   k) At the point halfway between the end and the center, at any given time $\frac{\partial^2 T}{\partial x^2}$ is a factor of $\sin(\pi/4) = 1/\sqrt{2}$ less than at the center, and so the initial rate of change of temperature is -7.71 C°/s.

A plot of temperature as a function of both position and time for $0 \le t \le 50$ s is shown below.

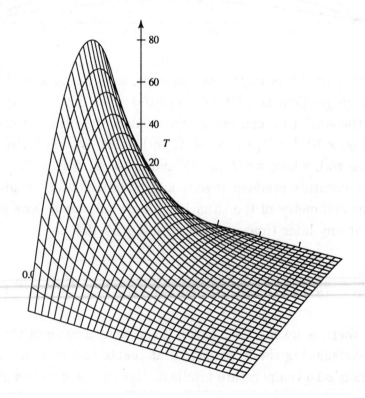

15-114: a) In hot weather, the moment of inertia I and the length d in Eq. (13-39) will both increase by the same factor, and so the period will be longer and the clock will run slow (lose time). Similarly, the clock will run fast (gain time) in cold weather. (An ideal pendulum is a special case of a physical pendulum.) b) $\frac{\Delta L}{L_0} = \alpha \Delta T = (1.2 \times 10^{-5} \ (\text{C}°)^{-1}) \times (10.0 \ \text{C}°) = 1.2 \times 10^{-4}$. c) See Problem 13-83; to avoid possible confusion, denote the pendulum period by τ. For this problem, $\frac{\Delta \tau}{\tau} = \frac{1}{2}\frac{\Delta L}{L} = +6.0 \times 10^{-5}$, so in one day the clock will gain $(86{,}400 \ \text{s})(6.0 \times 10^{-5}) = 5.2$ s to two figures. d) $\left|\frac{\Delta \tau}{\tau}\right| = (1/2)\alpha \Delta T < (86{,}400)^{-1}$, so

$$\Delta T < 2\left((1.2 \times 10^{-5} \ (\text{C}°)^{-1})(86{,}400)\right)^{-1} = 1.93 \ \text{C}°.$$

15-115: The rate at which heat is aborbed at the blackened end is the heat current in the rod,

$$A\sigma\left(T_{\text{s}}^4 - T_2^4\right) = \frac{kA}{L}(T_2 - T_1),$$

where $T_1 = 20.00$ K and T_2 is the temperature of the blackened end of the rod. If this were to be solved exactly, the equation would be a quartic, very likely not worth the trouble. Following the hint, approximate T_2 on the left side of the above expression as T_1 to obtain

$$T_2 = T_1 + \frac{\sigma L}{k}\left(T_{\text{s}}^2 - T_1^4\right) = T_1 + (6.79 \times 10^{-12} \ \text{K}^{-3})\left(T_{\text{s}}^4 - T_1^4\right) = T_1 + 0.424 \ \text{K}.$$

This approximation for T_2 is indeed only slightly larger than T_1, and is a good estimate of the temperature. Using this for T_2 in the original expression to find a better value of ΔT gives the same ΔT to eight figures, and further iterations are not worthwhile.

A numerical program used to find roots of the quartic equation returns a value for ΔT that differed from that found above in the eighth place; this, of course, is more precision than is warranted in this problem.

15-116: a) The rates are:

(i) 280 W,

(ii) $(54 \ \text{J/h} \cdot \text{C}° \cdot \text{m}^2)(1.5 \ \text{m}^2)(11 \ \text{C}°)/(3600 \ \text{s/h}) = 0.248$ W,

(iii) $(1400 \ \text{W/m}^2)(1.5 \ \text{m}^2) = 2.10 \times 10^3$ W,

(iv) $(5.67 \times 10^{-8} \ \text{W/m}^2 \cdot \text{K}^4)(1.5 \ \text{m}^2)((320 \ \text{K})^4 - (309 \ \text{K})^4) = 116$ W.

The total is 2.50 kW, with the largest portion due to radiation from the sun.

b) $\dfrac{P}{\rho L_{\text{v}}} = \dfrac{2.50 \times 10^3 \ \text{W}}{(1000 \ \text{kg/m}^3)(2.42 \times 10^6 \ \text{J/kg} \cdot \text{K})} = 1.03 \times 10^{-6} \ \text{m}^3/\text{s} = 3.72 \ \text{L/h}.$

c) Redoing the above calculations with $e = 0$ and the decreased area gives a power of 945 W and a corresponding evaporation rate of 1.4 L/h. Wearing reflective clothing helps a good deal. Large areas of loose weave clothing also facilitate evaporation.

Chapter 16   Thermal Properties of Matter

In doing the numerical calculations for the exercises and problems for this chapter, the values of the ideal-gas constant have been used with the precision given on page 501 of the text,

$$R = 8.3145 \text{ J/mol·K} = 0.08206 \text{ L·atm/mol·K}.$$

Use of values of these constants with either greater or less precision may introduce differences in the third figures of some answers.

16-1:   a) $n = m_{tot}/M = (0.225 \text{ kg})/(4.00 \times 10^{-3} \text{ kg/mol}) = 56.3 \text{ mol}$.   b) Of the many ways to find the pressure, Eq. (16-3) gives

$$p = \frac{nRT}{V} = \frac{(56.3 \text{ mol})(0.08206 \text{ L·atm/mol·K})(294.15 \text{ K})}{(20.0 \text{ L})}$$

$$= 67.2 \text{ atm} = 6.81 \times 10^6 \text{ Pa}.$$

16-2:   a) The final temperature is four times the initial Kelvin temperature, or $4(314.15 \text{ K}) - 273.15 = 983°\text{C}$ to the nearest degree.

b) $m_{tot} = nM = \dfrac{MpV}{RT} = \dfrac{(4.00 \times 10^{-3} \text{ kg/mol})(1.30 \text{ atm})(2.60 \text{ L})}{(0.08206 \text{ L·atm/mol·K})(314.15 \text{ K})} = 5.24 \times 10^{-4} \text{ kg}.$

16-3:   For constant temperature, Eq. (16-6) becomes

$$p_2 = p_1(V_1/V_2) = (3.40 \text{ atm})(0.110/0.390) = 0.96 \text{ atm}.$$

16-4:   a) Decreasing the pressure by a factor of one-third decreases the Kelvin temperature by a factor of one-third, so the new Celsius temperature is $1/3(293.15 \text{ K}) - 273.15 = -175°\text{C}$ rounded to the nearest degree. b) The net effect of the two changes is to keep the pressure the same while decreasing the Kelvin temperature by a factor of one-third, resulting in a decrease in volume by a factor of one-third, to 1.00 L.

16-5:   From Eq. (16-6),

$$T_2 = T_1\left(\frac{p_2 V_2}{p_1 V_1}\right) = (300.15 \text{ K})\left(\frac{(2.821 \times 10^6 \text{ Pa})(46.2 \text{ cm}^3)}{(1.01 \times 10^5 \text{ Pa})(499 \text{ cm}^3)}\right) = 776 \text{ K} = 503°\text{C}.$$

16-6:   a) $m_{tot} = \dfrac{MpV}{RT} = \dfrac{(32.0 \times 10^{-3} \text{ kg/mol})(4.013 \times 10^5 \text{ Pa})(0.0750 \text{ m}^3)}{(8.3145 \text{ J/mol·K})(310.15 \text{ K})} = 0.373 \text{ kg}.$

b) Using the final pressure of 2.813×10^5 Pa and temperature of 295.15 K, $m' = 0.275$ kg, so the mass lost is 0.098 kg, where extra figures were kept in the intermediate calculation of m_{tot}.

16-7: From Eq. (16-6),

$$p_2 = p_1 \left(\frac{T_2}{T_1}\right)\left(\frac{V_1}{V_2}\right) = (1.50 \times 10^5 \text{ Pa})\left(\frac{430.15 \text{ K}}{300.15 \text{ K}}\right)\left(\frac{0.750 \text{ m}^3}{0.48 \text{ m}^3}\right) = 3.36 \times 10^5 \text{ Pa.}$$

16-8: a) $\quad n = \dfrac{pV}{RT} = \dfrac{(1.00 \text{ atm})(140 \times 10^3 \text{ L})}{(0.08206 \text{ L·atm/mol·K})(295.15 \text{ K})} = 5.78 \times 10^3 \text{ mol.}$

b) $(32.0 \times 10^{-3} \text{ kg/mol})(5.78 \times 10^3 \text{ mol}) = 185 \text{ kg.}$

16-9: $V_2 = V_1(T_2/T_1) = (0.600 \text{ L})(77.3/292.15) = 0.159 \text{ L.}$

16-10: a) $nRT/V = 7.28 \times 10^6$ Pa while Eq. (16-7) gives 5.87×10^6 Pa. b) The van der Waals equation, which accounts for the attraction between molecules, gives a pressure that is 20% lower. c) 7.28×10^5 Pa, 7.13×10^5 Pa, 2.1%. d) As n/V decreases, the formulas and the numerical values are the same.

16-11: At constant temperature, $p_2 = p_1(V_1/V_2) = (1.0 \text{ atm})(6.0/5.7) = 1.1 \text{ atm.}$

16-12: a) $\frac{V_2}{V_1} = \frac{p_1 T_2}{p_2 T_1} = (3.50)(\frac{296 \text{ K}}{277 \text{ K}}) = 3.74.$ b) Lungs cannot withstand such a volume change; breathing is a good idea.

16-13: a) $\quad T_2 = \dfrac{p_2 V}{nR} = \dfrac{(100 \text{ atm})(3.10 \text{ L})}{(11.0 \text{ mol})(0.08206 \text{ L·atm/mol·K})} = 343 \text{ K} = 70.3°\text{C.}$

b) This is a very small temperature increase and the thermal expansion of the tank may be neglected; in this case, neglecting the expansion means not including expansion in finding the highest safe temperature, and including the expansion would tend to relax safe standards.

16-14: Repeating the calculation of Example 16-4 (and using the same numerical values for R and the temperature gives) $p = (0.537)p_{\text{atm}} = 5.44 \times 10^4$ Pa.

16-15: $p = \rho RT/M = (0.364 \text{ kg/m}^3)(8.3145 \text{ J/mol·K})(273.15 \text{ K} - 56.5 \text{ K})/(28.8 \times 10^{-3} \text{kg/mol}) = 2.28 \times 10^4$ Pa.

16-16:

$$N = nN_A = \frac{pV}{RT}N_A$$

$$= \frac{(9.119 \times 10^{-9} \text{ Pa})(1.00 \times 10^{-6} \text{ m}^3)}{(8.3145 \text{ J/mol·K})(300 \text{ K})}\left(6.023 \times 10^{23} \text{ molecules/mol}\right)$$

$$= 2.20 \times 10^6 \text{ molecules.}$$

16-17: a)

$$p = \frac{nRT}{V} = \frac{N}{V}\frac{RT}{N_a} = \left(80 \times 10^3 \frac{\text{molecules}}{\text{L}}\right)\frac{(0.08206 \text{ L·atm/mol·K})(7500 \text{ K})}{(6.023 \times 10^{23} \text{ molecules/mol})}$$

$$= 8.2 \times 10^{-17}\text{atm,}$$

about 8.2×10^{-12} Pa. This is much lower, by a factor of a thousand, than the pressures considered in Exercise 16-16. b) Variations in pressure of this size are not likely to affect the motion of a starship.

16-18: Since this gas is at standard conditions, the volume will be $V = (22.4 \times 10^{-3} \text{ m}^3)$ $\frac{N}{N_A} = 2.23 \times 10^{-16}$ m^3, and the length of a side of a cube of this volume is $(2.23 \times 10^{-16} \text{ m}^3)^{\frac{1}{3}} = 6.1 \times 10^{-6}$ m.

16-19: $\frac{1000 \text{ g}}{18.0 \text{ g/mol}} = 55.6$ mol, which is $(55.6 \text{ mol})(6.023 \times 10^{23} \text{ molecules/mol}) = 3.35 \times 10^{25}$ molecules.

16-20: a) The volume per molecule is

$$\frac{V}{N} = \frac{nRT/p}{nN_A} = \frac{RT}{N_A p}$$

$$= \frac{(8.3145 \text{ J/mol·K})(300.15 \text{ K})}{(6.023 \times 10^{23} \text{ molecules/mol})(1.013 \times 10^5 \text{ Pa})}$$

$$= 4.091 \times 10^{-26} \text{ m}^3.$$

If this volume were a cube of side L,

$$L = (4.091 \times 10^{-26} \text{ m}^3)^{\frac{1}{3}} = 3.45 \times 10^{-9} \text{ m},$$

which is (b) a bit more than ten times the size of a molecule.

16-21: a) $V = m/\rho = nM/\rho = (5.00 \text{ mol})(18.0 \text{ g/mol})/(1.00 \text{ g/cm}^3) = 90.0 \text{ cm}^3 = 9.00 \times 10^{-5}$ m^3.   b) See Exercise 16-20;

$$\left(\frac{V}{N}\right)^{1/3} = \left(\frac{V}{nN_A}\right)^{1/3} = \left(\frac{9.00 \times 10^{-5} \text{ m}^3}{(5.00 \text{ mol})(6.023 \times 10^{23} \text{ molecules/mol})}\right)^{1/3}$$

$$= 3.10 \times 10^{-10} \text{ m}.$$

c) This is comparable to the size of a water molecule.

16-22: a) From Eq. (16-16), the average kinetic energy depends only on the temperature, not on the mass of individual molecules, so the average kinetic energy is the same for the molecules of each element. b) Equation (16-19) also shows that the rms speed is proportional to the inverse square root of the mass, and so

$$\frac{v_{\text{rms Kr}}}{v_{\text{rms Ne}}} = \sqrt{\frac{20.18}{83.80}} = 0.491, \quad \frac{v_{\text{rms Rn}}}{v_{\text{rms Ne}}} = \sqrt{\frac{20.18}{222.00}} = 0.301 \quad \text{and}$$

$$\frac{v_{\text{rms Rn}}}{v_{\text{rms Kr}}} = \sqrt{\frac{83.80}{222.00}} = 0.614.$$

16-23: a) At the same temperature, the average speeds will be different for the different isotopes; a stream of such isotopes would tend to separate into two groups.   b) $\sqrt{\frac{0.352}{0.349}} = 1.004$.

16-24: (Many calculators have statistics functions that are preprogrammed for such calculations as part of a statistics application. The results presented here were done on such a calculator.)   a) With the multiplicity of each score denoted by n_1, the average is

$\left(\frac{1}{150}\right)\sum n_i x_i = 54.6.$ b) $\left[\left(\frac{1}{150}\right)\sum n_i x_i^2\right]^{1/2} = 61.1.$ (Extra significant figures are warranted because the sums are known to higher precision.)

16-25: a) $\frac{3}{2}kT = (3/2)(1.381 \times 10^{-23} \text{ J/K})(300 \text{ K}) = 6.21 \times 10^{-21} \text{ J}.$

b) $\dfrac{2K_{\text{ave}}}{m} = \dfrac{2(6.21 \times 10^{-21} \text{ J})}{(32.0 \times 10^{-3} \text{ kg/mol})/(6.023 \times 10^{23} \text{ molecules/mol})} = 2.34 \times 10^5 \text{ m}^2/\text{s}^2.$

c) $v_{\text{rms}} = \sqrt{\dfrac{3RT}{M}} = \sqrt{\dfrac{3(8.3145 \text{ J/mol·K})(300 \text{ K})}{(32.0 \times 10^{-3} \text{ kg/mol})}} = 4.84 \times 10^2 \text{ m/s},$

which is of course the square root of the result of part (b).

d) $mv_{\text{rms}} = \left(\dfrac{M}{N_A}\right)v_{\text{rms}} = \dfrac{(32.0 \times 10^{-3} \text{ kg/mol})}{(6.023 \times 10^{23} \text{ molecules/mol})}(4.84 \times 10^2 \text{ m/s})$

$$= 2.57 \times 10^{-23} \text{ kg·m/s}$$

This may also be obtained from

$$\sqrt{2mK_{\text{ave}}} = \sqrt{\dfrac{2(6.21 \times 10^{-21} \text{ J})(32.0 \times 10^{-3} \text{ kg/mol})}{(6.023 \times 10^{23} \text{ molecules/mol})}}$$

e) The average force is the change in momentum of the atom, divided by the time between collisions. The magnitude of the momentum change is twice the result of part (d) (assuming an elastic collision), and the time between collisions is twice the length of a side of the cube, divided by the speed. Numerically,

$$F_{\text{ave}} = \frac{2mv_{\text{rms}}}{2L/v_{\text{rms}}} = \frac{mv_{\text{rms}}^2}{L} = \frac{2K_{\text{rms}}}{L} = \frac{2(6.21 \times 10^{-21} \text{ J})}{(0.100 \text{ m})} = 1.24 \times 10^{-19} \text{ N}.$$

f) $p_{\text{ave}} = F_{\text{ave}}/L^2 = 1.24 \times 10^{-17} \text{ Pa}.$

g) $p/p_{\text{ave}} = (1.013 \times 10^5 \text{ Pa})/(1.24 \times 10^{-17} \text{ Pa}) = 8.15 \times 10^{21} \text{ molecules}.$

h) $N = nN_A = \dfrac{pV}{RT}N_A$

$$= \left(\frac{(1.00 \text{ atm})(1.00 \text{ L})}{(0.08206 \text{ L·atm/mol·K})(300 \text{ K})}\right)(6.023 \times 10^{23} \text{ molecules/mol})$$

$$= 2.45 \times 10^{22}.$$

i) The result of part (g) was obtained by assuming that all of the molecules move in the same direction, and that there was a force on only two of the sides of the cube.

16-26: This is the same calculation done in Example 16-9, but with $T = 300$ K and $p = 3.50 \times 10^{-13}$ atm, giving $\lambda = 1.6 \times 10^5$ m.

16-27: The rms speeds will be the same if the Kelvin temperature is proportional to the molecular mass; $T_{N_2} = T_{H_2}(M_{N_2}/M_{H_2}) = (293.15 \text{ K})(28.0/2.02) = 4.06 \times 10^3 \text{ K} = 3.79 \times 10^{3}$°C.

16-28: a) $\sqrt{\frac{3kT}{m}} = \sqrt{\frac{3(1.381 \times 10^{-23} \text{ J/K})(300 \text{ K})}{(3.00 \times 10^{-16} \text{ kg})}} = 6.44 \times 10^{-3}$ m/s. b) If the particle is in thermal equilibrium with its surroundings, its motion will depend only on the surrounding temperature, not the mass of the individual particles.

16-29: a) The six degrees of freedom would mean a heat capacity at constant volume of $6\left(\frac{1}{2}\right)R = 3R = 24.9$ J/mol·K. $\frac{3R}{M} = \frac{3(8.3145 \text{ J/mol·K})}{(18.0 \times 10^{-3} \text{ kg/mol})} = 1.39 \times 10^3$ J/kg·K, b) vibrations do contribute to the heat capacity.

16-30: a) $C_v = (C)$ (molar mass), so $(833 \text{ J/kg·°C})(0.018 \text{ kg/mol}) = 15.0$ J/mol·°C at -180°C, $(1640 \text{ J/kg·°C})(0.018 \text{ kg/mol}) = 29.5$ J/mol·°C at -60°C, $(2060 \text{ J/kg·°C}) \times (0.018 \text{ kg/mol}) = 37.1$ J/mol·°C at -5.0°C. b) Vibrational degrees of freedom become more important. c) C_V exceeds $3R$ because H_2O also has rotational degrees of freedom.

16-31: a) Using Eq. (16-26), $Q = (2.50 \text{ mol})(20.79 \text{ J/mol·K})(30.0 \text{ K}) = 1.56$ kJ. b) From Eq. (16-25), $\frac{3}{5}$ of the result of part (a), 936 J.

16-32: a)
$$c = \frac{C_V}{M} = \frac{20.76 \text{ J/mol·K}}{28.0 \times 10^{-3} \text{ kg/mol}} = 741 \text{ J/kg·K},$$

which is $\frac{741}{4190} = 0.177$ times the specific heat capacity of water.

b) $m_N C_N \Delta T_N = m_w C_w \Delta T_w$, or $m_N = \frac{m_w C_w}{C_N}$. Inserting the given data and the result from part (a) gives $m_N = 5.65$ kg. To find the volume, use $pV = nRT$, or $V = \frac{nRT}{p} = \frac{[(5.65 \text{ kg})/(0.028 \text{ kg/mol})](0.08206 \text{ L·atm/mol·K})(293 \text{ K})}{1 \text{ atm}} = 4855$ L.

16-33: From Table (16-2), the speed is $(1.60)v_{rms}$, and so

$$v_{rms}^2 = \frac{3kT}{m} = \frac{3RT}{M} = \frac{v^2}{(1.60)^2}$$

(see Exercise 16-36), and so the temperature is

$$T = \frac{Mv^2}{3(1.60)^2 R} = \frac{(28.0 \times 10^{-3} \text{ kg/mol})}{3(1.60)^2(8.3145 \text{ J/mol·K})}v^2 = (4.385 \times 10^{-4} \text{ K·s}^2/\text{m}^2)v^2.$$

a) $(4.385 \times 10^{-4} \text{ K·s}^2/\text{m}^2)(1500 \text{ m/s})^2 = 987$ K.
b) $(4.385 \times 10^{-4} \text{ K·s}^2/\text{m}^2)(1000 \text{ m/s})^2 = 438$ K.
c) $(4.385 \times 10^{-4} \text{ K·s}^2/\text{m}^2)(500 \text{ m/s})^2 = 110$ K.

16-34: Making the given substitution $\epsilon = \frac{1}{2}mv^2$,

$$f(v) = 4\pi \left(\frac{m}{2\pi kT}\right)^{3/2} \frac{2\epsilon}{m} e^{-\epsilon/kT} = \frac{8\pi}{m} \left(\frac{m}{2\pi kT}\right)^{3/2} \epsilon e^{-\epsilon/kT}.$$

16-35: Express Eq. (16-33) as $f = A\epsilon e^{-\epsilon/kT}$, with A a constant. Then,

$$\frac{df}{d\epsilon} = A\left[e^{-\epsilon/kT} - \frac{\epsilon}{kT}e^{-\epsilon/kT}\right] = Ae^{-\epsilon/kT}\left[1 - \frac{\epsilon}{kT}\right].$$

Thus, f will be a maximum when the term in square brackets is zero, or $\epsilon = \frac{1}{2}mv^2 = kT$, which is Eq. (16-34).

16-36: Note that $\frac{k}{m} = \frac{R/N_A}{M/N_A} = \frac{R}{M}$.

    a) $\sqrt{2(8.3145 \text{ J/mol·K})(300 \text{ K})/(44.0 \times 10^{-3} \text{ kg/mol})} = 3.37 \times 10^2 \text{ m/s}$.

    b) $\sqrt{8(8.3145 \text{ J/mol·K})(300 \text{ K})/(\pi(44.0 \times 10^{-3} \text{ kg/mol}))} = 3.80 \times 10^2 \text{ m/s}$.

    c) $\sqrt{3(8.3145 \text{ J/mol·K})(300 \text{ K})/(44.0 \times 10^{-3} \text{ kg/mol})} = 4.12 \times 10^2 \text{ m/s}$.

16-37: Ice crystals will form if $T = 0.0°C$; using this in the given relation for temperature as a function of altitude gives $y = 2.5 \times 10^3 \text{ m} = 2.5 \text{ km}$.

16-38: a) The pressure must be above the triple point, $p_1 = 610 \text{ Pa}$. If $p < p_1$, the water cannot exist in the liquid phase, and the phase transition is from solid to vapor (sublimation). b) p_2 is the critical pressure, $p_2 = p_c = 221 \times 10^5 \text{ Pa}$. For pressures below p_2 but above p_1, the phase transition is the most commonly observed sequence, solid to liquid to vapor, or ice to water to steam.

16-39: The temperature of $0.00°C$ is just below the triple point of water, and so there will be no liquid. Solid ice and water vapor at $0.00°C$ will be in equilibrium.

16-40: The atmospheric pressure is below the triple point pressure of water, and there can be no liquid water on Mars. The same holds true for CO_2.

16-41: a) $\Delta V = \beta V_o \Delta T = (3.6 \times 10^{-5}/°C)(11 \text{ L})(21°C) = 0.0083 \text{ L}$

             $\Delta V = -kV_o \Delta p = (6.25 \times 10^{-12}/\text{Pa})(11 \text{ L})(2.1 \times 10^7 \text{ Pa}) = -0.0014 \text{ L}$

So the total change in volume is $\Delta V = 0.0083 \text{ L} - 0.0014 \text{ L} = 0.0069 \text{ L}$.   b) Yes; ΔV is much less than the original volume of 11.0 L.

16-42: $m = nM = \dfrac{MpV}{RT}$

$$= \frac{(28.0 \times 10^{-3} \text{ kg/mol})(2.026 \times 10^{-8} \text{ Pa})(3000 \times 10^{-6} \text{ m}^3)}{(8.3145 \text{ J/mol·K})(295.15 \text{ K})}$$

$$= 6.94 \times 10^{-16} \text{ kg}.$$

16-43: $\Delta m = \Delta nM = \dfrac{\Delta p V M}{RT}$

$$= \frac{(1.05 \times 10^6 \text{ Pa})((1.00 \text{ m})\pi(0.060 \text{ m})^2)(44.10 \times 10^{-3} \text{ kg/mol})}{(8.3145 \text{ J/mol·K})(295.15 \text{ K})} = 0.213 \text{ kg}.$$

16-44: a) The height h' at this depth will be proportional to the volume, and hence inversely proportional to the pressure and proportional to the Kelvin temperature;

$$h' = h\frac{p}{p'}\frac{T'}{T} = h\frac{p_{\text{atm}}}{p_{\text{atm}} + \rho g y}\frac{T'}{T}$$

$$= (2.30 \text{ m})\frac{(1.013 \times 10^5 \text{ Pa})}{(1.013 \times 10^5 \text{ Pa}) + (1030 \text{ kg/m}^3)(9.80 \text{ m/s}^2)(73.0 \text{ m})}\left(\frac{280.15 \text{ K}}{300.15 \text{ K}}\right)$$

$$= 0.26 \text{ m},$$

so $\Delta h = h - h' = 2.04$ m. b) The necessary gauge pressure is the term $\rho g y$ from the above calculation, $p_g = 7.37 \times 10^5$ Pa.

16-45:   The change in the height of the column of mercury is due to the pressure of the air. The mass of the air is

$$m_{N_2} = nM = \frac{PV}{RT}M = \frac{\rho_{Hg}g\Delta hV}{RT}M$$

$$= \left(\frac{\begin{array}{c}(13.6 \times 10^3 \text{ kg/m}^3)(9.80 \text{ m/s}^2)(0.060 \text{ m}) \\ \times ((0.900 \text{ m} - 0.690 \text{ m}))(0.620 \times 10^{-4} \text{ m}^2)\end{array}}{(8.3145 \text{ J/mol·K})(293.15 \text{ K})} \right) (28.0 \text{ g/mol})$$

$$= 1.23 \times 10^{-3} \text{ g.}$$

16-46:   The density ρ' of the hot air must be $\rho' = \rho - \frac{m}{V}$, where ρ is the density of the ambient air and m is the load. The density is inversely proportional to the temperature, so

$$T' = T\frac{\rho}{\rho'} = \frac{\rho}{\rho - (m/V)} = T\left(1 - \frac{m}{\rho V}\right)^{-1}$$

$$= (288.15 \text{ K})\left(1 - \frac{(290 \text{ kg})}{(1.23 \text{ kg/m}^3)(500 \text{ m}^3)}\right)^{-1} = 545 \text{ K,}$$

which is 272°C.

16-47: $p_2 = p_1\left(\frac{V_1 T_2}{V_2 T_1}\right) = (2.72 \text{ atm})\left(\frac{(0.0150 \text{ m}^3)(318.15 \text{ K})}{(0.0159 \text{ m}^3)(278.15 \text{ K})}\right) = 2.94 \text{ atm,}$

so the gauge pressure is 1.92 atm.

16-48:   (Neglect the thermal expansion of the flask.) a) $p_2 = p_1(T_2/T_1) = (1.013 \times 10^5 \text{ Pa})(300/380) = 8.00 \times 10^4$ Pa.

b) $m_{tot} = nM = \left(\frac{p_2 V}{RT_2}\right)M$

$$= \left(\frac{(8.00 \times 10^4 \text{ Pa})(1.50 \text{ L})}{(8.3145 \text{ J/mol·K})(300 \text{ K})}\right)(30.1 \text{ g/mol}) = 1.45 \text{ g.}$$

16-49:   a) At atmospheric pressure, the volume of hydrogen will increase by a factor of $\frac{1.30 \times 10^6}{1.01 \times 10^5}$, so the number of tanks is

$$\frac{750 \text{ m}^3}{(1.90 \text{ m}^3)((1.30 \times 10^6)/(1.01 \times 10^5))} = 31.$$

b) The difference between the weight of the air displaced and the weight of the hydrogen is

$$(\rho_{\text{air}} - \rho_{\text{H}_2})Vg = \left(\rho_{\text{air}} - \frac{pM_{\text{H}_2}}{RT}\right)Vg$$

$$= \left(1.23\ \text{kg/m}^3 - \frac{(1.01 \times 10^5\ \text{Pa})(2.02 \times 10^{-3}\ \text{kg/mol})}{(8.3145\ \text{J/mol·K})(288.15\ \text{K})}\right)$$

$$\times\ (9.80\ \text{m/s}^2)(750\ \text{m}^3)$$

$$= 8.42 \times 10^3\ \text{N}.$$

c) Repeating the above calculation with $M = 4.00 \times 10^{-3}$ kg/mol gives a weight of 7.80×10^3 N.

16-50: If the original height is h and the piston descends a distance y, the final pressure of the air will be $p_{\text{atm}}\left(\frac{h}{h-y}\right)$. This must be the same as the pressure at the bottom of the mercury column, $p_{\text{atm}} + (\rho g)y$. Equating these two, performing some minor algebra and solving for y gives

$$y = h - \frac{p_{\text{atm}}}{\rho g} = (0.900\ \text{m}) - \frac{(1.013 \times 10^5\ \text{Pa})}{(13.6 \times 10^3\ \text{kg/m}^3)(9.80\ \text{m/s}^2)} = 0.140\ \text{m}.$$

16-51: a) The tank is given as being "large," so the speed of the water at the top of the surface in the tank may be neglected. The efflux speed is then obtained from

$$\frac{1}{2}\rho v^2 = \rho g \Delta h + \Delta p, \quad \text{or}$$

$$v = \sqrt{2\left(g\Delta h + \frac{\Delta p}{\rho}\right)} = \sqrt{2\left((9.80\ \text{m/s}^2)(2.50\ \text{m}) + \frac{(3.20 \times 10^5\ \text{Pa})}{(1000\ \text{kg/m}^3)}\right)}$$

$$= 26.2\ \text{m/s}.$$

b) Let $h_0 = 3.50$ m and $p_0 = 4.20 \times 10^5$ Pa. In the above expression for v, $\Delta h = h - 1.00$ m and $\Delta p = p_0\left(\frac{4.00\ \text{m} - h_0}{4.00\ \text{m} - h}\right) - p_a$. Repeating the calculation for $h = 3.00$ m gives $v = 16.1$ m/s and with $h = 2.00$ m, $v = 5.44$ m/s. c) Setting $v^2 = 0$ in the above expression gives a quadratic equation in h which may be re-expressed as

$$(h - 1.00\ \text{m}) = \frac{p_a}{\rho g} - \frac{p_0}{\rho g}\frac{0.50\ \text{m}}{4.00\ \text{m} - h}.$$

Denoting $\frac{p_a}{\rho g} = y = 10.204$ m and $\frac{p_0(0.50\ \text{m})}{\rho g} = z^2 = 21.43$ m^2, this quadratic becomes

$$h^2 - (5.00\ \text{m} + y)h + ((4.00\ \text{m})y + (4.00\ \text{m}^2) - z^2) = 0,$$

which has as its solutions $h = 1.737$ m and $h = 13.47$ m. The larger solution is unphysical (the height is greater than the height of the tank), and so the flow stops when $h = 1.74$ m.

Although use of the quadratic formula is correct, for this problem it is more efficient for those with programmable calculators to find the solution to the quadratic by iteration. Using $h = 2.00$ m (the lower height in part (b)) gives convergence to three figures after four iterations. (The larger root is not obtained by a convergent iteration.)

16-52: a) $\dfrac{N}{\Delta t} = \dfrac{nN_A}{\Delta t} = \dfrac{pVN_A}{RT\Delta t} = \dfrac{(1.00 \text{ atm})(14.5 \text{ L})(6.023 \times 10^{23} \text{ molecules/mol})}{(0.08206 \text{ L·atm/mol·K})(293.15 \text{ K})(3600 \text{ s})}$

$$= 1.01 \times 10^{20} \text{ molecule.}$$

b) $\dfrac{(14.5 \text{ L})/(60 \text{ min})}{(0.5 \text{ L})(0.210 - 0.163)} = 10/\text{min.}$

c) The density of the air has decreased by a factor of 0.72, and so the respiration rate must increase by a factor of $\frac{1}{0.72}$, to 14 breaths/min. If the breathing rate is not increased, one would experience "shortness of breath."

16-53: $3N = 3nN_A = 3(m/M)N_A = 3\dfrac{(6.023 \times 10^{23} \text{ molecules/mol})(50 \text{ kg})}{(18.0 \times 10^{-3} \text{ kg/mol})}$

$$= 5.0 \times 10^{27} \text{ atoms.}$$

16-54: The volume of gas per molecule (see Problem 16-20) is $\frac{RT}{N_A p}$, and the volume of a molecule is about $V_0 = \frac{4}{3}\pi(2.0 \times 10^{-10} \text{ m})^3 = 3.4 \times 10^{-29} \text{ m}^3$. Denoting the ratio of these volumes as f,

$$p = f\frac{RT}{N_A V_0} = f\frac{(8.3145 \text{ J/mol·K})(300 \text{ K})}{(6.023 \times 10^{23} \text{ molecules/mol})(3.4 \times 10^{-29} \text{ m}^3)} = (1.2 \times 10^8 \text{ Pa})f.$$

"Noticeable deviations" is a subjective term, but f on the order of unity gives a pressure of 10^8 Pa. Deviations from ideality are likely to be seen at values of f substantially lower than this.

16-55: a) Dividing both sides of Eq. (16-7) by the product RTV gives the result. b) The algorithm described is best implemented on a programmable calculator or computer; for a calculator, the numerical procedure is an iteration of

$$x = \left[\frac{(9.8 \times 10^5)}{(8.3145)(400.15)} + \frac{(0.448)}{(8.3145)(400.15)}x^2\right][1 - (4.29 \times 10^{-5})x].$$

Starting at $x = 0$ gives a fixed point at $x = 3.03 \times 10^2$ after four iterations. The number density is 3.03×10^2 mol/m^2.   c) The ideal-gas equation is the result after the first iteration, 295 mol/m^3. The van der Waals density is larger. The term corresponding to a represents the attraction of the molecules, and hence more molecules will be in a given volume for a given pressure.

16-56: a) $U = mgh = \frac{M}{N_A}gh = (\frac{28.0 \times 10^{-3} \text{ kg/mol}}{6.023 \times 10^{23} \text{ molecules/mol}})(9.80 \text{ m/s}^2)(400 \text{ m}) = 1.82 \times 10^{-22}$ J.

b) Setting $U = \frac{3}{2}kT$, $T = \frac{2}{3}\frac{1.82 \times 10^{-22} \text{ J}}{1.38 \times 10^{-23} \text{ J/K}} = 8.80$ K.   c) It is possible, but not at all likely for a molecule to rise to that altitude. This altitude is much larger than the mean free path.

16-57:   a), b) (See figure.) The solid curve is $U(r)$, in units of U_0, and with $x = r/R_0$. The dashed curve is $F(r)$ in units of U_0/R_0. Note that $r_1 < r_2$.

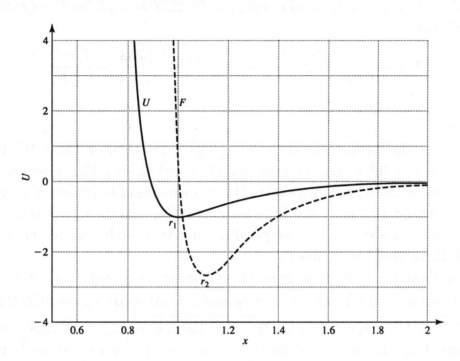

c) When $U = 0$, $\left(\frac{R_0}{r_1}\right)^{12} = 2\left(\frac{R_0}{r_1}\right)^6$, or $r_1 = R_0/2^{1/6}$. Setting $F = 0$ in Eq. (13-26) gives $r_2 = R_0$, and $\frac{r_1}{r_2} = 2^{-1/6}$.   d) $U(r_2) = U(R_0) = -U_0$, so the work required is U_0.

16-58:   a) $\frac{3}{2}nRT = \frac{3}{2}pV = \frac{3}{2}(1.01 \times 10^5 \text{ Pa})(5.00 \times 10^{-3} \text{ m}^3) = 758$ J.   b) The mass of the gas is $\frac{MpV}{RT}$, and so the ratio of the energies is

$$\frac{\frac{1}{2}\frac{MpV}{RT}v^2}{\frac{3}{2}pV} = \frac{1}{3}\frac{Mv^2}{RT} = \frac{1}{3}\frac{(2.016 \times 10^{-3} \text{ kg/mol})(30.0 \text{ m/s})^2}{(8.3145 \text{ J/mol·K})(300 \text{ K})} = 2.42 \times 10^{-4} = 0.0242\%.$$

16-59:   a) From Eq. (16-19),

$$v_{\text{rms}} = \sqrt{3(8.3145 \text{ J/mol·K})(300.15 \text{ K})/(28.0 \times 10^{-3} \text{ kg/mol})} = 517 \text{ m/s}.$$

b) $v_{\text{rms}}/\sqrt{3} = 299$ m/s.

16-60:   a) $\sqrt{\dfrac{3kT}{m}} = \sqrt{\dfrac{3(1.38 \times 10^{-23} \text{ J/K})(5800 \text{ K})}{(1.67 \times 10^{-27} \text{ kg})}} = 1.20 \times 10^4$ m/s.

b) $\sqrt{\dfrac{2GM}{R}} = \sqrt{\dfrac{2(6.673 \times 10^{-11} \text{ N·m}^2/\text{kg}^2)(1.99 \times 10^{30} \text{ kg})}{(6.96 \times 10^8 \text{ m})}} = 6.18 \times 10^5$ m/s.

c) The escape speed is about 50 times the rms speed, and any of Fig. (16-17), Eq. (16-32) or Table (16-2) will indicate that there is a negligibly small fraction of molecules with the escape speed.

16-61:   a) To escape, the total energy must be positive, $K + U > 0$. At the surface of the earth, $U = -GmM/R = -mgR$, so to escape $K > mgR$.   b) Setting the average kinetic energy equal to the expression found in part (a), $(3/2)kT = mgR$, or $T = (2/3)(mgR/k)$. For nitrogen, this is

$$T = \frac{2}{3} \frac{(28.0 \times 10^{-3} \text{ kg/mol})(9.80 \text{ m/s}^2)(6.38 \times 10^6 \text{ m})}{(6.023 \times 10^{23} \text{ molecules/mol})(1.381 \times 10^{-23} \text{ J/K})}$$

$$= 1.40 \times 10^5 \text{ K}$$

and for hydrogen the escape temperature is $\left(\frac{2.02}{28.0}\right)$ times this, or 1.01×10^4 K.   c) For nitrogen, $T = 6.36 \times 10^3$ K and for hydrogen, $T = 459$ K.   d) The escape temperature for hydrogen on the moon is comparable to the temperature of the moon, and so hydrogen would tend to escape until there would be none left. Although the escape temperature for nitrogen is higher than the moon's temperature, nitrogen would escape, and continue to escape, until there would be none left.

16-62:   (See Example 12-5 for calculation of the escape speeds)   a) Jupiter: $v_{rms} = \sqrt{3(8.3145 \text{ J/mol·K})(140 \text{ K})/(2.02 \times 10^{-3} \text{ kg/mol})} = 1.31 \times 10^3 \text{ m/s} = (0.0221)v_e$. Earth: $v_{rms} = \sqrt{3(8.3145 \text{ J/mol·K})(220 \text{ K})/(2.02 \times 10^{-3} \text{ kg/mol})} = 1.65 \times 10^3 \text{ m/s} = (0.146)v_e$.

b) Escape from Jupiter is not likely for any molecule, while escape from earth is possible for some and hence possible for all.

c) $v_{rms} = \sqrt{3(8.3145 \text{ J/mol·K})(200 \text{ K})/(32.0 \times 10^{-3} \text{ kg/mol})} = 395$ m/s. The radius of the asteroid is $R = (3M/4\pi\rho)^{1/3} = 4.68 \times 10^5$ m, and the escape speed is $\sqrt{2GM/R} = 542$ m/s, so there can be no such atmosphere.

16-63:   a) From Eq. (16-19),

$$m = \frac{3kT}{v_{rms}^2} = \frac{3(1.381 \times 10^{-23} \text{ J/K})(300 \text{ K})}{(0.001 \text{ m/s})^2} = 1.24 \times 10^{-14} \text{ kg}.$$

b) $mN_A/M = (1.24 \times 10^{-14} \text{ kg})(6.023 \times 10^{23} \text{ molecules/mol})/(18.0 \times 10^{-3} \text{ kg/mol}) = 4.16 \times 10^{11}$ molecules.

c)
$$D = 2r = 2\left(\frac{3V}{4\pi}\right)^{1/3} = 2\left(\frac{3m/\rho}{4\pi}\right)^{1/3}$$

$$= 2\left(\frac{3(1.24 \times 10^{-14} \text{ kg})}{4\pi(920 \text{ kg/m}^3)}\right)^{1/3} = 2.95 \times 10^{-6} \text{ m},$$

which is too small to see.

16-64: From $x = A\cos\omega t$, $v = -\omega A \sin\omega t$,

$$U_{\text{ave}} = \frac{1}{2}kA^2(\cos^2\omega t)_{\text{ave}}, \quad K_{\text{ave}} = \frac{1}{2}m\omega^2 A^2(\sin^2\omega t)_{\text{ave}}.$$

Using $(\sin^2\theta)_{\text{ave}} = (\cos^2\theta)_{\text{ave}} = \frac{1}{2}$ and $m\omega^2 = k$ shows that $K_{\text{ave}} = U_{\text{ave}}$.

16-65: a) In the same manner that Eq. (16-27) was obtained, the heat capacity of the two-dimensional solid would be $2R = 16.6$ J/mol·K. b) The heat capacity would behave qualitatively like those in Fig. (16-15), and heat capacity would decrease with decreasing temperature.

16-66: a) The two degrees of freedom associated with the rotation for a diatomic molecule account for two-fifths of the total kinetic energy, so $K_{\text{rot}} = nRT = (1.00)(8.3145 \text{ J/mol·K})(300 \text{ K}) = 2.49 \times 10^3$ J.

b) $$I = 2m(L/2)^2 = 2\left(\frac{16.0 \times 10^{-3} \text{ kg/mol}}{6.023 \times 10^{23} \text{ molecules/mol}}\right)(6.05 \times 10^{-11} \text{ m})^2$$

$$= 1.94 \times 10^{-46} \text{ kg·m}^2.$$

c) Using the results of parts (a) and (b),

$$\omega_{\text{rms}} = \sqrt{\frac{2K_{\text{rot}}/N_A}{I}} = \sqrt{\frac{2(2.49 \times 10^3 \text{ J})}{(1.94 \times 10^{-46} \text{ kg·m}^2)(6.023 \times 10^{23} \text{ molecules/mol})}}$$

$$= 6.52 \times 10^{12} \text{ rad/s},$$

much larger than that of machinery.

16-67: For CO_2, the contribution to C_V other than vibration is

$$\frac{5}{2}R = 20.79 \text{ J/mol·K}, \quad \text{and} \quad C_V - \frac{5}{2}R = 0.270C_V$$

For both SO_2 and H_2S, the contribution to C_V other than vibration is

$$\frac{6}{2}R = 24.94 \text{ J/mol·K},$$

and the respective fractions of C_V are 0.205 and 0.039.

16-68: a) $$\int_0^\infty f(v)\,dv = 4\pi\left(\frac{m}{2\pi kT}\right)^{3/2}\int_0^\infty v^2 e^{-mv^2/2kT}\,dv$$

$$= 4\pi\left(\frac{m}{2\pi kT}\right)^{3/2}\left(\frac{1}{4(m/2kT)}\right)\sqrt{\frac{\pi}{m/2kT}} = 1,$$

where the tabulated integral (given in Problem 16-69) has been used. b) $f(v)\,dv$ is the probability that a particle has speed between v and $v + dv$; the probability that the particle has some speed is unity, so the sum (integral) of $f(v)\,dv$ must be 1.

16-69: With $n = 2$ and $a = m/2kT$, the integral is

$$4\pi \left(\frac{m}{2\pi kT}\right)^{3/2} \left(\frac{3}{2^3(m/2kT)^2}\right) \sqrt{\frac{\pi}{(m/2kT)}} = \frac{3kT}{m},$$

which is Eq. (16-16).

16-70:

$$\int_0^\infty f(v)\,dv = 4\pi \left(\frac{m}{2\pi kT}\right)^{3/2} \int_0^\infty v^3 e^{-mv^2/2kT}\,dv.$$

Making the suggested change of variable, $v^2 = x$, $2v\,dv = dx$, $v^3\,dv = (1/2)x\,dx$, the integral becomes

$$\int_0^\infty vf(v)\,dv = 2\pi \left(\frac{m}{2\pi kT}\right)^{3/2} \int_0^\infty x e^{-mx/2kT}\,dx$$

$$= 2\pi \left(\frac{m}{2\pi kT}\right)^{3/2} \left(\frac{2kT}{m}\right)^2$$

$$= \frac{2}{\sqrt{\pi}}\sqrt{\frac{2kT}{m}} = \sqrt{\frac{8kT}{\pi m}},$$

which is Eq. (16-35).

16-71: a) See Problem 16-68. Because $f(v)\,dv$ is the probability that a particle has a speed between v and $v + dv$, $f(v)\,dv$ is the fraction of the particles that have speed in that range. The number of particles with speeds between v and $v + dv$ is therefore $dN = Nf(v)\,dv$ and

$$\Delta N = N \int_v^{v+\Delta v} f(v)\,dv.$$

b) $v_{mp} = \sqrt{\frac{2kT}{m}}$, and

$$f(v_{mp}) = 4\pi \left(\frac{m}{2\pi kT}\right)^{3/2} \left(\frac{2kT}{m}\right) e^{-1} = \frac{4}{e\sqrt{\pi}v_{mp}}.$$

For oxygen gas at 300 K, $v_{mp} = 3.95 \times 10^2$ m/s, and $f(v)\Delta v = 0.0421$, keeping an extra figure. c) Increasing v by a factor of 7 changes f by a factor of $7^2 e^{-48}$, and $f(v)\Delta v = 2.94 \times 10^{-21}$. d) Multiplying the temperature by a factor of 2 increases the most probable speed by a factor of $\sqrt{2}$, and the answers are decreased by $\sqrt{2}$; 0.0297 and 2.08×10^{-21}. e) Similarly, when the temperature is one-half what it was in parts (b) and (c), the fractions increase by $\sqrt{2}$ to 0.0595 and 4.15×10^{-21}. f) At lower temperatures, the distribution is more sharply peaked about the maximum (the most probable speed), as is shown in Fig. (16-17).

16-72: a) $(0.60)(2.34 \times 10^3 \text{ Pa}) = 1.40 \times 10^3 \text{ Pa}$.

b) $$m = \frac{MpV}{RT} = \frac{(18.0 \times 10^{-3} \text{ kg/mol})(1.40 \times 10^3 \text{ Pa})(1.00 \text{ m}^3)}{(8.3145 \text{ J/mol·K})(293.15 \text{ K})} = 10 \text{ g}.$$

16-73: The partial pressure of water in the room is the vapor pressure at which condensation occurs. The relative humidity is $\frac{1.81}{4.25} = 42.6\%$.

16-74: a) The partial pressure is $(0.35)(3.78 \times 10^3 \text{ Pa}) = 1.323 \times 10^3$ Pa. This is close to the vapor pressure at 12°C, which would be at an altitude $(30°C - 12°C)/(0.6°C/100 \text{ m}) = 3$ km above the ground (more precise interpolation is not warranted for this estimate).

b) The vapor pressure will be the same as the water pressure at around 24°C, corresponding to an altitude of about 1 km.

16-75: a) From Eq. (16-21),

$$\lambda = (4\pi\sqrt{2}r^2(N/V))^{-1} = (4\pi\sqrt{2}(5.0 \times 10^{-11} \text{ m})^2(50 \times 10^6 \text{ m}^{-3}))^{-1}$$
$$= 4.5 \times 10^{11} \text{ m}.$$

b) $\sqrt{3(8.3145 \text{ J/mol·K})(20 \text{ K})/(1.008 \times 10^{-3} \text{ kg/mol})} = 703$ m/s, and the time between collisions is then $(4.5 \times 10^{11} \text{ m})/(703 \text{ m/s}) = 6.4 \times 10^8$ s, about 20 yr. Collisions are not very important. c) $p = (N/V)kT = (50 \times 10^6 \text{ m}^{-3})(1.381 \times 10^{-23} \text{ J/K})(20 \text{ K}) = 1.4 \times 10^{-14}$ Pa.

d)
$$v_e = \sqrt{\frac{2GM}{R}} = \sqrt{\frac{2G(Nm/V)(4\pi R^3/3)}{R}} = \sqrt{(8\pi/3)G(N/V)mR}$$
$$= \sqrt{(8\pi/3)(6.673 \times 10^{-11} \text{ N·m}^2/\text{kg}^2)(50 \times 10^6 \text{ m}^{-3})(1.67 \times 10^{-27} \text{ kg})}$$
$$\times (10 \times 9.46 \times 10^{15} \text{ m})$$
$$= 650 \text{ m/s}.$$

This is lower than v_{rms}, and the cloud would tend to evaporate. e) In equilibrium (clearly not *thermal* equilibrium), the pressures will be the same; from $pV = NkT$,

$$kT_{\text{ISM}}(N/V)_{\text{ISM}} = kT_{\text{nebula}}(N/V)_{\text{nebula}}$$

and the result follows. f) With the result of part (e),

$$T_{\text{ISM}} = T_{\text{nebula}}\left(\frac{(V/N)_{\text{nebula}}}{(V/N)_{\text{ISM}}}\right) = (20 \text{ K})\left(\frac{50 \times 10^6 \text{ m}^3}{(200 \times 10^{-6} \text{ m}^3)^{-1}}\right) = 2 \times 10^5 \text{ K},$$

more than three times the temperature of the sun. This indicates a high average kinetic energy, but the thinness of the ISM means that a ship would not burn up.

16-76: a) Following Example 16-4, $\frac{dP}{dy} = -\frac{pM}{RT}$, which in this case becomes

$$\frac{dp}{p} = -\frac{Mg}{R}\frac{dy}{T_0 - \alpha y},$$

which integrates to

$$\ln\left(\frac{p}{p_0}\right) = \frac{Mg}{R\alpha}\ln\left(1 - \frac{\alpha y}{T_0}\right), \quad \text{or} \quad p = p_0\left(1 - \frac{\alpha y}{T_0}\right)^{\frac{Mg}{R\alpha}}$$

b) Using the first equation above, for sufficiently small α, $\ln(1 - \frac{\alpha y}{T_0}) \approx -\frac{\alpha y}{T_0}$, and this gives the expression derived in Example 16-4.

c)
$$\left(1 - \frac{(0.6 \times 10^{-2} \ C°/m)(8863 \ m)}{(288 \ K)}\right) = 0.8154,$$

$$\frac{Mg}{R\alpha} = \frac{(28.8 \times 10^{-3})(9.80 \ m/s^2)}{(8.3145 \ J/mol \cdot K)(0.6 \times 10^{-2} \ C°/m)} = 5.6576$$

(the extra significant figures are needed in exponents to reduce roundoff error), and $p_0(0.8154)^{5.6576} = 0.315$ atm, which is 0.95 of the result found in Example 16-4. Note: for calculators without the x^y function, the pressure in part (c) must be found from $p = p_0 \exp((5.6576) \ln(0.8154))$.

16-77: a) A positive slope $\frac{\partial p}{\partial V}$ would mean that an increase in pressure causes an increase in volume, or that decreasing volume results in a decrease in pressure, which cannot be the case for any real gas. b) See Fig. (16-4). From part (a), p cannot have a positive slope along an isotherm, and so can have no extremes (maxima or minima) along an isotherm. When $\frac{\partial p}{\partial V}$ vanishes along an isotherm, the point on the curve in a p-V diagram must be an inflection point, and $\frac{\partial^2 p}{\partial V^2} = 0$.

c)
$$p = \frac{nRT}{V - nb} - \frac{an^2}{V^2}$$

$$\frac{\partial p}{\partial V} = -\frac{nRT}{(V - nb)^2} + \frac{2an^2}{V^3}$$

$$\frac{\partial^2 p}{\partial V^2} = \frac{2nRT}{(V - nb)^3} - \frac{6an^2}{V^4}.$$

Setting the last two of these equal to zero gives

$$V^3 nRT = 2an^2(V - nb)^2, \quad V^4 nRT = 3an^2(V - nb)^3.$$

d) Following the hint, $V = (3/2)(V - nb)$, which is solved for $(V/n)_c = 3b$. Substituting this into either of the last two expressions in part (c) gives $T_c = 8a/27Rb$.

e)
$$p_c = \frac{RT}{(V/n)_c - b} - \frac{a}{(V/n)_c} = \frac{R\left(\frac{8a}{27Rb}\right)}{2b} - \frac{a}{9b^2} = \frac{a}{27b^2}.$$

f)
$$\frac{RT_c}{p_c(V/n)_c} = \frac{\frac{8}{27}\frac{a}{b}}{\frac{a}{27b^2}3b} = \frac{8}{3}.$$

g) H_2: 3.28. N_2: 3.44. H_2O: 4.35. h) While all are close to 8/3, the agreement is not good enough to be useful in predicting critical point data. The van der Waals equation models certain gases, and is not accurate for substances near critical points.

16-78: a) $v_{av} = \frac{1}{2}(v_1 + v_2)$ and $v_{rms} = \frac{1}{\sqrt{2}}\sqrt{v_1^2 + v_2^2}$, and

$$v_{rms}^2 - v_{av}^2 = \frac{1}{2}\left(v_1^2 + v_2^2\right) - \frac{1}{4}\left(v_1^2 + v_2^2 + 2v_1v_2\right)$$

$$= \frac{1}{4}\left(v_1^2 + v_2^2 - 2v_1v^2\right)$$

$$= \frac{1}{4}(v_1 - v_2)^2.$$

This shows that $v_{rms} \geq v_{av}$, with equality holding if and only if the particles have the same speeds.

b) $v_{rms}'^2 = \frac{1}{N+1}(Nv_{rms}^2 + u^2)$, $v_{av}' = \frac{1}{N+1}(Nv_{av} + u)$, and the given forms follow immediately.

c) The algebra is similar to that in part (a); it helps somewhat to express

$$v_{av}'^2 = \frac{1}{(N+1)^2}\left(N((N+1)-1)v_{av}^2 + 2Nv_{av}u + ((N+1)-N)u^2\right)$$

$$= \frac{N}{N+1}v_{av}^2 + \frac{N}{(N+1)^2}\left(-v_{av}^2 + 2v_{av}u - u^2\right) + \frac{1}{N+1}u^2.$$

Then,

$$v_{rms}'^2 - v_{av}'^2 = \frac{N}{(N+1)}\left(v_{rms}^2 - v_{av}^2\right) + \frac{N}{(N+1)^2}\left(v_{av}^2 - 2v_{av}u + u^2\right)$$

$$= \frac{N}{N+1}\left(v_{rms}^2 - v_{av}^2\right) + \frac{N}{(N+1)^2}(v_{av} - u)^2.$$

If $v_{rms} > v_{av}$, then this difference is necessarily positive, and $v'_{rms} > v'_{av}$.

d) The result has been shown for $N = 1$, and it has been shown that validity for N implies validity for $N + 1$; by induction, the result is true for all N.

Chapter 17   The First Law of Thermodynamics

17-1: a)

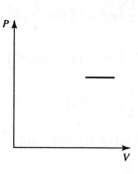

b) $p\Delta V = nR\Delta T = (2.00\ \text{mol})(8.3145\ \text{J/mol·K})(80\ \text{C}°) = 1.33 \times 10^3\ \text{J}.$

17-2: a)

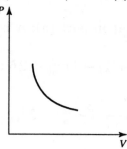

b) If the pressure is reduced by 60%, the final volume is (5/2) of its original value. From Eq. (17-4),

$$W = nRT \ln \frac{V_2}{V_1} = (3)(8.3145\ \text{J/mol·K})(400.15\ \text{K}) \ln \left(\frac{5}{2}\right) = 9.15 \times 10^3\ \text{J}.$$

17-3: a)

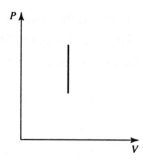

b) At constant volume, $dV = 0$ and so $W = 0$.

17-4: a)

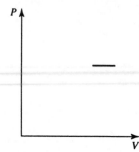

b) $p\Delta V = (1.50 \times 10^5\ \text{Pa})(0.0600\ \text{m}^3 - 0.0900\ \text{m}^3) = -4.50 \times 10^3\ \text{J}.$

17-5: a)

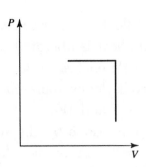

b) In the first process, $W_1 = p\Delta V = 0$. In the second process, $W_2 = p\Delta V = (5.00 \times 10^5 \text{ Pa})(-0.080 \text{ m}^3) = -4.00 \times 10^4 \text{ J}$.

17-6: a) $W_{13} = p_1(V_2 - V_1)$, $W_{32} = 0$, $W_{24} = p_2(V_1 - V_2)$ and $W_{41} = 0$. The total work done by the system is $W_{13} + W_{32} + W_{24} + W_{41} = (p_1 - p_2)(V_2 - V_1)$, which is the area in the p-V plane enclosed by the loop. b) For the process in reverse, the pressures are the same, but the volume changes are all the negatives of those found in part (a), so the total work is negative of the work found in part (a).

17-7: $Q = 254 \text{ J}$, $W = -73 \text{ J}$ (work is done *on* the system), and so $\Delta U = Q - W = 327 \text{ J}$.

17-8: a) $p\Delta V = (1.80 \times 10^5 \text{ Pa})(0.210 \text{ m}^2) = 3.78 \times 10^4 \text{ J}$.

b) $\Delta U = Q - W = 1.15 \times 10^5 \text{ J} - 3.78 \times 10^4 \text{ J} = 7.72 \times 10^4 \text{ J}$.

c) The relations $W = p\Delta V$ and $\Delta U = Q - W$ hold for any system.

17-9: a) $p\Delta V = (2.30 \times 10^5 \text{ Pa})(-0.50 \text{ m}^3) = -1.15 \times 10^5 \text{ J}$. b) $Q = \Delta U + W = -1.40 \times 10^5 \text{ J} + (-1.15 \times 10^5 \text{ J}) = -2.55 \times 10^5 \text{ J}$ (heat flows out of the gas). c) No; the first law of thermodynamics is valid for any system.

17-10: a) The greatest work is done along the path that bounds the largest area above the V-axis in the p-V plane (see Fig. (17-7)), which is path 1. The least work is done along path 3. b) $W > 0$ in all three cases; $Q = \Delta U + W$, so $Q > 0$ for all three, with the greatest Q for the greatest work, that along path 1. When $Q > 0$, heat is absorbed.

17-11: a) The energy is

$$(2.0 \text{ g})(4.0 \text{ kcal/g}) + (17.0 \text{ g})(4.0 \text{ kcal/g}) + (7.0 \text{ g})(9.0 \text{ kcal/g}) = 139 \text{ kcal},$$

and the time required is $(139 \text{ kcal})/(510 \text{ kcal/h}) = 0.273 \text{ h} = 16.4 \text{ min}$. b) $v = \sqrt{2K/m} = \sqrt{2(139 \times 10^3 \text{ cal})(4.186 \text{ J/cal})/(60 \text{ kg})} = 139 \text{ m/s} = 501 \text{ km/h}$.

17-12: a) The container is said to be well-insulated, so there is no heat transfer. b) Stirring requires work. The stirring needs to be irregular so that the stirring mechanism moves against the water, not with the water. c) The work mentioned in part (b) is work done *on* the system, so $W < 0$, and since no heat has been transferred, $\Delta U = -W > 0$.

17-13: The work done is positive from a to b and negative from b to a; the net work is the area enclosed and is positive around the clockwise path. For the closed path $\Delta U = 0$, so $Q = W > 0$. A positive value for Q means heat is absorbed. b) $|Q| = 7200 \text{ J}$, and from part (a), $Q > 0$ and so $Q = W = 7200$ J.   c) For the counterclockwise path, $Q = W < 0$. $W = -7200$ J, so $Q = -7200$ J and heat is liberated, with $|Q| = 7200$ J.

17-14:   a), b) The clockwise loop (I) encloses a larger area in the p-V plane than the counterclockwise loop (II). Clockwise loops represent positive work and counterclockwise loops negative work, so $W_{\text{I}} > 0$ and $W_{\text{II}} < 0$. Over one complete cycle, the net work $W_{\text{I}} + W_{\text{II}} > 0$, and the net work done by the system is positive. c) For the complete cycle, $\Delta U = 0$ and so $W = Q$. From part (a), $W > 0$ so $Q > 0$, and heat flows into the system.   d) Consider each loop as beginning and ending at the intersection point of the loops. Around each loop, $\Delta U = 0$, so $Q = W$; then, $Q_{\text{I}} = W_{\text{I}} > 0$ and $Q_{\text{II}} = W_{\text{II}} < 0$. Heat flows into the system for loop I and out of the system for loop II.

17-15:   a) Yes; heat has been transferred from the gasses to the water (and very likely the can), as indicated by the temperature rise of the water. For the system of the gasses, $Q < 0$. b) The can is given as being constant-volume, so the gasses do no work. Neglecting the thermal expansion of the water, no work is done.   c) $\Delta U = Q - W = Q < 0$.

17-16:   a) $p\Delta V = (2.026 \times 10^5 \text{ Pa})(0.824 \text{ m}^3 - 1.00 \times 10^{-3} \text{ m}^3) = 1.67 \times 10^5$ J.

b)
$$\Delta U = Q - W = mL_{\text{v}} - W$$
$$= (1.00 \text{ kg})(2.20 \times 10^6 \text{ J/kg}) - 1.67 \times 10^5 \text{ J} = 2.03 \times 10^6 \text{ J}.$$

17-17:   a) Using Equation (17-12), $\mathrm{d}T = \dfrac{\mathrm{d}Q}{nC_V} = \dfrac{645 \text{ J}}{(0.185 \text{ mol})(20.76 \text{ J/mol·K})} = 167.9$ K, or $T = 948$ K.

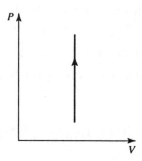

b) Using Equation (17-14), $\mathrm{d}T = \dfrac{\mathrm{d}Q}{nC_p} = \dfrac{645 \text{ J}}{(0.185 \text{ mol})(29.07 \text{ J/mol·K})} = 119.9$ K, or $T = 900$ K.

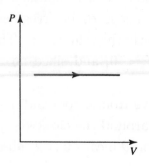

17-18: a) $nC_V\Delta T = (0.0100 \text{ mol})(12.47 \text{ J/mol·K})(40.0 \text{ C°}) = 4.99 \text{ J}.$

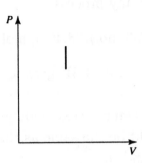

b) $nC_p\Delta T = (0.0100 \text{ mol})(20.78 \text{ J/mol·K})(40.0 \text{ C°}) = 8.31 \text{ J}.$

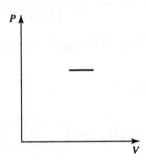

c) In the first process, $W = 0$ but in the second $W > 0$. ΔU is the same for both, and so Q is larger for the second.   d) For an ideal gas, $\Delta U = nC_V\Delta T = 4.99 \text{ J}$ for both parts (a) and (b).

17-19: a) For an isothermal process,

$$W = nRT \ln(V_2/V_1) = (0.150 \text{ mol})(8.3145 \text{ J/mol·K})(350.15 \text{ K}) \ln(1/4)$$
$$= -605 \text{ J}.$$

b) For an isothermal process for an ideal gas, $\Delta T = 0$ and $\Delta U = 0$.   c) For a process with $\Delta U = 0$, $Q = W = -605 \text{ J}$; 605 J are liberated.

17-20: For an isothermal process, $\Delta U = 0$, so $W = Q = -335 \text{ J}$.

17-21: For an ideal gas $\gamma = C_p/C_V = 1 + R/C_V$, and so $C_V = R/(\gamma - 1) = (8.3145 \text{ J/mol·K})/(0.127) = 65.5 \text{ J/mol·K}$ and $C_p = C_V + R = 73.8 \text{ J/mol·K}$.

17-22: a)

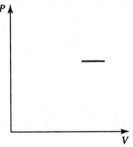

b) $$pV_2 - pV_1 = nR(T_2 - T_1)$$
$$= (0.250 \text{ mol})(8.3145 \text{ J/mol·K})(100.0 \text{ K}) = 208 \text{ J}.$$

c) The work is done on the piston.

d) Since Eq. (17-13) holds for any process,

$$\Delta U = nC_V\Delta T = (0.250 \text{ mol})(28.46 \text{ J/mol·K})(100.0 \text{ K}) = 712 \text{ J}.$$

e) Either $Q = nC_P\Delta T$ or $Q = \Delta U + W$ gives $Q = 924 \times 10^3$ J to three significant figures.

f) The lower pressure would mean a correspondingly larger volume, and the net result would be that the work done would be the same as that found in part (b).

17-23: a) $C_p = R/(1 - (1/\gamma))$, and so

$$Q = nC_p\Delta T = \frac{(2.40 \text{ mol})(8.3145 \text{ J/mol·K})(5.0 \text{ C°})}{1 - 1/1.220} = 553 \text{ J}.$$

b) $nC_V\Delta T = nC_P\Delta T/\gamma = (553 \text{ J})/(1.220) = 454$ J. (An extra figure was kept for these calculations.)

17-24: a) See also Exercise 17-26;

$$p_2 = p_1 \left(\frac{V_1}{V_2}\right)^\gamma = (1.50 \times 10^5 \text{ Pa}) \left(\frac{0.0800 \text{ m}^3}{0.0400 \text{ m}^3}\right)^{\frac{5}{3}} = 4.76 \times 10^5 \text{ Pa}.$$

b) This result may be substituted into Eq. (17-26), or, substituting the above form for p_2,

$$W = \frac{1}{\gamma - 1}p_1V_1 \left(1 - \left(\frac{V_1}{V_2}\right)^{\gamma-1}\right)$$

$$= \frac{3}{2}(1.50 \times 10^5 \text{ Pa})(0.0800 \text{ m}^3) \left(1 - \left(\frac{0.0800}{0.0400}\right)^{\frac{2}{3}}\right) = -1.06 \times 10^4 \text{ J}.$$

c) From Eq. (17-22), $(T_2/T_1) = (V_2/V_1)^{\gamma-1} = (0.0800/0.0400)^{2/3} = 1.59$, and since the final temperature is higher than the initial temperature, the gas is heated (see the note in Section 17-9 regarding "heating" and "cooling.")

17-25: a)

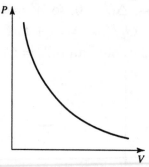

b) (Use $\gamma = 1.400$, as in Example 17-6.) From Eq. (17-22),

$$T_2 = T_1(V_1/V_2)^{\gamma-1} = (293.15 \text{ K})(11.1)^{0.400} = 768 \text{ K} = 495°C$$

and from Eq. (17-24), $p_2 = p_1(V_1/V_2)^\gamma = (1.00 \text{ atm})(11.1)^{1.400} = 29.1$ atm.

17-26: Equations (17-22) and (17-24) may be re-expressed as

$$\frac{T_2}{T_1} = \left(\frac{V_1}{V_2}\right)^{\gamma-1}, \quad \frac{p_2}{p_1} = \left(\frac{V_1}{V_2}\right)^{\gamma}.$$

a) $\gamma = \frac{5}{3}$, $p_2 = (4.00 \text{ atm})(2/3)^{\frac{5}{3}} = 2.04$ atm, $T_2 = (350 \text{ K})(2/3)^{\frac{2}{3}} = 267$ K.

b) $\gamma = \frac{7}{5}$, $p_2 = (4.00 \text{ atm})(2/3)^{\frac{7}{5}} = 2.27$ atm, $T_2 = (350 \text{ K})(2/3)^{\frac{2}{5}} = 298$ K.

17-27: a)

b) From Eq. (17-25), $W = nC_V\Delta T = (0.450 \text{ mol})(12.47 \text{ J/mol·K})(40.0 \text{ C}°) = 224$ J. For an adiabatic process, $Q = 0$ and there is no heat flow. $\Delta U = Q - W = -W = -224$ J.

17-28: a) $T = \frac{pV}{nR} = \frac{(1.00 \times 10^5 \text{ Pa})(2.50 \times 10^{-3} \text{ m}^3)}{(0.1 \text{ mol})(8.3145 \text{ J/mol·K})} = 301$ K.

b) i) Isothermal: If the expansion is *isothermal*, the process occurs at constant temperature and the final temperature is the same as the initial temperature, namely 301 K.

ii) Isobaric:

$$T = \frac{pV}{nR} = \frac{(1.00 \times 10^5 \text{ Pa})(5.00 \times 10^{-3} \text{ m}^3)}{(0.100 \text{ mol})(8.3145 \text{ J/mol·K})}$$

$$T = 601 \text{ K}.$$

iii) Adiabatic: Using Equation (17-22), $T_2 = \frac{T_1 V_1^{\gamma-1}}{V_2^{\gamma-1}} = \frac{(301 \text{ K})(V_1^{.67})}{(2V_1^{.67})} = (301 \text{ K})(\frac{1}{2})^{.67} = 189$ K.

17-29: See Exercise 17-24. a) $p_2 = p_1(V_1/V_2)^{\gamma} = (1.10 \times 10^5 \text{ Pa})((5.00 \times 10^{-3} \text{ m}^3)/(1.00 \times 10^{-2} \text{ m}^3))^{1.29} = 4.50 \times 10^4$ Pa. b) Using Equation (17-26),

$$W = \frac{(p_1 V_1 - p_2 V_2)}{\gamma - 1}$$
$$= \frac{[(1.1 \times 10^5 \text{ N/m}^3)(5.0 \times 10^{-3} \text{ m}^3) - (4.5 \times 10^4 \text{ N/m}^3)(1.0 \times 10^{-2} \text{ m}^3)]}{(1.29 - 1)},$$

and thus $W = 345$ J. c) $(T_2/T_1) = (V_2/V_1)^{\gamma-1} = ((5.00 \times 10^{-3} \text{ m}^3)/(1.00 \times 10^{-2} \text{ m}^3))^{0.29} = 0.818$. The final temperature is lower than the initial temperature, and the gas is cooled.

17-30: a) The product pV increases, and even for a non-ideal gas, this indicates a temperature increase. b) The work is the area in the p-V plane bounded by the blue line representing the process and the verticals at V_a and V_b. The area of this trapezoid is

$$\frac{1}{2}(p_b + p_a)(V_b - V_a) = \frac{1}{2}(2.40 \times 10^5 \text{ Pa})(0.0400 \text{ m}^3) = 4800 \text{ J}.$$

17-31: a) $\Delta U = Q - W = (90.0 \text{ J}) - (60.0 \text{ J}) = 30.0 \text{ J}$ for any path between a and b. If $W = 15.0 \text{ J}$ along path abd, then $Q = \Delta U + W = 30.0 \text{ J} + 15.0 \text{ J} = 45.0 \text{ J}$. b) Along the return path, $\Delta U = -30.0 \text{ J}$, and $Q = \Delta U + W = (-30.0 \text{ J}) + (-35.0 \text{ J}) = -65.0 \text{ J}$; the negative sign indicates that the system liberates heat. c) In the process db, $dV = 0$ and so the work done in the process ad is 15.0 J; $Q_{ad} = (U_d - U_a) + W_{ad} = (8.00 \text{ J}) + (15.0 \text{ J}) = 23.0 \text{ J}$. In the process db, $W = 0$ and so $Q_{db} = U_b - U_d = 30.0 \text{ J} - 8.0 \text{ J} = 22.0 \text{ J}$.

17-32: For each process, $Q = \Delta U + W$. No work is done in the processes ab and dc, and so $W_{bc} = W_{abc}$ and $W_{ad} = W_{adc}$, and the heat flow for each process is: for ab, $Q = 90 \text{ J}$: for bc, $Q = 440 \text{ J} + 450 \text{ J} = 890 \text{ J}$: for ad, $Q = 180 \text{ J} + 120 \text{ J} = 300 \text{ J}$: for dc, $Q = 350 \text{ J}$. Since $Q > 0$ for each process, heat is absorbed in each process. Note that the arrows representing the processes all point in the direction of increasing temperature (increasing U).

17-33: We will need to use Equations (17-3), $W = p(V_2 - V_1)$ and (17-4), $\Delta U = Q - W$.

a) The work done by the system during the process: Along ab or cd, $W = 0$. Along bc, $W_{bc} = P_c(V_c - V_a)$. Along ad, $W_{ad} = p_a(V_c - V_a)$.

b) The heat flow into the system during the process: $Q = \Delta U + W$.

$\Delta U_{ab} = U_b - U_a$, so $Q_{ab} = U_b - U_a + 0$.

$\Delta U_{bc} = U_c - U_b$, so $Q_{bc} = (U_c - U_b) + p_c(V_c - V_a)$.

$\Delta U_{ad} = U_d - U_a$, so $Q_{ad} = (U_d - U_a) + p_a(V_c - V_a)$.

$\Delta U_{dc} = U_c - U_d$, so $Q_{dc} = (U_c - U_d) + 0$.

c) From state a to state c along path abc:

$$W_{abc} = p_c(V_c - V_a). \quad Q_{abc} = U_b - U_a + (U_c - U_b) + p_c(V_c - V_a) = (U_c - U_a) + p_c(V_c - V_a)$$

From state a to state c along path adc:

$$W_{abc} = p_a(V_c - V_a). \quad Q_{adc} = (U_c - U_a) + p_a(V_c - V_a)$$

Assuming $P_c > P_a$, $Q_{abc} > Q_{adc}$, and $W_{abc} > W_{adc}$.

d) To understand this difference, start from the relationship $Q = W + \Delta U$. The internal energy change ΔU is path independent and so it is the same for path abc and path adc. The work done by the system is the area *under* the path in the pV-plane and is *not* the same for the two paths. Indeed, it is larger for path abc. Since ΔU is the same and W is different, Q must be different for the two paths. The heat flow Q is path dependent.

17-34: a) $n = \frac{Q}{C_P \Delta T} = \frac{(+2.5 \times 10^4 \text{ J})}{(29.07 \text{ J/mol·K})(40.0 \text{ K})} = 21.5$ mol.

b) $\Delta U = nC_V \Delta T = Q\frac{C_V}{C_P} = (-2.5 \times 10^4 \text{ J})\frac{20.76}{29.07} = -1.79 \times 10^4 \text{ J}$.

c) $W = Q - \Delta U = -7.15 \times 10^3 \text{ J}$.

d) ΔU is the same for both processes, and if $dV = 0$, $W = 0$ and $Q = \Delta U = -1.79 \times 10^4$ J.

17-35: $\Delta U = 0$, and so $Q = W = p\Delta V$ and

$$\Delta V = \frac{W}{p} = \frac{(-2.15 \times 10^5 \text{ J})}{(9.50 \times 10^5 \text{ Pa})} = -0.226 \text{ m}^3,$$

with the negative sign indicating a decrease in volume.

17-36: a)

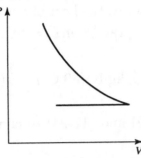

b) At constant temperature, the product pV is constant, so $V_2 = V_1(p_1/p_2) = (1.5 \text{ L})\left(\frac{1.00 \times 10^5 \text{ Pa}}{2.50 \times 10^4 \text{ Pa}}\right) = 6.00$ L. The final pressure is given as being the same as $p_3 = p_2 = 2.5 \times 10^4$ Pa. The final volume is the same as the initial volume, so $T_3 = T_1(p_3/p_1) = 75.0$ K.   c) Treating the gas as ideal, the work done in the first process is

$$nRT \ln(V_2/V_1) = p_1 V_1 \ln(p_1/p_2)$$

$$= (1.00 \times 10^5 \text{ Pa})(1.5 \times 10^{-3} \text{ m}^3) \ln\left(\frac{1.00 \times 10^5 \text{ Pa}}{2.50 \times 10^4 \text{ Pa}}\right)$$

$$= 208 \text{ J},$$

keeping an extra figure. For the second process,

$$p_2(V_3 - V_2) = p_2(V_1 - V_2) = p_2 V_1(1 - (p_1/p_2))$$

$$= (2.50 \times 10^4 \text{ Pa})(1.5 \times 10^{-3} \text{ m}^3)\left(1 - \frac{1.00 \times 10^5 \text{ Pa}}{2.50 \times 10^4 \text{ Pa}}\right) = -113 \text{ J}.$$

d) Heat at constant volume.

17-37: a) The fractional change in volume is

$$\Delta V = V_0 \beta \Delta T = (1.20 \times 10^{-2} \text{ m}^3)(1.20 \times 10^{-3} \text{ K}^{-1})(30.0 \text{ K}) = 4.32 \times 10^{-4} \text{ m}^3.$$

b) $p\Delta V = (F/A)\Delta V = ((3.00 \times 10^4 \text{ N})/0.0200 \text{ m}^2))(4.32 \times 10^{-4} \text{ m}^3) = 648$ J.

c) $Q = mC_p\Delta T = V_0 \rho C_p \Delta T = (1.20 \times 10^{-2} \text{ m}^3)(791 \text{ kg/m}^3)(2.51 \times 10^3 \text{ J/kg·K})$ $(30.0 \text{ K}) = 7.15 \times 10^5$ J.

d) $\Delta U = Q - W = 7.15 \times 10^5$ J to three figures.   e) Under these conditions, there is no substantial difference between c_V and c_p.

17-38: a) $\beta \Delta T V_0 = (5.1 \times 10^{-5} \text{ (C°)}^{-1})(70.0 \text{ C°})(2.00 \times 10^{-2})^3 = 2.86 \times 10^{-8}$ m^3.

b) $p\Delta V = 2.88 \times 10^{-3}$ J.

c)
$$Q = mC\Delta T = \rho V_0 C \Delta T$$
$$= (8.9 \times 10^3 \text{ kg/m}^3)(8.00 \times 10^{-6} \text{ m}^3)(390 \text{ J/kg·K})(70.0 \text{ C}°)$$
$$= 1944 \text{ J}.$$

d) To three figures, $\Delta U = Q = 1940$ J.   e) Under these conditions, the difference is not substantial.

17-39:   For a mass m of ejected spray, the heat of reaction L is related to the temperature rise and the kinetic energy of the spray by $mL = mC\Delta T - (1/2)mv^2$, or

$$L = C\Delta T - \frac{1}{2}v^2 = (4190 \text{ J/kg·K})(80 \text{ C}°) - \frac{1}{2}(19 \text{ m/s})^2 = 3.4 \times 10^5 \text{ J/kg}.$$

17-40:   Solving Equations (17-22) and (17-24) to eliminate the volumes,

$$p_1^{\gamma-1}T_1^\gamma = p_2^{\gamma-1}T_2^\gamma, \quad \text{or} \quad T_1 = T_2 \left(\frac{p_1}{p_2}\right)^{1-\frac{1}{\gamma}}.$$

Using $\gamma = \frac{7}{5}$ for air, $T_1 = (273.15 \text{ K})(\frac{1.60 \times 10^6}{2.80 \times 10^5})^{\frac{2}{7}} = 449$ K, which is 176°C.

17-41:   a) As the air moves to lower altitude its density increases; under an adiabatic compression, the temperature rises. If the wind is fast-moving, Q is not as likely to be significant, and modeling the process as adiabatic (no heat loss to the surroundings) is more accurate.   b) See Problems 17-43 and 17-40: The temperature at the higher pressure is $T_2 = (258.15 \text{ K})((8.12 \times 10^4 \text{ Pa})/(5.60 \times 10^4 \text{ Pa}))^{2/7} = 287.1$ K, which is 13.9°C and so the temperature would rise by 11.9 C°.

17-42:   a)

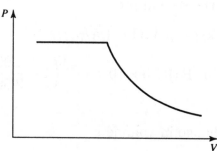

b) The work done is

$$W = p_0(2V_0 - V_0) + \frac{C_V}{R}(p_0(2V_0) - p_3(4V_0)).$$

$p_3 = p_0(2V_0/4V_0)^\gamma$ and so

$$W = p_0 V_0 \left[1 + \frac{C_V}{R}(2 - 2^{2-\gamma})\right]$$

Note that p_0 is the absolute pressure.   c) The most direct way to find the temperature is to find the ratio of the final pressure and volume to the original and treat the air as an ideal gas;

$$T_3 = T_0 \frac{p_3 V_3}{p_1 V_1} = T_0 \left(\frac{V_2}{V_3}\right)^{\gamma} \left(\frac{V_3}{V_1}\right) = T_0 \left(\frac{1}{2}\right)^{\gamma} 4 = T_0 (2)^{2-\gamma}$$

d) Since $n = \frac{p_0 V_0}{RT_0}$, $Q = \frac{p_0 V_0}{RT_0}(C_V + R)(2T_0 - T_0) = p_0 V_0 \left(\frac{C_V}{R} + 1\right)$. This amount of heat flows into the gas.

17-43:   a) For constant cross-section area, the volume is proportional to the length, and Eq. (17-24) becomes $L_2 = L_1 (p_1/p_2)^{1/\gamma}$ and the distance the piston has moved is

$$L_1 - L_2 = L_1 \left(1 - \left(\frac{p_1}{p_2}\right)^{1/\gamma}\right) = (0.250 \text{ m}) \left(1 - \left(\frac{1.01 \times 10^5 \text{ Pa}}{5.21 \times 10^5 \text{ Pa}}\right)^{1/1.400}\right)$$

$$= 0.173 \text{ m}.$$

b) Raising both sides of Eq. (17-22) to the power γ and both sides of Eq. (17-24) to the power $\gamma - 1$, dividing to eliminate the terms $V_1^{\gamma(\gamma-1)}$ and $V_2^{\gamma(\gamma-1)}$ and solving for the ratio of the temperatures,

$$T_2 = T_1 \left(\frac{p_2}{p_1}\right)^{1-(1/\gamma)} = (300.15 \text{ K}) \left(\frac{5.21 \times 10^5 \text{ Pa}}{1.01 \times 10^5 \text{ Pa}}\right)^{1-(1/1.400)} = 480 \text{ K} = 206°\text{C}.$$

Using the result of part (a) to find L_2 and then using Eq. (17-22) gives the same result.

c) Of the many possible ways to find the work done, the most straightforward is to use the result of part (b) in Eq. (17-25),

$$W = nC_V \Delta T = (20.0 \text{ mol})(20.8 \text{ J/mol·K})(179.0 \text{ C}°) = 7.45 \times 10^4 \text{ J},$$

where an extra figure was kept for the temperature difference.

17-44:   a)

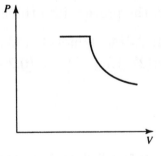

b) The final temperature is the same as the initial temperature, and the density is proportional to the absolute pressure. The mass needed to fill the cylinder is then

$$m = \rho_0 V \frac{p}{p_a} = (1.23 \text{ kg/m}^3)(575 \times 10^{-6} \text{ m}^3)\frac{1.45 \times 10^5 \text{ Pa}}{1.01 \times 10^5 \text{ Pa}} = 1.02 \times 10^{-3} \text{ kg}.$$

The increase in power is proportional to the increase in pressure; the percentage increase is $\frac{1.45}{1.01} - 1 = 0.44 = 44\%$.   c) The temperature of the compressed air is not the same as the original temperature; the density is proportional to the pressure, and for the process, modeled as abiabatic, the volumes are related to the pressures by Eq. (17-24), and the mass of air needed to fill the cylinder is

$$m = \rho_0 V \left(\frac{p}{p_a}\right)^{1/\gamma} = (1.23 \text{ kg/m}^3)(575 \times 10^{-6} \text{ m}^3)\left(\frac{1.45 \times 10^5 \text{ Pa}}{1.01 \times 10^5 \text{ Pa}}\right)^{1/1.40}$$

$$= 9.16 \times 10^{-4} \text{ kg},$$

an increase of $(1.45/1.01)^{1/1.40} - 1 = 0.29 = 29\%$.

17-45:   a) For an isothermal process for an ideal gas, $\Delta T = 0$ and $\Delta U = 0$, so $Q = W = 300$ J.   b) For an adiabatic process, $Q = 0$, and $\Delta U = -W = -300$ J.   c) For isobaric, $W = pdV = nRdT$, or $dT = \frac{W}{nR}$. Then, $Q = nC_pdT$ and substituting for dT gives $Q = nC_p\frac{W}{nR} = C_p\frac{W}{R}$, or $Q = \frac{5}{2}R\frac{W}{R} = \frac{5}{2}(300$ J$)$. Thus, $Q = 750$ J. To find ΔU, use $\Delta U = nC_vdT$. Substituting for dT and C_v, $\Delta U = n(\frac{3}{2}R)\frac{W}{nR} = \frac{3}{2}W = 450$ J.

17-46:   a)

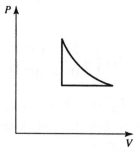

b) The isobaric process doubles the temperature to 710 K, and this must be the temperature of the isothermal process. c) After the isothermal process, the oxygen is at its original volume but twice the original temperature, so the pressure is twice the original pressure, 4.80×10^5 Pa. d) Break the process into three steps.

$W_1 = -nRT_\circ = -(0.25 \text{ mol})(8.3145 \text{ J/mol} \cdot \text{K})(335 \text{ K}) = -738$ J;

$W_2 = nRT \ln(p_1/p_2) = nR(2T_\circ) \ln(1/2) = (0.250 \text{ mol})(8.3145 \text{ J/mol} \cdot \text{K})(710 \text{ K})$ $(.693) = 1023$ J;

$W_3 = 0$ (because $dV = 0$).

Thus, $W = 285$ J.

17-47:   a) During the expansion, the Kelvin temperature doubles and $\Delta T = 300$ K. $W = p\Delta V = nR\Delta T = (0.250 \text{ mol})(8.3145 \text{ J/mol·K})(355 \text{ K}) = 738$ J, $Q = nC_p\Delta T = (0.250 \text{ mol})(29.17 \text{ J/mol·K})(355 \text{ K}) = 2590$ J and $\Delta U = nC_V\Delta T = Q - W = 1850$ J. b) The final cooling is isochoric; $dV = 0$ and so $W = 0$. The temperature change is $\Delta T = -355$ K, and $Q = \Delta U = nC_V\Delta T = -1850$ J.   c) For the isothermal compression, $\Delta T = 0$ and so $\Delta U = 0$.

17-48: a)

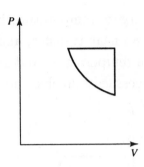

b) At constant pressure, halving the volume halves the Kelvin temperature, and the temperature at the beginning of the adiabatic expansion is 150 K.   c) The volume doubles during the adiabatic expansion, and from Eq. (17-22), the temperature at the end of the expansion is $(150 \text{ K})(1/2)^{0.40} = 114 \text{ K}$.   c) The minimum pressure occurs at the end of the adiabatic expansion. During the heating the volume is held constant, so the minimum pressure is proportional to the Kelvin temperature, $p_{min} = (1.80 \times 10^5 \text{ Pa})(113.7 \text{ K}/300 \text{ K}) = 6.82 \times 10^4 \text{ Pa}$.

17-49: a) $W = p\Delta V = nR\Delta T = (0.150 \text{ mol})(8.3145 \text{ J/mol·K})(-150 \text{ K}) = -187 \text{ J}$, $Q = nC_p\Delta T = (0.150 \text{ mol})(29.07 \text{ J/mol·K})(-150 \text{ K}) = -654 \text{ J}$, $\Delta U = Q - W = -467 \text{ J}$.

b) From Eq. (17-24), using the expression for the temperature found in Problem 17-48,

$$W = \frac{1}{0.40}(0.150 \text{ mol})(8.3145 \text{ J/mol·K})(150 \text{ K})\left(1 - (1/2)^{0.40}\right) = 113 \text{ J},$$

$Q = 0$ for an adiabatic process, and $\Delta U = Q - W = -W = -113 \text{ J}$.   c) $dV = 0$, so $W = 0$. Using the temperature change as found in Problem 17-48 and part (b), $Q = nC_V\Delta T = (0.150 \text{ mol})(20.76 \text{ J/mol·K})(300 \text{ K} - 113.7 \text{ K}) = 580 \text{ J}$, and $\Delta U = Q - W = Q = 580 \text{ J}$.

17-50: a) $W = nRT \ln(\frac{V_2}{V_1}) = nRT \ln(3) = 3.29 \times 10^3 \text{ J}$.

b) See Problem 17-24(b); $nC_V T_1 (1 - (1/3)^{\frac{2}{3}}) = 2.33 \times 10^3 \text{ J}$.

c) $V_2 = 3V_1$, so $p\Delta V = 2pV_1 = 2nRT_1 = 6.00 \times 10^3 \text{ J}$.

d)

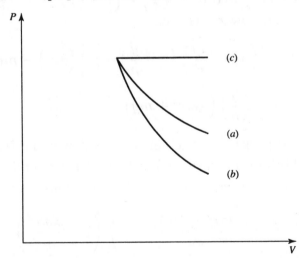

The most work done is in the isobaric process, as the pressure is maintained at its original value. The least work is done in the abiabatic process. e) The isobaric process involves the most work and the largest temperature increase, and so requires the most heat. Adiabatic processes involve no heat transfer, and so the magnitude is zero. f) The isobaric process doubles the Kelvin temperature, and so has the largest change in internal energy. The isothermal process necessarily involves no change in internal energy.

17-51:   a)

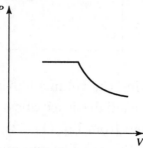

   b) No heat is supplied during the adiabatic expansion; during the isobaric expansion, the heat added is $nC_p\Delta T$. The Kelvin temperature doubles, so $\Delta T = 300.15$ K and $Q = (0.350 \text{ mol})(34.60 \text{ J/mol·K})(300.15 \text{ K}) = 3.63 \times 10^3$ J.   c) For the entire process, $\Delta T = 0$ and so $\Delta U = 0$.   d) If $\Delta U = 0$, $W = Q = 3.63 \times 10^3$ J.   e) During the isobaric expansion, the volume doubles. During the adiabatic expansion, the temperature decreases by a factor of two, and from Eq. (17-22) the volume changes by a factor of $2^{1-\gamma} = 2^{-0.33}$, and the final volume is $(14 \times 10^{-3} \text{ m}^3)2^{-0.33} = 0.0113$ m^3.

17-52:   a) The difference between the pressure, multiplied by the area of the piston, must be the weight of the piston. The pressure in the trapped gas is $p_0 + \frac{mg}{A} = p_0 + \frac{mg}{\pi r^2}$.
b) When the piston is a distance $h + y$ above the cylinder, the pressure in the trapped gas is

$$\left(p_0 + \frac{mg}{\pi r^2}\right)\left(\frac{h}{h+y}\right)$$

and for values of y small compared to h, $\frac{h}{h+y} = \left(1 + \frac{y}{h}\right)^{-1} \sim 1 - \frac{y}{h}$. The net force, taking the positive direction to be upward, is then

$$F = \left[\left(p_0 + \frac{mg}{\pi r^2}\right)\left(1 - \frac{y}{h}\right) - p_0\right](\pi r^2) - mg$$

$$= -\left(\frac{y}{h}\right)\left(p_0\pi r^2 + mg\right).$$

This form shows that for positive h, the net force is down; the trapped gas is at a lower pressure than the equilibrium pressure, and so the net force tends to restore the piston to equilibrium.   c) The angular frequency of small oscillations would be

$$\omega^2 = \frac{(p_0\pi r^2 + mg)/h}{m} = \frac{g}{h}\left(1 + \frac{p_0\pi r^2}{mg}\right).$$

If the displacements are not small, the motion is not simple harmonic. This can be seen be considering what happens if $y \sim -h$; the gas is compressed to a very small volume, and the force due to the pressure of the gas would become unboundedly large for a finite displacement, which is not characteristic of simple harmonic motion. If $y \gg h$ (but not so large that the piston leaves the cylinder), the force due to the pressure of the gas becomes small, and the restoring force due to the atmosphere and the weight would tend toward a constant, and this is not characteristic of simple harmonic motion.

17-53: a) Solving for p as a function of V and T and integrating with respect to V,

$$p = \frac{nRT}{V - nb} - \frac{an^2}{V^2}$$

$$W = \int_{V_1}^{V_2} p\,dV = nRT \ln\left[\frac{V_2 - nb}{V_1 - nb}\right] + an^2\left[\frac{1}{V_2} - \frac{1}{V_1}\right].$$

When $a = b = 0$, $W = nRT \ln(V_2/V_1)$, as expected. b) Using the expression found in part (a),

i) $W = (1.80 \text{ mol})(8.3145 \text{ J/mol·K})(300 \text{ K})$

$$\times \ln\left[\frac{(4.00 \times 10^{-3} \text{ m}^3) - (1.80 \text{ mol})(6.38 \times 10^{-5} \text{ m}^2/\text{mol})}{(2.00 \times 10^{-3} \text{ m}^3) - (1.80 \text{ mol})(6.38 \times 10^{-5} \text{ m}^2/\text{mol})}\right]$$

$$+(0.554 \text{ J·m}^3/\text{mol}^2)(1.80 \text{ mol})^2\left[\frac{1}{4.00 \times 10^{-3} \text{ m}^3} - \frac{1}{2.00 \times 10^{-3} \text{ m}^3}\right]$$

$$= 2.80 \times 10^3 \text{ J}.$$

ii) $nRT \ln(2) = 3.11 \times 10^3 \text{ J}.$

c) 300 J to two figures, larger for the ideal gas. For this case, the difference due to nonzero a is more than that due to nonzero b. The presence of a nonzero a indicates that the molecules are attracted to each other and so do not do as much work in the expansion.

Chapter 18   The Second Law of Thermodynamics

18-1:   a) $2200 \text{ J} + 4300 \text{ J} = 6500 \text{ J}$.   b) $\frac{2200}{6500} = 0.338 = 33.8\%$.

18-2:   a) $9000 \text{ J} - 6400 \text{ J} = 2600 \text{ J}$.   b) $\frac{2600 \text{ J}}{9000 \text{ J}} = 0.289 = 28.9\%$.

18-3:   a) $\frac{3700}{16,100} = 0.230 = 23.0\%$.

   b) $16,100 \text{ J} - 3700 \text{ J} = 12,400 \text{ J}$.

   c) $\frac{16,100 \text{ J}}{4.60 \times 10^4 \text{ J/kg}} = 0.350 \text{ g}$.

   d) $(3700 \text{ J})(60.0 \text{ /s}) = 222 \text{ kW} = 298 \text{ hp}$.

18-4:   a) $Q = \frac{1}{e}Pt = \frac{(180 \times 10^3 \text{ W})(1.00 \text{ s})}{(0.280)} = 6.43 \times 10^5 \text{ J}$.

   b) $Q - Pt = 6.43 \times 10^5 \text{ J} - (180 \times 10^3 \text{ W})(1.00 \text{ s}) = 4.63 \times 10^5 \text{ J}$.

18-5:   a) $e = \frac{330 \text{ MW}}{1300 \text{ MW}} = 0.25 = 25\%$.   b) $1300 \text{ MW} - 330 \text{ MW} = 970 \text{ MW}$.

18-6:   Solving Eq. (18-6) for r,

$$(1 - \gamma) \ln r = \ln(1 - e) \quad \text{or}$$

$$r = (1 - e)^{\frac{1}{1-\gamma}} = (0.350)^{-2.5} = 13.8.$$

If the first equation is used (for instance, using a calculator without the x^y function), note that the symbol "e" is the ideal efficiency, not the base of natural logarithms.

18-7:   a) $T_b = T_a r^{\gamma-1} = (295.15 \text{ K})(9.5)^{0.40} = 726 \text{ K} = 453°C$.

   b) $p_b = p_a r^\gamma = (8.50 \times 10^4 \text{ Pa})(9.50)^\gamma = 1.99 \times 10^6 \text{ Pa}$.

18-8:   a) From Eq. (18-6), $e = 1 - r^{1-\gamma} = 1 - (8.8)^{-0.40} = 0.58 = 58\%$.

   b) $1 - (9.6)^{-0.40} = 60\%$, an increase of 2%. If more figures are kept for the efficiencies, the difference is 1.4%.

18-9:   a) $|W| = \frac{|Q_C|}{K} = \frac{3.40 \times 10^4 \text{ J}}{2.10} = 1.62 \times 10^4 \text{ J}$.

   b) $|Q_H| = |Q_C| + |W| = |Q_C|\left(1 + \frac{1}{K}\right) = 5.02 \times 10^4 \text{ J}$.

18-10: $P = \dfrac{W}{\Delta t} = \dfrac{|Q_C|}{K \Delta t} = \dfrac{1}{K}\left(\dfrac{\Delta m}{\Delta t}\right)(L_f + c_p |\Delta T|)$

$$= \frac{1}{2.8}\left(\frac{8.0 \text{ kg}}{3600 \text{ s}}\right)\left((1.60 \times 10^5 \text{ J/kg}) + (485 \text{ J/kg·K})(2.5 \text{ K})\right) = 128 \text{ W}.$$

18-11:   a) $\dfrac{1.44 \times 10^5 \text{ J} - 9.80 \times 10^4 \text{ J}}{60.0 \text{ s}} = 767 \text{ W}$.   b) $EEF = H/P$, or

$$EEF = \frac{(9.8 \times 10^4 \text{ J})/(60 \text{ s})}{[(1.44 \times 10^5 \text{ J})/(60 \text{ s}) - (9.8 \times 10^4 \text{ J})/(60 \text{ s})]}(3.413) = \frac{1633 \text{ W}}{767 \text{ W}}(3.413) = 7.27.$$

18-12:   a) $|Q_C| = m(L_f + c_{ice}|\Delta T_{ice}| + c_{water}|\Delta T_{water}|)$

$$= (1.80 \text{ kg})\left(\begin{matrix} 334 \times 10^3 \text{ J/kg} + (2100 \text{ J/kg·K})(5.0 \text{ K}) \\ + (4190 \text{ J/kg·K})(25.0 \text{ K}) \end{matrix}\right)$$

$$= 8.09 \times 10^5 \text{ J}.$$

b) $W = \frac{|Q_C|}{K} = \frac{8.08 \times 10^5 \text{ J}}{2.40} = 3.37 \times 10^5 \text{ J}.$

c) $|Q_H| = W + |Q_C| = 3.37 \times 10^5 \text{ J} + 8.08 \times 10^5 \text{ J} = 1.14 \times 10^6 \text{ J}$ (note that $|Q_H| = |Q_C|(1 + \frac{1}{K})$.)

18-13: a) $|Q_H| - |Q_C| = 550 \text{ J} - 335 \text{ J} = 215 \text{ J}.$

b) $T_C = T_H(|Q_C|/|Q_H|) = (620 \text{ K})(335 \text{ J}/550 \text{ J}) = 378 \text{ K}.$

c) $1 - (|Q_C|/|Q_H|) = 1 - (335 \text{ J}/550 \text{ J}) = 39\%.$

18-14: a) From Eq. (18-13), the rejected heat is $(\frac{300 \text{ K}}{520 \text{ K}})(6450 \text{ J}) = 3.72 \times 10^3 \text{ J}.$

b) $6450 \text{ J} - 3.72 \times 10^3 \text{ J} = 2.73 \times 10^3 \text{ J}.$

c) From either Eq. (18-4) or Eq. (18-14), $e = 0.423 = 42.3\%.$

18-15: a)
$$|Q_H| = |Q_C|\frac{T_H}{T_C} = mL_f\frac{T_H}{T_C}$$
$$= (85.0 \text{ kg})(334 \times 10^3 \text{ J/kg})\frac{(297.15 \text{ K})}{(273.15 \text{ K})} = 3.088 \times 10^7 \text{ J},$$

or 3.09×10^7 J to two figures.   b) $|W| = |Q_H| - |Q_C| = |Q_H|(1 - (T_C/T_H)) = (3.09 \times 10^7 \text{ J}) \times (1 - (273.15/297.15)) = 2.49 \times 10^6 \text{ J}.$

18-16: a) From Eq. (18-13), $(\frac{320 \text{ K}}{270 \text{ K}})(415 \text{ J}) = 492 \text{ J}.$ b) The work per cycle is $492 \text{ J} - 415 \text{ J} = 77 \text{ J}$, and $P = (2.75) \times \frac{77 \text{ J}}{1.00 \text{ s}} = 212 \text{ W}$, keeping an extra figure.   c) $T_C/(T_H - T_C) = (270 \text{ K})/(50 \text{ K}) = 5.4.$

18-17: For all cases, $|W| = |Q_H| - |Q_C|$.   a) The heat is discarded at a higher temperature, and a refrigerator is required; $|W| = |Q_C|((T_H/T_C) - 1) = (5.00 \times 10^3 \text{ J})((298.15/263.15) - 1) = 665 \text{ J}.$ b) Again, the device is a refrigerator, and $|W| = |Q_C|((273.15/263.15) - 1) = 190 \text{ J}.$ c) The device is an engine; the heat is taken from the hot resevoir, and the work done by the engine is $|W| = (5.00 \times 10^3 \text{ J})((248.15/263.15) - 1) = 285 \text{ J}.$

18-18: The claimed efficiency of the engine is $\frac{1.51 \times 10^8 \text{ J}}{2.60 \times 10^8 \text{ J}} = 58\%$. While the most efficient engine that can operate between those temperatures has efficiency $e_{\text{Carnot}} = 1 - \frac{250 \text{ K}}{400 \text{ K}} = 38\%$. The proposed engine would violate the second law of thermodynamics, and is not likely to find a market among the prudent.

18-19: a) Combining Eq. (18-14) and Eq. (18-15),

$$K = \frac{T_C/T_H}{1 - (T_C/T_H)} = \frac{1-e}{(1-(1-e))} = \frac{1-e}{e}.$$

b) As $e \to 1$, $K \to 0$; a perfect ($e = 1$) engine exhausts no heat ($Q_C = 0$), and this is useless as a refrigerator. As $e \to 0$, $K \to \infty$; a useless ($e = 0$) engine does no work ($W = 0$), and a refrigerator that requires no energy input is very good indeed.

18-20: a) $\frac{Q}{T_C} = \frac{mL_f}{T_C} = \frac{(0.350 \text{ kg})(334 \times 10^3 \text{ J/kg})}{(273.15 \text{ K})} = 428 \text{ J/K}.$

b) $\frac{-1.17 \times 10^5 \text{ J}}{298.15 \text{ K}} = -392 \text{ J/K}.$

c) $\Delta S = 428 \text{ J/K} + (-392 \text{ J/K}) = 36 \text{ J/K}.$ (If more figures are kept in the intermediate calculations, or if $\Delta S = Q((1/273.15 \text{ K}) - (1/298.15 \text{ K}))$ is used, $\Delta S = 35.6 \text{ J/K}.$

18-21: The final temperature will be

$$\frac{(1.00 \text{ kg})(20.0°C) + (2.00 \text{ kg})(80.0°C)}{(3.00 \text{ kg})} = 60°C,$$

and so the entropy change is

$$(4190 \text{ J/kg·K}) \left[(1.00 \text{ kg}) \ln \left(\frac{333.15 \text{ K}}{293.15 \text{ K}} \right) + (2.00 \text{ kg}) \ln \left(\frac{333.15 \text{ K}}{353.15 \text{ K}} \right) \right] = 47.4 \text{ J/K}.$$

18-22: For an isothermal expansion, $\Delta T = 0$, $\Delta U = 0$ and $Q = W$. The change of entropy is $\frac{Q}{T} = \frac{1850 \text{ J}}{293.15 \text{ K}} = 6.31 \text{ J/K}$.

18-23: The entropy change is $dS = \frac{dQ}{T}$, and $dQ = mL_v$. Thus,

$$dS = \frac{-mL_v}{T} = \frac{-(0.13 \text{ kg})(2.09 \times 10^4 \text{ J/kg})}{(4.216 \text{ K})} = -644 \text{ J/K}.$$

18-24: a) $\Delta S = \frac{Q}{T} = \frac{mL_v}{T} = \frac{(1.00 \text{ kg})(2256 \times 10^3 \text{ J/kg})}{(373.15 \text{ K})} = 6.05 \times 10^3 \text{ J/K}$. Note that this is the change of entropy of the water as it changes to steam.   b) The magnitude of the entropy change is roughly five times the value found in Example 18-5. Water is less ordered (more random) than ice, but water is far less random than steam; a consideration of the density changes indicates why this should be so.

18-25: a) $\Delta S = \frac{Q}{T} = \frac{mL_v}{T} = \frac{(18.0 \times 10^{-3} \text{ kg})(2256 \times 10^3 \text{ J/kg})}{(373.15 \text{ K})} = 109 \text{ J/K}$.

b) $N_2:$ $\dfrac{(28.0 \times 10^{-3} \text{ kg})(201 \times 10^3 \text{ J/kg})}{(77.34 \text{ K})} = 72.8 \text{ J/K}$

$Ag:$ $\dfrac{(107.9 \times 10^{-3} \text{ kg})(2336 \times 10^3 \text{ J/kg})}{(2466 \text{ K})} = 102.2 \text{ J/K}$

$Hg:$ $\dfrac{(200.6 \times 10^{-3} \text{ kg})(272 \times 10^3 \text{ J/kg})}{(630 \text{ K})} = 86.6 \text{ J/K}$

c) The results are the same order of magnitude, all around 100 J/K. The entropy change is a measure of the increase in randomness when a certain number (one mole) goes from the liquid to the vapor state. The entropy per particle for any substance in a vapor state is expected to be roughly the same, and since the randomness is much higher in the vapor state (see Exercise 18-24), the entropy change per molecule is roughly the same for these substances.

18-26: a) The final temperature, found using the methods of Chapter 15, is

$$T = \frac{(3.50 \text{ kg})(390 \text{ J/kg·K})(100 \text{ C°})}{(3.50 \text{ kg})(390 \text{ J/kg·K}) + (0.800 \text{ kg})(4190 \text{ J/kg·K})} = 28.94°C,$$

or 28.9°C to three figures.   b) Using the result of Example 18-10, the total change in entropy is (making the conversion to Kelvin temperature)

$$\Delta S = (3.50 \text{ kg})(390 \text{ J/kg·K}) \ln \left(\frac{302.09 \text{ K}}{373.15 \text{ K}} \right)$$

$$+ (0.800 \text{ kg})(4190 \text{ J/kg·K}) \ln \left(\frac{302.09 \text{ K}}{273.15 \text{ K}} \right)$$

$$= 49.2 \text{ J/K}.$$

(This result was obtained by keeping even more figures in the intermediate calculation. Rounding the Kelvin temperatures to the nearest 0.01 K gives the same result.

18-27: As in Example 18-8,

$$\Delta S = nR \, \ln \left(\frac{V_2}{V_1} \right) = (2.00 \text{ mol})(8.3145 \text{ J/mol·K}) \ln \left(\frac{0.0420 \text{ m}^3}{0.0280 \text{ m}^3} \right) = 6.74 \text{ J/K}.$$

18-28: a) On the average, each half of the box will contain half of each type of molecule, 250 of nitrogen and 50 of oxygen. b) See Example 18-11. The total change in entropy is

$$\Delta S = kN_1 \ln(2) + kN_2 \ln(2) = (N_1 + N_2)k \ln(2)$$
$$= (600)(1.381 \times 10^{-23} \text{ J/K}) \ln(2) = 5.74 \times 10^{-21} \text{ J/K}.$$

c) See also Exercise 18-30. The probability is $(1/2)^{500} \times (1/2)^{100} = (1/2)^{600} = 2.4 \times 10^{-181}$, and is not likely to happen.

The numerical result for part (c) above may not be obtained directly on some standard calculators. For such calculators, the result may be found by taking the log base ten of 0.5 and multiplying by 600, then adding 181 and then finding 10 to the power of the sum. The result is then $10^{-181} \times 10^{0.87} = 2.4 \times 10^{-181}$.

18-29: a) No; the velocity distribution is a function of the mass of the particles, the number of particles and the temperature, none of which change during the isothermal expansion. b) As in Example 18-11, $w_1 = \frac{1}{3}^N w_2$ (the volume has increased, and $w_2 < w_1$); $\ln(w_2/w_1) = \ln \left(3^N \right) = N \ln(3)$, and

$$\Delta S = kN \ln(3) = knN_A \ln(3) = nR \ln(3) = 18.3 \text{ J/K}.$$

c) As in Example 18-8, $\Delta S = nR \, \ln(V_2/V_1) = nR \, \ln(3)$, the same as the expression used in part (b), and $\Delta S = 18.3$ J/K.

18-30: For those with a knowledge of elementary probability, all of the results for this exercise are obtained from

$$P(k) = \binom{n}{k} p^k (1-p)^{n-k} = \frac{4!}{k!(4-k)!} \left(\frac{1}{2} \right)^4,$$

where $P(k)$ is the probability of obtaining k heads, $n = 4$ and $p = 1 - p = \frac{1}{2}$ for a fair coin. This is of course consistent with Fig. (18-15).

a) $\frac{4!}{4!0!}(1/2)^4 = \frac{4!}{0!4!}(1/2)^4 = \frac{1}{16}$ for all heads or all tails.   b) $\frac{4!}{3!1!}(1/2)^4 = \frac{4!}{1!3!} = \frac{1}{4}$.

c) $\frac{4!}{2!2!}(1/2)^4 = \frac{3}{8}$.   d) $2 \times \frac{1}{16} + 2 \times \frac{1}{4} + \frac{3}{8} = 1$. The number of heads must be one of 0, 1, 2, 3 or 4, and there must be unit probability of one and only one of these possibilities.

18-31:   The area is the average power divided by the useful power per unit area S, or

$$A = \frac{P_{ave}}{S} = \frac{(1.50 \times 10^{10} \text{ J})}{(31 \text{ da})(86,400 \text{ s/da})(0.60)(65.7 \text{ W/m}^2)} = 142 \text{ m}^2,$$

which should fit on the roof.

18-32:   a) $mL_c(1 - e) = (4500 \text{ kg})(2.70 \times 10^7 \text{ J/kg})(0.80) = 9.72 \times 10^{10} \text{ J}$.

   b) From $Q = mc\Delta T = \rho V c \Delta T$,

$$V = \frac{Q}{\rho c \Delta T} = \frac{(9.72 \times 10^{10} \text{ J})}{(1.00 \times 10^3 \text{ kg/m}^3)(4190 \text{ J/kg·K})(22.0 \text{ K})} = 1.05 \times 10^3 \text{ m}^3.$$

The length of a side of a cube with this volume is $(1.05 \times 10^3 \text{ m}^3)^{\frac{1}{3}} = 10.2 \text{ m}$.

18-33:   See Problem 15-109.

a) $$\frac{V}{t} = \frac{(Q/\rho_{water}c\Delta T)}{t} = \frac{1}{\rho_{water}c\Delta T}\left(\frac{Q}{t}\right)$$

$$= \frac{(0.60)(150 \text{ W/m}^2)(8.0 \text{ m}^2)}{(1000 \text{ kg/m}^3)(4190 \text{ J/kg·K})(40.0 \text{ K})} = 4.30 \times 10^{-6} \text{ m}^3/\text{s},$$

which is 15.5 L/hr, or 16 L/hr to two figures.

   b) $(15.5 \text{ L/hr})(24 \text{ hr/da})/(75 \text{ L/person·da}) = 4.9$ folks, so 4 inhabitants can be accommodated pleasantly, 5 in a pinch.

18-34:   a) $\frac{(0.85 \times 10^9 \text{ W})}{(0.60)(200 \text{ W/m}^2)} = 7.08 \times 10^6 \text{ m}^2 = 7.08 \text{ km}^2 = 2.73 \text{ mi}^2$.

   b) $\frac{(0.85 \times 10^9 \text{ W})}{(0.30)(200 \text{ W/m}^2)} = 1.42 \times 10^7 \text{ m}^2 = 14.2 \text{ km}^2 = 5.47 \text{ mi}^2$.

18-35:   a) $$V = \frac{m}{\rho} = \frac{Q}{\rho c \Delta T} = \frac{(4.00 \times 10^9 \text{ J})}{(1000 \text{ kg/m}^3)(4190 \text{ J/kg·K})(28.0 \text{ K})} = 34.1 \text{ m}^3.$$

   b) The heat added to (and later removed from) the Glauber salt is related to the mass (and hence the volume) by

$$Q = m(c_{solid}\Delta T_{solid} + L_f + c_{liquid}\Delta T_{liquid}),$$

and the needed volume is

$$V = \frac{m}{\rho} = \frac{(4.00 \times 10^9 \text{ J})}{(1600 \text{ kg/m}^3)\left(\begin{array}{c}(1930 \text{ J/kg·K})(11.0 \text{ C}°) + (2.42 \times 10^5 \text{ J/kg}) \\ + (2850 \text{ J/kg·K})(17.0 \text{ C}°)\end{array}\right)}$$

$$= 8.02 \text{ m}^3.$$

   c) Glauber salts take up far less space.

18-36: a) $\frac{P_{\text{out}}}{e} = \frac{1100 \text{ MW}}{0.350} = 3140$ MW.

   b) $\frac{P_{\text{out}}/e}{L_c} = \frac{(11.00 \times 10^8 \text{ W})/(0.350)}{(3.00 \times 10^7 \text{ J/kg})} = 104.8$ kg/s. The coal burned in one day is $(104.8 \text{ kg/s}) \times$ (86,400 s) $= 9.05 \times 10^6$ kg. Note that extra figures were needed to avoid roundoff error.

   c) The input minus the output is 3140 MW $-$ 1100 MW $=$ 2040 MW.

   d) See Problem 15-109; $p_{\text{dis}} = c\Delta T \rho \frac{dV}{dt}$, so

$$\frac{dV}{dt} = \frac{p_{\text{dis}}}{c\Delta T \rho} = \frac{(2.04 \times 10^9 \text{ W})}{(4190 \text{ J/kg·K})(4.0 \text{ K})(1.00 \times 10^3 \text{ kg/m}^3)} = 120 \text{ m}^3/\text{s}.$$

   e) $v = \frac{1}{A} \frac{dV}{dt} = \frac{(120 \text{ m}^3/\text{s})}{(100 \text{ m})(5.0 \text{ m})} = 0.24$ m/s.

18-37: a) The product of the efficiencies of the individual steps is the efficiency of the overall process. The efficiency is the fraction of the energy input that is used, and so the efficiencies must be multiplied. b) $(0.40) \times (1.0 - 10) \times (1 - 0.10) \times (0.80) \times (0.90) = 0.23 = 23\%$. c) The overall efficiency is somewhat larger.

18-38: a) Solving Eq. (18-14) for T_H, $T_H = T_C \frac{1}{1-e}$, so the temperature change

$$T_H' - T_H = T_C \left(\frac{1}{1 - e'} - \frac{1}{1 - e} \right) = (183.15 \text{ K}) \left(\frac{1}{0.55} - \frac{1}{0.600} \right) = 27.8 \text{ K}.$$

   b) Similarly, $T_C = T_H(1 - e)$, and if $T_H' = T_H$,

$$T_C' - T_C = T_C \frac{e' - e}{1 - e} = (183.15 \text{ K}) \left(\frac{0.050}{0.600} \right) = 15.3 \text{ K}.$$

18-39: The initial volume is $V_1 = \frac{nRT_1}{p_1} = 8.62 \times 10^{-3}$ m^3. a) At point 1, the pressure is given as atmospheric, and $p_1 = 1.01 \times 10^5$ Pa, with the volume found above, $V_1 = 8.62 \times 10^{-3}$ m^3. $V_2 = V_1 = 8.62 \times 10^{-3}$ m^3, and $p_2 = \frac{T_2}{T_1} p_1 = 2p_1 = 2.03 \times 10^5$ Pa (using $p_a = 1.013 \times 10^5$ Pa). $p_3 = p_1 = 1.01 \times 10^5$ Pa and $V_3 = V_1 \frac{T_3}{T_1} = 1.31 \times 10^{-2}$ m^3. b) Process 1-2 is isochoric, $dV = 0$ so $W = 0$. $\Delta U = Q = nC_V \Delta T = (0.350 \text{ mol})(5/2) \times (8.3145 \text{ J/mol·K})(300 \text{ K}) = 2.18 \times 10^3$ J. The process 2-3 is adiabatic, $Q = 0$, and $\Delta U = -W = nC_V \Delta T = (0.350 \text{ mol})(5/2)(8.3145 \text{ J/mol·K})(-145 \text{ K}) = -1060$ J $(W > 0)$. The process 3-1 is isobaric; $W = p\Delta V = nR\Delta T = (0.350 \text{ mo;})(8.3145 \text{ J/mol·K})(-155 \text{ K}) = -451$ J, $\Delta U = nC_V \Delta T = n(5/2)(8.3145 \text{ J/mol·K})(-155 \text{ K}) = -1130$ J and $Q = nC_p \Delta T = (0.350 \text{ mol})(7/2)(8.3145 \text{ J/mol·K})(-155 \text{ K}) = -1580$ J $= \Delta U + W$. c) The net work done is 1060 J $-$ 451 J $=$ 609 J. d) Keeping extra figures in the calculations for the process 1-2, the heat flow into the engine for one cycle is 2183 J $-$ 1579 J $=$ 604 J. e) $e = \frac{604 \text{ J}}{2183 \text{ J}} = 0.277 = 27.7\%$. For a Carnot-cycle engine operating between 300 K and 600 K, the thermal efficiency is $1 - \frac{300}{600} = 0.500 = 50\%$.

18-40: a) $e = 1 - \frac{279.15 \text{ K}}{300.15 \text{ K}} = 7.0\%$. b) $\frac{P_{\text{out}}}{e} = \frac{210 \text{ kW}}{0.070} = 3.0$ MW, 3.0 MW $-$ 210 kW $= (\frac{1}{e} - 1)(210 \text{ kW}) = 2.8$ MW.

   c) $\frac{dm}{dt} = \frac{d|Q_C|/dt}{c\Delta T} = \frac{(2.8 \times 10^6 \text{ W})(3600 \text{ s/hr})}{(4190 \text{ J/kg·K})(4 \text{ K})} = 6 \times 10^5 \text{ kg/hr} = 6 \times 10^5 \text{ L/hr}.$

18-41: There are many equivalent ways of finding the efficiency; the method presented here saves some steps. The temperature at point 3 is $T_3 = 4T_0$, and so

$$Q_H = \Delta U_{13} + W_{13} = nC_V(T_3 - T_0) + (2p_0)(2V_0 - V_0) = \frac{5}{2}nRT_0(3) + 2p_0V_0 = \frac{19}{2}p_0V_0,$$

where $nRT_0 = p_0V_0$ has been used for an ideal gas. The work done by the gas during one cycle is the area enclosed by the blue square in Fig. (18-43), $W = p_0V_0$, and so the efficiency is $e = \frac{W}{Q_H} = \frac{2}{19} = 10.5\%$.

18-42: a) $p_2 = p_1 = 2.00$ atm, $V_2 = V_1\frac{T_2}{T_1} = (4.00 \text{ L})(3/2) = 6.00$ L. $V_3 = V_2 = 6.00$ L, $p_3 = p_2\frac{T_3}{T_2} = p_2(5/9) = 1.111$ atm, $p_4 = p_3\frac{V_3}{V_4} = p_3(3/2) = 1.67$ atm. As a check, $p_1 = p_4\frac{T_1}{T_4} = p_4(6/5) = 2.00$ atm. To summarize,

$$(p_1, V_1) = (2.00 \text{ atm}, 4.00 \text{ L}) \qquad (p_2, V_2) = (2.00 \text{ atm}, 6.00 \text{ L})$$
$$(p_3, V_3) = (1.111 \text{ atm}, 6.00 \text{ L}) \qquad (p_4, V_4) = (1.67 \text{ atm}, 4.00 \text{ L}).$$

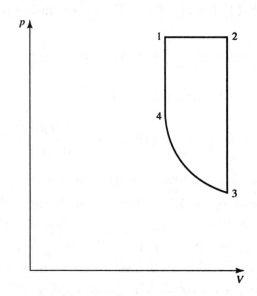

b) The number of moles of oxygen is $n = \frac{p_1V_1}{RT_1}$, and the heat capacities are those in Table (17-1). The product p_1V_1 has the value $x = 810.4$ J; using this and the ideal gas law,

i: $$Q = nC_P\Delta T = \frac{C_P}{R}x\left(\frac{T_2}{T_1} - 1\right) = (3.508)(810.4 \text{ J})(1/2) = 1422 \text{ J},$$

$$W = p_1\Delta V = x\left(\frac{T_2}{T_1} - 1\right) = (810.4 \text{ J})(1/2) = 405 \text{ J}.$$

ii: $$Q = nC_V\Delta T = \frac{C_V}{R}x\left(\frac{T_3 - T_2}{T_1}\right) = (2.508)(810.4 \text{ J})(-2/3) = -1355 \text{ J}, \quad W = 0.$$

iii: $$W = nRT_3\ln\left(\frac{V_4}{V_3}\right) = x\frac{T_3}{T_1}\ln\left(\frac{V_4}{V_3}\right) = (810.4 \text{ J})(5/6)\ln(2/3) = -274 \text{ J}, \quad Q = W$$

iv: $$Q = nC_V\Delta T = \frac{C_V}{R}x\left(1 - \frac{T_4}{T_1}\right) = (2.508)(810.4 \text{ J})(1/6) = 339 \text{ J}, \quad W = 0.$$

In the above, the terms are given to to nearest integer number of joules to reduce roundoff error.

c) The net work done in the cycle is 405 J − 274 J = 131 J.

d) Heat is added in steps i and iv, and the added heat is 1422 J + 339 J = 1761 J and the efficiency is $\frac{131 \text{ J}}{1761 \text{ J}} = 0.075$, or 7.5%. The efficiency of a Carnot-cycle engine operating between 250 K and 450 K is $1 - \frac{250}{450} = 0.44 = 44\%$.

18-43: a) $\Delta U = 1657$ kJ − 1005 kJ $= 6.52 \times 10^5$ J, $W = p\Delta V = (363 \times 10^3 \text{ Pa}) \times (0.4513 \text{ m}^3 - 0.2202 \text{ m}^3) = 8.39 \times 10^4$ J, and so $Q = \Delta U + W = 7.36 \times 10^5$ J.

b) Similarly,

$$
\begin{aligned}
Q_H &= \Delta U - p\Delta V \\
&= (1171 \text{ kJ} - 1969 \text{ kJ}) + (2305 \times 10^3 \text{ Pa})(0.00946 \text{ m}^3 - 0.0682 \text{ m}^3) \\
&= -9.33 \times 10^5 \text{ J}.
\end{aligned}
$$

c) The work done during the adiabatic processes must be found indirectly (the coolant is not ideal, and is not always a gas). For the entire cycle, $\Delta U = 0$, and so the net work done by the coolant is the sum of the results of parts (a) and (b), -1.97×10^5 J. The work done by the motor is the negative of this, 1.97×10^5 J. d) $K = \frac{|Q_c|}{|W|} = \frac{7.36 \times 10^5 \text{ J}}{1.97 \times 10^5 \text{ J}} = 3.74$.

18-44: For a monatomic ideal gas, $C_P = \frac{5}{2}R$ and $C_V = \frac{3}{2}R$.

a) *ab*: The temperature changes by the same factor as the volume, and so

$$Q = nC_P\Delta T = \frac{C_P}{R}p_a(V_a - V_b) = (2.5)(3.00 \times 10^5 \text{ Pa})(0.300 \text{ m}^3) = 2.25 \times 10^5 \text{ J}.$$

The work $p\Delta V$ is the same except for the factor of $\frac{5}{2}$, so $W = 0.90 \times 10^5$ J. $\Delta U = Q - W = 1.35 \times 10^5$ J.

bc: The temperature now changes in proportion to the pressure change, and $Q = \frac{3}{2}(p_c - p_b)V_b = (1.5)(-2.00 \times 10^5 \text{ Pa})(0.800 \text{ m}^3) = -2.40 \times 10^5$ J, and the work is zero ($\Delta V = 0$). $\Delta U = Q - W = -2.40 \times 10^5$ J.

ca: The easiest way to do this is to find the work done first; W will be the negative of area in the *p-V* plane bounded by the line representing the process *ca* and the verticals from points *a* and *c*. The area of this trapezoid is $\frac{1}{2}(3.00 \times 10^5 \text{ Pa} + 1.00 \times 10^5 \text{ Pa})(0.800 \text{ m}^3 - 0.500 \text{ m}^3) = 6.00 \times 10^4$ J, and so the work is -0.60×10^5 J. ΔU must be 1.05×10^5 J (since $\Delta U = 0$ for the cycle, anticipating part (b)), and so Q must be $\Delta U + W = 0.45 \times 10^5$ J.

b) See above; $Q = W = 0.30 \times 10^5$ J, $\Delta U = 0$.

c) The heat added, during processes *ab* and *ca*, is 2.25×10^5 J $+ 0.45 \times 10^5$ J $= 2.70 \times 10^5$ J and the efficiency is $\frac{W}{Q_H} = \frac{0.30 \times 10^5}{2.70 \times 10^5} = 0.111 = 11.1\%$.

18-45: a) *ab*: For the isothermal process, $\Delta T = 0$ and $\Delta U = 0$. $W = nRT_1 \ln(V_b/V_a) = nRT_1 \ln(1/r) = -nRT_1 \ln(r)$, and $Q = W = -nRT_1 \ln(r)$. *bc*: For the isochoric process, $dV = 0$ and $W = 0$; $Q = \Delta U = nC_V\Delta T = nC_V (T_2 - T_1)$. *cd*: As in the process *ab*, $\Delta U = 0$ and $W = Q = nRT_2 \ln(r)$. *da*: As in process *bc*, $dV = 0$, $W = 0$ and $\Delta U = W = nC_V (T_1 - T_2)$. b) The values of Q for the processes are the negatives of each other.

c) The net work for one cycle is $W_{net} = nR(T_2 - T_1)\ln(r)$, and the heat added (neglecting the heat exchanged during the isochoric expansion and compression, as mentioned in part (b)) is $Q_{cd} = nRT_2 \ln(r)$, and the efficiency is $\frac{W_{net}}{Q_{cd}} = 1 - (T_1/T_2)$. This is the same as the efficiency of a Carnot-cycle engine operating between the two temperatures.

18-46: The efficiency of the first engine is $e_1 = \frac{T_H - T'}{T_H}$ and that of the second is $e_2 = \frac{T' - T_C}{T'}$, and the overall efficiency is

$$e = e_1 e_2 = \left[\frac{T_H - T'}{T_H}\right]\left[\frac{T' - T_C}{T'}\right].$$

The first term in the product is necessarily less than the original efficiency since $T' > T_C$, and the second term is less than 1, and so the overall efficiency has been reduced.

18-47: a) The cylinder described contains a mass of air $m = \rho(\pi d^2/4)L$, and so the total kinetic energy is $K = \rho(\pi/8)d^2 L v^2$. This mass of air will pass by the turbine in a time $t = L/v$, and so the maximum power is

$$P = \frac{K}{t} = \rho(\pi/8)d^2 v^3.$$

Numerically, the product $\rho_{air}(\pi/8) \approx 0.5 \text{ kg/m}^3 = 0.5 \text{ W·s}^4/\text{m}^5$.

b) $\quad v = \left(\frac{P/e}{kd^2}\right)^{1/3} = \left(\frac{(3.2 \times 10^6 \text{ W})/(0.25)}{(0.5 \text{ W·s}^4/\text{m}^5)(97 \text{ m})^2}\right)^{1/3} = 14 \text{ m/s} = 50 \text{ km/h}.$

c) Wind speeds tend to be higher in mountain passes.

18-48: a) $\quad (105 \text{ km/h})\left(\frac{1 \text{ gal}}{25 \text{ mi}}\right)\left(\frac{1 \text{ mi}}{1.609 \text{ km}}\right)\left(\frac{3.788 \text{ L}}{1 \text{ gal}}\right) = 9.89 \text{ L/h}.$

b) From Eq. (18-6), $e = 1 - r^{1-\gamma} = 1 - (8.5)^{-0.40} = 0.575 = 57.5\%$.

c) $\quad \left(\frac{9.89 \text{ L/h}}{3600 \text{ s/hr}}\right)(0.740 \text{ kg/L})(4.60 \times 10^7 \text{ J/kg})(0.575) = 5.38 \times 10^4 \text{ W} = 72.1 \text{ hp}.$

d) Repeating the calculation gives $1.4 \times 10^4 \text{ W} = 19 \text{ hp}$, about 8% of the maximum power.

18-49: (Extra figures are given in the numerical answers for clarity.) a) The efficiency is $e = 1 - r^{-0.40} = 0.611$, so the work done is $Q_H e = 122 \text{ J}$ and $|Q_C| = 78 \text{ J}$. b) Denote the length of the cylinder when the piston is at point a by L_0 and the stroke as s. Then, $\frac{L_0}{L_0 - s} = r$, $L_0 = \frac{r}{r-1}s$ and the volume is

$$L_0 A = \frac{r}{r-1}sA = \frac{10.6}{9.6}(86.4 \times 10^{-3} \text{ m})\pi(41.25 \times 10^{-3} \text{ m})^2 = 5.10 \times 10^{-4} \text{ m}^3.$$

c) The calculations are presented symbolically, with numerical values substituted at the end. At point a, the pressure is $p_a = 8.50 \times 10^4 \text{ Pa}$, the volume is $V_a = 5.10 \times 10^{-4} \text{ m}^3$ as found in part (b) and the temperature is $T_a = 300 \text{ K}$. At point b, the volume is $V_b = V_a/r$, the pressure after the adiabatic compression is $p_b = p_a r^\gamma$ and the temperature is $T_b = T_a r^{\gamma-1}$. During the burning of the fuel, from b to c, the volume remains constant and so $V_c = V_b = V_a/r$. The temperature has changed by an amount

$$\Delta T = \frac{Q_H}{nC_V} = \frac{Q_H}{(p_a V_a/RT_a)C_V} = \frac{RQ_H}{p_a V_a C_V}T_a =$$

$$= \frac{(8.3145 \text{ J/mol·K})(200 \text{ J})}{(8.50 \times 10^4 \text{ Pa})(5.10 \times 10^{-4} \text{ m}^3)(20.5 \text{ J/mol·K})}T_a = f T_a,$$

where f is a dimensionless constant equal to 1.871 to four figures. The temperature at c is then $T_c = T_b + fT_a = T_a \left(r^{\gamma-1} + f\right)$. The pressure is found from the volume and temperature, $p_c = p_a r \left(r^{\gamma-1} + f\right)$. Similarly, the temperature at point d is found by considering the temperature change in going from d to a, $\left|\frac{Q_C}{nC_V}\right| = (1-e)\frac{Q_H}{nC_V} = (1-e)fT_a$, so $T_d = T_a \left(1 + (1-e)f\right)$. The process from d to a is isochoric, so $V_d = V_a$, and $p_d = p_a(1+(1-e)f)$. As a check, note that $p_d = p_c r^{-\gamma}$. To summarize,

	p	V	T
a	p_a	V_a	T_a
b	$p_a r^\gamma$	V_a/r	$T_a r^{\gamma-1}$
c	$p_a r \left(r^{\gamma-1} + f\right)$	V_a/r	$T_a \left(r^{\gamma-1} + f\right)$
d	$p_a \left(1 + (1-e)f\right)$	V_a	$T_a \left(1 + (1-e)f\right)$

Using numerical values (and keeping all figures in the intermediate calculations),

	p	V	T
a	8.50×10^4 Pa	5.10×10^{-4} m³	300 K
b	2.32×10^6 Pa	4.81×10^{-5} m³	771 K
c	4.00×10^6 Pa	4.81×10^{-5} m³	1330 K
d	1.47×10^5 Pa	5.10×10^{-4} m³	518 K

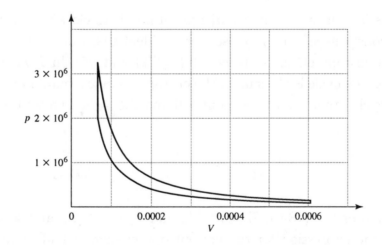

d) With the temperature found at point c in part (c), the efficiency is $1 - \frac{300 \text{ K}}{1333 \text{ K}} = 0.775 = 77.5\%$.

18-50: See Figure (18-12(c)), and Example 18-8.

a) For the isobaric expansion followed by the isochoric process, follow a parth from T to $2T$ to T. Use $dQ = nC_V \, dT$ or $dQ = nC_p \, dT$ to get $\Delta S = nC_p \ln 2 + nC_V \ln \frac{1}{2} = n(C_p - C_V) \ln 2 = nR \ln 2$.

b) For the isochoric cooling followed by the isobaric expansion, follow a path from T to $T/2$ to T. Then $\Delta S = nC_V \ln \frac{1}{2} + nC_p \ln 2 = n(C_p - C_V) \ln 2 = nR \ln 2$.

18-51: The much larger mass of water suggests that the final state of the system will be water at a temperature between $0°C$ and $60.0°C$. This temperature would be

$$T = \frac{\left(\begin{array}{l}(0.600 \text{ kg})(4190 \text{ J/kg·K})(45.0 \text{ C}°) \\[4pt] - (0.0500 \text{ kg})((2100 \text{ J/kg·K})(15.0 \text{ C}°) \\[4pt] + 334 \times 10^3 \text{ J/kg}\end{array}\right)}{(0.650 \text{ kg})(4190 \text{ J/kg·K})} = 34.83°C,$$

keeping an extra figure. The entropy change of the system is then

$$\Delta S = (0.600 \text{ kg})(4190 \text{ J/kg·K}) \ln \left(\frac{307.98}{318.15}\right)$$

$$+ (0.0500 \text{ kg}) \left[\begin{array}{l}(2100 \text{ J/kg·K}) \ln \left(\dfrac{273.15}{258.15}\right) \\[10pt] + \dfrac{334 \times 10^3 \text{ J/kg}}{273.15 \text{ K}} \\[10pt] + (4190 \text{ J/kg·K}) \ln \left(\dfrac{307.98}{273.15}\right)\end{array}\right] = 10.5 \text{ J/K}.$$

(Some precision is lost in taking the logarithms of numbers close to unity.)

18-52: a) For constant-volume processes for an ideal gas, the result of Example 18-10 may be used; the entropy changes are $nC_V \ln(T_c/T_b)$ and $nC_V \ln(T_a/T_d)$. b) The total entropy change for one cycle is the sum of the entropy changes found in part (a); the other processes in the cycle are adiabatic, with $Q = 0$ and $\Delta S = 0$. The total is then

$$\Delta S = nC_V \ln \frac{T_c}{T_b} + nC_V \ln \frac{T_a}{T_d} = nC_V \ln \left(\frac{T_c T_a}{T_b T_d}\right).$$

From the derivation of Eq. (18-6), $T_b = r^{\gamma-1} T_a$ and $T_c = r^{\gamma-1} T_d$, and so the argument of the logarithm in the expression for the net entropy change is 1 identically, and the net entropy change is zero. c) The system is not isolated, and a zero change of entropy for an irreversible system is certainly possible.

18-53: a)

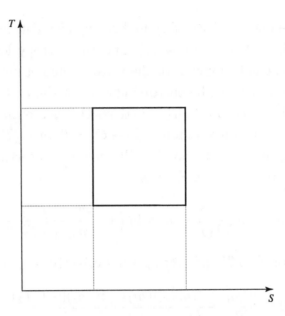

b) From Eq. (18-17), $dS = \frac{dQ}{T}$, and so $dQ = T\,dS$, and

$$Q = \int dQ = \int T\,dS,$$

which is the area under the curve in the TS plane.   c) Q_H is the area under the rectangle bounded by the horizontal part of the rectangle at T_H and the verticals. $|Q_C|$ is the area bounded by the horizontal part of the rectangle at T_C and the verticals. The net work is then $Q_H - |Q_C|$, the area bounded by the rectangle that represents the process. The ratio of the areas is the ratio of the lengths of the vertical sides of the respective rectangles, and the efficiency is $e = \frac{W}{Q_H} = \frac{T_H - T_C}{T_H}$.   d) As explained in Problem 18-45, the substance that mediates the heat exchange during the isochoric expansion and compression does not leave the system, and the diagram is the same as in part (a). As found in that problem, the ideal efficiency is the same as for a Carnot-cycle engine.

18-54:   a) $\Delta S = \frac{Q}{T} = -\frac{mL_f}{T} = -\frac{(0.160 \text{ kg})(334 \times 10^3 \text{ J/kg})}{(373.15 \text{ K})} = -143 \text{ J/K}.$

b) $\Delta S = \frac{Q}{T} = \frac{mL_f}{T} = \frac{(0.160 \text{ kg})(334 \times 10^3 \text{ J/kg})}{(273.15 \text{ K})} = 196 \text{ J/K}.$

c) From the time equilbrium has been reached, there is no net heat exchange between the rod and its surroundings (as much heat leaves the end of the rod in the ice as enters at the end of the rod in the boiling water), so the entropy change of the copper rod is zero.   d) $196 \text{ J/K} - 143 \text{ J/K} = 53 \text{ J/K}.$

18-55:   a) $\Delta S = mc \ln(T_2/T_1)$

$$= (250 \times 10^{-3} \text{ kg})(4190 \text{ J/kg·K}) \ln(338.15 \text{ K}/293.15 \text{ K})$$

$$= 150 \text{ J/K}.$$

b) $\Delta S = \frac{-mc\Delta T}{T_{\text{element}}} = \frac{-(250 \times 10^{-3} \text{ kg})(4190 \text{ J/kg · K})(338.15 \text{ K} - 293.15 \text{ K})}{393.15 \text{ K}} = -120 \text{ J/K}.$ c) The sum of the results of parts (a) and (b) is $\Delta S_{\text{system}} = 30 \text{ J/K}.$ d) Heating a liquid is not reversible. Whatever the energy source for the heating element, heat is being delivered at a higher temperature than that of the water, and the entropy loss of the source will be less in magnitude than the entropy gain of the water. The net entropy change is positive.

18-56: a) As in Example 18-10, the entropy change of the first object is $m_1 c_1 \ln(T/T_1)$ and that of the second is $m_2 c_2 \ln(T'/T_2)$, and so the net entropy change is as given. Neglecting heat transfer to the surroundings, $Q_1 + Q_2 = 0$, $m_1 c_1 (T - T_1) + m_2 c_2 (T' - T_2) = 0$, which is the given expression. b) Solving the energy-conservation relation for T' and substituting into the expression for ΔS gives

$$\Delta S = m_1 c_1 \ln\left(\frac{T}{T_1}\right) + m_2 c_2 \ln\left(1 - \frac{m_1 c_1}{m_2 c_2}\left(\frac{T}{T_2} - \frac{T_1}{T_2}\right)\right).$$

Differentiating with respect to T and setting the derivative equal to 0 gives

$$0 = \frac{m_1 c_1}{T} + \frac{(m_2 c_2)(m_1 c_1/m_2 c_2)(-1/T_2)}{\left(1 - (m_1 c_1/m_2 c_2)\left(\frac{T}{T_2} - \frac{T_1}{T_2}\right)\right)}.$$

This may be solved for

$$T = \frac{m_1 c_1 T_1 + m_2 c_2 T_2}{m_1 c_1 + m_2 c_2},$$

which is the same as T' when substituted into the expression representing conservation of energy.

Those familiar with Lagrange multipliers can use that technique to obtain the relations

$$\frac{\partial}{\partial T}\Delta S = \lambda \frac{\partial Q}{\partial T}, \quad \frac{\partial}{\partial T'}\Delta S = \lambda \frac{\partial Q}{\partial T'}$$

and so conclude that $T = T'$ immediately; this is equivalent to treating the differentiation as a related rate problem, as

$$\frac{d}{dT'}\Delta S = \frac{m_1 c_1}{T} + \frac{m_2 c_2}{T'}\frac{dT'}{dT} = 0,$$

and using $\frac{dT'}{dT} = -\frac{m_1 c_1}{m_2 c_2}$ gives $T = T'$ with a great savings of algebra.

c) The final state of the system will be that for which no further entropy change is possible. If $T < T'$, it is possible for the temperatures to approach each other while increasing the total entropy, but when $T = T'$, no further spontaneous heat exchange is possible.

18-57: a) For an ideal gas, $C_P = C_V + R$, and taking air to be diatomic, $C_P = \frac{7}{2}R$, $C_V = \frac{5}{2}R$ and $\gamma = \frac{7}{5}$. Referring to Fig. (18-4), $Q_H = n\frac{7}{2}R(T_c - T_b) = \frac{7}{2}(p_c V_c - p_b V_b)$. Similarly, $Q_C = n\frac{5}{2}R(p_a V_a - p_d V_d)$. What needs to be done is to find the relations between the product of the pressure and the volume at the four points.

For an ideal gas, $\frac{p_c V_c}{T_c} = \frac{p_a V_a}{T_a}$, so $p_c V_c = p_a V_a (\frac{T_c}{T_a})$. For a compression ratio r, and given that for the Diesel cycle the process ab is adiabatic,

$$p_b V_b = p_a V_a \left(\frac{V_a}{V_b}\right)^{\gamma-1} = p_a V_a r^{\gamma-1}.$$

Similarly, $p_d V_d = p_c V_c (\frac{V_c}{V_a})^{\gamma-1}$. Note that the last result uses the fact that process da is isochoric, and $V_d = V_a$; also, $p_c = p_b$ (process bc is isobaric), and so $V_c = V_b(\frac{T_c}{T_a})$. Then,

$$\frac{V_c}{V_a} = \frac{T_c}{T_b} \cdot \frac{V_b}{V_a} = \frac{T_c}{T_a} \cdot \frac{T_a}{T_b} \cdot \frac{V_a}{V_b}$$

$$= \frac{T_c}{T_a} \cdot \left(\frac{T_a V_a^{\gamma-1}}{T_b V_b^{\gamma-1}}\right) \left(\frac{V_a}{V_b}\right)^{-\gamma}$$

$$= \frac{T_c}{T_a} r^{\gamma}.$$

Combining the above results,

$$p_d V_d = p_a V_a \left(\frac{T_c}{T_a}\right)^{\gamma} r^{\gamma-\gamma^2}.$$

Substitution of the above results into Eq. (18-4) gives

$$e = 1 - \frac{5}{7} \left[\frac{\left(\frac{T_c}{T_a}\right)^{\gamma} r^{\gamma-\gamma^2} - 1}{\left(\frac{T_c}{T_a}\right) - r^{\gamma-1}} \right]$$

$$= 1 - \frac{1}{1.4} \left[\frac{(5.022) r^{-0.56} - 1}{(3.167) - r^{0.40}} \right],$$

where $\frac{T_c}{T_a} = 3.167$, $\gamma = 1.4$ have been used. Substitution of $r = 21.0$ yields $e = 0.708 = 70.8\%$.

Chapter 19 Mechanical Waves

19-1: a) The period is twice the time to go from one extreme to the other, and $v = f\lambda = \lambda/T = (6.00 \text{ m})/(5.0 \text{ s}) = 1.20 \text{ m/s}$, or 1.2 m/s to two figures. b) The amplitude is half the total vertical distance, 0.310 m. c) The amplitude does not affect the wave speed; the new amplitude is 0.150 m. d) For the waves to exist, the water level cannot be level (horizontal), and the boat would tend to move along a wave toward the lower level, alternately in the direction of and opposed to the direction of the wave motion.

19-2: a) $f = \frac{344 \text{ m/s}}{10 \times 12.0 \times 10^{-2} \text{ m}} = 287 \text{ Hz}$.

b) Ten times the result of part (a), 2.87 kHz. c) 28.7 kHz.

19-3: a) $\lambda = v/f = (344 \text{ m/s})/(784 \text{ Hz}) = 0.439 \text{ m}$.

b) $f = v/\lambda = (344 \text{ m/s})/(6.55 \times 10^{-5} \text{ m}) = 5.25 \times 10^6 \text{ Hz}$.

19-4: Denoting the speed of light by c, $\lambda = \frac{c}{f}$, and

a) $\frac{3.00 \times 10^8 \text{ m/s}}{540 \times 10^3 \text{ Hz}} = 556 \text{ m}$.   b) $\frac{3.00 \times 10^8 \text{ m/s}}{104.5 \times 10^6 \text{ Hz}} = 2.87 \text{ m}$.

19-5: a) $\lambda_{\max} = (344 \text{ m/s})/(20.0 \text{ Hz}) = 17.2 \text{ m}$, $\lambda_{\min} = (344 \text{ m/s})/(20,000 \text{ Hz}) = 1.72 \text{ cm}$.

b) $\lambda_{\max} = (1480 \text{ m/s})/(20.0 \text{ Hz}) = 74.0 \text{ m}$, $\lambda_{\min} = (1480 \text{ m/s})/(20,000 \text{ Hz}) = 74.0 \text{ mm}$.

19-6: Comparison with Eq. (19-4) gives a) 6.50 mm,   b) 28.0 cm,   c) $f = \frac{1}{T} = \frac{1}{0.0360 \text{ s}} = 27.8 \text{ Hz}$ and from Eq. (19-1),   d) $v = (0.280 \text{ m})(27.8 \text{ Hz}) = 7.78 \text{ m/s}$,   e) $+x$ direction.

19-7: a) $f = v/\lambda = (8.00 \text{ m/s})/(0.320 \text{ m}) = 25.0 \text{ Hz}$, $T = 1/f = 1/(25.0 \text{ Hz}) = 4.00 \times 10^{-2} \text{ s}$, $k = 2\pi/\lambda = (2\pi)/(0.320 \text{ m}) = 19.6 \text{ rad/m}$.

b) $$y(x, t) = (0.0700 \text{ m}) \sin 2\pi \left(t(25.0 \text{ Hz}) + \frac{x}{0.320 \text{ m}} \right).$$

c) $(0.0700 \text{ m}) \sin [2\pi((0.150 \text{ s})(25.0 \text{ Hz}) + (0.360 \text{ m})/(0.320 \text{ m}))] = -4.95 \text{ cm}$.   d) The argument in the square brackets in the expression used in part (c) is $2\pi(4.875)$, and the displacement will next be zero when the argument is 10π; the time is then $T(5 - x/\lambda) = (1/25.0 \text{ Hz})(5 - (0.360 \text{ m})/(0.320 \text{ m})) = 0.1550 \text{ s}$ and the elapsed time is 0.0050 s,   e) $T/2 = 0.02 \text{ s}$.

19-8: a)

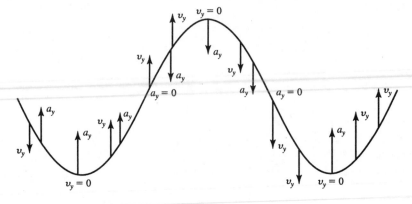

b)

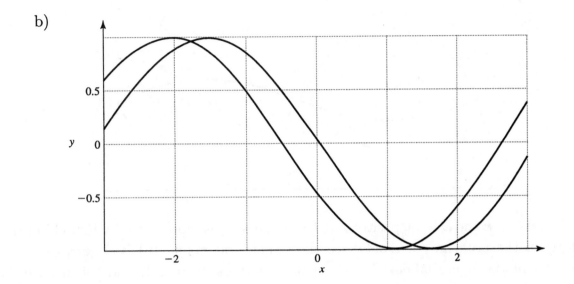

19-9: a)

$$\frac{\partial y}{\partial x} = Ak\cos(\omega t + kx) \qquad \frac{\partial^2 y}{\partial x^2} = -Ak^2\sin(\omega t + kx)$$

$$\frac{\partial y}{\partial t} = A\omega\cos(\omega t + kx) \qquad \frac{\partial^2 y}{\partial t^2} = -A\omega^2\sin(\omega t + kx),$$

and so $\frac{\partial^2 y}{\partial x^2} = \frac{k^2}{\omega^2}\frac{\partial^2 y}{\partial t^2}$, and $y(x,\ t)$ is a solution of Eq. (19-12) with $v = \omega/k$.

b)

$$\frac{\partial y}{\partial x} = -Ak\sin(\omega t + kx) \qquad \frac{\partial^2 y}{\partial x^2} = -Ak^2\cos(\omega t + kx)$$

$$\frac{\partial y}{\partial t} = -A\omega\sin(\omega t + kx) \qquad \frac{\partial^2 y}{\partial t^2} = -A\omega^2\cos(\omega t + kx),$$

and so $\frac{\partial^2 y}{\partial x^2} = \frac{k^2}{\omega^2}\frac{\partial^2 y}{\partial t^2}$, and $y(x,\ t)$ is a solution of Eq. (19-12) with $v = \omega/k$.   c) Both waves are moving in the $-x$-direction, as explained in the discussion preceding Eq. (19-8).

d)   Taking   derivatives   yields $v_y(x,t) = -\omega A\sin(\omega t + kx)$ and $a_y(x,t) = -\omega^2 A\cos(\omega t - kx)$.

19-10:   a) The relevant expressions are

$$y(x,\ t) = A\sin(\omega t - kx)$$

$$v_y = \frac{\partial y}{\partial t} = \omega A\cos(\omega t - kx)$$

$$a_y = \frac{\partial^2 y}{\partial t^2} = \frac{\partial v_y}{\partial t} = -\omega^2 A\sin(\omega t - kx).$$

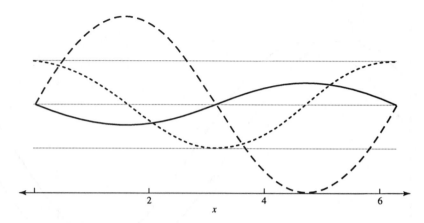

b) (Take A, k and ω to be positive. At $t = 0$, the wave is represented by Fig. (19-6(a)); point (i) in the problem corresponds to the origin, and points (ii)-(vii) correspond to the points in the figure labeled 1-7.)   (i) $v_y = \omega A \cos(0) = \omega A$, and the particle is moving upward (in the positive y-direction). $a_y = -\omega^2 A \sin(0) = 0$, and the particle is instantaneously not accelerating.   (ii) $v_y = \omega A \cos(-\pi/4) = \omega A/\sqrt{2}$, and the particle is moving up. $a_y = -\omega^2 A \sin(-\pi/4) = \omega^2 A/\sqrt{2}$, and the particle is speeding up.   (iii) $v_y = \omega A \cos(-\pi/2) = 0$, and the particle is instantaneously at rest. $a_y = -\omega^2 A \sin(-\pi/2) = \omega^2 A$, and the particle is speeding up.   (iv) $v_y = \omega A \cos(-3\pi/4) = -\omega A/\sqrt{2}$, and the particle is moving down. $a_y = -\omega^2 A \sin(-3\pi/4) = \omega^2 A/\sqrt{2}$, and the particle is slowing down (v_y is becoming less negative).   (v) $v_y = \omega A \cos(-\pi) = -\omega A$ and the particle is moving down. $a_y = -\omega^2 A \sin(-\pi) = 0$, and the particle is instantaneously not accelerating.   (vi) $v_y = \omega A \cos(-5\pi/4) = -\omega A/\sqrt{2}$ and the particle is moving down. $a_y = -\omega^2 A \sin(-5\pi/4) = -\omega^2 A/\sqrt{2}$ and the particle is speeding up (v_y and a_y have the same sign).   (vii) $v_y = \omega A \cos(-3\pi/2) = 0$, and the particle is instantaneously at rest. $a_y = -\omega^2 A \sin(-3\pi/2) = -\omega^2 A$ and the particle is speeding up.   (vii) $v_y = \omega A \cos(-7\pi/4) = \omega A/\sqrt{2}$, and the particle is moving upward. $a_y = -\omega^2 A \sin(-7\pi/4) = -\omega^2 A/\sqrt{2}$ and the particle is slowing down (v_y and a_y have opposite signs).

19-11:   Reading from the graph, a) $A = 4.0$ mm, b) $T = 0.040$ s.   c) A displacement of 0.090 m corresponds to a time interval of 0.025 s; that is, the part of the wave represented by the point where the red curve crosses the origin corresponds to the point where the blue curve crosses the t-axis $(y = 0)$ at $t = 0.025$ s, and in this time the wave has traveled 0.090 m, and so the wave speed is 3.6 m/s and the wavelength is $vT = (3.6$ m/s$)(0.040$ s$) =$ 0.14 m.   d) 0.090 m/0.015 s $= 6.0$ m/s and the wavelength is 0.24 m.   d) No; there could be many wavelengths between the places where $y(t)$ is measured.

19-12:   a)
$$A \sin 2\pi \left(\frac{t}{T} - \frac{x}{\lambda} \right) = -A \sin \frac{2\pi}{\lambda} \left(x - \frac{\lambda}{T} t \right)$$

$$= -A \sin \frac{2\pi}{\lambda} (x - vt),$$

where $\frac{\lambda}{T} = \lambda f = v$ has been used.

b)

$$v_y = \frac{\partial y}{\partial t} = \frac{2\pi v}{\lambda} A \cos \frac{2\pi}{\lambda}(x - vt).$$

c) The speed is the greatest when the cosine is 1, and that speed is $2\pi v A/\lambda$. This will be equal to v if $A = \lambda/2\pi$, less than v if $A < \lambda/2\pi$ and greater than v if $A > \lambda/2\pi$.

19-13: a) $t = 0$:

x(cm)	0.00	1.50	3.00	4.50	6.00	7.50	9.00	10.50	12.00
y(cm)	0.000	−0.212	−0.300	−0.212	0.000	0.212	0.300	0.212	0.000

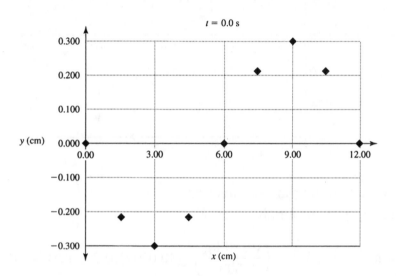

b) i) $t = 0.400$ s:

x(cm)	0.00	1.50	3.00	4.50	6.00	7.50	9.00	10.50	12.00
y(cm)	0.285	0.136	−0.093	−0.267	−0.285	−0.136	0.093	0.267	0.285

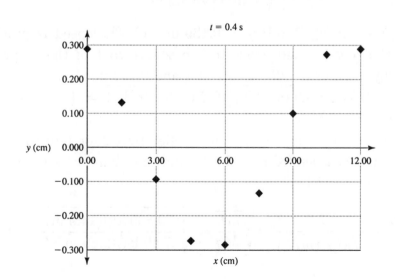

ii) $t = 0.800$ s:

x(cm)	0.00	1.50	3.00	4.50	6.00	7.50	9.00	10.50	12.00
y(cm)	0.176	0.296	0.243	0.047	−0.176	−0.296	−0.243	−0.047	0.176

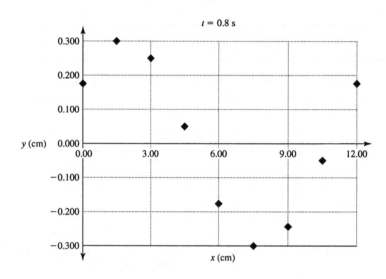

iii) The wave is traveling in $+x$-direction.

19-14:   Solving Eq. (19-13) for the force F,

$$F = \mu v^2 = \mu(f\lambda)^2 = \left(\frac{0.120 \text{ kg}}{2.50 \text{ m}}\right)((40.0 \text{ Hz})(0.750 \text{ m}))^2 = 43.2 \text{ N}.$$

19-15:   a) Neglecting the mass of the string, the tension in the string is the weight of the pulley, and the speed of a transverse wave on the string is

$$v = \sqrt{\frac{F}{\mu}} = \sqrt{\frac{(1.50 \text{ kg})(9.80 \text{ m/s}^2)}{(0.0550 \text{ kg/m})}} = 16.3 \text{ m/s}.$$

b) $\lambda = v/f = (16.3 \text{ m/s})/(120 \text{ Hz}) = 0.136$ m.   c) The speed is proportional to the square root of the tension, and hence to the square root of the suspended mass; the answers change by a factor of $\sqrt{2}$, to 23.1 m/s and 0.192 m.

19-16:   a) $v = \sqrt{F/\mu} = \sqrt{(140.0 \text{ N})(10.0 \text{ m})/0.800 \text{ kg}} = 41.8$ m/s.   b) $\lambda = v/f = (41.8 \text{ m/s})/(1.20 \text{ Hz}) = 34.9$ m.   c) The speed is larger by a factor of $\sqrt{2}$, and so for the same wavelength, the frequency must be multiplied by $\sqrt{2}$, or 1.70 Hz.

19-17:   Denoting the suspended mass by M and the string mass by m, the time for the pulse to reach the other end is

$$t = \frac{L}{v} = \frac{L}{\sqrt{Mg/(m/L)}} = \sqrt{\frac{mL}{Mg}} = \sqrt{\frac{(0.800 \text{ kg})(14.0 \text{ m})}{(7.50 \text{ kg})(9.80 \text{ m/s}^2)}} = 0.390 \text{ s}.$$

19-18: a) The tension at the bottom of the rope is due to the weight of the load, and the speed is the same 88.5 m/s as found in Example 19-4. b) The tension at the middle of the rope is $(21.0 \text{ kg})(9.80 \text{ m/s}^2) = 205.8$ N (keeping an extra figure) and the speed of the rope is 90.7 m/s. c) The tension at the top of the rope is $(22.0 \text{ kg})(9.80 \text{ m/s}^2) = 215.6$ m/s and the speed is 92.9 m/s. (See Challenge Problem (19-46) for the effects of varying tension on the time it takes to send signals.)

19-19: a) Using Equation (19-21), $B = v^2\rho = (\lambda f)^2$, so $B = [(8 \text{ m})(400/\text{s})]^2 \times (1300 \text{ kg/m}^3) = 1.33 \times 10^{10}$ Pa.

b) Using Equation (19-22), $Y = v^2\rho = (L/t)^2\rho = [(1.5 \text{ m})/(3.9 \times 10^{-4} \text{ s})]^2 \times (6400 \text{ kg/m}^3) = 9.47 \times 10^{10}$ Pa.

19-20: a) The time for the wave to travel to Caracas was 9 min 39 s = 579 s and the speed was 1.085×10^4 m/s (keeping an extra figure). Similarly, the time for the wave to travel to Kevo was 680 s for a speed of 1.278×10^4 m/s, and the time to travel to Vienna was 767 s for a speed of 1.258×10^4 m/s. The average speed for these three measurements is 1.21×10^4 m/s. Due to variations in density, or reflections (a subject addressed in later chapters), not all waves travel in straight lines with constant speeds. b) From Eq. (19-21), $B = v^2\rho$, and using the given value of $\rho = 3.3 \times 10^3$ kg/m^3 and the speeds found in part (a), the values for the bulk modulus are, respectively, 3.9×10^{11} Pa, 5.4×10^{11} Pa and 5.2×10^{11} Pa. These are larger, by a factor of 2 or 3, than the largest values in Table (11-1).

19-21: (The water temperature is not specified; using $v_{\text{water}} = 1482$ m/s at 20°C, as given in Table (19-1), gives the following result.) The sound wave travels in water for the same time as the wave travels a distance $22.0 \text{ m} - 1.20 \text{ m} = 20.8$ m in air, and so the depth of the diver is

$$(20.8 \text{ m})\frac{v_{\text{water}}}{v_{\text{air}}} = (20.8 \text{ m})\frac{1482 \text{ m/s}}{344 \text{ m/s}} = 89.6 \text{ m}.$$

This is the depth of the diver; the distance from the horn is 90.8 m.

19-22: a), b), c) Using Eq. (19-27),

$$v_{\text{H}_2} = \sqrt{\frac{(1.41)(8.3145 \text{ J/mol·K})(300.15 \text{ K})}{(2.02 \times 10^{-3} \text{ kg/mol})}} = 1.32 \times 10^3 \text{ m/s}$$

$$v_{\text{He}} = \sqrt{\frac{(1.67)(8.3145 \text{ J/mol·K})(300.15 \text{ K})}{(4.00 \times 10^{-3} \text{ kg/mol})}} = 1.02 \times 10^3 \text{ m/s}$$

$$v_{\text{Ar}} = \sqrt{\frac{(1.67)(8.3145 \text{ J/mol·K})(300.15 \text{ K})}{(39.9 \times 10^{-3} \text{ kg/mol})}} = 323 \text{ m/s}.$$

d) Repeating the calculation of Example 19-7 at $T = 300.15$ K gives $v_{\text{air}} = 348$ m/s, and so $v_{\text{H}_2} = 3.80\, v_{\text{air}}$, $v_{\text{He}} = 2.94\, v_{\text{air}}$ and $v_{\text{Ar}} = 0.928\, v_{\text{air}}$.

19-23:   Solving Eq. (19-27) for the temperature,

$$T = \frac{Mv^2}{\gamma R} = \frac{(28.8 \times 10^{-3} \text{ kg/mol})\left(\left(\frac{850 \text{ km/h}}{0.85}\right)\left(\frac{1 \text{ m/s}}{3.6 \text{ km/hr}}\right)\right)^2}{(1.40)(8.3145 \text{ J/mol·K})} = 191 \text{ K},$$

or $-82\ °\text{C}$.   b) See the results of Problem 16-78, the variation of atmospheric pressure with altitude, assuming a non-constant temperature. If we know the altitude we can use the result of Problem 16-78, $p = p_0\left(1 - \frac{\alpha y}{T_0}\right)^{\left(\frac{Mg}{R\alpha}\right)}$. Since $T = T_0 - \alpha y$, for $T = 191$ K, $\alpha = .6 \times 10^{-2}°\text{C/m}$, and $T_0 = 273$ K, $y = 13{,}667$ m (44,840 ft.). Although a very high altitude for commercial aircraft, some military aircraft fly this high. This result assumes a uniform decrease in temperature that is solely due to the increasing altitude. Then, if we use this altitude, the pressure can be found:

$$p = p_0\left(1 - \frac{(.6 \times 10^{-2}°\text{C/m})(13{,}667 \text{ m})}{273 \text{ K}}\right)^{\left(\frac{28.8 \times 10^{-3} \text{ kg/mol})(9.8 \text{ m/s}^2)}{(8.315 \text{ J/mol·K})(.6 \times 10^{-2 0} \text{ C/m})}\right)},$$

and $p = p_0(.70)^{5.66} = .13p_0$, or about .13 atm. Using an altitude of 13,667 m in the equation derived in Example 16-4 gives $p = .18p_0$, which overestimates the pressure due to the assumption of an isothermal atmosphere.

19-24:   (This is the same as Problem 11-86(a), reproduced here.)

For constant temperature ($\Delta T = 0$),

$$\Delta(pV) = (\Delta p)V + p(\Delta V) = 0 \quad \text{and} \quad B = -\frac{(\Delta p)V}{(\Delta V)} = p.$$

19-25:   Table 19-1 suggests that the speed of longitudinal waves in brass is much higher than in air, and so the sound that travels through the metal arrives first. The time difference is

$$\Delta t = \frac{L}{v_{\text{air}}} - \frac{L}{v_{\text{Brass}}} = \frac{80.0 \text{ m}}{344 \text{ m/s}} - \frac{80.0 \text{ m}}{\sqrt{(0.90 \times 10^{11} \text{ Pa})/(8600 \text{ kg/m}^3)}} = 0.208 \text{ s}.$$

19-26:
$$\sqrt{\frac{(1.40)(8.3145 \text{ J/mol·K})(300.15 \text{ K})}{(28.8 \times 10^{-3} \text{ kg/mol})}} - \sqrt{\frac{(1.40)(8.3145 \text{ J/mol·K})(260.15 \text{ K})}{(28.8 \times 10^{-3} \text{ kg/mol})}}$$

$$= 24 \text{ m/s}.$$

(The result is known to only two figures, being the difference of quantities known to three figures.)

19-27:   The mass per unit length μ is related to the density (assumed uniform) and the cross-section area A by $\mu = A\rho$, so combining Eq. (19-13) and Eq. (19-22) with the given relations between the speeds,

$$\frac{Y}{\rho} = 900\,\frac{F}{A\rho} \quad \text{so} \quad F/A = \frac{Y}{900}.$$

19-28: a) Using Eq. (19-33),

$$P_{\text{ave}} = \frac{1}{2}\sqrt{\mu F}\,\omega^2 A^2$$

$$= \frac{1}{2}\sqrt{\left(\frac{3.00 \times 10^{-3}\text{ kg}}{0.80\text{ m}}\right)(25.0\text{ N})(2\pi(120.0\text{ Hz}))^2(1.6\times 10^{-3}\text{ m})^2}$$

$$= 0.223\text{ W},$$

or 0.22 W to two figures. b) Halving the amplitude quarters the average power, to 0.056 W.

19-29: a) $\quad \lambda = \dfrac{v}{f} = \dfrac{\sqrt{Y/\rho}}{f} = \dfrac{\sqrt{(11.0\times 10^{10}\text{ Pa})/(8.9\times 10^3\text{ kg/m}^3)}}{220\text{ Hz}} = 16.0\text{ m}.$

b) Solving Eq. (19-35) for the amplitude A (as opposed to the area $a = \pi r^2$) in terms of the product $p = Ia$,

$$A = \sqrt{\frac{(2P/a)}{\sqrt{\rho Y}\,\omega^2}}$$

$$= \sqrt{\frac{2(6.50\times 10^{-6}\text{ W})/(\pi(0.800\times 10^{-2}\text{ m})^2)}{\sqrt{(8.9\times 10^3\text{ kg/m}^3)(11.0\times 10^{10}\text{ Pa})}(2\pi(220\text{ Hz}))^2}} = 3.29\times 10^{-8}\text{ m}.$$

c) $\omega A = 2\pi f A = 2\pi(220\text{ Hz})(3.289\times 10^{-8}\text{ m}) = 4.55\times 10^{-5}\text{ m/s}.$

19-30: a) See Exercise 19-29. The amplitude is

$$A = \sqrt{\frac{2I}{\sqrt{\rho B}\,\omega^2}}$$

$$= \sqrt{\frac{2(3.00\times 10^{-6}\text{ W/m}^2)}{\sqrt{(1000\text{ kg/m}^3)(2.18\times 10^9\text{ Pa})}(2\pi(3400\text{ Hz}))^2}} = 9.44\times 10^{-11}\text{ m}.$$

The wavelength is

$$\lambda = \frac{v}{f} = \frac{\sqrt{B/\rho}}{f} = \frac{\sqrt{(2.18\times 10^9\text{ Pa})/(1000\text{ kg/m}^3)}}{3400\text{ Hz}} = 0.434\text{ m}.$$

c) Repeating the above with $B = \gamma p = 1.40\times 10^5$ Pa and the density of air gives $A = 5.66\times 10^{-9}$ m and $\lambda = 0.100$ m. The amplitude is larger in air, by a factor of about 60. For a given frequency, the much less dense air molecules must have a larger amplitude to transfer the same amount of energy.

19-31: a) $f = v/\lambda = (36.0 \text{ m/s})/(1.80 \text{ m}) = 20.0$ Hz, $\omega = 2\pi f = 126$ rad/s, $k = \omega/v = 2\pi/\lambda = 3.49$ rad/m.

b) $y(x, t) = A\sin(\omega t - kx) = (2.50 \text{ mm})\sin[(126 \text{ rad/s})t - (3.49 \text{ rad/m})x]$.

c) At $x = 0$, $y(0, t) = A\sin\omega t = (2.50 \text{ mm})\sin[(126 \text{ rad/s})t]$. From this form it may be seen that at $x = 0$, $t = 0$, $\frac{\partial y}{\partial t} > 0$.   d) At $x = 1.35$ m $= 3\lambda/4$, $kx = 3\pi/2$ and

$$y(3\lambda/4, t) = A\sin[\omega t - 3\pi/2].$$

e) See Exercise 19-12; $\omega A = 0.315$ m/s. f) From the result of part (d), $y = 0$ mm. $v_y = -0.315$ m/s.

19-32: a) From comparison with Eq. (19-4), $A = 0.75$ cm, $\lambda = \frac{2}{0.400/\text{cm}} = 5.00$ cm, $f = 125$ Hz, $T = \frac{1}{f} = 0.00800$ s and $v = \lambda f = 6.25$ m/s.

b)

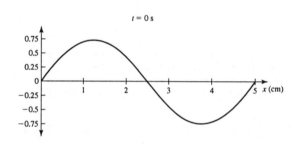

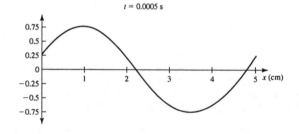

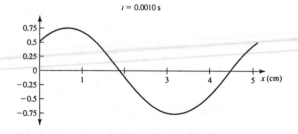

c) To stay with a wavefront as t increases, x decreases and so the wave is moving in the $-x$-direction. d) From Eq. (19-13), the tension is $F = \mu v^2 = (0.50 \text{ kg/m})(6.25 \text{ m/s})^2 = 19.5$ N.

e) $P_{\text{ave}} = \frac{1}{2}\sqrt{\mu F}\omega^2 A^2 = 54.2$ W.

19-33: a) Speed in each segment is $v = \sqrt{F/\mu}$. The time to travel through a segment is $t = L/v$. The travel times then, are $t_1 = L\sqrt{\frac{\mu_1}{F}}$, $t_2 = L\sqrt{\frac{4\mu_1}{F}}$, and $t_3 = L\sqrt{\frac{\mu_1}{4F}}$. Adding gives $t_{\text{total}} = L\sqrt{\frac{\mu_1}{F}} + 2L\sqrt{\frac{\mu_1}{F}} + \frac{1}{2}L\sqrt{\frac{\mu_1}{F}} = \frac{7}{2}L\sqrt{\frac{\mu_1}{F}}$.

b) No, because the tension is uniform throughout each piece.

19-34: The maximum vertical acceleration must be at least g. Because $a = \omega^2 A$, $g = \omega^2 A_{\text{min}}$ and thus $A_{\text{min}} = g/\omega^2$. Using $\omega = 2\pi f = 2\pi v/\lambda$ and $v = \sqrt{F/\mu}$, this becomes $A_{\text{min}} = \frac{g\lambda^2\mu}{4\pi^2 F}$.

19-35: a) See Exercise 19-10; $a_y = \frac{\partial^2 y}{\partial t^2} = -\omega^2 y$, and so $k' = \Delta m\omega^2 = \Delta x\mu\omega^2$.

b)
$$\omega^2 = (2\pi f)^2 = \left(\frac{2\pi v}{\lambda}\right)^2 = \frac{4\pi^2 F}{\mu\lambda^2}$$

and so $k' = (4\pi^2 F/\lambda^2)\Delta x$. The effective force constant k' is independent of amplitude, as for a simple harmonic oscillator, and is proportional to the tension that provides the restoring force. The factor of $1/\lambda^2$ indicates that the curvature of the string creates the restoring force on a segment of the string. More specifically, one factor of $1/\lambda$ is due to the curvature, and a factor of $1/(\lambda\mu)$ represents the mass in one wavelength, which determines the frequency of the overall oscillation of the string. The mass $\Delta m = \mu\Delta x$ also contains a factor of μ, and so the effective spring constant per unit length is independent of μ.

19-36: a), b)

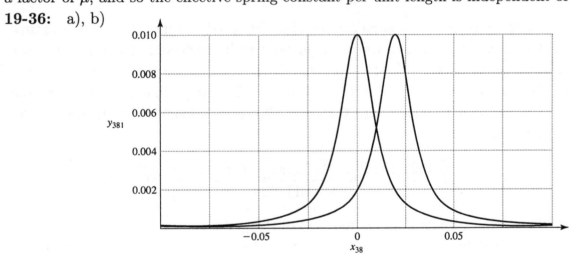

c) The displacement is a maximum when the term in parentheses in the denominator is zero; the denominator is the sum of two squares and is minimized when $x = vt$, and the maximum displacement is A. At $x = 4.50$ cm, the displacement is a maximum at $t = (4.50 \times 10^{-2} \text{ m})/(20.0 \text{ m/s}) = 2.25 \times 10^{-3}$ s. The displacement will be half of the maximum when $(x - vt)^2 = A^2$, or $t = (x \pm A)/v = 1.75 \times 10^{-3}$ s and 2.75×10^{-3} s.

d) Of the many ways to obtain the result, the method presented saves some algebra and minor calculus, relying on the chain rule for partial derivatives. Specifically, let $u = u(x, t) = x - vt$, so that if $f(x, t) = g(u)$, $\frac{\partial f}{\partial x} = \frac{dg}{du}\frac{\partial u}{\partial x} = \frac{dg}{du}$ and $\frac{\partial f}{\partial t} = \frac{dg}{du}\frac{\partial u}{\partial t} = -\frac{dg}{du}v$. (In this form it may be seen that any function of this form satisfies the wave equation; see Problem 19-39.) In this case, $y(x, t) = A^3 \left(A^2 + u^2\right)^{-1}$, and so

$$\frac{\partial y}{\partial x} = \frac{-2A^3 u}{\left(A^2 + u^2\right)^2}, \quad \frac{\partial^2 y}{\partial x^2} = -\frac{2A^3 \left(A^2 - 3u^2\right)}{\left(A^2 + u^2\right)^3}$$

$$\frac{\partial y}{\partial t} = v\frac{2A^3 u}{\left(A^2 + u^2\right)^2}, \quad \frac{\partial^2 y}{\partial t^2} = -v^2\frac{2A^3 \left(A^2 - 3u^2\right)}{\left(A^2 + u^2\right)^2},$$

and so the given form for $y(x, t)$ is a solution to the wave equation with speed v.

19-37: a) (1): The curve appears to be horizontal, and $v_y = 0$. As the wave moves, the point will begin to move downward, and $a_y < 0$. (2): As the wave moves in the $+x$-direction (to the right in Fig. (19-39)), the particle will move upward. The portion of the curve to the left of the point is steeper, so $a_y > 0$. (3) The point is moving down, and will increase its speed as the wave moves; $v_y < 0$, $a_y < 0$. (4) The curve appears to be horizontal, and $v_y = 0$. As the wave moves, the point will move away from the x-axis, and $a_y > 0$. (5) The point is moving downward, and will increase its speed as the wave moves; $v_y < 0$, $a_y < 0$. (6) The particle is moving upward, but the curve that represents the wave appears to have no curvature, so $v_y > 0$ and $a_y = 0$. b) The accelerations, which are related to the curvatures, will not change. The transverse velocities will all change sign.

19-38: The speed of light is so large compared to the speed of sound that the travel time of the light from the lightning or the radio signal may be neglected. Then, the distance from the storm to the dorm is $(344 \text{ m/s})(4.43 \text{ s}) = 1523.92 \text{ m}$ and the distance from the storm to the ballpark is $(344 \text{ m/s})(3.00 \text{ s}) = 1032 \text{ m}$. The angle that the direction from the storm to the ballpark makes with the north direction is found from these distances using the law of cosines;

$$\theta = \arccos\left[\frac{(1523.92 \text{ m})^2 - (1032 \text{ m})^2 - (1120 \text{ m})^2}{-2(1032 \text{ m})(1120 \text{ m})}\right] = 90.07°,$$

so the storm can be considered to be due west of the park.

19-39: a) As time goes on, someone moving with the wave would need to move in such a way that the wave appears to have the same shape. If this motion can be described by $x = vt + c$, with c a constant (not the speed of light), then $y(x, t) = f(c)$, and the waveform is the same to such an observer. b) See Problem 19-36. The derivation is completed by taking the second partials,

$$\frac{\partial^2 y}{\partial x^2} = \frac{1}{v^2}\frac{d^2 f}{du^2}, \quad \frac{\partial^2 y}{\partial t^2} = \frac{d^2 f}{du^2},$$

so $y(x, t) = f(t - x/v)$ is a solution to the wave equation with wave speed v.   c) This is of the form $y(x, t) = f(u)$, with $u = t - x/v$ and

$$f(u) = De^{-C^2(t-(B/C)x)^2},$$

and the result of part (b) may be used to determine the speed $v = C/B$ immediately.

19-40:   a)

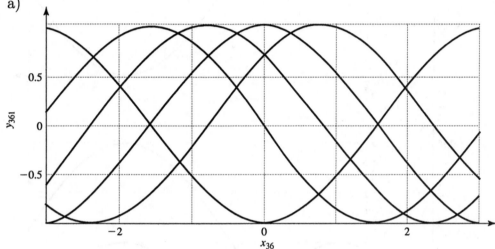

b) $\frac{\partial y}{\partial t} = \omega A \cos(\omega t - kx + \phi)$.   c) No; $\phi = \pi/4$ or $\phi = 3\pi/4$ would both give $A/\sqrt{2}$. If the particle is known to be moving downward, the result of part (b) shows that $\cos\phi < 0$, and so $\phi = 3\pi/4$.   d) To identify ϕ uniquely, the quadrant in which ϕ is must be known. In physical terms, the signs of both the position and velocity, and the magnitude of either, are necessary to determine ϕ (within additive multiples of 2π).

19-41:   a) $\sqrt{\mu F} = F\sqrt{\mu/F} = F/v = Fk/\omega$ and subsituting this into Eq. (19-33) gives the result.

  b) Quadrupling the tension from F to $F' = 4F$ increases the speed $v = \sqrt{F/\mu}$ by a factor of 2, so the new frequency ω' and new wave number k' are related to ω and k by $(\omega'/k') = 2(\omega/k)$. For the average power to be the same, we must have $Fk\omega = F'k'\omega'$, so $k\omega = 4k'\omega'$ and $k'\omega' = k\omega/4$.

  Multiplying the first and second equations together gives

$$\omega'^2 = \omega^2/2, \quad \text{so} \quad \omega' = \omega/\sqrt{2}.$$

Thus, the frequency must decrease by a factor of $\sqrt{2}$. Dividing the second equation by the first equation gives

$$k'^2 = k^2/8, \quad \text{so} \quad k' = k/\sqrt{8}.$$

19-42:   a), d)

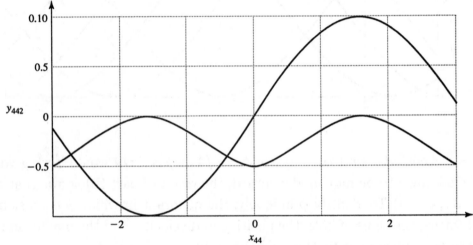

b) The power is a maximum where the displacement is zero, and the power is a minimum of zero when the magnitude of the displacement is a maximum.   c) The direction of the energy flow is always in the same direction.   d) In this case, $\frac{\partial y}{\partial x} = kA\cos(\omega t + kx)$, and so Eq. (19-30) becomes

$$P(x,\,t) = -Fk\omega A^2 \cos(\omega t + kx).$$

The power is now negative (energy flows in the $-x$-direction), but the qualitative relations of part (b) are unchanged.

19-43: $$v_1^2 = \frac{F_1}{\mu}, \quad v_2^2 = \frac{F_2}{\mu} = \frac{F_1 - YA\alpha\Delta T}{\mu}.$$

Solving for α,

$$\alpha = \frac{v_1^2 - v_2^2}{Y(A/\mu)\Delta T} = \frac{v_1^2 - v_2^2}{(Y/\rho)\Delta T}.$$

19-44: a) Consider the derivation of the speed of a longitudinal wave in Section 19-6. Instead of the bulk modulus B, the quantity of interest is the change in force per fractional length change. The force constant k' is the change in force per length change, so the force change per fractional length change is $k'L$, the applied force at one end is $F = (k'L)(v_y/v)$ and the longitudinal impulse when this force is applied for a time t is $k'Ltv_y/v$. The change in longitudinal momentum is $((vt)m/L)v_y$ and equating the expressions, canceling a factor of t and solving for v gives $v^2 = L^2 k'/m$.

An equivalent method is to use the result of Problem 11-70(a), which relates the force constant k' and the "Young's modulus" of the Slinky™, $k' = YA/L$, or $Y = k'L/A$. The mass density is $\rho = m/(AL)$, and Eq. (19-22) gives the result immediately.

b) $(2.00 \text{ m})\sqrt{(1.50 \text{ N/m})/(0.250 \text{ kg})} = 4.90 \text{ m/s}$.

19-45: a)
$$u_k = \frac{\Delta K}{\Delta x} = \frac{(1/2)\Delta m v_y^2}{\Delta m/\mu} = \frac{1}{2}\mu \left(\frac{\partial y}{\partial t}\right)^2.$$

b) $\frac{\partial y}{\partial t} = \omega A \cos(\omega t - kx)$ and so

$$u_k = \frac{1}{2}\mu\omega^2 A^2 \cos^2(\omega t - kx).$$

c) The piece has width Δx and height $\Delta x \frac{\partial y}{\partial x}$, and so the length of the piece is

$$\left((\Delta x)^2 + \left(\Delta x \frac{\partial y}{\partial x}\right)^2\right)^{1/2} = \Delta x \left(1 + \left(\frac{\partial y}{\partial x}\right)^2\right)^{1/2}$$

$$\approx \Delta x \left[1 + \frac{1}{2}\left(\frac{\partial y}{\partial x}\right)^2\right],$$

where the relation given in the hint has been used.

d)
$$u_p = F \frac{\Delta x \left[1 + \frac{1}{2}\left(\frac{\partial y}{\partial x}\right)^2\right] - \Delta x}{\Delta x} = \frac{1}{2}F\left(\frac{\partial y}{\partial x}\right)^2.$$

e) $\frac{\partial y}{\partial x} = -kA\cos(\omega t - kx)$, and so

$$u_p = \frac{1}{2}Fk^2 A^2 \cos^2(\omega t - kx)$$

and f) comparison with the result of part (c), with $k^2 = \omega^2/v^2 = \omega^2\mu/F$, shows that for a sinusoidal wave $u_k = u_p$. g) In this graph, u_k and u_p coincide, as shown in part (f).

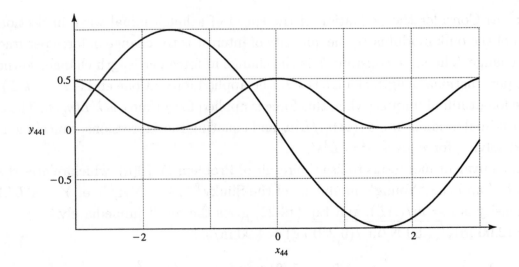

At $y = 0$, the string is stretched the most, and is moving the fastest, so u_k and u_p are maximized. At the extremes of y, the string is unstretched and is not moving, so u_k and u_p are both at their minimum of zero.

h) $$u_k + u_p = Fk^2 A^2 \cos^2(\omega t - kx) = Fk(\omega/v)A^2 \cos^2(\omega t - kx) = \frac{P}{v}.$$

The energy density (which is itself a wave; see Problem 19-39) travels with the wave, and the rate at which the energy is transported is the product of the density per unit length and the speed.

19-46:   a) The tension is the difference between the diver's weight and the buoyant force,

$$F = (m - \rho_{\text{water}}V)g = \left(120 \text{ kg} - (1000 \text{ kg/m}^3)(0.0800 \text{ m}^3)(9.80 \text{ m/s}^2)\right) = 392 \text{ N}.$$

b) The increase in tension will be the weight of the cable between the diver and the point at x, minus the buoyant force. This increase in tension is then

$$(\mu x - \rho(Ax))g = \left(1.10 \text{ kg/m} - (1000 \text{ kg/m}^3)\pi(1.00 \times 10^{-2} \text{ m})^2\right)(9.80 \text{ m/s}^2)x$$

$$= (7.70 \text{ N/m})x.$$

The tension as a function of x is then $F(x) = (392 \text{ N}) + (7.70 \text{ N/m})x$.   c) Denote the tension as $F(x) = F_0 + ax$, where $F_0 = 392 \text{ N}$ and $a = 7.70 \text{ N/m}$. Then, the speed of transverse waves as a function of x is $v = \frac{dx}{dt} = \sqrt{(F_0 + ax)/\mu}$ and the time t needed for a wave to reach the surface is found from

$$t = \int dt = \int \frac{dx}{dx/dt} = \int \frac{\sqrt{\mu}}{\sqrt{F_0 + ax}}\, dx.$$

Let the length of the cable be L, so

$$t = \sqrt{\mu} \int_0^L \frac{dx}{\sqrt{F_0 + ax}} = \sqrt{\mu}\,\frac{2}{a}\sqrt{F_0 + ax}\Big|_0^L$$

$$= \frac{2\sqrt{\mu}}{a}\left(\sqrt{F_0 + aL} - \sqrt{F_0}\right)$$

$$= \frac{2\sqrt{1.10 \text{ kg/m}}}{7.70 \text{ N/m}}\left(\sqrt{392 \text{ N} + (7.70 \text{ N/m})(100 \text{ m})} - \sqrt{392 \text{ N}}\right) = 3.89 \text{ s.}$$

19-47: The tension in the rope will vary with radius r. The tension at a distance r from the center must supply the force to keep the mass of the rope that is further out than r accelerating inward. The mass of this piece is $m\frac{L-r}{L}$, and its center of mass moves in a circle of radius $\frac{L+r}{2}$, and so

$$T(r) = \left[m\,\frac{L-r}{L}\right]\omega^2\left[\frac{L+r}{2}\right] = \frac{m\omega^2}{2L}\left(L^2 - r^2\right).$$

An equivalent method is to consider the net force on a piece of the rope with length dr and mass $dm = dr\, m/L$. The tension must vary in such a way that $T(r) - T(r + dr) = -\omega^2 r\, dm$, or $\frac{dT}{dr} = -(m\omega^2/L)r\, dr$. This is integrated to obtain $T(r) = -(m\omega^2/2L)r^2 + C$, where C is a constant of integration. The tension must vanish at $r = L$, from which $C = (m\omega^2 L/2)$ and the previous result is obtained.

The speed of propagation as a function of distance is

$$v(r) = \frac{dr}{dt} = \sqrt{\frac{T(r)}{\mu}} = \sqrt{\frac{TL}{m}} = \frac{\omega}{\sqrt{2}}\sqrt{L^2 - r^2},$$

where $\frac{dr}{dt} > 0$ has been chosen for a wave traveling from the center to the edge. Separating variables and integrating, the time t is

$$t = \int dt = \frac{\sqrt{2}}{\omega}\int_0^L \frac{dr}{\sqrt{L^2 - r^2}}.$$

The integral may be found in a table, or in Appendix B. The integral is done explicitly by letting $r = L\sin\theta$, $dr = L\cos\theta\, d\theta$, $\sqrt{L^2 - r^2} = L\cos\theta$, so that

$$\int \frac{dr}{\sqrt{L^2 - r^2}} = \theta = \arcsin\frac{r}{L}, \quad \text{and}$$

$$t = \frac{\sqrt{2}}{\omega}\arcsin(1) = \frac{\pi}{\omega\sqrt{2}}.$$

19-48: The intensity will be the product of the net force per unit area and the longitudinal speed. The net force on a section of the rod will be equal to the Young's modulus times the area times the rate at which the stress changes with respect to distance. That is, Eq. (11-10) may be re-expressed as

$$F_y = -aY\frac{\partial y}{\partial x},$$

where a is the cross-section area. The minus sign indicates that F_y is positive if y decreases with increasing x, and $\frac{\partial y}{\partial x}$ replaces $\Delta l/l_0$, the fractional change in elongation. Thus, Eq. (19-29) becomes

$$I = \frac{P}{a} = -Y\frac{\partial y}{\partial x}\frac{\partial y}{\partial t},$$

Eq. (19-31) becomes

$$I = \sqrt{\rho Y}\,\omega^2 A^2 \cos^2(\omega t - kx)$$

and Eq. (19-33) becomes Eq. (19-35).

Chapter 20   Wave Interference and Normal Modes

20-1:   a) The wave form for the given times, respectively, is shown.

b)

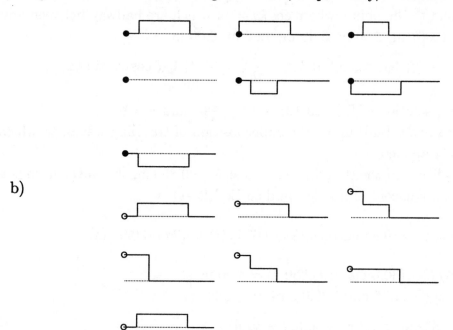

20-2:   a) The wave form for the given times, respectively, is shown.

b)

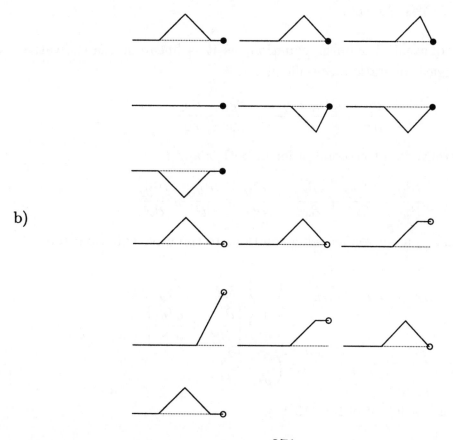

20-3: a) The nodes correspond to the places where $y = 0$ for all t in Eq. (20-1); that is, $\sin kx_{node} = 0$ or $kx_{node} = n\pi$, n an integer. With $k = 0.75\pi$ rad/m, $x_{node} = (1.333$ m$)n$ and for $n = 0\ldots5$, $x_{node} = 0$, 1.333 m, 2.67 m, 4.00 m, 5.33 m, 6.67 m. Values of n greater than 5 or less than 0 correspond to values of x that are not part of the string. b) The antinodes correspond to the points where $\cos kx = 0$, which are halfway between any two adjacent nodes, at 0.667 m, 2.00 m, 3.33 m, 4.67 m, 6.00 m.

20-4: a) $\quad \dfrac{\partial^2 y}{\partial x^2} = -k^2[A_{SW} \cos \omega t] \sin kx, \quad \dfrac{\partial^2 y}{\partial t^2} = -\omega^2[A_{SW} \cos \omega t] \sin kx,$

so for $y(x, t)$ to be a solution of Eq. (19-12), $-k^2 = \frac{-\omega^2}{v^2}$, and $v = \frac{\omega}{k}$.

b) A standing wave is built up by the superposition of traveling waves, to which the relationship $v = \omega/k$ applies.

20-5: a) The amplitude of the standing wave is $A_{SW} = 0.85$ cm, the wavelength is twice the distance between adjacent antinodes, and so Eq. (20-1) is

$$y(x, t) = (0.85 \text{ cm}) \cos((2\pi/0.075 \text{ s})t) \sin(2\pi x/30.0 \text{ cm}).$$

b) $v = \lambda f = \lambda/T = (30.0 \text{ cm})/(0.0750 \text{ s}) = 4.00$ m/s.

c) $(0.850 \text{ cm}) \sin(2\pi(10.5 \text{ cm})/(30.0 \text{ cm})) = 0.688$ cm.

20-6: $\quad y_1 + y_2 = A\left[\sin(\omega t + kx) - \sin(\omega t - kx)\right]$

$$= A\left[\sin \omega t \cos kx + \cos \omega t \sin kx - \sin \omega t \cos kx + \cos \omega t \sin kx\right]$$

$$= 2A \cos \omega t \sin kx.$$

20-7: The wave equation is a linear equation, as it is linear in the derivatives, and differentiation is a linear operation. Specifically,

$$\frac{\partial y}{\partial x} = \frac{\partial (y_1 + y_2)}{\partial x} = \frac{\partial y_1}{\partial x} + \frac{\partial y_2}{\partial x}.$$

Repeating the differentiation to second order in both x and t,

$$\frac{\partial^2 y}{\partial x^2} = \frac{\partial^2 y_1}{\partial x^2} + \frac{\partial^2 y_2}{\partial x^2}, \quad \frac{\partial^2 y}{\partial t^2} = \frac{\partial^2 y_1}{\partial t^2} + \frac{\partial^2 y_2}{\partial t^2}.$$

The functions y_1 and y_2 are given as being solutions to the wave equation; that is,

$$\frac{\partial^2 y}{\partial x^2} = \frac{\partial^2 y_1}{\partial x^2} + \frac{\partial^2 y_2}{\partial x^2} = \left(\frac{1}{v^2}\right)\frac{\partial^2 y_1}{\partial t^2} + \left(\frac{1}{v^2}\right)\frac{\partial^2 y_2}{\partial t^2}$$

$$= \left(\frac{1}{v^2}\right)\left[\frac{\partial^2 y_1}{\partial t^2} + \frac{\partial^2 y_2}{\partial t^2}\right]$$

$$= \left(\frac{1}{v^2}\right)\frac{\partial^2 y}{\partial t^2}$$

and so $y = y_1 + y_2$ is a solution of Eq. (19-12).

20-8:   a) From Eq. (20-8),

$$f_1 = \frac{1}{2L}\sqrt{\frac{FL}{m}} = \frac{1}{2(0.400 \text{ m})}\sqrt{\frac{(800 \text{ N})(0.400 \text{ m})}{(3.00 \times 10^{-3} \text{ kg})}} = 408 \text{ Hz.}$$

b) $\frac{10,000 \text{ Hz}}{408 \text{ Hz}} = 24.5$, so the 24th harmonic may be heard, but not the 25th.

20-9:   a) In the fundamental mode, $\lambda = 2L = 1.60$ m and so $v = f\lambda = (60.0 \text{ Hz})(1.60 \text{ m}) = 96.0$ m/s.

b) $F = v^2\mu = v^2 m/L = (96.0 \text{ m/s})^2(0.0400 \text{ kg})/(0.800 \text{ m}) = 461$ N.

20-10:   The condition that $x = L$ is a node becomes $k_n L = n\pi$. The wave number and wavelength are related by $k_n\lambda_n = 2\pi$, and so $\lambda_n = 2L/n$.

20-11:   a) The product of the frequency and the string length is a constant for a given string, equal to half of the wave speed, so to play a note with frequency 587 Hz, $x = (60.0 \text{ cm})(440 \text{ Hz})/(587 \text{ Hz}) = 45.0$ cm.   b) No; the fundamental is the lowest possible frequency.

20-12:   a) (i) $x = \frac{\lambda}{2}$ is a node, and there is no motion. (ii) $x = \frac{\lambda}{4}$ is an antinode, and $v_{\text{max}} = A(2\pi f) = 2\pi fA$, $a_{\text{max}} = (2\pi f)v_{\text{max}} = 4\pi^2 f^2 A$. (iii) $\cos\frac{\pi}{4} = \frac{1}{\sqrt{2}}$, and this factor multiplies the results of (ii), so $v_{\text{max}} = \sqrt{2}\pi fA$, $a_{\text{max}} = 2\sqrt{2}\pi^2 f^2 A$.   b) The amplitude is $A\sin kx$, or (i) 0, (ii) A, (iii) $A/\sqrt{2}$.   c) The time between the extremes of the motion is the same for any point on the string (although the period of the zero motion at a node might be considered indeterminate) and is $\frac{1}{2f}$.

20-13:   a) $\lambda_1 = 2L = 3.00$ m, $f_1 = \frac{v}{2L} = \frac{(48.0 \text{ m/s})}{2(1.50 \text{ m})} = 16.0$ Hz.   b) $\lambda_3 = \lambda_1/3 = 1.00$ m, $f_2 = 3f_1 = 48.0$ Hz.   c) $\lambda_4 = \lambda_1/4 = 0.75$ m, $f_3 = 4f_1 = 64.0$ Hz.

20-14:   a) For the fundamental mode, the wavelength is twice the length of the string, and $v = f\lambda = 2fL = 2(245 \text{ Hz})(0.635 \text{ m}) = 311$ m/s.   b) The frequency of the fundamental mode is proportional to the speed and hence to the square root of the tension; $(245 \text{ Hz})\sqrt{1.01} = 246$ Hz.   c) The frequency will be the same, 245 Hz. The wavelength will be $\lambda_{\text{air}} = v_{\text{air}}/f = (344 \text{ m/s})/(245 \text{ Hz}) = 1.40$ m, which is larger than the wavelength of the standing wave on the string by a factor of the ratio of the speeds.

20-15:   a) Refer to Fig. (20-15).   i) The fundamental has a displacement node at $\frac{L}{2} = 0.600$ m, the first overtone mode has displacement nodes at $\frac{L}{4} = 0.300$ m and $\frac{3L}{4} = 0.900$ m, and the second overtone mode has displacement nodes at $\frac{L}{6} = 0.200$ m, $\frac{L}{2} = 0.600$ m and $\frac{5L}{6} = 1.000$ m.   ii) Fundamental: 0, $L = 1.200$ m. First: 0, $\frac{L}{2} = 0.600$ m, $L = 1.200$ m. Second: 0, $\frac{L}{3} = 0.400$ m, $\frac{2L}{3} = 0.800$ m, $L = 1.200$ m.

b) Refer to Fig. (20-16); distances are measured from the right end of the pipe in the figure.   Pressure nodes at: Fundamental: $L = 1.200$ m. First overtone: $L/3 = 0.400$ m, $L = 1.200$ m. Second overtone: $L/5 = 0.240$ m, $3L/5 = 0.720$ m, $L = 1.200$ m.   Displacement nodes at: Fundamental: 0. First overtone: 0, $2L/3 = 0.800$ m. Second overtone: 0, $2L/5 = 0.480$ m, $4L/5 = 0.960$ m.

20-16:   a) $f_1 = \frac{v}{2L} = \frac{(344 \text{ m/s})}{2(0.450 \text{ m})} = 382$ Hz, $2f_1 = 764$ Hz, $f_3 = 3f_1 = 1147$ Hz, $f_4 = 4f_1 = 1529$ Hz.   b) $f_1 = \frac{v}{4L} = 191$ Hz, $f_3 = 3f_1 = 573$ Hz, $f_5 = 5f_1 = 956$ Hz,

$f_7 = 7f_1 = 1338$ Hz. Note that the symbol "f_1" denotes different frequencies in the two parts. The frequencies are not always exact multiples of the fundamental, due to rounding.

c) Open: $\frac{20,000}{f_1} = 52.3$, so the 52nd harmonic is heard. Stopped; $\frac{20,000}{f_1} = 105 = (2 \times 52 + 1)$, so the 52nd is heard.

20-17: $f_1 = \frac{(344 \text{ m/s})}{4(0.17 \text{ m})} = 506$ Hz, $f_2 = 3f_1 = 1517$ Hz, $f_3 = 5f_1 = 2529$ Hz.

20-18:   a) The fundamental frequency is proportional to the square root of the ratio $\frac{\gamma}{M}$ (see Eq. (19-27)), so

$$f_{He} = f_{air}\sqrt{\frac{\gamma_{He}}{\gamma_{air}} \cdot \frac{M_{air}}{M_{He}}} = (262 \text{ Hz})\sqrt{\frac{(5/3)}{(7/5)} \cdot \frac{28.8}{4.00}} = 767 \text{ Hz},$$

b) No; for a fixed wavelength, the frequency is proportional to the speed of sound in the gas.

20-19:

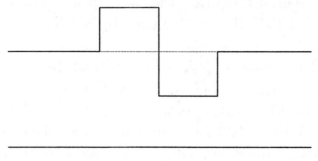

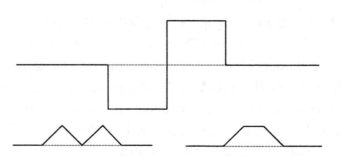

20-20:

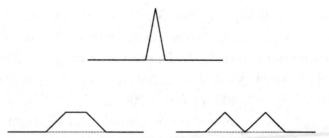

20-21:   a) For constructive interference, the path difference $d = 2.00$ m must be equal to an integer multiple of the wavelength, so $\lambda_n = d/n$ (see Example 20-7),

$$f_n = \frac{v}{\lambda_n} = \frac{vn}{d} = n\left(\frac{v}{d}\right) = n\frac{344 \text{ m/s}}{2.00 \text{ m}} = n(172 \text{ Hz}).$$

Therefore, the lowest frequency is 172 Hz.

b) Repeating the above with the path difference an odd multiple of half a wavelength, $f_n = \left(n + \frac{1}{2}\right)$ (172 Hz). Therefore, the lowest frequency is 86 Hz ($n = 0$).

20-22: The difference in path length is $\Delta x = (L-x)-x = L-2x$, or $x = (L-\Delta x)/2$. For destructive interference, $\Delta x = (n + (1/2))\lambda$, and for constructive interference, $\Delta x = n\lambda$. The wavelength is $\lambda = v/f = (344$ m/s$)/(206$ Hz$) = 1.670$ m (keeping an extra figure), and so to have $0 \leq x \leq L$, $-4 \leq n \leq 3$ for destructive interference and $-4 \leq n \leq 4$ for constructive interference. Note that neither speaker is at a point of constructive or destructive interference.

a) The points of destructive interference would be at $x = .58$ m, 1.42 m.

b) Constructive interference would be at the points $x = .17$ m, 1.00 m, 1.83 m.

c) The positions are very sensitive to frequency, the amplitudes of the waves will not be the same (except possibly at the middle), and exact cancellation at any frequency is not likely. Also, treating the speakers as point sources is a poor approximation for these dimensions (see Exercise 19-2).

20-23: a) For a stopped pipe, the wavelength of the fundamental standing wave is $4L = 0.56$ m, and so the frequency is $f_1 = (344$ m/s$)/(0.56$ m$) = 0.614$ kHz. b) The length of the column is half of the original length, and so the frequency of the fundamental mode is twice the result of part (a), or 1.23 kHz.

20-24: For a string fixed at both ends, Equation (20-6), $f_n = \frac{nv}{2L}$, is useful. It is important to remember the second *overtone* is the third *harmonic*. Solving for v, $v = \frac{nf_nL}{2}$, and inserting the data, $v = \frac{(2)(.635 \text{ m})(588/\text{s})}{3}$, and $v = 249$ m/s.

20-25: a) To show this relationship is valid, take the second time derivative:

$$\frac{\partial^2 y(x,t)}{\partial t^2} = \frac{\partial^2}{\partial t^2}[(A_{SW}\sin kx)\cos\omega t],$$

$$\frac{\partial^2 y(x,t)}{\partial t^2} = -\omega\frac{\partial}{\partial t}[(A_{SW}\sin kx)\sin\omega t]$$

$$\frac{\partial^2 y(x,t)}{\partial t^2} = -\omega^2[(A_{SW}\sin kx)\cos\omega t],$$

$$\frac{\partial^2 y(x,t)}{\partial t^2} = -\omega^2 y(x,t), \text{ Q.E.D.}$$

The displacement of the harmonic oscillator is periodic in both time and space.

b) Yes, the traveling wave is also a solution of the wave equation. See Section 19-4, *Mathematical Description of a Wave.*

20-26: a) The wave moving to the left is inverted and reflected; the reflection means that the wave moving to the left is the same function of $-x$, and the inversion means that the function is $-f(-x)$. More rigorously, the wave moving to the left in Fig. (20-3) is obtained from the wave moving to the right by a rotation of 180°, so both the coordinates (f and x) have their signs changed. b) The wave that is the sum is $f(x) - f(-x)$ (an inherently odd function), and for any f, $f(0) - f(-0) = 0$. c) The wave is reflected but not inverted (see the discussion in part (a) above), so the wave moving to the left in Fig. (20-4) is $+f(-x)$.

$$\text{d)} \quad \frac{dy}{dx} = \frac{d}{dx}\left(f(x) + f(-x)\right) = \frac{d\,f(x)}{dx} + \frac{d\,f(-x)}{dx} = \frac{d\,f(x)}{dx} + \frac{d\,f(-x)}{d(-x)}\frac{d\,(-x)}{dx}$$

$$= \frac{d\,f}{dx} - \frac{d\,f}{dx}\bigg|_{x=-x}.$$

At $x = 0$, the terms are the same and the derivative is zero. (See Exercise 20-2 for a situation where the derivative of f is not finite, so the string is not always horizontal at the boundary.)

20-27: a) $y(x,\,t) = y_1(x,\,t) + y_2(x,\,t)$

$$= A\left[\sin(\omega t + kx) + \sin(\omega t - kx)\right]$$

$$= A\left[\sin\omega t\,\cos kx + \cos\omega t\,\sin kx + \sin\omega t\,\cos kx - \cos\omega t\,\sin kx\right]$$

$$= (2A)\sin\omega t\,\cos kx.[-28pt]$$

b) At $x = 0$, $y(0,\,t) = (2A)\sin\omega t$, and so $x = 0$ is an antinode.   c) The maximum displacement is, from part (b), $A_{\text{SW}} = 2A$, the maximum speed is $\omega A_{\text{SW}} = 2\omega A$ and the magnitude of the maximum acceleration is $\omega^2 A_{\text{SW}} = 2\omega^2 A$.

20-28: a) $\lambda = v/f = (192.0\text{ m/s})/(240.0\text{ Hz}) = 0.800$ m, and the wave amplitude is $A_{\text{SW}} = 0.400$ cm. The amplitude of the motion at the given points is (i) $(0.400\text{ cm})\sin(\pi) = 0$ (a node), (ii) $(0.400\text{ cm})\sin(\pi/2) = 0.400$ cm (an antinode) and (iii) $(0.400\text{ cm})\sin(\pi/4) = 0.283$ cm.   b) The time is half of the period, or $1/(2f) = 2.08 \times 10^{-3}$ s.   c) In each case, the maximum velocity is the amplitude multiplied by $\omega = 2\pi f$ and the maximum acceleration is the amplitude multiplied by $\omega^2 = 4\pi^2 f^2$, or (i) 0, 0;   (ii) 6.03 m/s, 9.10×10^3 m/s^2;   (iii) 4.27 m/s, 6.43×10^3 m/s^2.

20-29: The plank is oscillating in its fundamental mode, so $\lambda = 2L = 10.0$ m, with a frequency of 2.00 Hz.   a) $v = f\lambda = 20.0$ m/s.   b) The plank would be in its first overtone, with twice the frequency, or 4 jumps/s.

20-30: a) The frequency of transverse standing wave in the string is (see Eq. (20-6)) $\frac{3v_t}{2(L/10)}$ and the frequency of the longitudinal standing wave in the air is (see Eq. (20-15)) $\frac{v_a}{4L}$. Equating these frequencies gives $v_a = 60v_t$.   b) The speed of transverse waves in the string is increased by a factor of $\sqrt{2}$ (assuming that in stretching the string, the mass per unit length is unchanged), and the fundamental frequency is increased by a factor of $\sqrt{2}$. There will be no integer multiple of this frequency that is the same as the frequency of vibration of the air in the can, and so there is no resonance and the amplitude of the standing waves on the string is greatly reduced.

20-31: a) For an open pipe, the difference between successive frequencies is the fundamental, in this case 392 Hz, and all frequencies would be integer multiples of this frequency. This is not the case, so the pipe cannot be an open pipe. For a stopped pipe, the difference between successive frequencies is twice the fundamental, and each frequency is an odd integer multiple of the fundamental. In this case, $f_1 = 196$ Hz, and

1372 Hz $= 7f_1$, 1764 Hz $= 9f_1$. b) $n = 7$ for 1372 Hz, $n = 9$ for 1764 Hz. c) $f_1 = v/4L$, so $L = v/4f_1 = (344 \text{ m/s})/(784 \text{ Hz}) = 0.439$ m.

20-32: The steel rod has standing waves much like a pipe open at both ends, as shown in Figure (20-15). An integral number of half wavelengths must fit on the rod, that is, $f_n = \frac{nv}{2L}$.

a) The ends of the rod are antinodes because the ends of the rod are free to ocsillate.

b) The fundamental can be produced when the rod is held at the middle because a node is located there.

c) $f_1 = \frac{(1)(5941 \text{ m/s})}{2(1.50 \text{ m})} = 1980$ Hz.

d) The next harmonic is $n = 2$, or $f_2 = 3961$ Hz. We would need to hold the rod at an $n = 2$ node, which is located at $L/4$ from either end, or at .375 m from either end.

20-33: The shower stall can be modeled as a pipe closed at both ends, and hence there are nodes at the two end walls. Figure (20-7) shows standing waves on a *string* fixed at both ends but the sequence of harmonics is the same, namely that an integral number of half wavelengths must fit in the stall.

a) The condition for standing waves is $f_n = \frac{nv}{2L}$, so the first three harmonics are $n = 1$, 2, 3.

b) A particular physics professor's shower has a length of $L = 1.48$ m. Using $f_n = \frac{nv}{2L}$, the resonant frequencies can be found when $v = 344$ m/s.

n	$f(\text{Hz})$
1	116
2	232
3	349

Note that the fundamental and second harmonic, which would have the greatest amplitude, are frequencies typically in the normal range of male singers. Hence, men do sing better in the shower! (For a further discussion of resonance and the human voice, see Thomas D. Rossing, *The Science of Sound*, Second Edition, Addison-Wesley, 1990, especially Chapters 4 and 17.)

20-34: a) The cross-section area of the string would be $a = (900 \text{ N})/(7.0 \times 10^8 \text{ Pa}) = 1.29 \times 10^{-6}$ m^2, corresponding to a radius of 0.640 mm (keeping extra figures). The length is the volume divided by the area,

$$L = \frac{V}{a} = \frac{m/\rho}{a} = \frac{(4.00 \times 10^{-3} \text{ kg})}{(7.8 \times 10^3 \text{ kg/m}^3)(1.29 \times 10^{-6} \text{ m}^2)} = 0.40 \text{ m}.$$

b) Using the above result in Eq. (20-8) gives $f_1 = 377$ Hz, or 380 Hz to two figures.

20-35: a) The fundamental has nodes only at the ends, $x = 0$ and $x = L$. b) For the second harmonic, the wavelength is the length of the string, and the nodes are at $x = 0$, $x = L/2$ and $x = L$.

c)

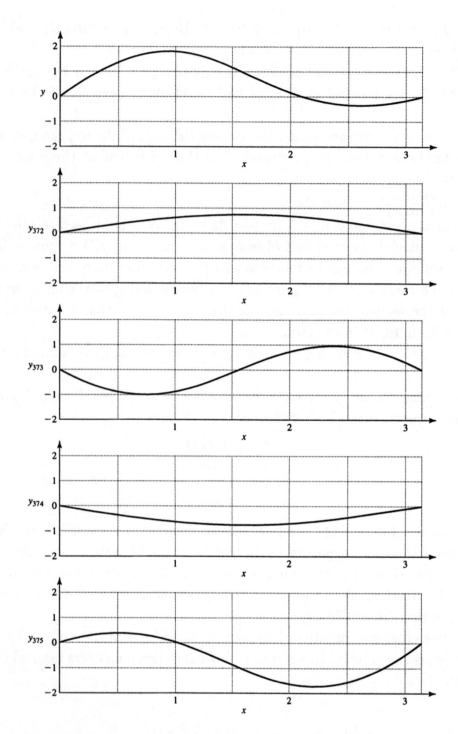

d) No; no part of the string, except for $x = L/2$, oscillates with a single frequency.

20-36: a) The new tension F' in the wire is

$$F' = F - B = w - \frac{(1/3w)\,\rho_{\text{water}}}{\rho_{\text{Al}}} = w\left(1 - \frac{1}{3}\frac{\rho_{\text{water}}}{\rho_{\text{Al}}}\right)$$

$$= w\left(1 - \frac{(1.00 \times 10^3 \ \text{kg/m}^3)}{3(2.7 \times 10^3 \ \text{kg/m}^3)}\right) = (0.8765)w = (0.8765)F.$$

The frequency will be proportional to the square root of the tension, and so $f' = (200 \text{ Hz})\sqrt{0.8765} = 187$ Hz.

b) The water does not offer much resistance to the transverse waves in the wire, and hence the node will be located a the point where the wire attaches to the sculpture and not at the surface of the water.

20-37: a) The second distance is midway between the first and third, and if there are no other distances for which resonance occurs, the difference between the first and third positions is the wavelength $\lambda = 0.750$ m. (This would give the first distance as $\lambda/4 = 18.75$ cm, but at the end of the pipe, where the air is not longer constrained to move along the tube axis, the pressure node and displacement antinode will not coincide exactly with the end.) The speed of sound in the air is then $v = f\lambda = (500 \text{ Hz})(0.750 \text{ m}) = 375$ m/s.

b) Solving Eq. (19-27) for γ,

$$\gamma = \frac{Mv^2}{RT} = \frac{(28.8 \times 10^{-3} \text{ mau/mol})(375 \text{ m/s})^2}{(8.3145 \text{ J/mol·K})(350.15 \text{ K})} = 1.39.$$

c) Since the first resonance should occur at $\tau/4 = 0.1875$ m but actually occurs at 0.18 m, the difference is 0.0075 m.

20-38: a) Considering the ear as a stopped pipe with the given length, the frequency of the fundamental is $f_1 = v/4L = (344 \text{ m/s})/(0.10 \text{ m}) = 3440$ Hz; 3500 Hz is near the resonant frequency, and the ear will be sensitive to this frequency. b) The next resonant frequencies would be 10,500 Hz and 17,500 Hz, but 7000 Hz is between these. A standing wave with frequency 7000 Hz with a node at the eardrum would have a displacement node at the open end, and resonance does not occur.

20-39: a) Solving Eq. (20-8) for the tension F,

$$F = 4L^2 f_1^2 \mu = 4mLf_1^2 = 4(14.4 \times 10^{-3} \text{ kg})(0.600 \text{ m})(65.4 \text{ Hz})^2 = 148 \text{ N}.$$

b) The tension must increase by a factor of $\left(\frac{73.4}{65.4}\right)^2$, and the percent increase is $(73.4/65.4)^2 - 1 = 26.0\%$.

20-40: a) From Eq. (20-8), with m the mass of the string and M the suspended mass,

$$f_1 = \sqrt{\frac{F}{4mL}} = \sqrt{\frac{Mg}{\pi d^2 L^2 \rho}} =$$

$$\sqrt{\frac{(420.0 \times 10^{-3} \text{ kg})(9.80 \text{ m/s}^2)}{\pi(225 \times 10^{-6} \text{ m})^2(0.45 \text{ m})^2(21.4 \times 10^3 \text{ kg/m}^3)}} = 77.3 \text{ Hz}$$

and the tuning fork frequencies for which the fork would vibrate are integer multiples of 77.3 Hz. b) The ratio $m/M \approx 9 \times 10^{-4}$, so the tension does not vary appreciably along the string.

20-41:   a) $L = \lambda/4 = v/4f = (344 \text{ m/s})/(4(349 \text{ Hz})) = 0.246$ m. The frequency will be proportional to the speed, and hence to the square root of the Kelvin temperature. The temperature necessary to have the frequency be higher is

$$(293.15 \text{ K})(1.060)^2 = 329.5 \text{ K},$$

which is 56.3°C.

20-42:   The wavelength is twice the separation of the nodes, so

$$v = \lambda f = 2Lf = \sqrt{\frac{\gamma RT}{M}}.$$

Solving for γ,

$$\gamma = \frac{M}{RT}(2Lf)^2 = \frac{(16.0 \times 10^{-3} \text{ kg})}{(8.3145 \text{ J/mol·K})(293.15 \text{ K})}(2(0.200 \text{ m})(1100 \text{ Hz}))^2 = 1.27.$$

20-43:   If the separation of the speakers is denoted h, the condition for destructive interference is

$$\sqrt{x^2 + h^2} - x = \beta\lambda,$$

where β is an odd multiple of one-half. Adding x to both sides, squaring, cancelling the x^2 term from both sides and solving for x gives

$$x = \frac{h^2}{2\beta\lambda} - \frac{\beta}{2}\lambda.$$

Using $\lambda = \frac{v}{f}$ and h from the given data yields 9.01 m ($\beta = \frac{1}{2}$), 2.71 m ($\beta = \frac{3}{2}$), 1.27 m ($\beta = \frac{5}{2}$), 0.53 m ($\beta = \frac{7}{2}$) and 0.026 m ($\beta = \frac{9}{2}$). These are the only allowable values of β that give positive solutions for x. (Negative values of x may be physical, depending on speaker design, but in that case the difference between path lengths is $\sqrt{x^2 + h^2} + x$.) b) Repeating the above for integral values of β, constructive interference occurs at 4.34 m, 1.84 m, 0.86 m, 0.26 m. Note that these are between, but not midway between, the answers to part (a).   c) If $h = \lambda/2$, there will be destructive interference at speaker B. If $\lambda/2 > h$, the path difference can never be as large as $\lambda/2$. (This is also obtained from the above expression for x, with $x = 0$ and $\beta = \frac{1}{2}$.) The minimum frequency is then $v/2h = (344 \text{ m/s})/(4.0 \text{ m}) = 86$ Hz.

20-44:   a) $\frac{\partial y}{\partial x} = kA_{SW} \cos kx \cos \omega t$, $\frac{\partial y}{\partial t} = -\omega A_{SW} \sin kx \sin \omega t$, and so the instantaneous power is

$$P = FA_{SW}^2 \omega k \, (\sin kx \, \cos kx)(\sin \omega t \, \cos \omega t)$$
$$= \frac{1}{4} FA_{SW}^2 \omega k \, \sin(2kx)\sin(2\omega t).$$

b) The average value of P is proportional to the average value of $\sin(2\omega t)$, and the average of the sine function is zero; $P_{av} = 0$.   c) The waveform is the solid line, and the

power is the dashed line. At time $t = \pi/2\omega$, $y = 0$ and $P = 0$ and the graphs coincide.
d) When the standing wave is at its maximum displacement at all points, all of the energy
is potential, and is concentrated at the places where the slope is steepest (the nodes).
When the standing wave has zero displacement, all of the energy is kinetic, concentrated
where the particles are moving the fastest (the antinodes). Thus, the energy must be
transferred from the nodes to the antinodes, and back again, twice in each cycle. Note
that $|P|$ is greatest midway between adjacent nodes and antinodes, and that P vanishes
at the nodes and antinodes.

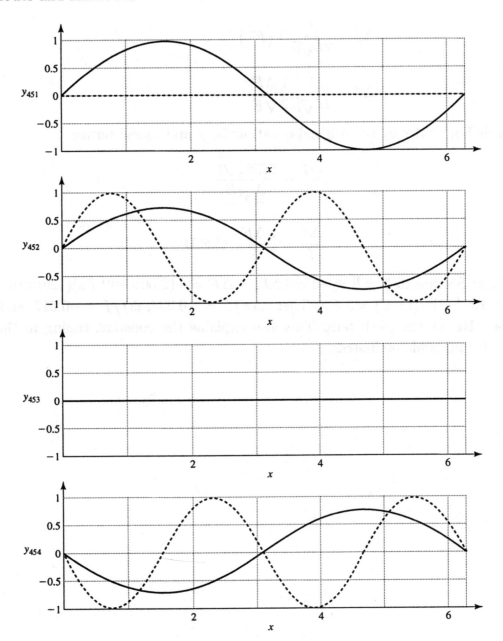

20-45: a) For a string, $f_n = \frac{n}{2L}\sqrt{\frac{F}{\mu}}$ and in this case, $n = 1$. Rearranging this and solving
for F gives $F = \mu 4L^2 f^2$. Note that $\mu = \pi r^2 \rho$, so $\mu = \pi(.203 \times 10^{-3} \text{ m})^2(7800 \text{ kg/m}^3) =$

1.01×10^{-3} kg/m. Substituting values, $F = (1.01 \times 10^{-3}$ kg/m$)4(.635$ m$)^2(247.0$ Hz$)^2 =$ 99.4 N.

b) To find the fractional change in the frequency we must take the ratio of Δf to f:

$$f = \frac{1}{2L}\sqrt{\frac{F}{\mu}},$$

$$\Delta(f) = \Delta\left(\frac{1}{2L}\sqrt{\frac{F}{\mu}}\right) = \Delta\left(\frac{1}{2L\sqrt{\mu}}F^{\frac{1}{2}}\right),$$

$$\Delta f = \frac{1}{2L\sqrt{\mu}}\Delta\left(F^{\frac{1}{2}}\right),$$

$$\Delta f = \frac{1}{2L\sqrt{\mu}}\frac{1}{2}\frac{\Delta F}{\sqrt{F}}.$$

Now divide both sides by the original equation for f and cancel terms:

$$\frac{\Delta f}{f} = \frac{\frac{1}{2L\sqrt{\mu}}\frac{1}{2}\frac{\Delta F}{\sqrt{F}}}{\frac{1}{2L}\sqrt{\frac{F}{\mu}}}$$

$$\frac{\Delta f}{f} = \frac{1}{2}\frac{\Delta F}{F}, \text{ Q.E.D.}$$

c) From Section 15-5, $\Delta F = -Y\alpha A\Delta T$, so $\Delta F = -(2.00\times10^{11}$ Pa$)(1.20\times10^{-5}/\text{C}°)\times (\pi(.203\times10^{-3}$ m$)^2)(11°\text{C}) = 3.4$ N. Then, $\Delta F/F = -0.034$, $\Delta f/f = -0.017$, and finally, $\Delta f = -4.2$ Hz, or the pitch falls. This also explains the constant tuning in the string sections of symphonic orchestras.

Chapter 21   Sound and Hearing

21-1: a) $\lambda = v/f = (344 \text{ m/s})/(1000 \text{ Hz}) = 0.344 \text{ m}$.   b) If $p \to 1000p_0$, then $A \to 1000A_0$. Therefore, the amplitude is 1.2×10^{-5} m.   c) Since $p_{max} = BkA$, increasing p_{max} while keeping A constant requires decreasing k, and increasing π, by the same factor. Therefore the new wavelength is $(0.688 \text{ m})(20) = 6.9$ m, $f_{new} = \frac{344 \text{ m/s}}{6.9 \text{ m}} = 50$ Hz.

21-2: $A = \frac{p_{max}v}{2\pi Bf} = \frac{(3.0 \times 10^{-2} \text{ Pa})(1480 \text{ m/s})}{2\pi(2.2 \times 10^9 \text{ Pa})(1000 \text{ Hz})}$, or $A = 3.21 \times 10^{-12}$ m. The much higher bulk modulus increases both the needed pressure amplitude and the speed, but the speed is proportional to the square root of the bulk modulus. The overall effect is that for such a large bulk modulus, large pressure amplitudes are needed to produce a given displacement.

21-3: From Eq. (21-5), $p_{max} = BkA = 2\pi BA/\lambda = 2\pi BAf/v$.

   a) $2\pi(1.42 \times 10^5 \text{ Pa})(2.00 \times 10^{-5} \text{ m})(150 \text{ Hz})/(344 \text{ m/s}) = 7.78$ Pa.

   b) $10 \times 7.78 \text{ Pa} = 77.8$ Pa.   c) $100 \times 7.78 \text{ Pa} = 778$ Pa.

   The amplitude at 1500 Hz exceeds the pain threshold, and at 15,000 Hz the sound would be unbearable.

21-4: From Eq. (21-8), $I = vp_{max}^2/2B$, and from Eq. (19-21), $v^2 = B/\rho$. Using Eq. (19-21) to eliminate v, $I = \left(\sqrt{B/\rho}\right)p_{max}^2/2B = p_{max}^2/2\sqrt{\rho B}$. Using Eq. (19-21) to eliminate B, $I = vp_{max}^2/2(v^2\rho) = p_{max}^2/2\rho v$.

21-5: a) $p_{max} = BkA = \frac{2\pi BfA}{v} = \frac{2\pi(1.42 \times 10^5 \text{ Pa})(150 \text{ Hz})(5.00 \times 10^{-6} \text{ m})}{(344 \text{ m/s})} = 1.95$ Pa.

   b) From Eq. (21-9), $I = p_{max}^2/2\rho v = (1.95 \text{ Pa})^2/(2 \times (1.2 \text{ kg/m}^3)(344 \text{ m/s})) = 4.58 \times 10^{-3}$ W/m^2.

   c) $10 \times \log\left(\frac{4.58 \times 10^{-3}}{10^{-12}}\right) = 96.6$ dB.

21-6: (a) The sound level is $\beta = (10 \text{ dB}) \log \frac{I}{I_0}$, so $\beta = (10 \text{ dB}) \log \frac{0.500 \,\mu\text{W/m}^2}{10^{-12} \,\text{W/m}^2}$, or $\beta = 57$ dB.

   b) First find v, the speed of sound at 20.0°C, from Table 19-1, $v = 344$ m/s. The density of air at that temperature is 1.20 kg/m^3. Using Equation (21-9), $I = \frac{p_{max}^2}{2\rho v} = \frac{(0.150 \text{ N/m}^2)^2}{2(1.20 \text{ kg/m}^3)(344 \text{ m/s})}$, or $I = 2.73 \times 10^{-5}$ W/m^2. Using this in Equation (21-11), $\beta = (10 \text{ dB}) \log \frac{2.73 \times 10^{-5} \text{ W/m}^2}{10^{-12} \text{ W/m}^2}$, or $\beta = 74.4$ dB.

21-7: a) As in Example 21-3, $I = \frac{(6.0 \times 10^{-5} \text{ Pa})^2}{2(1.20 \text{ kg/m}^3)(344 \text{ m/s})} = 4.4 \times 10^{-12}$ W/m^2. $\beta = 6.40$ dB.

21-8: a) $10 \times \log\left(\frac{4I}{I}\right) = 6.0$ dB.   b) The number must be multiplied by four, for an increase of 12 kids.

21-9: Mom is five times further away than Dad, and so the intensity she hears is $\frac{1}{25} = 5^{-2}$ of the intensity that he hears, and the difference in sound intensity levels is $10 \times \log(25) = 14$ dB.

21-10: a) The intensity is proportional to the reciprocal of the square of the distance, so a decrease in intensity by a factor of 25 corresponds to a quintupled distance, or 75.0 m.

   b) Using $4\pi r^2 I$ for either the 75.0-m or 15.0-m distance gives $P = 707$ W.

21-11: $\beta = (10 \text{ dB})\log\frac{I}{I_0}$, or $13 \text{ dB} = (10 \text{ dB})\log\frac{I}{I_0}$. Thus, $I/I_0 = 20.0$, or the intensity has increased a factor of 20.0.

21-12:   a) Since $f_{\text{beat}} = f_a - f_b$, the possible frequencies are $440.0 \text{ Hz} \pm 1.5 \text{ Hz} = 438.5 \text{ Hz}$ or 441.5 Hz.   b) The tension is proportional to the square of the frequency. Therefore $T \propto f^2$ and $\Delta T \propto 2f\Delta f$. So $\frac{\Delta T}{T} = \frac{2\Delta f}{f}$. i) $\frac{\Delta T}{T} = \frac{2(1.5 \text{ Hz})}{440 \text{ Hz}} = 6.82 \times 10^{-3}$. ii) $\frac{\Delta T}{T} = \frac{2(-1.5 \text{ Hz})}{440 \text{ Hz}} = -6.82 \times 10^{-3}$.

21-13:   a) A frequency of $\frac{1}{2}(108 \text{ Hz} + 112 \text{ Hz}) = 110 \text{ Hz}$ will be heard, with a beat frequency of $112 \text{ Hz} - 108 \text{ Hz} = 4$ beats per second.   b) The maximum amplitude is the sum of the amplitudes of the individual waves, $2(1.5 \times 10^{-8} \text{ m}) = 3.0 \times 10^{-8} \text{ m}$. The minimum amplitude is the difference, zero.

21-14:   Solving Eq. (21-17) for v, with $v_L = 0$, gives

$$v = \frac{f_L}{f_S - f_L}v_S = \left(\frac{1240 \text{ Hz}}{1200 \text{ Hz} - 1240 \text{ Hz}}\right)(-25.0 \text{ m/s}) = 775 \text{ m/s},$$

or 780 m/s to two figures (the difference in frequency is known to only two figures). Note that $v_S < 0$, since the source is moving toward the listener.

21-15:   Redoing the calculation with $+20.0 \text{ m/s}$ for v_S and -20.0 m/s for v_L gives 267 Hz.

21-16:   a) From Eq. (21-17), with $v_S = 0$, $v_L = -15.0 \text{ m/s}$, $f'_A = 375 \text{ Hz}$.

b) With $v_S = 35.0 \text{ m/s}$, $v_L = 15.0 \text{ m/s}$, $f'_B = 371 \text{ Hz}$.

c) $f'_A - f'_B = 4 \text{ Hz}$ (keeping an extra figure in f'_A). The difference between the frequencies is known to only one figure.

21-17:   In terms of wavelength, Eq. (21-17) is

$$\lambda_L = \frac{v + v_S}{v + v_L}\lambda_S.$$

a) $v_L = 0$, $v_S = -25.0$ m and $\lambda_L = \left(\frac{319}{344}\right)(344 \text{ m/s})/(400 \text{ Hz}) = 0.798 \text{ m}$. This is, of course, the same result as obtained directly from Eq. (21-15). $v_S = 25.0 \text{ m/s}$ and $\lambda_L = (369 \text{ m/s})/(400 \text{ Hz}) = 0.922 \text{ m}$. The frequencies corresponding to these wavelengths are   c) 431 Hz and   d) 373 Hz.

21-18:   a) In terms of the period of the source, Eq. (21-15) becomes

$$v_S = v - \frac{\lambda}{T_S} = 0.32 \text{ m/s} - \frac{0.12 \text{ m}}{1.6 \text{ s}} = 0.25 \text{ m/s}.$$

b) Using the result of part (a) in Eq. (21-16), or solving Eq. (21-15) for v_S and substituting into Eq. (21-16) (making sure to distinguish the symbols for the different wavelengths) gives $\lambda = 0.91 \text{ m}$.

21-19:   a) $v_L = 18.0 \text{ m/s}$, $v_S = -30.0 \text{ m/s}$, and Eq. (21-17) gives $f_L = \left(\frac{362}{314}\right)(262 \text{ Hz}) = 302 \text{ Hz}$.   b) $v_L = -18.0 \text{ m/s}$, $v_S = 30.0 \text{ m/s}$ and $f_L = 228 \text{ Hz}$.

21-20: a) In Eq. (21-19), $v/v_S = 1/1.70 = 0.588$ and $\alpha = \arcsin(0.588) = 36.0°$. b) As in Example 21-13,

$$t = \frac{(950 \text{ m})}{(1.70)(344 \text{ m/s})(\tan(36.0°))} = 2.23 \text{ s.}$$

21-21: a) Mathematically, the waves given by Eq. (21-1) and Eq. (21-4) are out of phase. Physically, at a displacement node, the air is most compressed or rarefied on either side of the node, and the pressure gradient is zero. Thus, displacement nodes are pressure antinodes. b) (This is the same as Fig. (21-2).) The solid curve is the pressure and the dashed curve is the displacement.

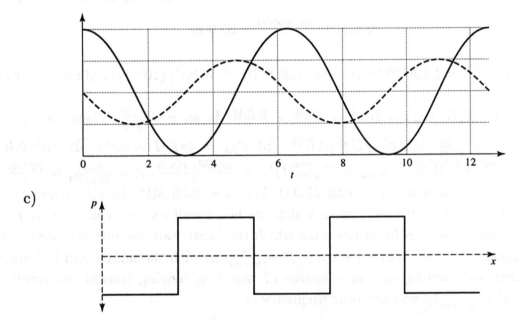

c)

The pressure amplitude is not the same. The pressure gradient is either zero or undefined. At the places where the pressure gradient is undefined mathematically (the "cusps" of the *y-x* plot), the particles go from moving at uniform speed in one direction to moving at the same speed in the other direction. In the limit that Fig. (21-19) is an accurate depiction, this would happen in a vanishing small time, hence requiring a very large force, which would result from a very large pressure gradient. d) The statement is true, but incomplete. The pressure is indeed greatest where the displacement is zero, but the pressure is equal to its largest value at points other than those where the displacement is zero.

21-22: The altitude of the plane when it passes over the end of the runway is $(1740 \text{ m} - 1200 \text{ m}) \tan 15° = 145 \text{ m}$, and so the sound intensity is $1/(1.45)^2$ of what the intensity would be at 100 m. The intensity level is then

$$100.0 \text{ dB} - 10 \times \log\left[(1.45)^2\right] = 96.8 \text{ dB,}$$

so the airliner is not in violation of the ordinance.

21-23:   a) Combining Eq. (21-9) and Eq. (21-11),

$$p_{max} = \sqrt{2\rho v I_0 10^{(\beta/10)}} = \sqrt{2(1.20 \text{ kg/m}^3)(344 \text{ m/s})(10^{-12} \text{ W/m}^2)10^{5.20}}$$
$$= 1.144 \times 10^{-2} \text{ Pa},$$

or 1.14×10^{-2} Pa to three figures.   b) From Eq. (21-5), and as in Example 21-1,

$$A = \frac{p_{max}}{Bk} = \frac{p_{max}v}{B2\pi f} = \frac{(1.144 \times 10^{-2} \text{ Pa})(344 \text{ m/s})}{2\pi(1.42 \times 10^5 \text{ Pa})(587 \text{ Hz})} = 7.51 \times 10^{-9} \text{ m}.$$

c) The distance is proportional to the reciprocal of the square root of the intensity, and hence to 10 raised to half of the sound intensity levels divided by 10. Specifically,

$$(5.00 \text{ m})10^{(5.20-3.00)/2} = 62.9 \text{ m}.$$

21-24:   a) $p = IA = I_0 10^{(\beta/10 \text{ dB})}A$.   b) $(1.00 \times 10^{-12} \text{ W/m}^2)(10^{5.50})(1.20\text{m}^2) = 3.79 \times 10^{-7}$ W.

21-25:   a) $A = \Delta R \cdot p_{max} = BkA = \frac{2\pi BA}{\lambda} = \frac{2\pi BAf}{v}$. In air $v = \sqrt{\frac{B}{\rho}}$. Therefore $p_{max} = 2\pi\sqrt{\rho B}f\Delta R, I = \frac{p_{max}^2}{2\sqrt{\rho B}} = 2\pi^2\sqrt{\rho B}f^2(\Delta R)^2$.   b) $P_{Tot} = 4\pi R^2 I = 8\pi^3\sqrt{\rho B}f^2R^2(\Delta R)^2$
c) $I = \frac{P_{Tot}}{4\pi d^2} = \frac{2\pi^2\sqrt{\rho B}f^2R^2(\Delta R)^2}{d^2}$, $p_{max} = (2\sqrt{\rho B}I)^{1/2} = \frac{2\pi\sqrt{\rho B}fR(\Delta R)}{d}$, $A = \frac{p_{max}}{2\pi\sqrt{\rho B}f} = \frac{R(\Delta R)}{d}$.

21-26:   (See also Problems 21-30 and 21-34). Let $f_0 = 2.00$ MHz be the frequency of the generated wave. The frequency with which the heart wall receives this wave is $f_H = \frac{v+v_H}{v}f_0$, and this is also the frequency with which the heart wall re-emits the wave. The detected frequency of this reflected wave is $f' = \frac{v}{v-v_H}f_H$, with the minus sign indicating that the heart wall, acting now as a source of waves, is moving toward the receiver. Combining, $f' = \frac{v+v_H}{v-v_H}f_0$, and the beat frequency is

$$f_{beat} = f' - f_0 = \left(\frac{v+v_H}{v-v_H} - 1\right)f_0 = \frac{2v_H}{v-v_H}f_0.$$

Solving for v_H,

$$v_H = v\left(\frac{f_{beat}}{2f_0 + f_{beat}}\right) = (1500 \text{ m/s})\left(\frac{85 \text{ Hz}}{2(2.00 \times 10^6 \text{ Hz}) + (85 \text{ Hz})}\right)$$
$$= 3.19 \times 10^{-2} \text{ m/s}.$$

Note that in the denominator in the final calculation, f_{beat} is negligible compared to f_0.
21-27:   a) $\lambda = v/f = (1482 \text{ m/s})/(22.0 \times 10^3 \text{ Hz}) = 6.74 \times 10^{-2}$ m.   b) See Problem 21-26 or Problem 21-30; the difference in frequencies is

$$\Delta f = f_S\left(\frac{2v_w}{v-v_w}\right) = (22.0 \times 10^3 \text{ Hz})\frac{2(4.95 \text{ m/s})}{(1482 \text{ m/s}) - (4.95 \text{ m/s})} = 147 \text{ Hz}.$$

21-28:   a) The maximum velocity of the siren is $\omega_P A_P = 2\pi f_P A_P$. You hear a sound with frequency $f_L = f_{siren}v/(v + v_S)$, where v_S varies between $+2\pi f_P A_P$ and $-2\pi f_P A_P$.

So $f_{L-max} = f_{siren}v/(v - 2\pi f_P A_P)$ and $f_{L-min} = f_{siren}v/(v + 2\pi f_P A_P)$. b) The maximum (minimum) frequency is heard when the platform is passing through equilibrium and moving up (down).

21-29: a) Let v_b be the speed of the bat, v_i the speed of the insect and f_i the frequency with which the sound waves both strike and are reflected from the insect. The frequencies at which the bat sends and receives the signals are related by

$$f_L = f_i \left(\frac{v + v_b}{v - v_i}\right) = f_S \left(\frac{v + v_i}{v - v_b}\right) \left(\frac{v + v_b}{v - v_i}\right).$$

Solving for v_i,

$$v_i = v \left[\frac{1 - \frac{f_S}{f_L}\left(\frac{v+v_b}{v-v_b}\right)}{1 + \frac{f_S}{f_L}\left(\frac{v+v_b}{v-v_b}\right)}\right] = v \left[\frac{f_L(v - v_b) - f_S(v + v_b)}{f_L(v - v_b) + f_S(v + v_b)}\right].$$

Letting $f_L = f_{refl}$ and $f_S = f_{bat}$ gives the result.

b) If $f_{bat} = 80.7$ kHz, $f_{refl} = 83.5$ kHz, and $v_{bat} = 3.9$ m/s, $v_{insect} = 2.0$ m/s.

21-30: (See Problems 21-26, 21-34, 21-27). a) In a time t, the wall has moved a distance $v_1 t$ and the wavefront that hits the wall at time t has traveled a distance vt, where $v = f_0\lambda_0$, and the number of wavecrests in the total distance is $\frac{(v+v_1)t}{\lambda_0}$. b) The reflected wave has traveled vt and the wall has moved $v_1 t$, so the wall and the wavefront are separated by $(v - v_1)t$. c) The distance found in part (b) must contain the number of reflected waves found in part (a), and the ratio of the quantities is the wavelength of the reflected wave, $\lambda_0 \frac{v-v_1}{v+v_1}$. d) The speed v divided by the result of part (c), expressed in terms of f_0 is $f_0\frac{v+v_1}{v-v_1}$. This is what is predicted by the problem-solving strategy. e) $f_0\frac{v+v_1}{v-v_1} - f_0 = f_0\frac{2v_1}{v-v_1}$.

21-31: a)

$$f_R = f_L\sqrt{\frac{c - v}{c + v}} = f_S\frac{\sqrt{1 - \frac{v}{c}}}{\sqrt{1 + \frac{v}{c}}} = f_S\left(1 - \frac{v}{c}\right)^{1/2}\left(1 + \frac{v}{c}\right)^{-1/2}.$$

b) For small x, the binomial theorem (see Appendix B) gives $(1 - x)^{1/2} \approx 1 - x/2$, $(1 + x)^{-1/2} \approx 1 - x/2$, so

$$f_L \approx f_S\left(1 - \frac{v}{2c}\right)^2 \approx f_S\left(1 - \frac{v}{c}\right)$$

where the binomial theorem has been used to approximate $(1 - x/2)^2 \approx 1 - x$.

The above result may be obtained without resort to the binomial theorem by expressing f_R in terms of f_S as

$$f_R = f_S\frac{\sqrt{1 - (v/c)}\sqrt{1 - (v/c)}}{\sqrt{1 + (v/c)}\sqrt{1 - (v/c)}} = f_S\frac{1 - (v/c)}{\sqrt{1 - (v/c)^2}}.$$

To first order in v/c, the square root in the denominator is 1, and the previous result is obtained. c) For an airplane, the approximation $v \ll c$ is certainly valid, and solving the expression found in part (b) for v,

$$v = c \left(\frac{f_S - f_R}{f_S} \right) = c \frac{f_{beat}}{f_S} = (3.00 \times 10^8 \text{ m/s}) \frac{46.0 \text{ Hz}}{2.43 \times 10^8 \text{ Hz}} = 56.8 \text{ m/s},$$

and the approximation $v \ll c$ is seen to be valid. Note that in this case, the frequency *difference* is known to three figures, so the speed of the plane is known to three figures.

21-32: a) As in Problem 21-31,

$$v = c \frac{f_S - f_R}{f_S} = (3.00 \times 10^8 \text{ m/s}) \frac{-0.018 \times 10^{14} \text{ Hz}}{4.568 \times 10^{14} \text{ Hz}} = -1.2 \times 10^6 \text{ m/s},$$

with the minus sign indicating that the gas is approaching the earth, as is expected since $f_R > f_S$. b) The radius is $(952 \text{ yr})(3.156 \times 10^7 \text{ s/yr})(1.2 \times 10^6 \text{ m/s}) = 3.6 \times 10^{16} \text{ m} = 3.8 \text{ ly}$. This may also be obtained from $(952 \text{ yr}) \frac{f_R - f_S}{f_S}$. c) The ratio of the width of the nebula to the distance from the earth is the ratio of the angular width (taken as 5 arc minutes) to an entire circle, which is 60×360 arc minutes. The distance to the nebula is then (keeping an extra figure in the intermediate calculation)

$$2 \times 3.75 \text{ ly} \frac{(60) \times (360)}{5} = 3.24 \times 10^4 \text{ ly},$$

so the explosion actually took place about 30,000 B.C.

21-33: a) The frequency is greater than 2800 MHz; the thunderclouds, moving toward the installation, encounter more wavefronts per time than would a stationary cloud, and so an observer in the frame of the storm would detect a higher frequency. Using the result of Problem 21-31, with $v = -42.0$ km/h,

$$f_R - f_S = f_S \frac{-v}{c} = (2800 \times 10^6 \text{ Hz}) \frac{(42.0 \text{ km/h})/(3.6 \text{ km/h/1 m/s})}{(3.00 \times 10^8 \text{ m/s})} = 109 \text{ Hz}.$$

b) The waves are being sent at a higher frequency than 2800 MHz from an approaching source, and so are received at a higher frequency. Repeating the above calculation gives the result that the waves are detected at the installation with a frequency 109 Hz greater than the frequency with which the cloud received the waves, or 218 Hz higher than the frequency at which the waves were originally transmitted at the receiver. Note that in doing the second calculation, $f_S = 2800 \text{ MHz} + 109 \text{ Hz}$ is the same as 2800 MHz to three figures.

21-34: a) (See also Example 21-12 and Problem 21-26.) The wall will receive and reflect pulses at a frequency $\frac{v}{v - v_w} f_0$, and the woman will hear this reflected wave at a frequency

$$\frac{v + v_w}{v} \cdot \frac{v}{v - v_w} f_0 = \frac{v + v_w}{v - v_w} f_0;$$

The beat frequency is

$$f_{beat} = f_0 \left(\frac{v + v_w}{v - v_w} - 1 \right) = f_0 \left(\frac{2v_w}{v - v_w} \right).$$

b) In this case, the sound reflected from the wall will have a lower frequency, and using $f_0(v - v_w)/(v + v_w)$ as the detected frequency (see Example 21-12; v_w is replaced by $-v_w$ in the calculation of part (a)),

$$f_{beat} = f_0 \left(1 - \frac{v - v_w}{v + v_w} \right) = f_0 \left(\frac{2v_w}{v + v_w} \right).$$

21-35: Refer to Equation (21-19) and Figure (21-16). The sound travels a distance vT and the plane travels a distance v_sT before the boom is found. So, $h^2 = (vT)^2 + (v_sT)^2$, or $v_sT = \sqrt{h^2 - v^2T^2}$. From Equation (21-19), $\sin \alpha = \frac{v}{v_S}$. Then, $v_S = \frac{hv}{\sqrt{h^2 - v^2T^2}}$.

21-36: a)

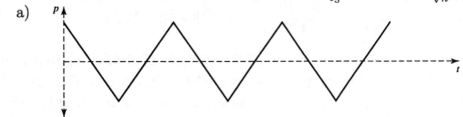

b) From Eq. (21-4), the function that has the given $p(x, 0)$ at $t = 0$ is given graphically as shown. Each section is a parabola, not a portion of a sine curve. The period is $\lambda/v = (0.200 \text{ m})/(344 \text{ m/s}) = 5.81 \times 10^{-4}$ s and the amplitude is equal to the area under the p-x curve between $x = 0$ and $x = 0.0500$ m divided by B, or 7.04×10^{-6} m.

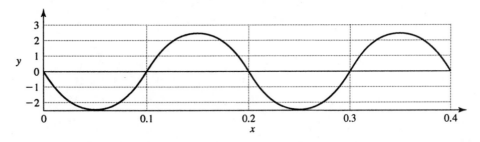

c) Assuming a wave moving in the $+x$-direction, $y(0, t)$ is as shown.

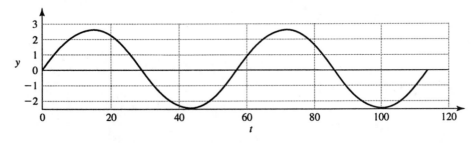

d) The maximum velocity of a particle occurs when a particle is moving throughout the origin, and the particle speed is $v_y = -\frac{\partial y}{\partial x} v = \frac{pv}{B}$. The maximum velocity is found from the maximum pressure, and $v_{y\,max} = (40\ \text{Pa})(344\ \text{m/s})/(1.42 \times 10^5\ \text{Pa}) = 9.69\ \text{cm/s}$. The maximum acceleration is the maximum pressure gradient divided by the density,

$$a_{max} = \frac{(80.0\ \text{Pa})/(0.100\ \text{m})}{(1.20\ \text{kg/m}^3)} = 6.67 \times 10^2\ \text{m/s}^2.$$

e) The speaker cone moves with the displacement as found in part (c); the speaker cone alternates between moving forward and backward with constant magnitude of acceleration (but changing sign). The acceleration as a function of time is a square wave with amplitude 667 m/s² and frequency $f = v/\lambda = (344\ \text{m/s})/(0.200\ \text{m}) = 1.72$ kHz.

21-37: Taking the speed of sound to be 344 m/s, the wavelength of the waves emitted by each speaker is 2.00 m. a) Point C is two wavelengths from speaker A and one and one-half from speaker B, and so the phase difference is $180° = \pi$ rad.

b)

$$I = \frac{P}{4\pi r^2} = \frac{8.00 \times 10^{-4}\ \text{W}}{4\pi (4.00\ \text{m})^2} = 3.98 \times 10^{-6}\ \text{W/m}^2,$$

and the sound intensity level is $(10\ \text{dB}) \log(3.98 \times 10^6) = 66.0$ dB. c) Repeating with $P = 6.00 \times 10^{-5}$ W and $r = 3.00$ m gives $I = 5.31 \times 10^{-7}$ W and $\beta = 57.2$ dB. d) With the result of part (a), the amplitudes, either displacement or pressure, must be subtracted. That is, the intensity is found by taking the square roots of the intensities found in part (b), subtracting, and squaring the difference. The result is that $I = 1.60 \times 10^{-6}$ W and $\beta = 62.1$ dB.